DAVID WAUGH

The New Wider World

CONTENTS

* indicates that a 'Places' mini-case study is included in this section.

Colour coding for Questions sections

ⓐ Based upon, or suited to, specimen GCSE questions at the Higher tier.

ⓐ Based upon, or suited to, specimen GCSE questions at the Foundation tier.

a Based upon, or suited to, specimen GCSE questions common to both tiers.

1 POPULATION

Distribution and density

Distribution describes the way in which people are spread out across the Earth's surface. This distribution is uneven and changes over periods of time. It is usual to show population distribution by means of a dot map (Figure 1.1). Notice how people are concentrated into certain parts of the world making those places very crowded. At the same time, other areas have relatively few people living there. These are said to be sparsely populated.

Figure 1.1
Dot map showing world population distribution

Figure 1.3
Factors affecting distribution and density of population

	PHYSICAL			
	DENSELY POPULATED	EXAMPLES	SPARSELY POPULATED	EXAMPLES
RELIEF	FLAT PLAINS AND LOW-LYING UNDULATING AREAS	BANGLADESH	HIGH, RUGGED MOUNTAINS	ANDES
	BROAD RIVER VALLEYS	GANGES VALLEY	WORN-DOWN SHIELD LANDS	CANADIAN SHIELD
	FOOTHILLS OF ACTIVE VOLCANOES	ETNA, PINATUBO		
CLIMATE	EVENLY DISTRIBUTED RAINFALL WITH NO TEMPERATURE EXTREMES	NORTH-WEST EUROPE	LIMITED ANNUAL RAINFALL	SAHARA DESERT
	AREAS WITH (I) HIGH SUNSHINE TOTALS (II) HEAVY SNOWFALL FOR TOURISM	(I) SPANISH COSTAS	LOW ANNUAL TEMPERATURES	GREENLAND
		(II) SWISS ALPINE VALLEYS	HIGH ANNUAL HUMIDITY	AMAZON RAINFOREST
	SEASONAL MONSOON RAINFALL	BANGLADESH	UNRELIABLE SEASONAL RAINFALL	SAHEL
VEGETATION	GRASSLANDS – EASY TO CLEAR/FARM	PARIS BASIN	FOREST	AMAZONIA, CANADIAN SHIELD
SOIL	DEEP FERTILE SILT LEFT BY RIVERS	NILE VALLEY AND DELTA	THIN SOILS IN MOUNTAINOUS OR GLACIATED AREAS	NORTHERN SCANDINAVIA
	VOLCANIC SOILS	ETNA	(I) LACKING HUMUS OR (II) AFFECTED BY LEACHING	(I) SAHEL (II) RAINFORESTS
NATURAL RESOURCES	MINERALS, e.g. COAL, IRON ORE	PENNSYLVANIA, JOHANNESBURG	LACKING MINERALS	ETHIOPIA
	ENERGY SUPPLIES, e.g. HEP	RHONE VALLEY	LACKING ENERGY SUPPLIES	NORTH-EAST BRAZIL
WATER SUPPLY	RELIABLE SUPPLIES	NORTH-WEST EUROPE	UNRELIABLE SUPPLIES	AFGHANISTAN
NATURAL ROUTES	GAPS THROUGH MOUNTAINS, CONFLUENCE OF VALLEYS	RHINE VALLEY, PARIS	MOUNTAIN BARRIER	HIMALAYAS

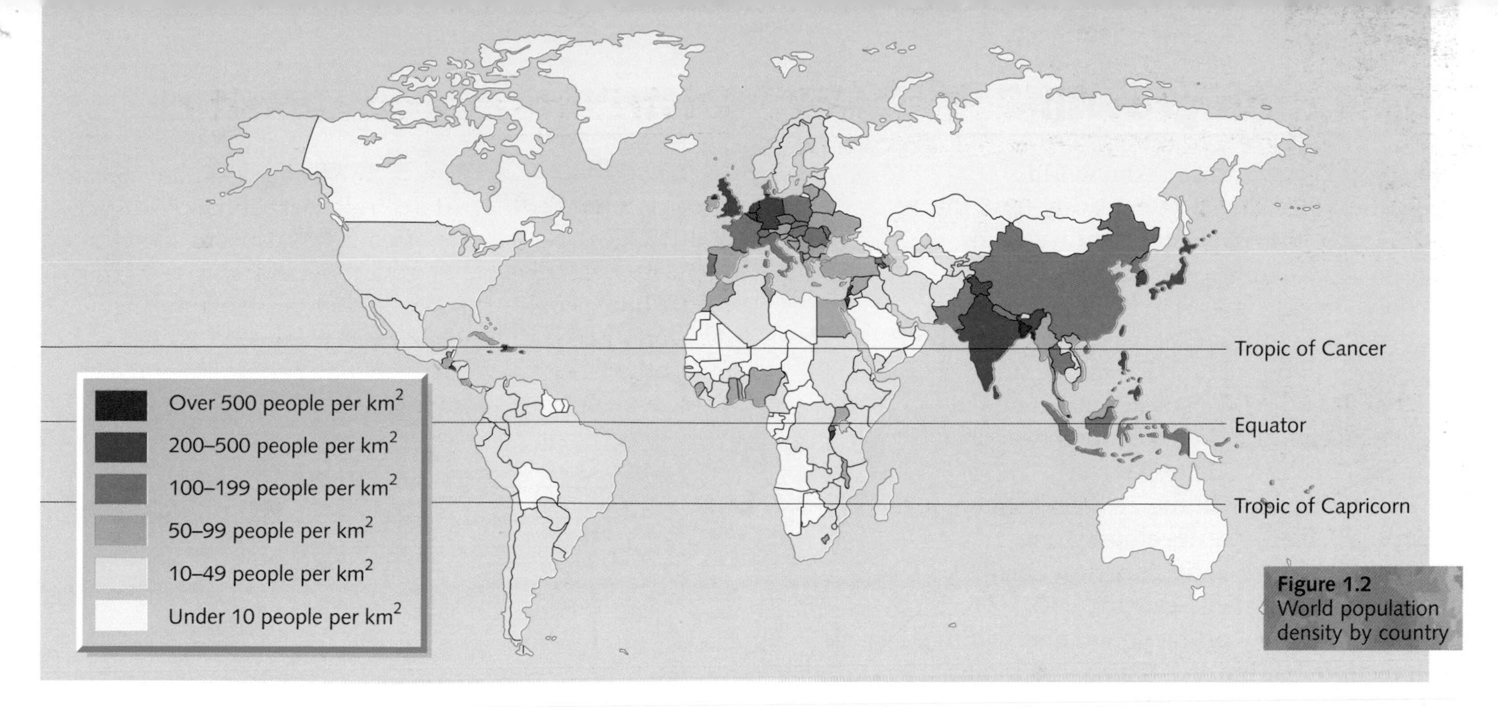

Figure 1.2 World population density by country

Density describes the number of people living in a given area, usually a square kilometre (km^2). Density is found by dividing the total population of a place by its area. Population density is usually shown by a choropleth map (Figure 1.2). A choropleth map is easy to read as it shows generalisations, but it does tend to hide concentrations. For example:

- Brazil appears on the world map in Figure 1.2 to have a low population density. However, on a larger-scale map (see Figure 1.15), several parts of the country are shown to have very high densities.
- Figure 1.2 suggests that the population of Egypt is evenly spread whereas in reality it is concentrated along the Nile Valley (Figure 1.1).

On global and continental scales, patterns of distribution and density are mainly affected by *physical* factors such as relief, climate, vegetation, soils, natural resources and water supply. At regional and more local scales, patterns are more likely to be influenced by *human* factors which may be economic, political or social. Figure 1.3 gives reasons, with specific examples, why some parts of the world are densely populated while others are sparsely populated. You should be aware, however, that:

- for a given place there are usually several reasons for its dense or sparse population, e.g. Nile Valley and Sahara Desert
- even within areas there are variations in density, e.g. parts of Japan have some of the highest densities in the world, yet less than one-fifth of the country is inhabited.

HUMAN				
	DENSELY POPULATED	EXAMPLES	SPARSELY POPULATED	EXAMPLES
ECONOMIC	PORTS	NEW YORK, SYDNEY	LIMITED FACILITIES FOR PORTS	BANGLADESH
	GOOD ROADS, RAILWAYS, AIRPORTS	GERMANY, CALIFORNIA	POOR TRANSPORT LINKS	HIMALAYAS
	INDUSTRIAL AREAS (TRADITIONAL)	PITTSBURGH, RUHR	LACK OF INDUSTRIAL DEVELOPMENT	SUDAN
	DEVELOPMENT OF TOURISM	BANFF (CANADA), JAMAICA	LACK OF TOURIST DEVELOPMENTS	IRAQ
	MONEY AVAILABLE FOR NEW HIGH-TECH INDUSTRIES	CALIFORNIA, SOUTH OF FRANCE	LACK OF MONEY FOR NEW INVESTMENTS	NEPAL, GAZA
POLITICAL	GOVERNMENT INVESTMENT	TOKYO REGION, NORTH ITALY	LACK OF GOVERNMENT INVESTMENT	DEM. REP. OF THE CONGO
	NEW TOWNS	SATELLITE TOWNS AROUND CAIRO, BRASILIA	DEPOPULATION OF RURAL AND OLD INDUSTRIAL AREAS	NORTH-EAST BRAZIL, BELGIAN COALFIELD
	RECLAMATION OF LAND	HONG KONG ISLAND, DUTCH POLDERS	LOSS OF LAND, E.G. DEFORESTATION, AND SOIL EROSION	AMAZONIA, APENNINES, SAHEL
SOCIAL	BETTER HOUSING OPPORTUNITIES	ARIZONA	POOR HOUSING OPPORTUNITIES	AFGHANISTAN, SOWETO
	EDUCATION, HEALTH FACILITIES, ENTERTAINMENT	SYDNEY, MILAN	LIMITED EDUCATION, HEALTH FACILITIES, ENTERTAINMENT	RWANDA
	RETIREMENT AREAS	SPANISH COSTAS, CANARY ISLANDS	POOR FACILITIES FOR RETIREMENT	ERITREA

Population growth

The annual growth rate of the world's population rose slowly but steadily until the beginning of the nineteenth century. Since then it has grown at a much faster rate and has only shown signs of slowing down in the last decade. Figure 1.4 is a bar graph showing the growth in population for this century in each continent. As well as emphasising the uneven distribution in population described on page 4, it shows that:

- The continents with the greatest increase in population are the developing ones of Asia, Africa and Latin America. One UN estimate suggests that by the year 2000, 39 per cent of the world's population will live in the two countries of China and India. Already in 1995 (Figure 1.5) the three developing continents were home to 84 per cent of the world's inhabitants.
- The continents with the slowest increase in population are the developed ones of Europe, the former USSR, North America and Oceania (Australasia). Estimates suggest that by the year 2000 several countries in North-west Europe will have a zero population growth, i.e. neither an increase nor a decrease.

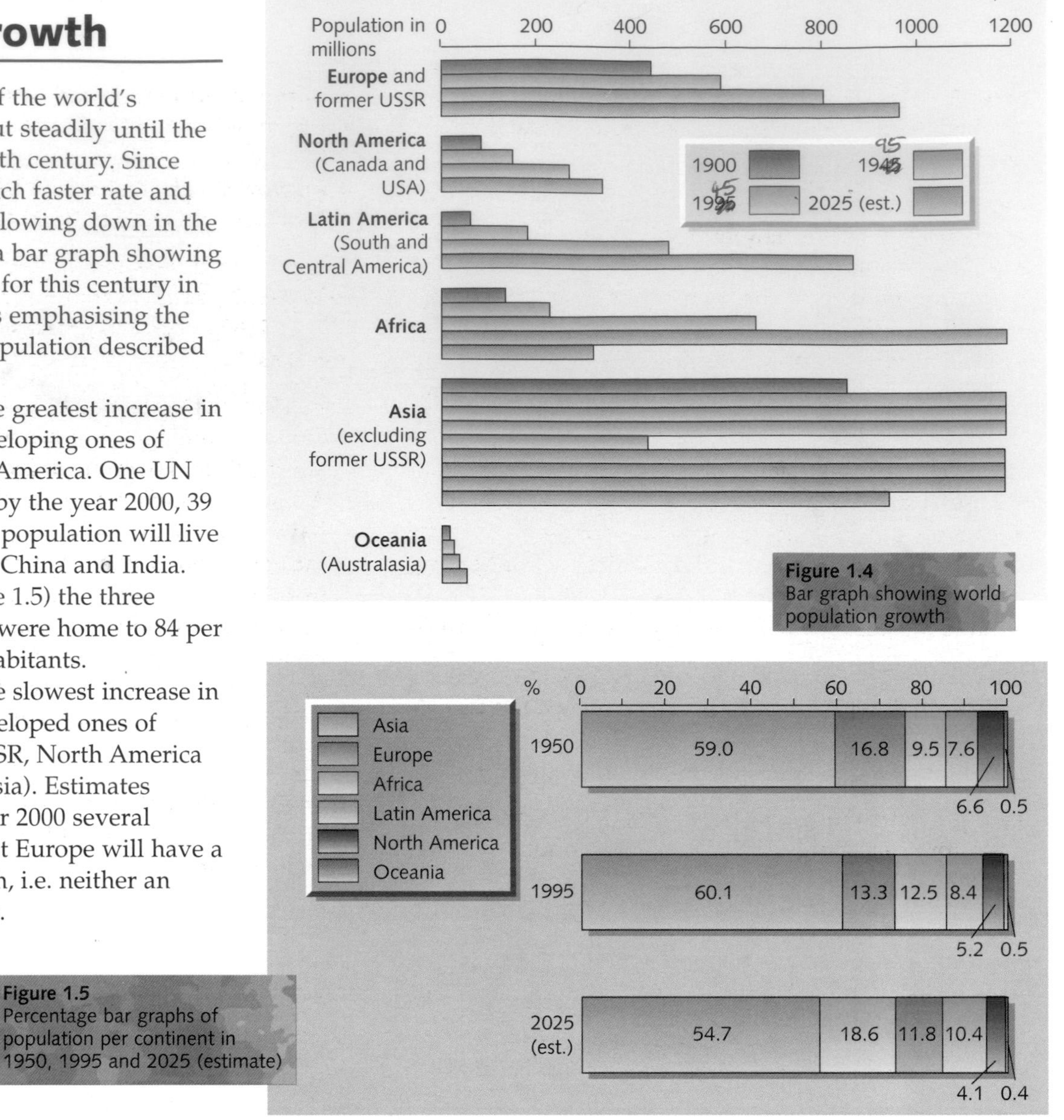

Figure 1.4
Bar graph showing world population growth

Figure 1.5
Percentage bar graphs of population per continent in 1950, 1995 and 2025 (estimate)

Figure 1.6
Demographic transition model

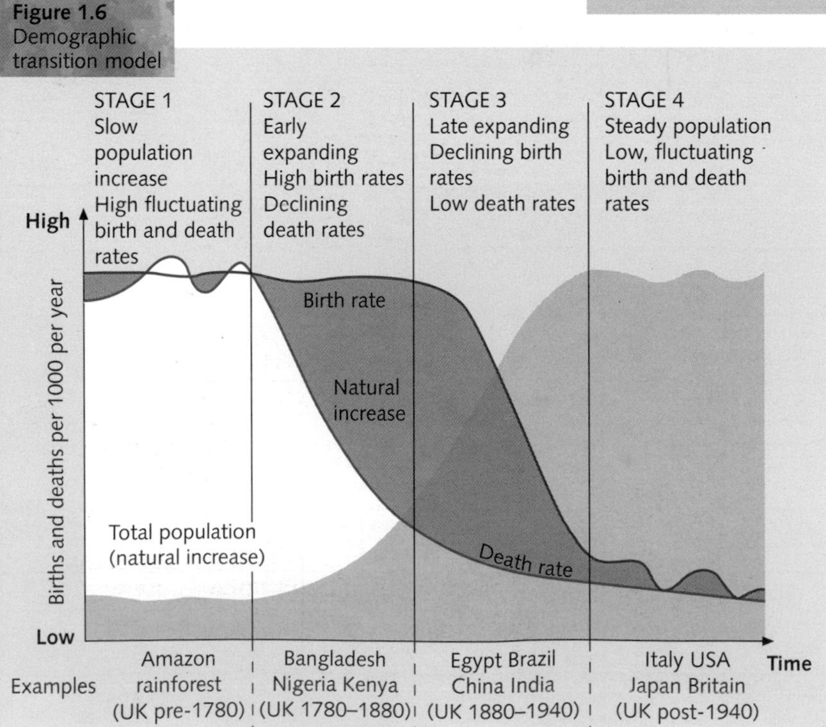

Population change

This depends on birth rate, death rate and migration. The *natural increase* or the annual growth rate is the difference between the *birth rate* (the average number of births per 1000 people) and the *death rate* (the average number of deaths per 1000 people). Based on growth rates in the industrial areas of Western Europe and North America, a model has been produced (Figure 1.6) which suggests that the population (or demographic) growth rate can be divided into four distinct stages. This demographic transition model has also been applied to developing countries despite the fact that the model assumes that the falling death rate (Stage 2) is a response to increased industrialisation – a process now accepted as being unlikely to take place in many of the less economically developed countries.

Stage 1 Here both birth rates and death rates fluctuate at a high level (about 35 per 1000) giving a small population growth.

Birth rates are high because:
- No birth control or family planning.
- So many children die in infancy that parents tend to produce more in the hope that several will live.
- Many children are needed to work on the land.
- Children are regarded as a sign of virility.
- Religious beliefs (e.g. Roman Catholics, Muslims and Hindus) encourage large families.

High death rates, especially among children, are due to:
- Disease and plague (bubonic, cholera, kwashiorkor).
- Famine, uncertain food supplies, poor diet.
- Poor hygiene – no piped, clean water and no sewage disposal.
- Little medical science – few doctors, hospitals, drugs.

Stage 2 Birth rates remain high, but death rates fall rapidly to about 20 per 1000 people giving a rapid population growth.

The fall in death rate results from:
- Improved medical care – vaccinations, hospitals, doctors, new drugs and scientific inventions.
- Improved sanitation and water supply.
- Improvements in food production (both quality and quantity).
- Improved transport to move food, doctors, etc.
- A decrease in child mortality.

Stage 3 Birth rates fall rapidly, to perhaps 20 per 1000 people, while death rates continue to fall slightly (15 per 1000 people) to give a slowly increasing population.

The fall in birth rate may be due to:
- Family planning – contraceptives, sterilisation, abortion and government incentives.
- A lower infant mortality rate therefore less need to have so many children.
- Increased industrialisation and mechanisation meaning fewer labourers are needed.
- Increased desire for material possessions (cars, holidays, bigger homes) and less for large families.
- Emancipation of women, enabling them to follow their own careers rather than being solely child bearers.

Stage 4 Both birth rates (16 per 1000) and death rates (12 per 1000) remain low, fluctuating slightly to give a steady population.

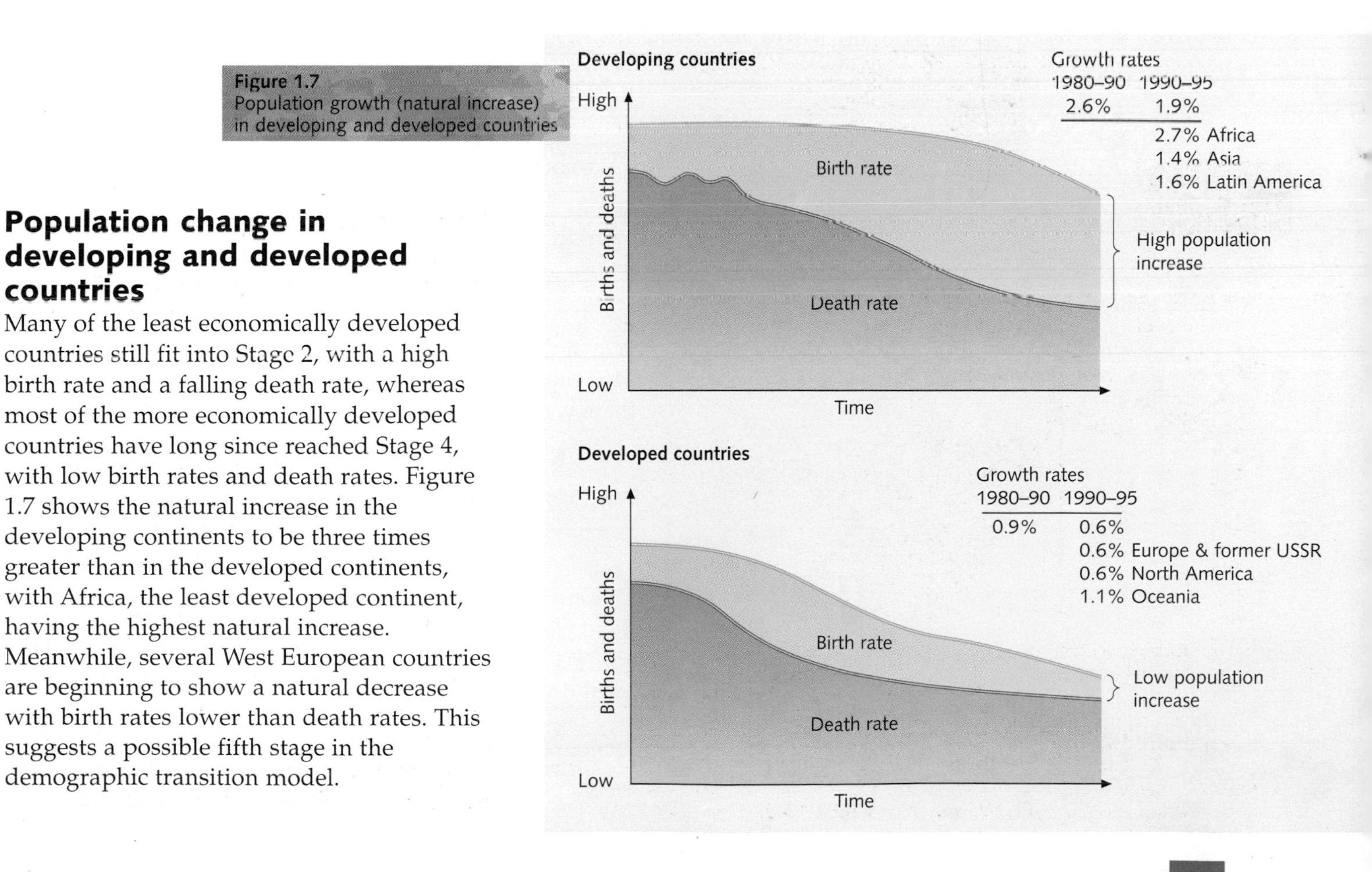

Figure 1.7 Population growth (natural increase) in developing and developed countries

Population change in developing and developed countries

Many of the least economically developed countries still fit into Stage 2, with a high birth rate and a falling death rate, whereas most of the more economically developed countries have long since reached Stage 4, with low birth rates and death rates. Figure 1.7 shows the natural increase in the developing continents to be three times greater than in the developed continents, with Africa, the least developed continent, having the highest natural increase. Meanwhile, several West European countries are beginning to show a natural decrease with birth rates lower than death rates. This suggests a possible fifth stage in the demographic transition model.

Population structures

The rate of natural increase, birth rate, death rate and life expectancy (*life expectancy* is the number of years that the average person born in a particular country can expect to live) all affect the population structure of a country. The population structure of a country can be shown by a population pyramid or, as it is sometimes known, an age–sex pyramid. The population is divided into five-year age groups (e.g. 5–9 years, 10–14 years), and also into males and females. The population pyramid for the United Kingdom is shown in Figure 1.8. The graph shows:

- A narrow pyramid indicating approximately equal numbers in each age group
- A low birth rate and a low death rate indicating a steady, or even a static, population growth
- More females than males live over 70 years
- There are more boys under 4 years of age than girls
- A relatively large proportion of the population in the pre- and post-reproductive age groups, and a relatively small number in the 15–64 age group which is the one that produces children and most of the national wealth.

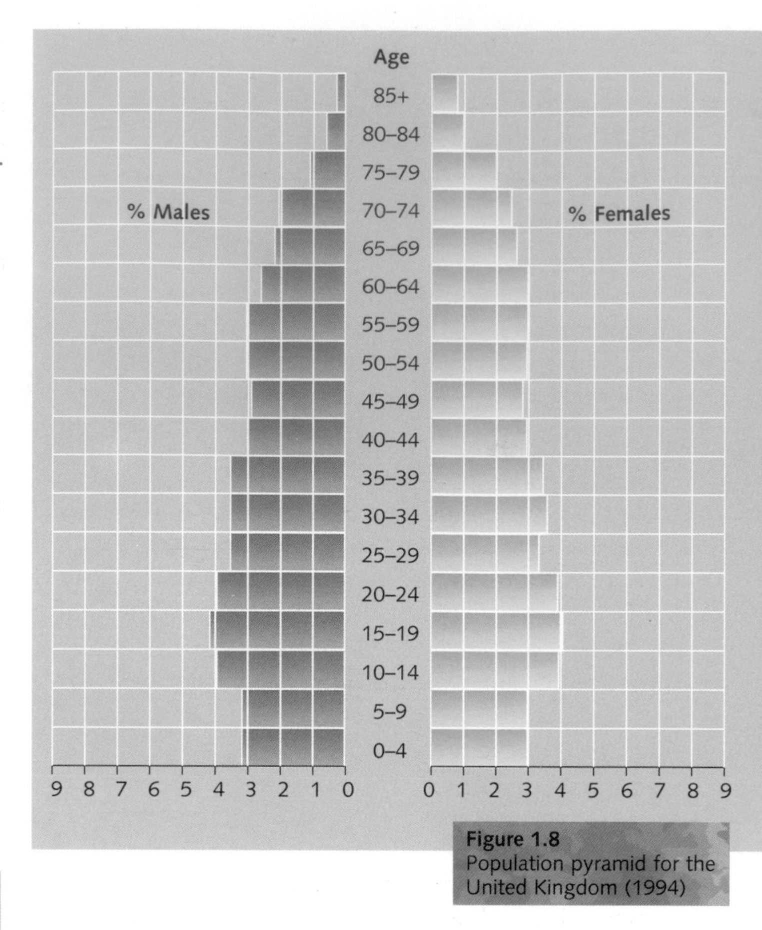

Figure 1.8
Population pyramid for the United Kingdom (1994)

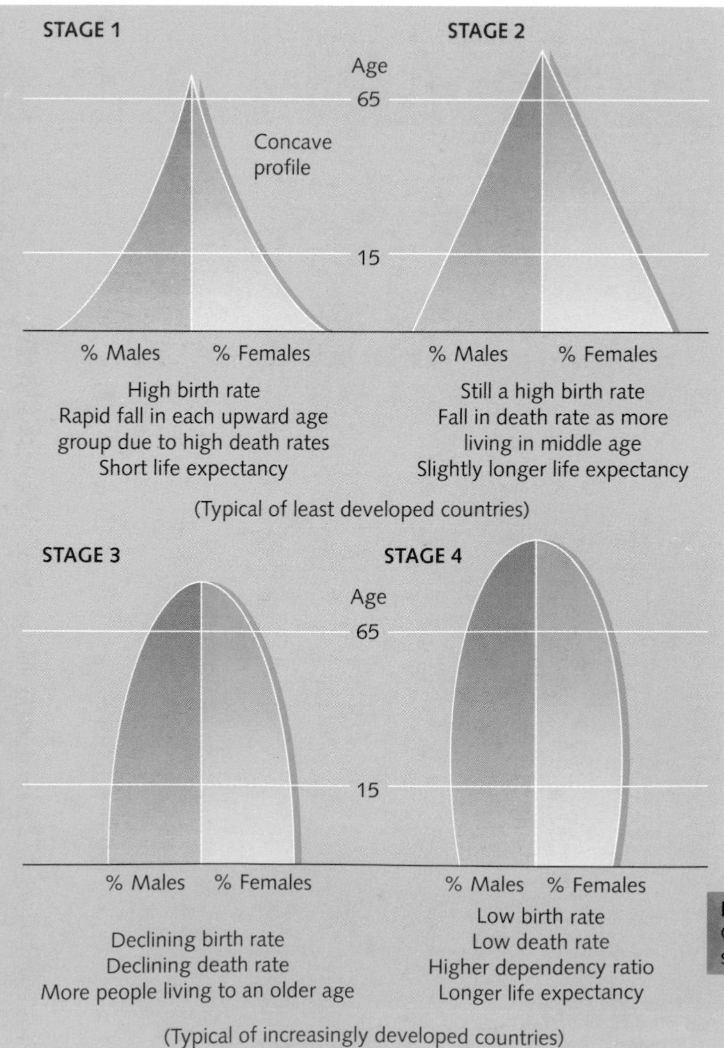

Figure 1.9
Changing population structures

This can be shown as the *dependency ratio* which can be expressed as:

$$\frac{\text{Non-economically active}}{\text{Economically active}} \quad \text{i.e.} \quad \frac{\text{Children (0–14) and and elderly (65+)}}{\text{Those of working age (15–64)}}$$

e.g. **UK 1971** (figures in millions)

$$\frac{13.387 + 7.307 \times 100}{31.616} = \text{Dependency ratio of } 65.45$$

That means that for every 100 people of working age, there were 65.45 people dependent upon them.

UK 1994 (figures in millions)

$$\frac{11.252 + 9.156 \times 100}{37.690} = \text{Dependency ratio of } 54.15$$

That means that for every 100 people of working age, there were 54.15 people dependent upon them.

Compared with 1971, the dependency ratio had fallen slightly. This was because despite a fall in Britain's birth rate, there had been an increase in the number of economically active people and an increase in life expectancy. (The dependency ratio does not take into account those of working age who are unemployed.) Most developed countries have a dependency ratio of between 50 and 70, whereas in developing countries the ratio is often over 100 due to the large numbers of children.

Population pyramids enable comparisons to be made between countries, and can help a country to plan for future service needs such as old people's homes if it has an ageing population or fewer schools if it has a declining, younger population. Unlike the demographic transition model (Figure 1.6), population pyramids include immigrants, but like that model they can produce four idealised types of graph representing different stages of development (Figure 1.9).

Stages of development

Three countries at different stages of economic development and which will be referred to in several later chapters are Kenya, India and Japan. The most recent population pyramids for these countries are given in Figure 1.10. How well, do you think, do the three pyramids fit with the models shown in Figure 1.9? Figure 1.10 confirms the point that population structures change over periods of time. Kenya is the least wealthy of the three. Its wide base (0–4 years) confirms that it has the highest birth rate and that its population declines rapidly due to a high infant mortality rate. Birth and infant mortality rates in India, which is slightly more wealthy, are both declining. This is shown by the narrower base (0–4 years) and relatively more people reaching child-bearing age (20–24 years). Both countries, and especially Kenya, have a rapidly narrowing pyramid indicating a high death rate and a short life expectancy (there are relatively few people aged over 65 years). The pyramid for Japan shows, by contrast, a low birth rate, a low death rate, a low infant mortality rate and a long life expectancy. The graphs also show that in Japan and India, in common with most other countries, there are slightly more male than female births; but that females have the longer life expectancy.

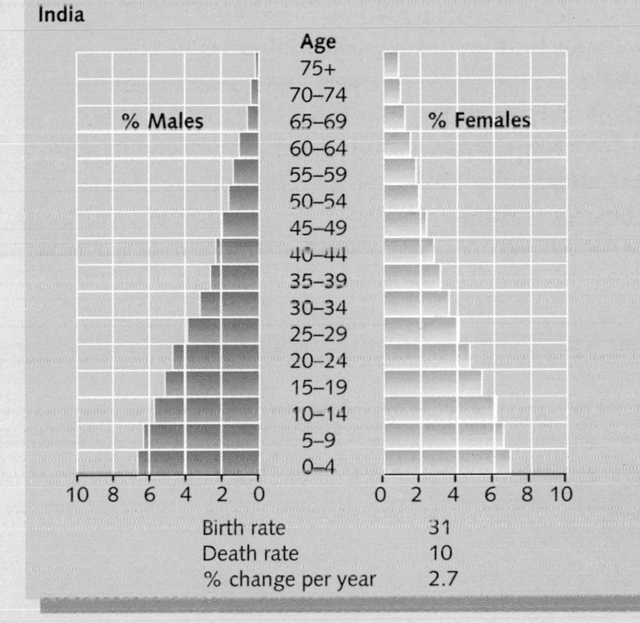

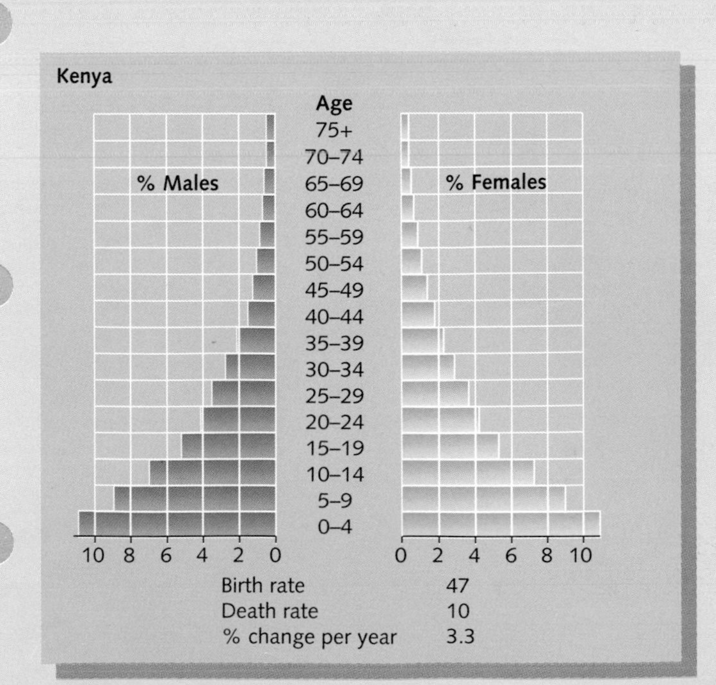

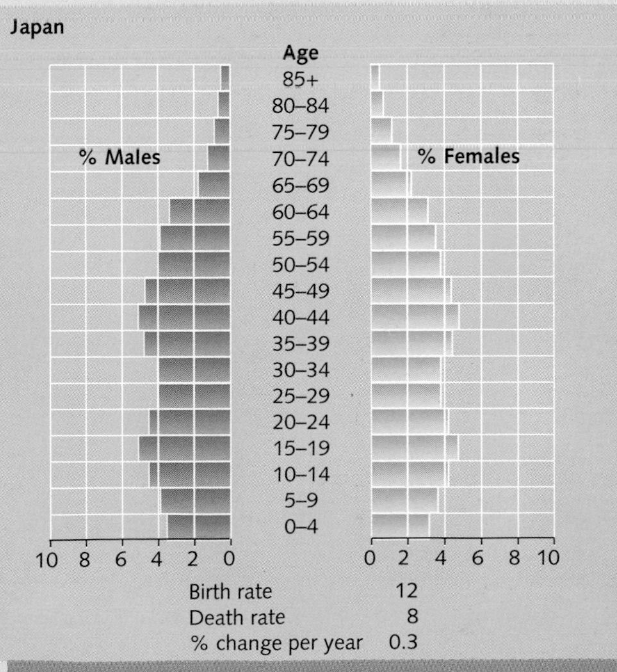

Figure 1.10
Population pyramids for Kenya, India and Japan

Places

Future trends

Between 1960 and 1970 the world's population grew, on average, by a record 2 per cent per year. The *growth rate*, which was even higher in economically less developed countries (Figure 1.4), caused increasing concern, as estimates suggested that the world's population would reach 7600 million by the year 2000. In 1996, however, the United Nations claimed that, taking the average for 1990 to 1995, the growth rate had fallen to 1.6 per cent a year. This means that the world's population is now not expected to exceed 6200 million by the turn of the century (Figure 1.11).

More recently, Earthscan has produced a report based on studies of fertility (i.e. birth rates), death rates and migration. This report suggests that the world's population will peak at around 10.6 billion in 2080 and then slowly decline. This decline is credited to:

- the global birth rate falling faster than previously predicted
- improvements in basic education and female literacy which have led to a reduction in the desired family size
- increased international migration to the more economically developed countries where people have fewer children
- diseases such as AIDS and malaria which are likely to reduce life expectancies, especially of people in sub-Saharan Africa, the region with the present highest birth rate.

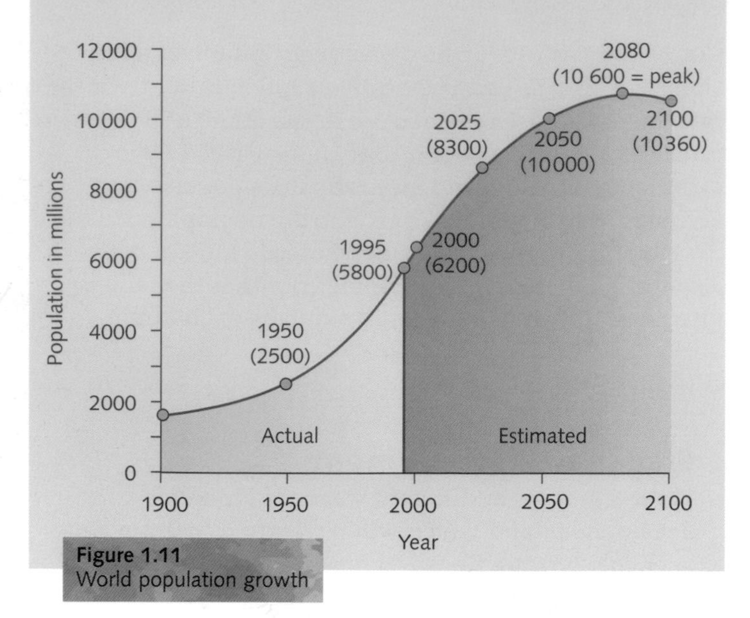

Figure 1.11
World population growth

Changing population structures

Children under 15

We have already seen (Figure 1.9) that countries at Stage 2 of the demographic transition model are economically less developed and have high birth rates. As high birth rates result in a high proportion of the total population, often over 40 per cent, being aged 15 or under (Figure 1.12), then:

- at present their youthful population will need child-health care and education – two services that these countries can ill-afford
- in the future, there will be more people reaching child-bearing age.

The consequence will be, as in most of Africa, a rapidly growing population.

Figure 1.12
Percentage of total population under 15 years

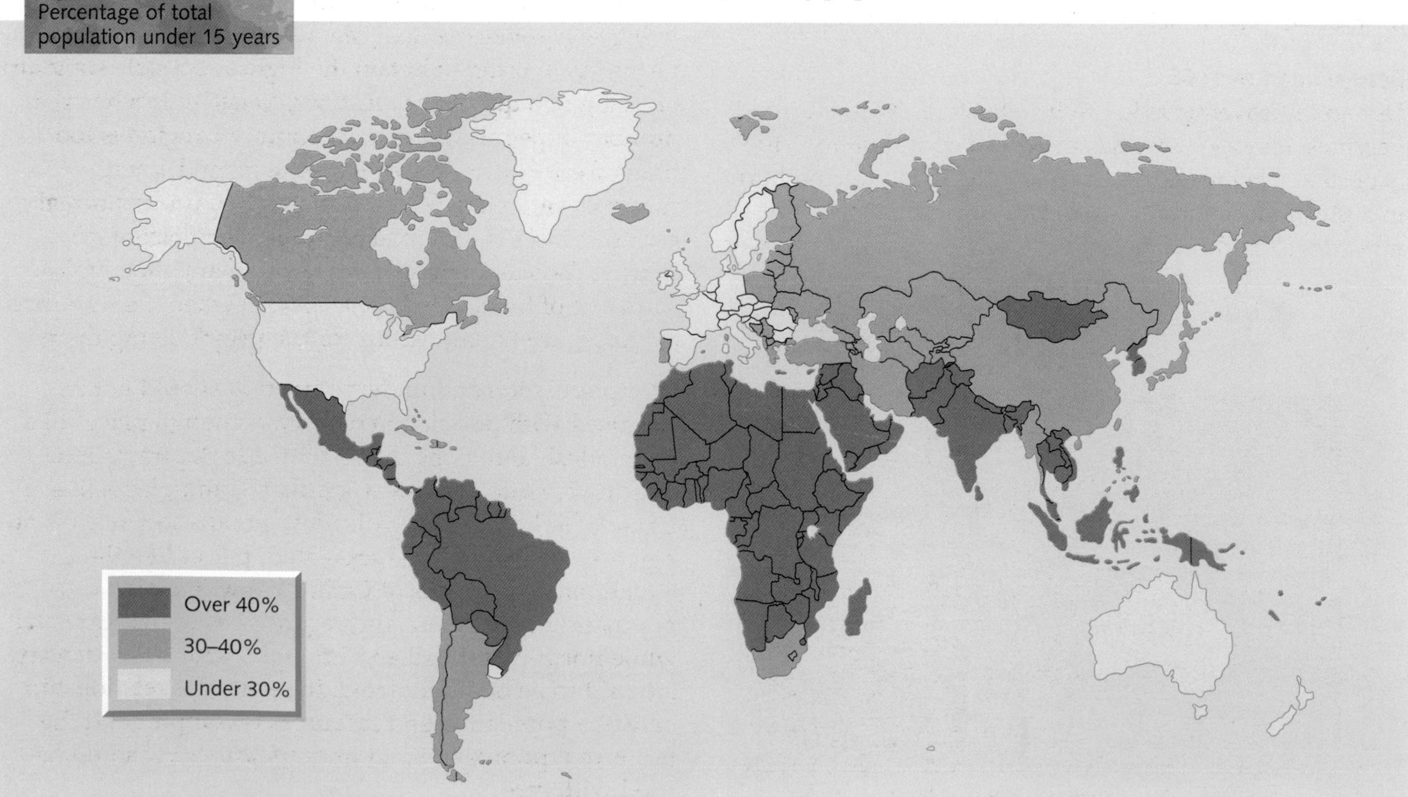

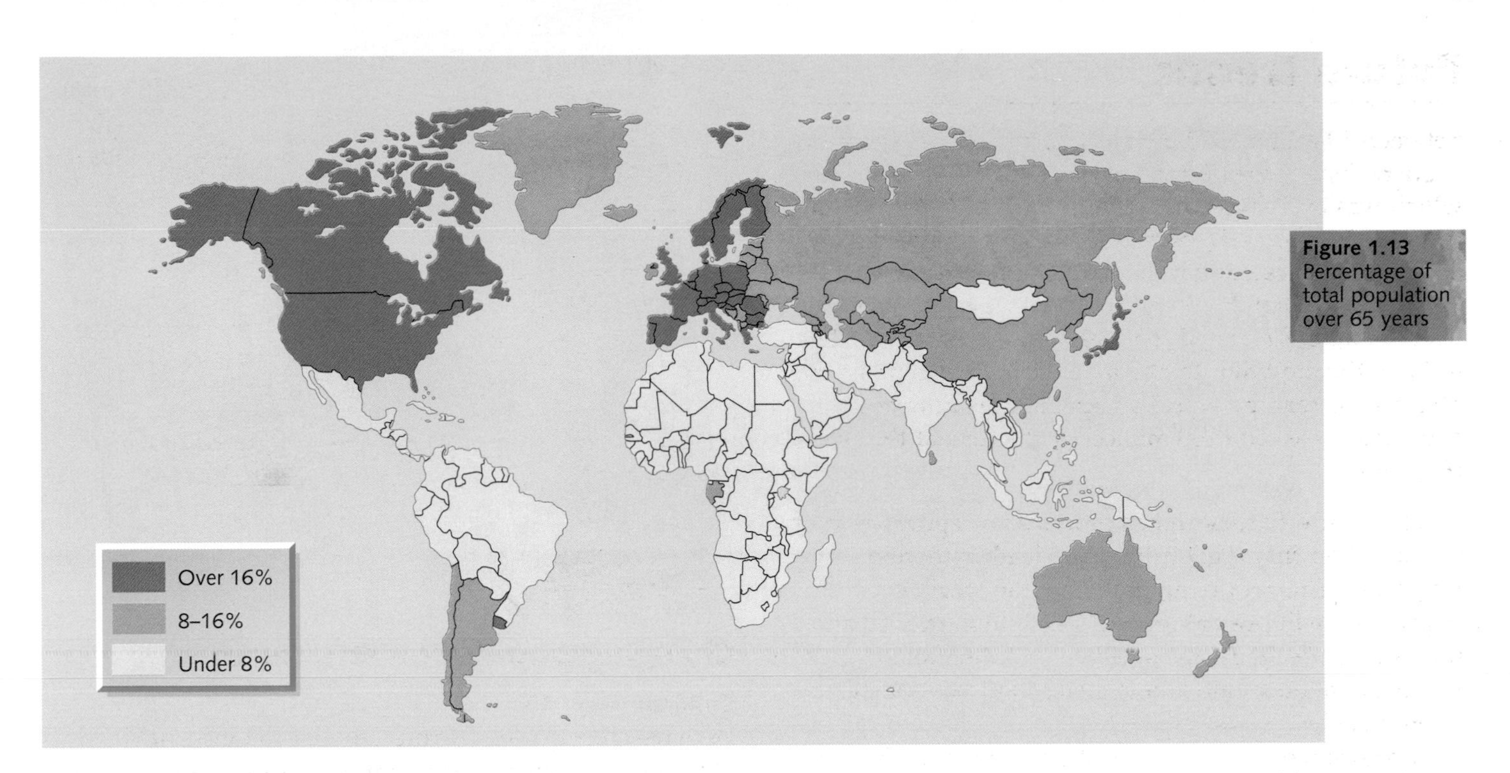

Figure 1.13 Percentage of total population over 65 years

In contrast, many of the economically more developed countries, with their low birth rates, have reached Stage 4. Here the problem is becoming 'too few' rather than 'too many' children. In 1997 the Population Institute claimed that there were already 79 countries in which too few children were being born to sustain long-term population growth. The worst affected are Germany, Italy and Russia, where deaths now exceed births. Countries where the *replacement rate* is not being met fear they will have too few consumers and skilled workers to keep their economy going, 'villages bereft of children and schools closed for lack of students', and a lack of security in providing pensions.

People aged over 65

Due to improvements in medical facilities, hygiene and vaccines, *life expectancy* has increased considerably – life expectancy being the number of years that a person born in a given place may be expected to live. Several of the most developed countries (Figure 1.13) already have over 16 per cent of their population of pensionable age, a figure likely to exceed 20 per cent in Japan (24 per cent), Germany (22 per cent) and France by the year 2020 (Figure 1.14). This is leading to an increase in demand for more money for pensions, medical care, residential homes and other social services. As a country's population continues to age, there will be fewer people in the economically active age group (page 8) to support them.

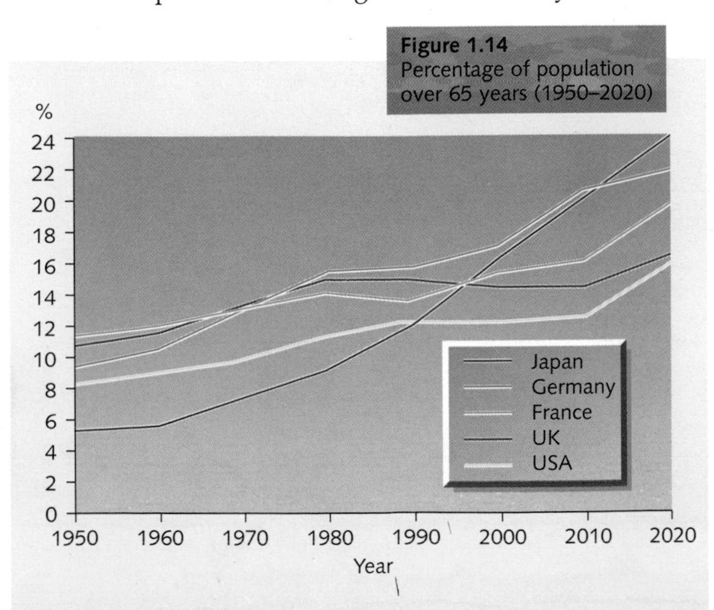

Figure 1.14 Percentage of population over 65 years (1950–2020)

Optimum, over- and underpopulation

Optimum population is, in theory, the number of people living in a given area that can maximise the use of resources in order to obtain the highest possible standard of living and quality of life. *Overpopulation* is when the number of people living in a country or region is too many for the resources available (e.g. food, energy, minerals and technology). It means that under normal circumstances there are likely to be insufficient jobs, houses, food and energy supplies to maintain a high standard of living. *Underpopulation* is when the resources available are too numerous for the people living there.

Overpopulation and underpopulation should not be confused with population density. Although places like Bangladesh and Hong Kong, with a high population density, are said to be overpopulated, and places like Canada and Australia, with a low population density, are regarded as being underpopulated, places like the Netherlands and parts of California, with a high population density, are also regarded as underpopulated, while north-east Brazil and Ethiopia, with a low density, are said to be overpopulated. In reality the relationship between population and resources is complex and the terms overpopulation and underpopulation should be used with care.

CASE STUDY 1

Population: Brazil

Distribution and density

In 1996 the average population density for Brazil was 19.3 per km^2 (i.e. 163 976 000 inhabitants living in 8 511 970 km^2). Although the distribution of population over the country is very uneven, the population density map (Figure 1.15) shows a relatively simple pattern. Over 90 per cent of Brazilians live near to the coast, mainly in the south-east of the country. Going inland, and towards the north and west, the density decreases very rapidly, with some of the more remote areas being virtually uninhabited.

The highest population densities occur either at irregular intervals along the coast or around the cities of São Paulo (Figure 1.16) and Belo Horizonte. Although the coastal climate is hot and wet, and flat land is limited due to mountains which frequently extend down to the sea, this region has a reliable water supply and a range of natural resources. Salvador and, later, Rio de Janeiro were the country's first two capital cities. Both had good natural harbours which encouraged trade, immigration, industry and, more recently, tourism (page 166). São Paulo, the world's second largest city, and Belo Horizonte grew up on the higher, cooler, healthier plateau of the eastern Brazilian Highlands. The rich soils around São Paulo were ideal for the growing of coffee (page 102). Later the presence of nearby minerals, such as iron ore, and energy supplies allowed the city to develop into a major industrial centre. The south-east region has the best transport system in Brazil, the greatest number of services, and has benefited most from government help.

Population density decreases with distance from the coast. This is because places further inland have fewer natural resources and cannot support as many people. Those areas shown as having an average density (between 0.5 and 4.9 on Figure 1.15) have a less favourable climate, a less reliable water supply, and fewer minerals and energy supplies than places to the south-east. Brasilia is an *anomaly*. An anomaly is something you would not normally expect. Brasilia exists only because in 1960 the Brazilian government chose this virtually uninhabited area as the site for the new federal capital. It was chosen in an attempt to spread out Brazil's population more evenly. Today Brasilia has a population of 1.6 million (page 65).

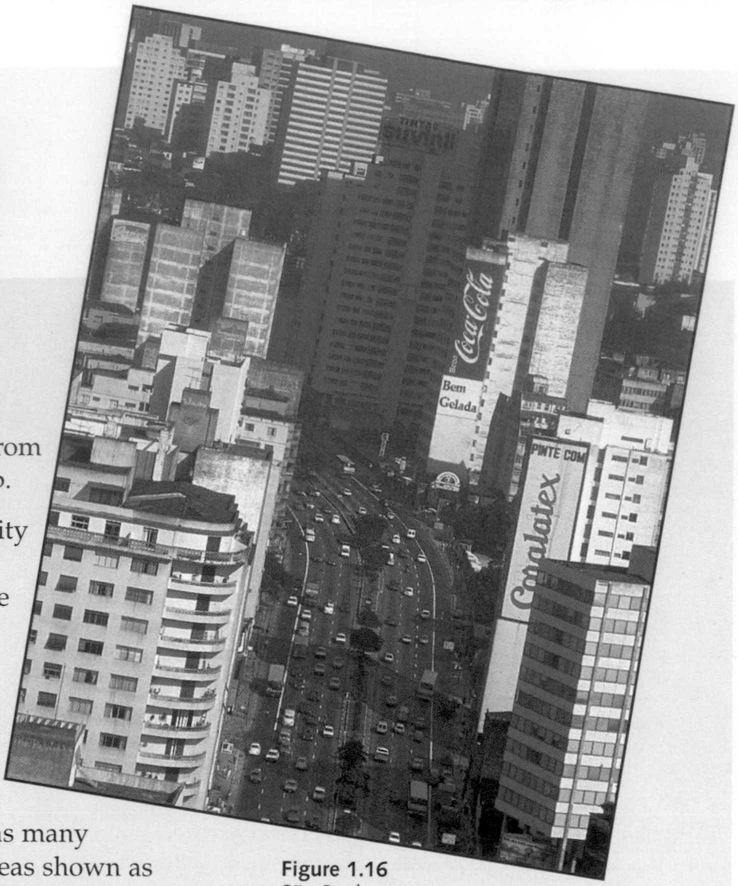

Figure 1.16
São Paulo

	Per km^2
North	2.6
North East	27.3
South East	69.2
South	38.4
West Central	5.9
All Brazil	**19.3**

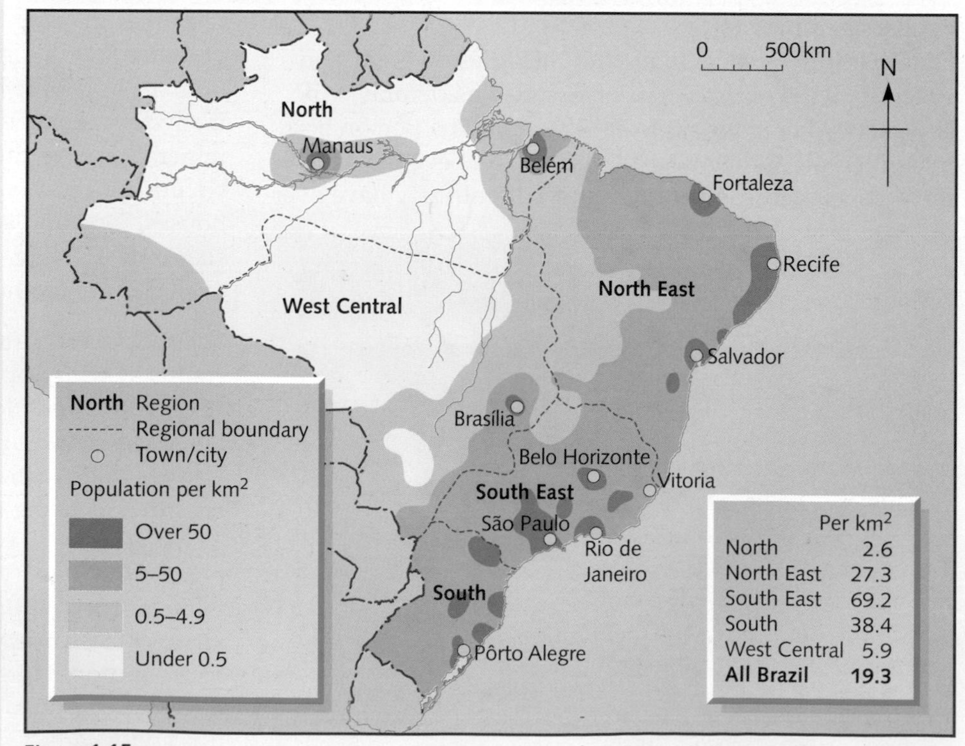

Figure 1.15
Population density in Brazil

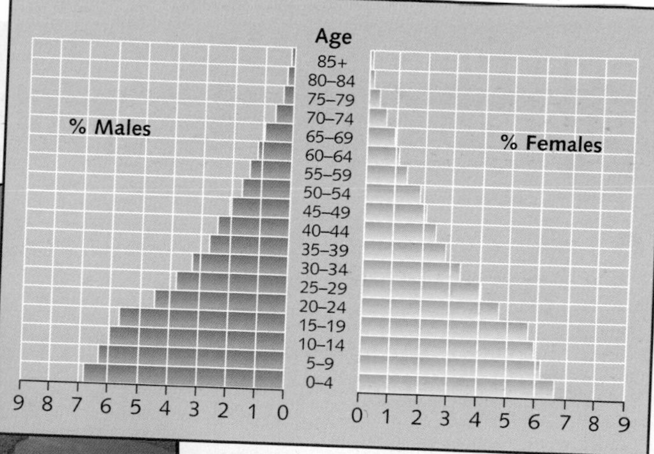

Figure 1.18
Population pyramid for Brazil

Figure 1.17
The Amazon and tropical rainforest

The sparsely populated region to the north-west is drained by the River Amazon and its numerous tributaries (Figure 1.17). The area, which is covered by tropical rainforest, is hot, wet and unhealthy. Soils are poor (page 100) and most areas lack known natural resources. Transport through the forest is difficult and much of the region lacks such basic services as health, education and electricity. Compared with south-east Brazil, birth, death and infant mortality rates are high and life expectancy is short. The region has, until recently, suffered from a lack of government investment. There are two anomalies in this region. The first is Manaus (population 1 million), which grew as the centre for rubber collecting and is now a large river port with a growing tourist industry. The other includes several places near to the mouth of the Amazon which have benefited either from the discovery of iron ore and bauxite or from the production of hydro-electricity (page 110).

Structure

Although figures for Brazil suggest that it has reached Stage 3 of the demographic transition model, parts of the interior still show the characteristics of Stage 1 (Amazon Indians) and Stage 2. Brazil's population pyramid, with its relatively wide base (Figure 1.18), shows that although both birth and infant mortality rates are still high, they are falling (birth rate 26 in 1991, 23 in 1996; infant mortality rate 59 in 1991, 57 in 1996). This means that, as shown by the widening pyramid, more people are reaching child-bearing age (20–24 year age group). Beyond that, the pyramid narrows, indicating that death rates are still high, although falling (10 in 1991; 8 in 1996), and life expectancy, although shorter than in a more developed country, is slowly increasing (64 for males; 69 for females in 1996).

Trends

Brazil's annual average population growth rate was 1.7 in 1996 compared with 2.1 in 1991 and 3.1 in 1960. The country still has a high proportion of its population aged 15 and under although this is now showing signs of a decline (39 per cent in 1991; 32.3 per cent in 1996). In contrast the proportion aged 65 and over is small but is beginning to increase (4.3 in 1991; 6.7 in 1996). The fertility rate of 2.8 (the number of children per female) hides the difference between the better-off urban areas in the south-east, and the more rural areas (Figure 1.19) and urban favelas (page 66).

"Families with children into the double figures are very common in north-east Brazil. Rural mothers in this state, Rio Grande do Norte, have an average of more than seven living children. A quarter of them have ten or more. And that is not counting the dead ones. You often meet women like Luisa Gomez, a slight, small 39-year-old. She married at 14. Since then she has been pregnant sixteen times, once every eighteen months. For half her adult life she has been pregnant, and for the other half breast-feeding the most recent addition. Only six of those sixteen are still alive. There were three stillbirths, and seven died in their first year. Ten wasted pregnancies. Seven and a half years of drain on an already weak organism for nothing. Worse than nothing, for all the anxiety, all the care, all the concern, and then the grief."

Figure 1.19
Having children in north-east Brazil

QUESTIONS

1

(Pages 4 and 5)

a i) What is meant by the terms: ■ population distribution and ■ population density? (2)

ii) How is population density calculated? (1)

b

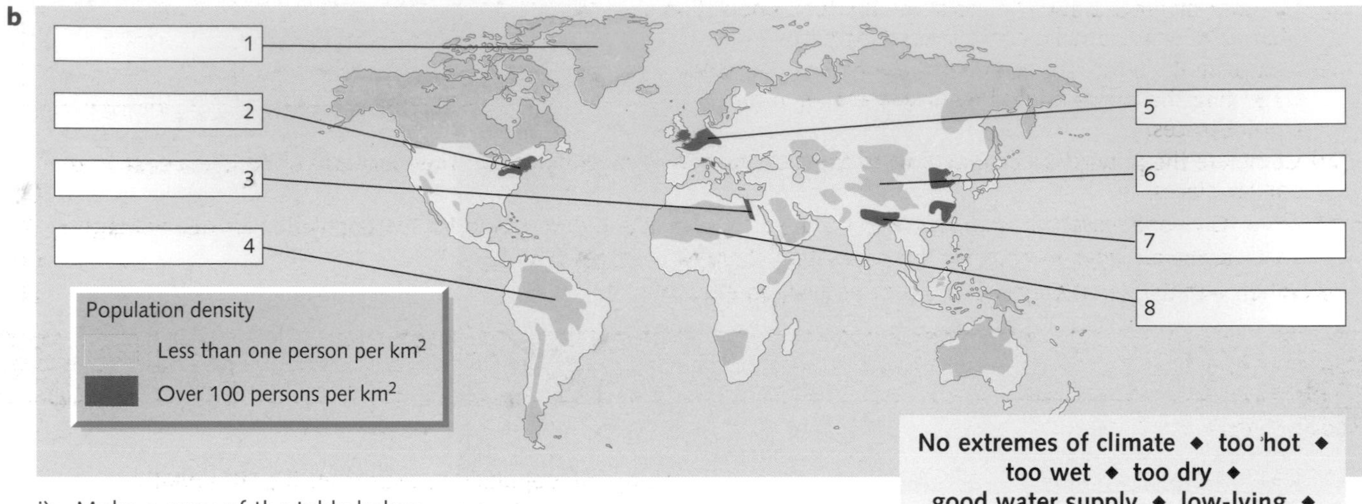

i) Make a copy of the table below.
Match the numbers on the map with the places named in the table. (8)

ii) Give reasons for the population density of each place.
The list (right) should help you with your answer.
Each label in the list may be used more than once. (16)

No extremes of climate ◆ too hot ◆ too wet ◆ too dry ◆ good water supply ◆ low-lying ◆ many resources ◆ few resources ◆ flat land ◆ fertile soil ◆ too mountainous ◆ good transport ◆ money for development

DENSELY POPULATED	MAP NUMBER	REASONS	SPARSELY POPULATED	MAP NUMBER	REASONS
EASTERN USA			HIMALAYAS		
GANGES VALLEY (INDIA/BANGLADESH)			SAHARA DESERT		
NORTH-WEST EUROPE			AMAZONIA		
NILE VALLEY (EGYPT)			GREENLAND		

2

(Page 6)

a i) What was the world's population in 1750? (1)

ii) What is it expected to be in the year 2000? (1)

b i) Name the three developing continents. (1)

ii) What is the expected population of the three developing continents in the year 2000? (1)

iii) Which continent will have the biggest share of the world's population in the year 2000? (1)

c i) Which three continents have more than doubled their population since 1945? (1)

ii) What proportion of the world's population (per cent) is expected to live in developing countries in the year 2000? (1)

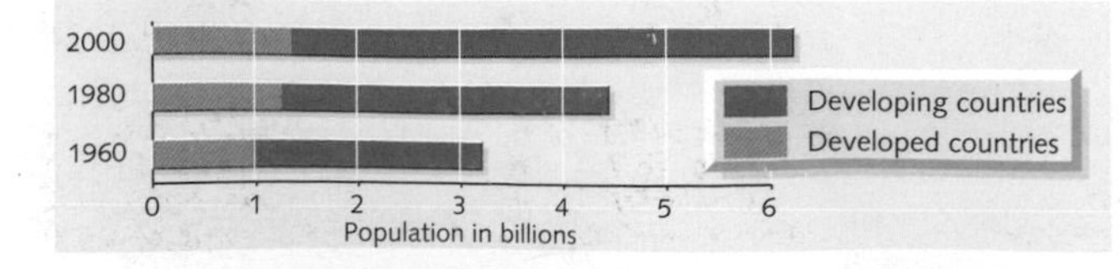

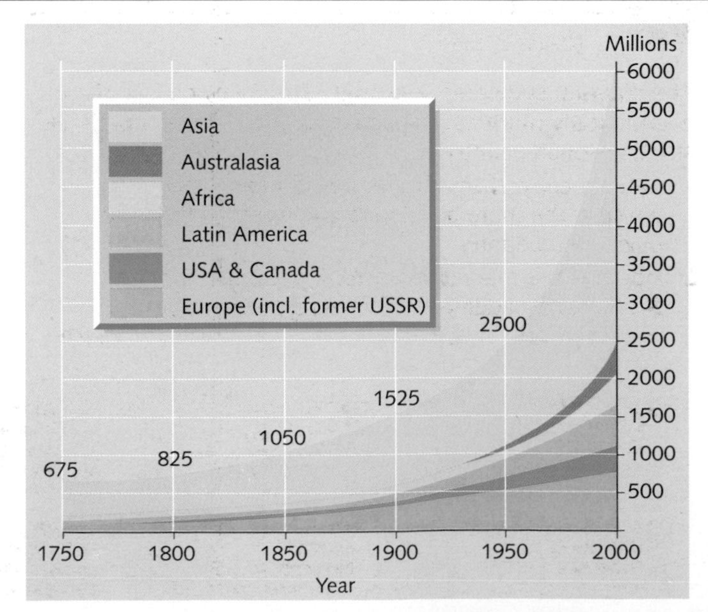

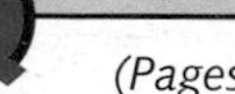

3

(Pages 6 and 7)

a What is meant by the following terms?
i) birth rate ii) death rate iii) natural increase (3)

b This diagram is the demographic transition model. It shows, in a simplified way, how population changes are likely to take place over a period of time. Make a copy of the figure and the table beneath it.

i) On your diagram label the birth rate and the death rate. (2)

ii) On your diagram draw a line to show the total population. Start at X. Shade and label the natural increase. (2)

iii) Complete the birth rate and death rate part of the table by adding the words: **high**, **low** or **decreasing** in the correct places. (3)

iv) Complete the population change part of the table by adding **slow decrease**, **slow increase** or **rapid increase** in the correct places. (3)

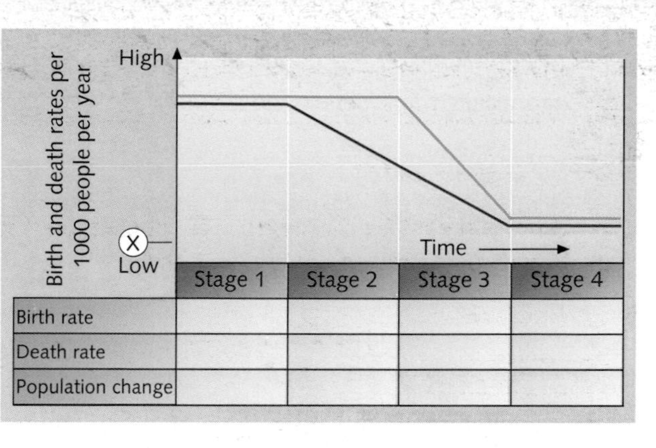

	Stage 1	Stage 2	Stage 3	Stage 4
Birth rate				
Death rate				
Population change				

c i) *Either* Give one reason for the population change in Stage 2 and one reason for the population change in Stage 3. (2)
Or Describe and give reasons for the population changes in Stages 2 and 3. (4)

ii) When was the UK at Stages 2 and 3 of population growth? (2)

4

(Page 8)

Age	% Males	% Females
75+	4.8	8.9
60–74	13.0	14.5
45–59	17.6	17.0
30–44	21.8	20.6
15–29	22.5	20.5
0–14	20.3	18.5

25 20 15 10 5 0 0 5 10 15 20 25

Using the population pyramid for the UK (1994):

a Complete the table. (6)

AGE GROUP	0–14	15–59	60 AND OVER
TOTAL %			

b i) On this graph, which three groups make up the dependant population? (3)

ii) Why are these groups called the dependant population? (2)

c In 1994 the dependency ratio for the UK was 54 and for India 86. Which is the more preferable ratio and why? (3)

5

(Pages 8 and 9)

The diagram compares population structures typical of an economically more developed country and an economically less developed country.

a Make a copy of the diagram. Add titles to say which is the more developed and which is the least developed country. (1)

b Describe the main differences in the population structures by adding the following labels in the correct places. (10)

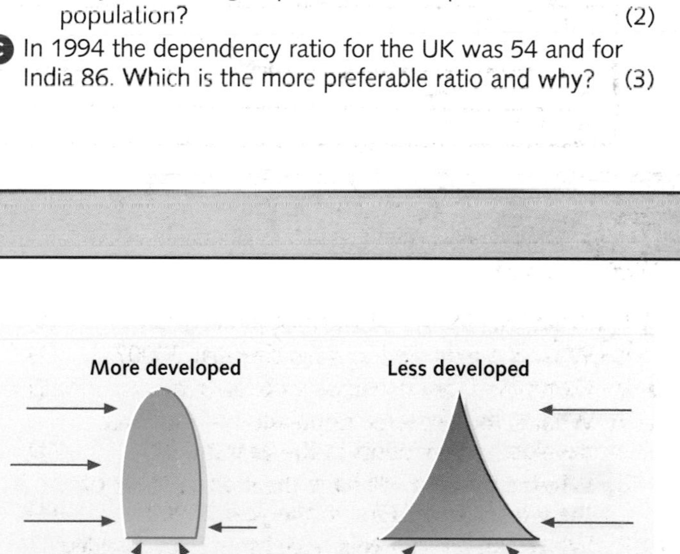

HIGH BIRTH-RATE, MANY CHILDREN	MANY ELDERLY, LONG LIFE EXPECTED	SQUAT-PYRAMID SHAPE	MANY PEOPLE DIE WHEN YOUNG	MANY MIDDLE AGED
NEARLY EQUAL NUMBERS AT EACH AGE	GENERALLY NARROW PYRAMID	MANY MORE YOUNG THAN OLD	LOW BIRTH RATE, FEW CHILDREN	FEW ELDERLY, SHORT LIFE EXPECTANCY

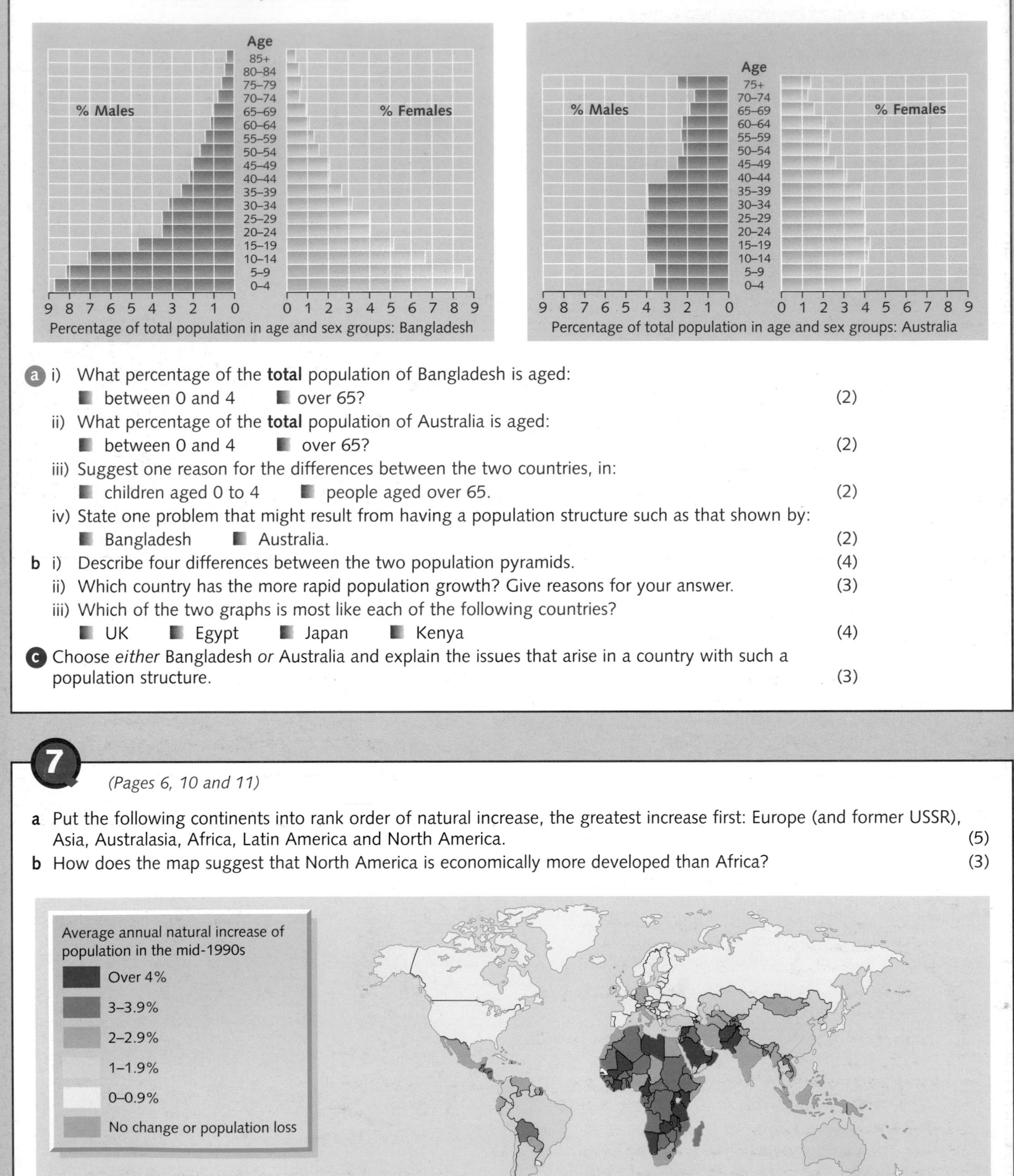

6

(Pages 8 and 9)

Study the population pyramids for Bangladesh (an economically less developed country) and Australia (an economically more developed country).

a i) What percentage of the **total** population of Bangladesh is aged:
- between 0 and 4
- over 65? (2)

ii) What percentage of the **total** population of Australia is aged:
- between 0 and 4
- over 65? (2)

iii) Suggest one reason for the differences between the two countries, in:
- children aged 0 to 4
- people aged over 65. (2)

iv) State one problem that might result from having a population structure such as that shown by:
- Bangladesh
- Australia. (2)

b i) Describe four differences between the two population pyramids. (4)

ii) Which country has the more rapid population growth? Give reasons for your answer. (3)

iii) Which of the two graphs is most like each of the following countries?
- UK
- Egypt
- Japan
- Kenya (4)

c Choose *either* Bangladesh *or* Australia and explain the issues that arise in a country with such a population structure. (3)

7

(Pages 6, 10 and 11)

a Put the following continents into rank order of natural increase, the greatest increase first: Europe (and former USSR), Asia, Australasia, Africa, Latin America and North America. (5)

b How does the map suggest that North America is economically more developed than Africa? (3)

8

(Pages 10 and 11)

a Give two reasons why many economically less developed countries have high birth rates. (2)

b i) State briefly why some people think that it is important to try to control the growth of world population. (3)

ii) With the help of the picture below, describe four methods by which the growth of world population might be controlled. (4)

c i) Why have developed countries been more successful than developing countries in lowering their birth rates? (3)

ii) What problems now face many developed countries as a result of their success in lowering the birth rate? (3)

9

(Page 11)

a i) Give two reasons why people living in economically more developed countries have a long life expectancy. (2)

ii) Why are governments in developed countries concerned about the increase in life expectancy? (2)

b Governments of countries with ageing populations may have to:

1 raise taxes 2 encourage more private saving

3 raise the present retirement age

4 encourage a higher birth rate.

Choose two of these policies and explain how they might help a government to cope with an ageing population. (4)

10

(Pages 4 to 7 and 10 to 13)

a Study Figure 1.18 which shows differences in population density in Brazil.

i) Describe the density of population in Brazil by locating places with: ■ a high density ■ an average density ■ a low density. (6)

ii) Give reasons for differences in population density. (9)

b Using this demographic transition graph for Brazil:

i) Give the 1996 birth and death rates. (2)

ii) Say which of the following periods was the one during which the population of Brazil grew most rapidly: 1920–1930; 1940–1950; 1960–1970; 1980–1990. (1)

iii) Give the total population in 1960 and in 1996. (2)

iv) Copy the graph and complete the line for total population between the years 1960 and 1996. (1)

v) Predict what you think Brazil's population might be in 2020. Give two reasons for your answer. (3)

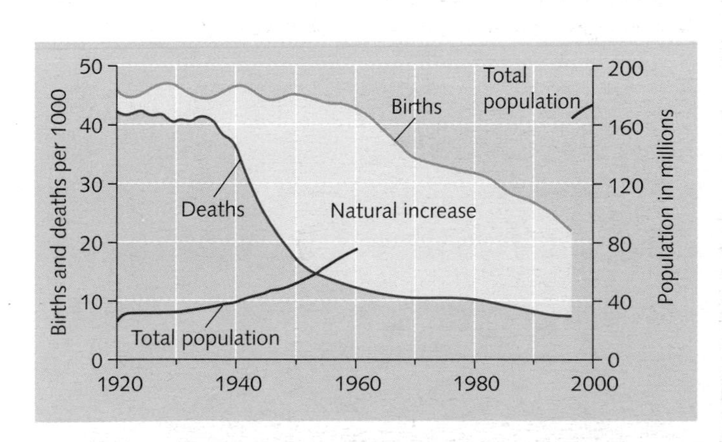

c Some parts of Brazil are overpopulated, others are underpopulated. What do the terms 'overpopulation' and 'underpopulation' mean? (2)

2 SETTLEMENT

A settlement is a place where people live. It can be large or small, permanent or temporary. Although each settlement is unique, it is likely to share similar characteristics with other settlements. Settlements can be grouped together (i.e. classified) using a variety of criteria. These criteria include location (*site and situation*), shape (*patterns*), major use (*function*), and position, or *rank*, as a service centre (*hierarchy*).

Site and situation

The location of a settlement is related to its site and situation.

Site describes the point at which the town (or hamlet, village, city) is located. Factors such as local relief, soil, water supply and resources were important in choosing the initial site of a settlement.

Situation describes where the settlement is located in relation to surrounding features such as other settlements, mountains, rivers and communications. It is the situation of a settlement that determines whether or not it will continue to grow to become a large town or city or whether it remains as a small hamlet or village. Paris, for example, had the site advantage of being located on an island in the River Seine which could be defended and which made bridging easier (Figure 2.1). However, the continued growth of Paris into Europe's largest city was due to its situation in the centre of a major farming area where several routes (rivers) converged.

Figure 2.2
Backbarrow, Cumbria: a wet-point site on a river

Figure 2.1
Ile de la Cité, Paris

Early settlements developed within a rural economy which aimed to be self-sufficient. Their sites were determined by a series of mainly physical factors. An ideal site was likely to have the benefit of several of these factors.

A **wet-point site**, especially in relatively dry areas, was essential as water is needed virtually every day throughout the year and is heavy to carry any distance. In early times, rivers were still sufficiently clean to give a safe, permanent supply. In lowland Britain, many early settlements were located at springs at the foot of chalk (Figure 2.2) or limestone escarpments.

A **dry-point** site, especially in relatively wet areas, was needed to avoid flooding or to be above unhealthy marshland, e.g. Ely in eastern England, which was built on a mound that acted as a natural island in the Fens.

Building materials, which ideally included stone, wood and clay, had to be obtained locally as these were heavy and bulky to move at a time when transport was poorly developed.

Defence against surrounding tribes was sometimes necessary. Good defensive sites may have been within a river meander, with the river giving protection on three sides, as at Durham and Shrewsbury (Figure 2.3), or on a hill with steep sides and commanding views as with Edinburgh and many Mediterranean settlements (Figure 2.4).

Figure 2.3
Shrewsbury

Figure 2.4
A Mediterranean hilltop village

A **fuel supply** was needed for heating and cooking, which was, in earlier times in Britain and still today in many developing countries, usually wood.

Food supplies were needed from land nearby, some of which was suitable for the rearing of animals and some for the growing of crops, e.g. sheep on the South Downs and crops in the clay vales (Figure 2.2).

Nodal points were where several valleys (natural routes) met to create a route centre, as at York and Paris, or where two rivers joined, as at Khartoum (Blue and White Nile) and St Louis (Mississippi and Missouri), or which controlled routes between hills, as at Guildford and Toulouse.

Bridging points may originally have been at a ford in the river, e.g. Oxford, or where the river was shallow and narrow enough to enable a bridge to be built, e.g. Paris.

Shelter and aspect In Britain it is an advantage to be sheltered from the strong prevailing south-westerly and cold northerly winds, and to have a south-facing aspect, as this gives most sunshine, heat and light (e.g. Torquay).

Figure 2.5 shows an area of land available for early settlement. What were the advantages and disadvantages of each of the five sites labelled A to E? Were any of the sites ideal? Finally, remember that many of these early site factors may no longer be important, e.g. in Britain most settlements now have a piped water supply, have one or more shops providing food, and do not need to be defended.

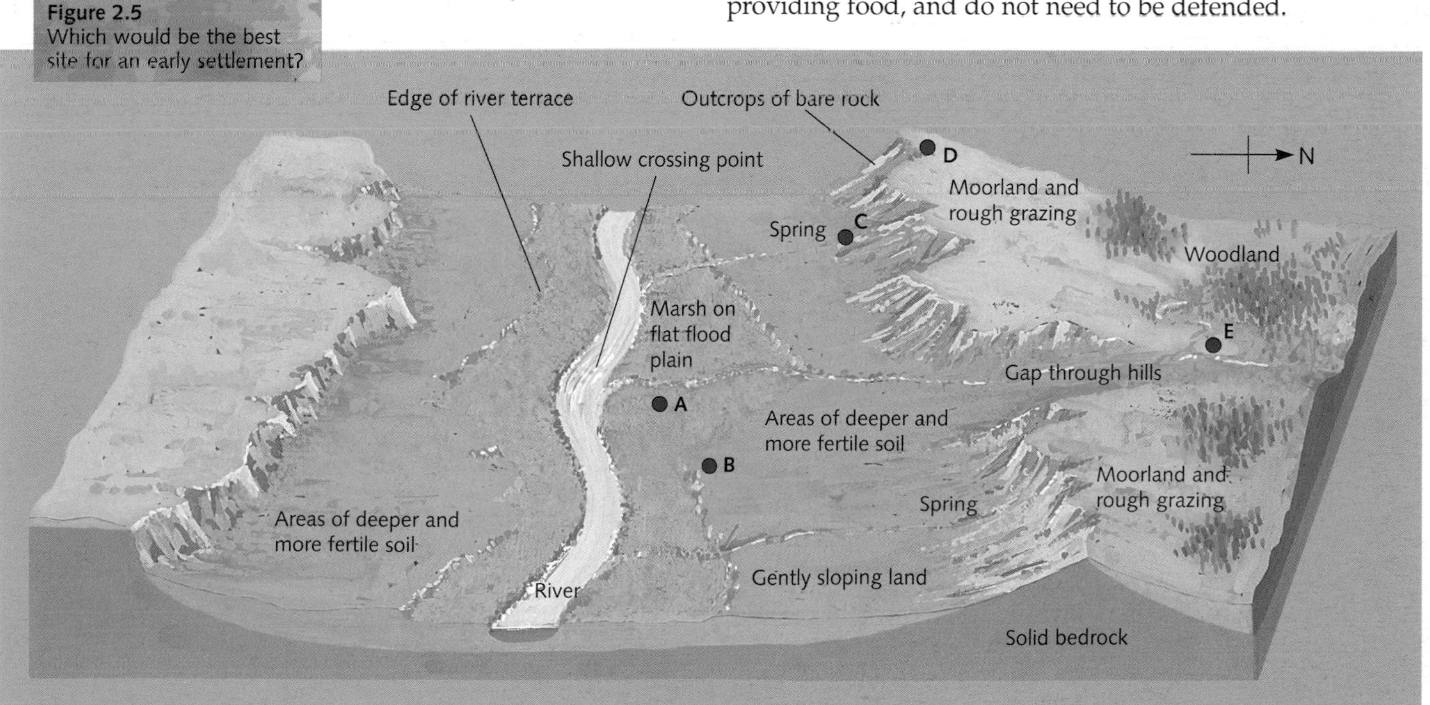

Figure 2.5
Which would be the best site for an early settlement?

Settlement types

Settlements are, for convenience, divided into rural and urban. However, it is often difficult to see and tell the difference between the two basic types. This difficulty is partly due to problems in defining the difference between the terms *rural* and *urban* and between what constitutes a village and a town. Figure 2.6 lists the usually accepted types of rural and urban settlement.

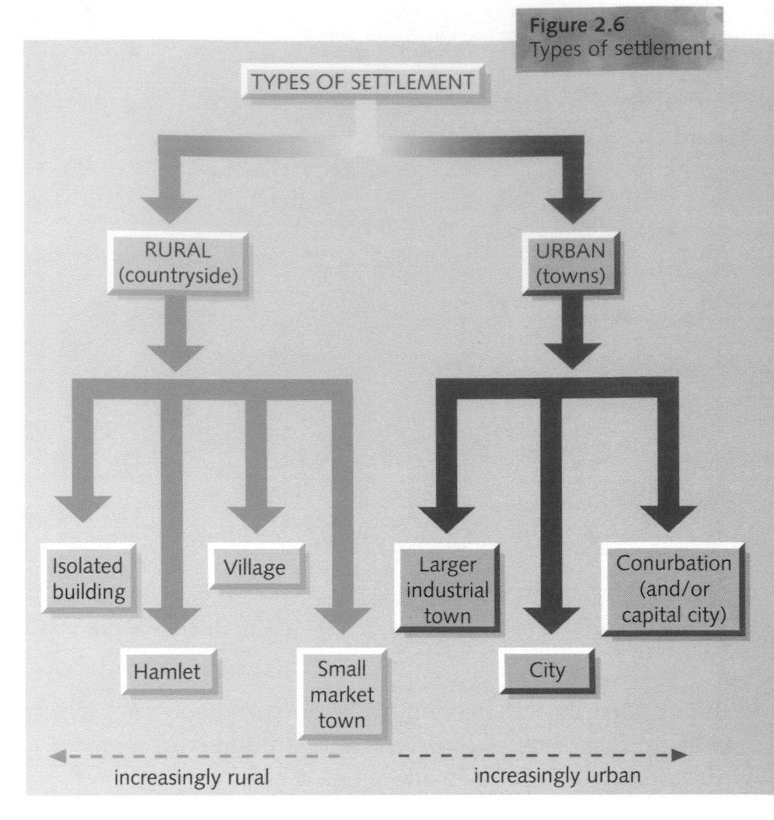

Figure 2.6
Types of settlement

Patterns

Geographers are interested in the patterns and shapes of villages and towns as well as in their main functions. Although villages have characteristic shapes, these vary from place to place both within Britain and across the world. Although it is unusual to find all the characteristic shapes within a small area, most can be seen in south Holderness to the east of Kingston upon Hull (Case Study 2, page 24).

Dispersed (Figure 2.7) This can either be:

- an *isolated*, individual building, or
- a group of two or three buildings, perhaps forming a hamlet, and separated from the next group by two or three kilometres.

Dispersed settlement occurs in an area of adverse physical difficulty where natural resources are insufficient to support more than a few people. Traditionally, most buildings are farms although increasingly some are being used as second homes or for holidays. In Britain, dispersed settlement occurs in mountainous parts of Scotland, Wales and northern England and in previously marshy areas such as the Fens (Figure 2.9).

Figure 2.7
Dispersed settlement in Northumberland

Nucleated (Figure 2.8) This is when several buildings were grouped together, initially often for defensive purposes and later for social and economic reasons. The nucleation of buildings into villages occurred where there was enough farmland for the inhabitants to be self-sufficient, as in the English Midlands and East Anglia, and where the water supply was reliable (Figure 2.2). Nucleated settlements often occur every 5 to 10 kilometres (Figure 2.9).

Linear or street (Figure 2.10) Linear settlement occurs where buildings are strung out along a line of communication. This may be a main road (e.g. leading out of a British city), a river valley (e.g. South Wales) or a canal or dyke (e.g. English Fens and the Netherlands).

Figure 2.8
A nucleated settlement: Hooton Pagnell, in Yorkshire

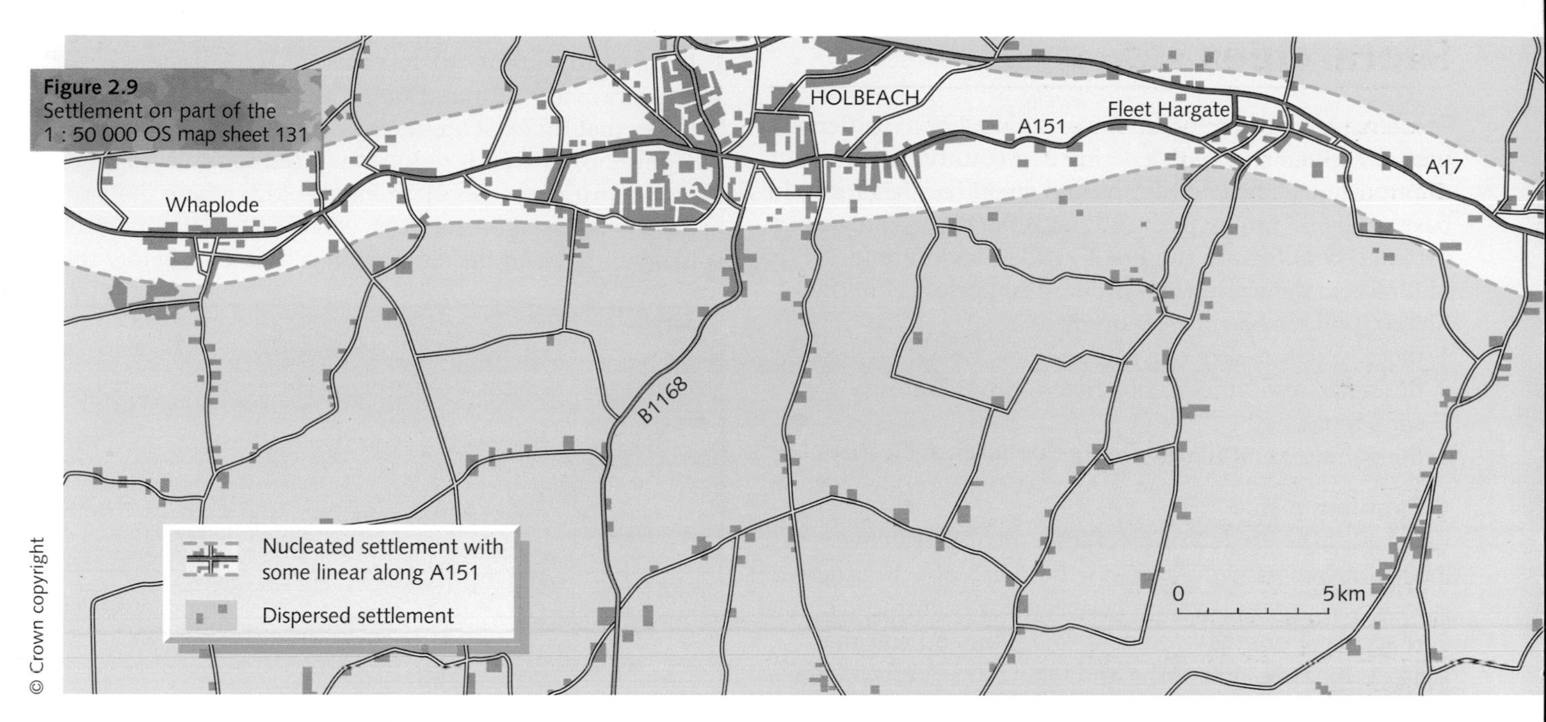

Figure 2.9
Settlement on part of the 1 : 50 000 OS map sheet 131

Planned These include suburbanised villages (page 44), surrounding large towns, and settlements on newly reclaimed or developed land (Dutch polders, page 45).

Figure 2.10
Linear settlement at Parson's Drove, Cambridgeshire

Functions

The function of a settlement relates to its economic and social development and refers to its main activities (Figure 2.11). Today, most larger settlements tend to be multi-functional (i.e. they have several functions) although one or two functions may be predominant. In some cases, the original function may no longer be applicable, e.g. British towns no longer have a defensive function. In other cases functions have changed over a period of time, e.g. a fishing village may now be a tourist resort, or a mining town may now have some high tech industry.

Figure 2.11
Types of function

		UK EXAMPLE	WORLD EXAMPLE
MARKET TOWNS	ORIGINALLY COLLECTING AND DISTRIBUTION CENTRES FOR SURROUNDING FARMING AREA. TODAY THEY MAY SERVICE AND PROCESS AGRICULTURAL MACHINERY AND PRODUCE.	YORK	WINNIPEG
MINING TOWNS	DEVELOPED TO EXPLOIT LOCAL MINERAL OR FUELS.	CORBY	PRUDHOE BAY
INDUSTRIAL–MANUFACTURING	WHERE RAW MATERIALS ARE PROCESSED INTO MANUFACTURED GOODS.	BIRMINGHAM	PITTSBURGH
PORTS	LOCATED ON COASTS, RIVERS AND LAKES FOR THE MOVEMENT OF GOODS AND PEOPLE FROM LAND TO SEA, OR VICE VERSA	SOUTHAMPTON	THUNDER BAY
ROUTE CENTRES	AT THE CONVERGENCE OF SEVERAL NATURAL ROUTES OR AT NODAL POINTS RESULTING FROM ECONOMIC DEVELOPMENT	CARLISLE	PARIS
COMMERCIAL	PROVIDING THE NEEDS OF INDUSTRY AND BUSINESS.	LONDON	HONG KONG
CULTURAL/RELIGIOUS	ATTRACTING PEOPLE, PERHAPS FOR A SHORT PERIOD, FOR EDUCATIONAL AND RELIGIOUS PURPOSES.	CAMBRIDGE	ROME
ADMINISTRATIVE	DEVELOPED TO CONTROL AREAS WHICH MAY VARY FROM A SMALL REGION (COUNTY TOWN) TO A COUNTRY (CAPITAL CITY).	EXETER	BRASILIA
RESIDENTIAL	WHERE THE MAJORITY OF RESIDENTS LIVE BUT DO NOT WORK.	TELFORD	MARNE-LA-VALLEE
TOURIST RESORTS	INCLUDE SPA TOWNS, COASTAL AND MOUNTAIN RESORTS.	BATH	ORLANDO

Hierarchies

The term *hierarchy* refers to the arrangement of settlements within a given area (e.g. a country or county) in an 'order of importance'. Isolated farms and small hamlets form the base of the hierarchy pyramid, with the largest and/or capital city at the top (Figure 2.12). Three different methods to determine the 'order of importance' in the hierarchy have been based upon:

1 the population size of a settlement
2 the range and number of services provided by a settlement
3 the sphere of influence, or market area, of a settlement.

1 Population size

Early attempts to determine a settlement hierarchy were based on size. However, no one has been able to produce a widely accepted division between, for example, a hamlet and a village or a village and a town. Indeed so-called villages in places like India and China are often as large as many British towns. Figure 2.12 lists the conventional hierarchy, as applied to Britain, in terms of types of settlement. However, the cut-off points for divisions based on population size and the distance between settlements, use generalised, arbitrary figures. Notice that the larger the settlement the fewer there are in number and the greater the distance between them.

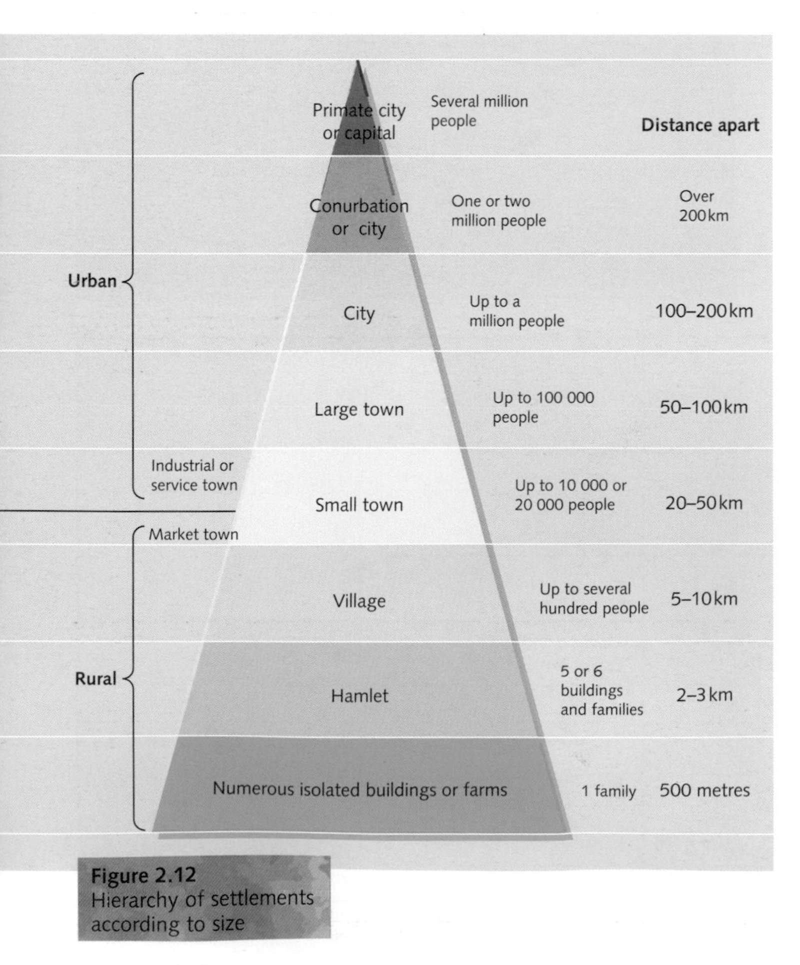

Figure 2.12
Hierarchy of settlements according to size

2 Range and number of services

Villages provide a limited range and number of services. Services that do exist are those likely to be used daily (the village shop) or which reduce the need to travel to other places (a primary school). In Figure 2.13, where the hierarchy is based on services, each place in the hierarchy is likely to have all the services of settlements below them.

HAMLET	PERHAPS NONE, OR PUBLIC TELEPHONE
VILLAGE	CHURCH, POST OFFICE, PUBLIC HOUSE, SHOP FOR DAILY GOODS, SMALL JUNIOR SCHOOL, VILLAGE HALL
SMALL TOWN	TOWN HALL, DOCTOR, SEVERAL CHURCHES/CHAPELS, CAFES AND RESTAURANTS, SMALL SECONDARY SCHOOL, RAILWAY STATION, SEVERAL SHOPS
LARGE TOWN	SEVERAL SHOPPING AREAS/ARCADES, HYPERMARKET, RAILWAY STATION, BUS STATION, HOTELS, BANKS, SMALL HOSPITAL, SMALL FOOTBALL TEAM
CITY	LARGE RAILWAY STATION, LARGE SHOPPING COMPLEX, CATHEDRAL, OPTICIANS AND JEWELLERS, LARGE HOSPITAL, LARGE FOOTBALL TEAM, UNIVERSITY, THEATRE, COUNTY HALL, AIRPORT
CAPITAL	CATHEDRALS, GOVERNMENT BUILDINGS, BANKING HQ, RAILWAY TERMINI, MUSEUMS AND ART GALLERIES, LARGE THEATRE, SHOPPING CENTRE, SEVERAL UNIVERSITIES, INTERNATIONAL AIRPORT

Figure 2.13
Hierarchy of settlements according to services

3 Sphere of influence

The sphere of influence, or *market area*, may be defined as the area served by a particular settlement. The area of the sphere of influence depends upon the size and services of a town and its surrounding settlements, the transport facilities available and the level of competition from rival settlements.

Figure 2.14
Market areas of Exeter

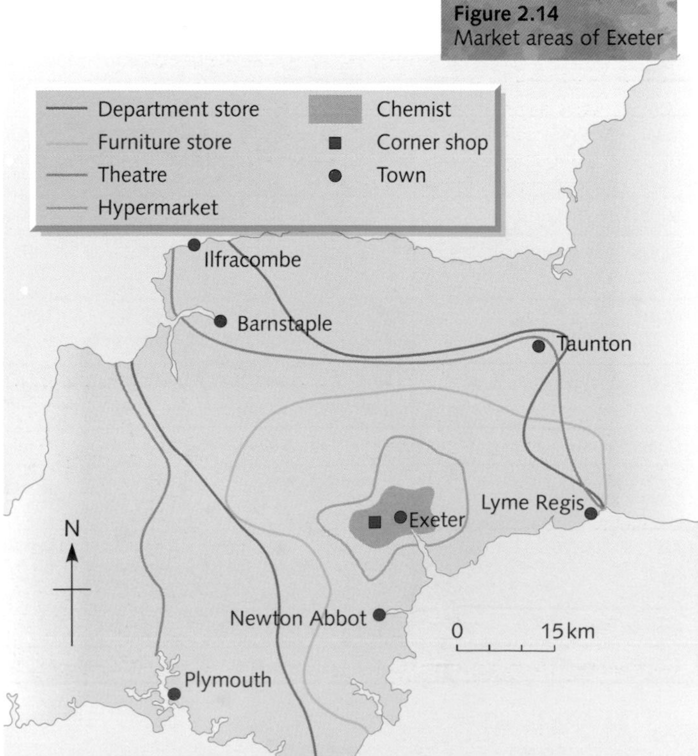

Two main ideas should be noted:

- A *threshold population* is the minimum number of people needed to ensure that demand is great enough for a special service to be offered to the people living in that area. For example, estimates suggest that 350 people are needed to make a village shop successful, 2500 for a single doctor to be available, and 10 000 people for a secondary school. Boots the Chemist prefers a threshold of 10 000 people in its market area, Marks & Spencer 50 000 people and Sainsbury's 60 000.
- *Range* is the maximum distance that people are prepared to travel to obtain a service. Figure 2.14 shows that people are not prepared to travel far to a corner shop or even to a chemist, but are prepared to travel much greater distances to shop at a hypermarket or, as these are visited less frequently, a furniture store or theatre.

Each settlement which provides a service is known as a *central place*. A central place provides goods and services for its own inhabitants and to people living in the surrounding area. The larger the settlement the more services it will provide and the more people it will serve. Large towns and cities will therefore have larger spheres of influence than smaller villages (Figure 2.15).

Changes in time

Few settlements remain constant in size. Villages near to large cities tend to increase in size as they become more suburbanised (page 44). Villages in more isolated areas, in contrast, tend to lose population. This means that villages increasing in size are likely to gain additional or improved services (a larger school, more shops) while villages declining in size will lose services (village shop, bus service).

Figure 2.15 Spheres of influence (market areas)

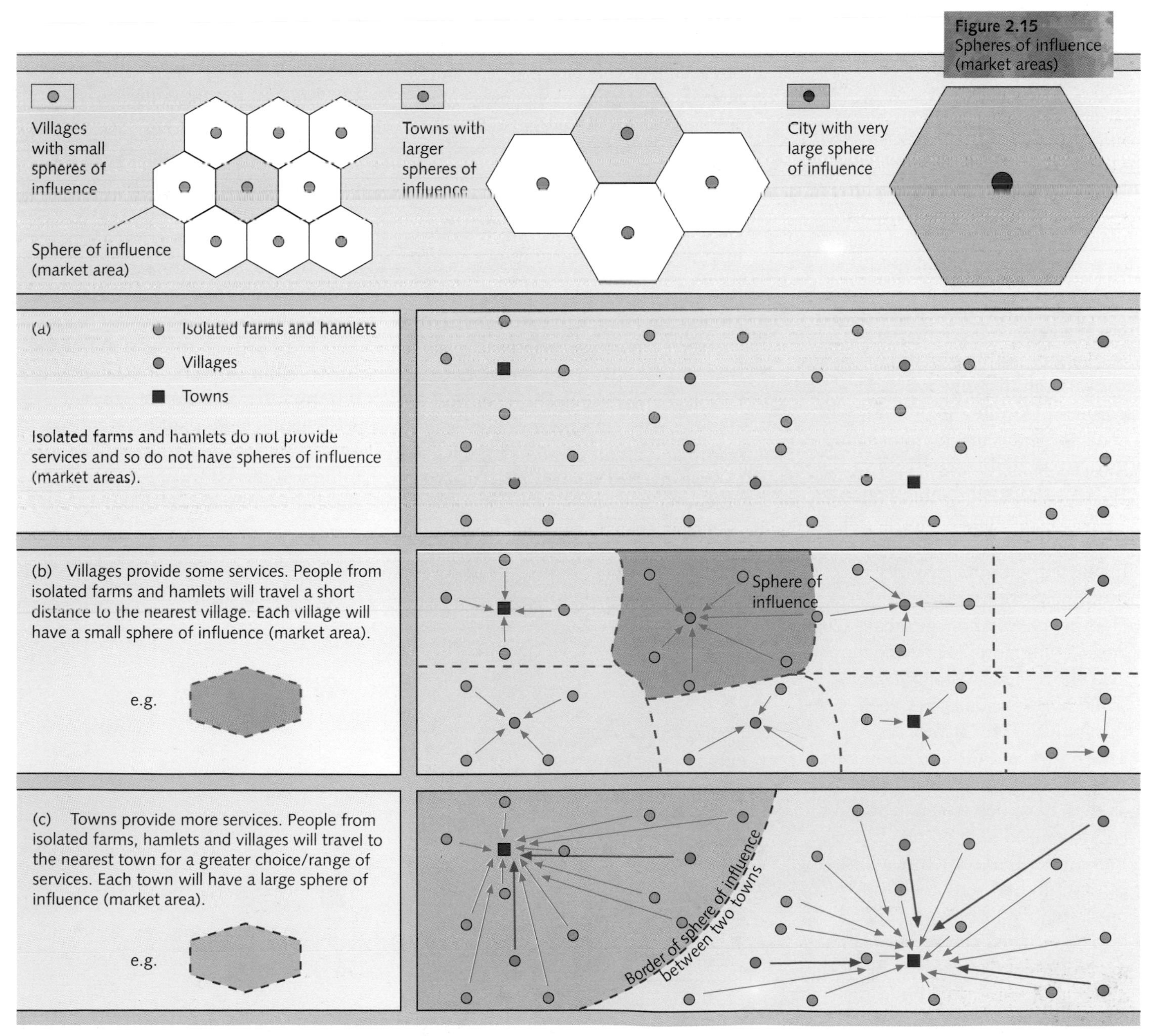

CASE STUDY 2

Settlement and the OS map

The map on page 25 is part of a 1 : 50 000 OS map (sheet 107). It shows that part of Holderness to the east of Kingston upon Hull.

Site

Although the map shows the area in the 1990s, some of the early site factors for the various settlements are still visible. Figure 2.16 shows, using map evidence only, some of the physical and human site factors for Hedon (1928). Other factors, either for Hedon or any other settlement under study, might be found by using older maps and documents or by making a special field visit.

Patterns

Settlement in this part of Holderness shows both dispersed and nucleated patterns. Despite the proximity of Hull, there are still many isolated farms, e.g. 209243, 250278 and 192326, and small hamlets such as Lelley (2032) and Wyton (1733). Most of the villages are, however, nucleated with the majority of buildings grouped together as at Hedon (1928), Thorngumbald (2026) and Keyingham (2425). The map also shows linear patterns along some of the roads as between Burstwick (224273) and Thorngumbald (214262). Several villages show evidence of planning as they have become increasingly suburbanised, e.g. Bilton (1533). Remember that sometimes it is difficult to fit a settlement into just one category. Burstwick (2227) has a fairly nucleated centre, some linear development leading out along the roads, and individual buildings (Stud Farm).

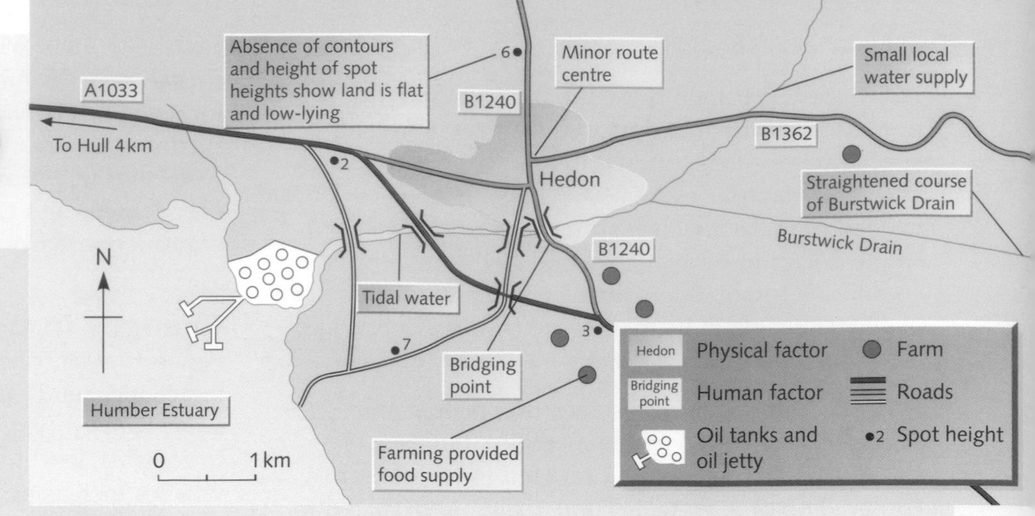

Figure 2.16
Annotated sketch map showing the site and situation of Hedon

Hierarchy

Size Although only a part of Kingston upon Hull is shown, it is obvious that this city is by far the largest settlement in the area. Other settlements are no bigger than villages. Of these Hedon appears to be the largest, followed by Thorngumbald and Keyingham (which seem about equal), Preston, Burton Pidsea and Burstwick.

Range and number of services
Hull again heads the hierarchy with – using map evidence only – more than ten schools, a college, two hospitals, numerous churches and chapels, parks, an information centre and a golf course. However, if the services for the other settlements are added up, Burstwick now comes second with five (two churches, a post office, school and public house) followed, with four each, by Hedon, Preston and Keyingham.

Sphere of influence Most people in the area are likely to visit Hull for most of their services. Outside of Hull, it would appear that only Burstwick and Preston have schools. That means, for example, that Burstwick will not only serve its own children, but also those from Thorngumbald, Keyingham and the surrounding smaller settlements (Figure 2.17). In contrast, most settlements have their own church and/or chapel and so the sphere of influence served by churches is smaller than that of schools.

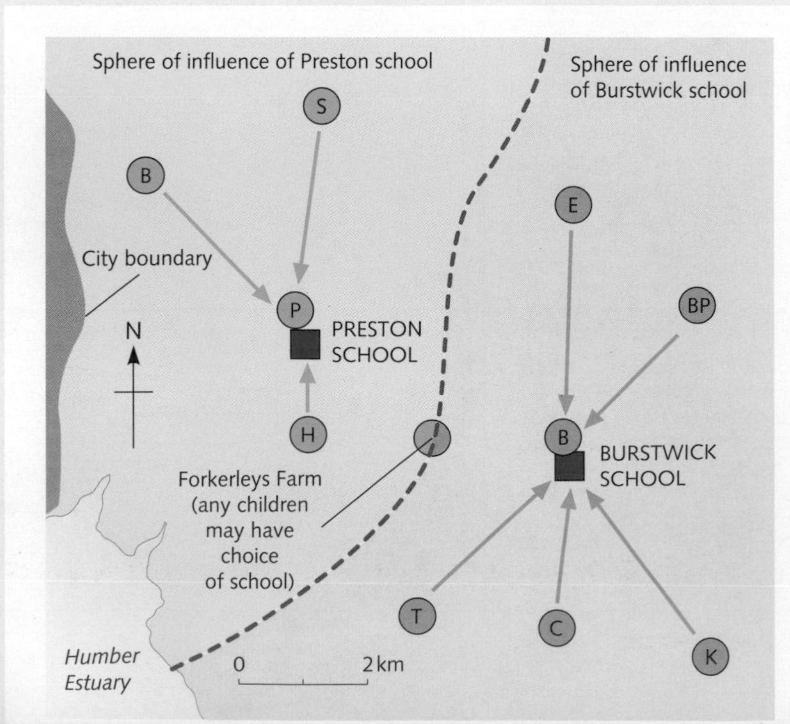

Figure 2.17
Spheres of influence of schools based on OS map evidence only

Figure 2.18
Part of the OS map sheet 107 Kingston upon Hull

QUESTIONS

(Pages 18 and 19)

Imagine that a group of settlers has sailed up the river shown in Figure 2.5. As their leader you have to choose the best site for a village. Your scouts have reported good possibilities at A, B, C, D and E.

a Copy the table. It lists several important factors that you will have to consider before choosing the best site.

i) Complete the table for each possible site. Give a score of 1 to 5 for each factor for each site.
1 = Very good; 2 = Good; 3 = Average; 4 = Poor; 5 = Very poor (5)

ii) State which you consider to be the best site (the one with the lowest score). Give reasons for your answer. (5)

b i) With help from the map, write a paragraph to describe the location of the chosen site. (4)

ii) What might be the main problems of living at this site? (2)

RESOURCE	SITE A	SITE B	SITE C	SITE D	SITE E
WATER					
CROP LAND					
GRAZING LAND					
FUEL					
BUILDING MATERIALS					
DEFENCE					
FLAT LAND THAT DOES NOT FLOOD					
TOTAL					

2

(Pages 20 and 21)

a Which of A, B and C is an example of:

- a dispersed settlement
- a nucleated settlement
- a linear settlement? (2)

b Give a definition for each of the settlements named in part (**a**). (3)

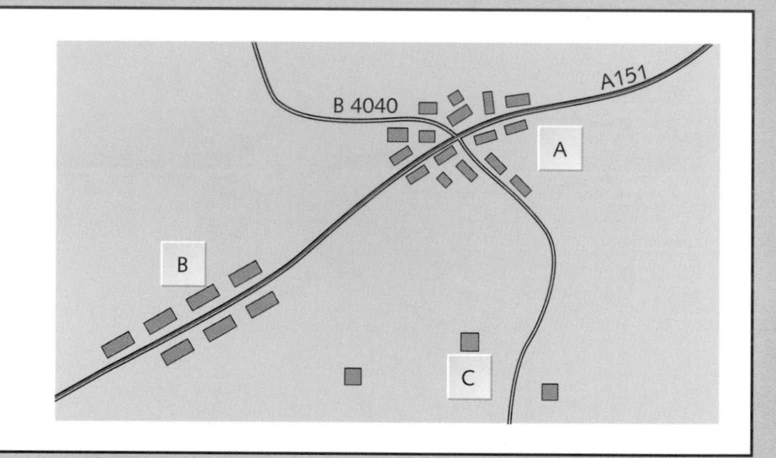

(Page 22)

a i) What is the population size of settlement A? (1)

ii) How many services does settlement A provide? (1)

iii) Of the four settlements labelled A to D, which is likely to be a small town, a large town, a village and a city? (3)

b i) Describe the relationship between population size and number of services in a settlement. (1)

ii) Give a reason for this relationship. (1)

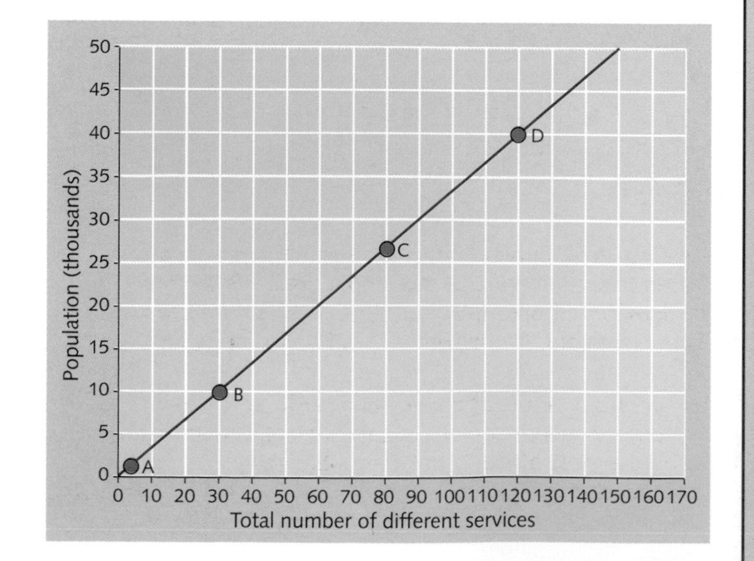

4

(Pages 22 and 23)

a i) Write out this list of settlements in order of size from largest to smallest:
town – village – city – conurbation – hamlet (2)

ii) Name two services that you might find in a village. (1)

iii) Name two services that you might find in a town, but **not** in a village. (1)

b Why does a village have fewer services than a town? (4)

c Complete the following sentences by using **one** of the terms in brackets.

i) Where something is organised in order of importance it is called a (service centre/hierarchy).

ii) The area from which people travel for a service is called a (market area/hierarchy).

iii) The minimum number of people in a market area needed to support a service is called the (range/threshold population).

iv) The market area is also known as the (range/sphere of influence). (4)

d The map below shows where people come from to use shops in towns A and B.

i) Copy the map and complete the market area for each town. (2)

ii) Why do market areas overlap? (2)

e i) What is the greatest range, in kilometres, of shoppers to:

- town A
- town B? (2)

ii) If a hypermarket was to be built at Y, how would this affect the market areas of towns A and B? (2)

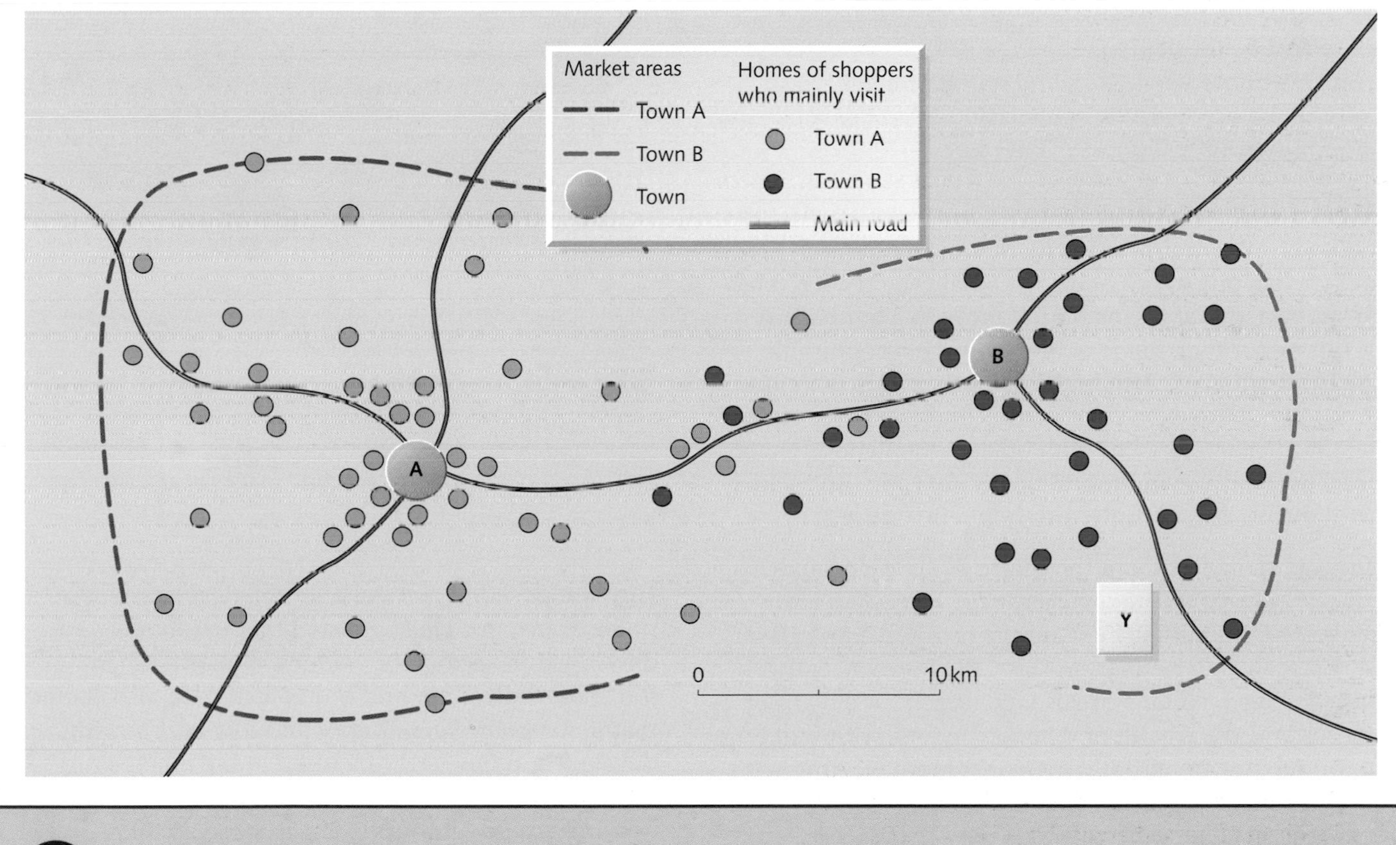

(Pages 24 and 25)

An Ordnance Survey map shows some of the services found in settlements. Locate on Figure 2.18 the ten settlements of Bilton, Wyton, Hedon, Preston, Burstwick, Thorngumbald, Camerton, Keyingham, Lelley and Burton Pidsea.

a Draw up a table and give each settlement one point for every church, chapel, school, public house, post office, hotel and public telephone shown on the OS map. (10)

b i) Rank the settlements in order, putting the one with most services (points) first and the one with fewest services (points) last. (5)

ii) Do you think there is a very close, close or not very close relationship between the number of services and the size of the ten settlements? (1)

iii) What other services may these settlements have that are not shown on the Ordnance Survey map? (2)

3 URBANISATION IN DEVELOPED COUNTRIES

Urban growth and land use

Urbanisation means an increase in the proportion of people living in towns and cities. Although towns were important even in the early civilisations of Mesopotamia and in the valleys of the Nile, Indus and Huang-He (China), most people tended to live and work in rural areas. It was not until the rapid growth of industry in the nineteenth century that large-scale urbanisation began in parts of western Europe and north-eastern USA. During the twentieth century, people continued to move to urban areas mainly for:

- more and better-paid jobs
- nearness to places of work and entertainment
- better housing, services (schools and hospitals) and shopping facilities.

Urban land use models

A model is a theoretical framework which may not actually exist, but which helps to explain the reality. It has been suggested that towns do not grow in a haphazard way, but rather they tend to develop with recognisable shapes and patterns. Although each urban area is unique, with its own distinctive pattern, it is likely to share certain generalised characteristics with other settlements. Two of the earliest land use models to be put forward, and which are still the easiest to apply, are shown in Figure 3.1.

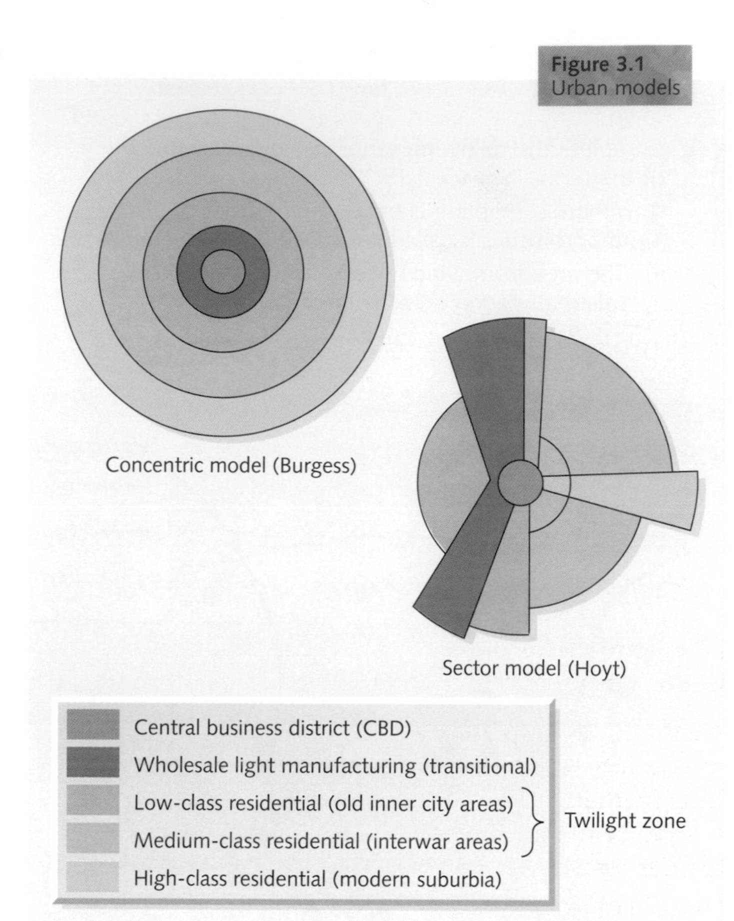

Figure 3.1 Urban models

- **Burgess** claimed that in the centre of all towns and cities there was a *central business district* (*CBD*). He suggested, initially using Chicago as his example, that towns grew outwards from this CBD in a concentric pattern. The resultant circles were based on the age of houses and the wealth of their occupants, with building becoming newer and the occupants more wealthy with increasing distance from the CBD.
- **Hoyt** proposed his model after the development of public transport. He suggested that urban areas developed in sectors, or wedges, alongside main transport routes into and out of a city. He also claimed that if, for example, industry and low-cost housing developed in one part of a town in the nineteenth century, then newer industry and modern low-cost housing would also locate in the same sector.

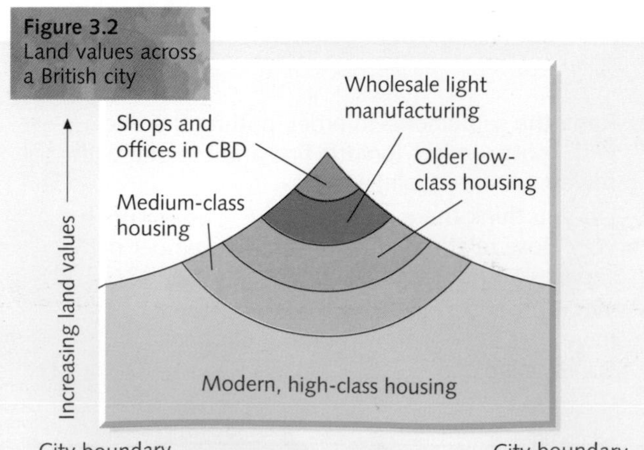

Figure 3.2 Land values across a British city

Urban land use and functional zones

Each of the zones shown in Figure 3.1 has a function. The four main types of function are shops and offices, industry, housing, and open space. The location of each zone and the distribution of each functional zone are related to several factors.

Land values and space Land values are highest and available sites more limited in the CBD where competition for land is greatest. As land values decrease rapidly towards the urban boundary then both the amount of space and the number of available sites increase (Figure 3.2).

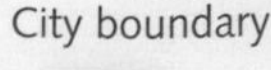

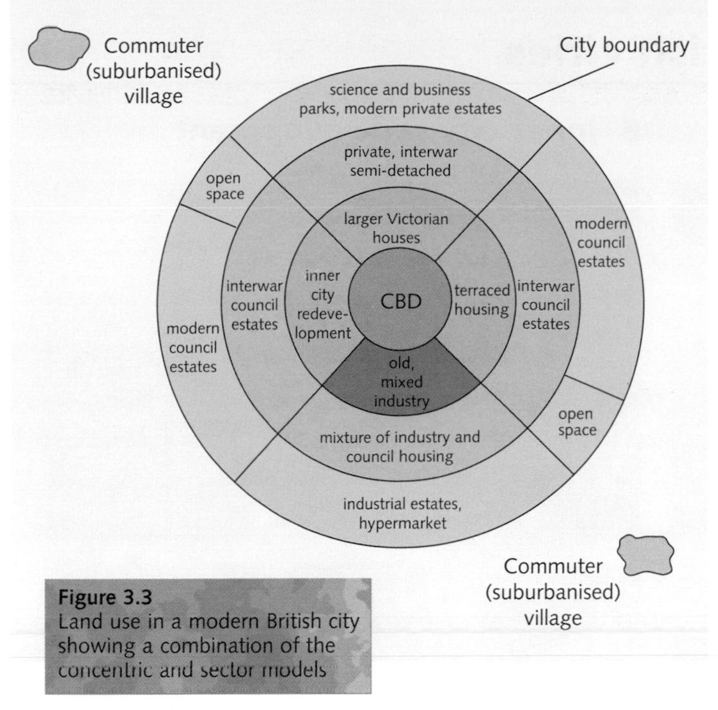

Figure 3.3
Land use in a modern British city showing a combination of the concentric and sector models

Age As towns developed outwards, the oldest buildings were near to the city centre (although many of these have now been replaced) and the newest ones in the outskirts.

Accessibility The CBD, where the main routes from the suburbs and surrounding towns meet, has been the easiest place to reach from all parts of the city, although this ease is now often reduced by increased congestion.

Wealth of the inhabitants The poorer members of the community tend to live in cheaper housing near to the CBD (with its shops) and the inner city (where most jobs used to be found). These people are less likely to be able to afford the higher transport (private or public) and housing costs of places nearer the city boundary.

Changes in demand Land use and functions change with time. For example:

- Nineteenth-century industry was located next to the CBD whereas modern industry prefers edge-of-city sites.
- The main land use demand in the nineteenth century was for industry and low-cost housing. Today it is for industry, shops and better-quality housing, all in a more pleasant environment, and open space.

Figure 3.3 is a more realistic model showing land use patterns and functional zones in a British city. Figure 3.4 is a transect, or cross-section, across a city.

It is possible to recognise on Figure 3.5 the CBD with its taller buildings; the inner city and twilight zone with factories and cleared land (left as waste or used for car parks); and, moving towards the top of the photo, inter-war and modern housing estates and an increasing amount of open space.

Figure 3.5
Changing land use in the British city of Peterborough

Figure 3.4
Transect across a typical British city

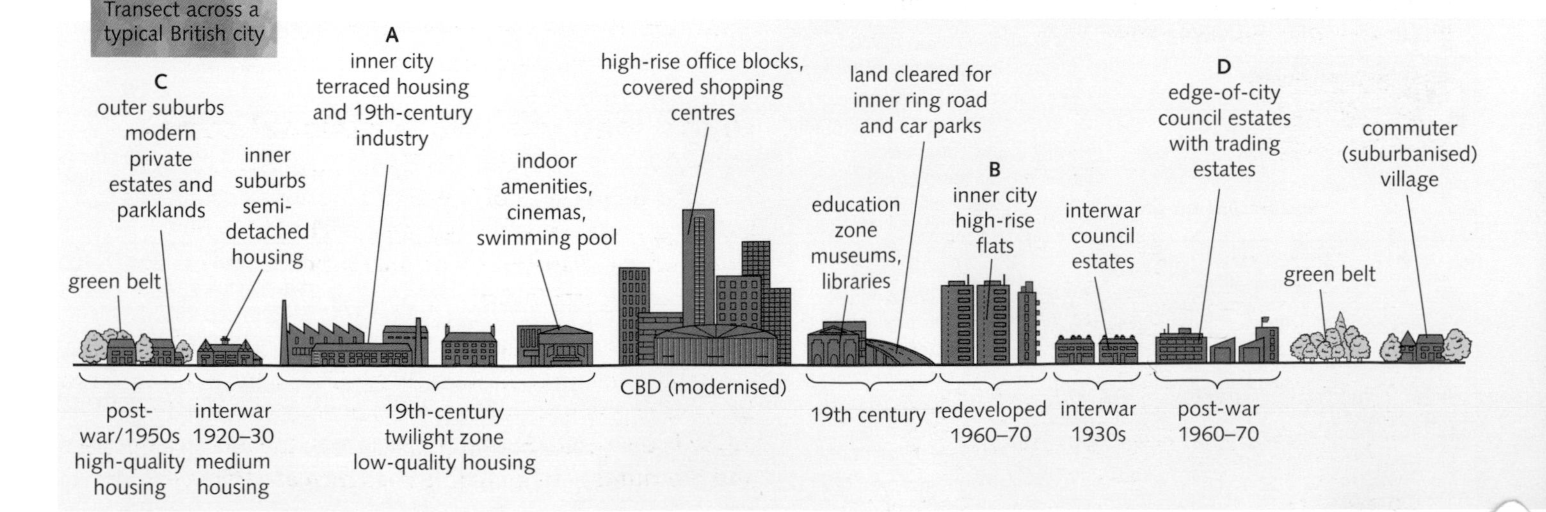

Residential environments in British cities

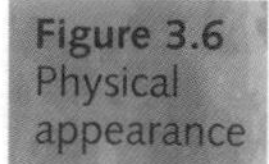

Figure 3.6 Physical appearance

A Old inner city area

B Inner city redevelopment

Figure 3.7 Land use

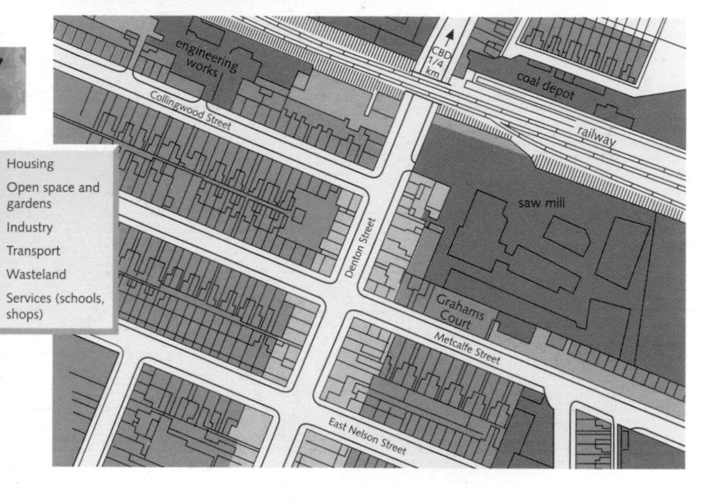

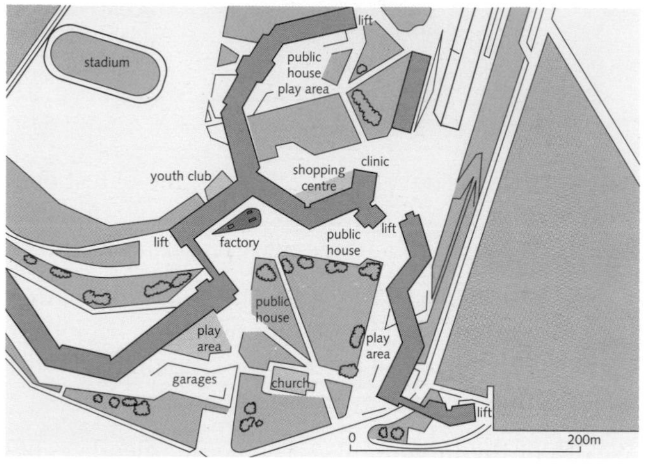

Figure 3.8 Description

The Industrial Revolution of the nineteenth century led to the growth of towns. The rapid influx of workers into these towns meant a big and immediate demand for cheap housing, and so builders constructed as many houses as possible in a small area, resulting in high-density housing with an overcrowded population. The houses were built in long, straight rows and in terraces. In those days of non-planning, few amenities were provided either in the house (e.g. no indoor WC, bathroom, sewerage or electricity) or around it (e.g. no open space and no gardens).

When in the 1950s and 1960s vast areas of inner cities were cleared by bulldozers, many of the displaced inhabitants either moved to council estates near the city boundary, or were rehoused in huge high-rise town blocks which were created on the sites of the old terraced houses. Although these high-rise buildings contained most modern amenities, the apartments had to be reached by lifts which led to narrow, dark corridors. Also, despite the areas of greenery between the flats, there was still a very high housing density.

Figure 3.9 Census data

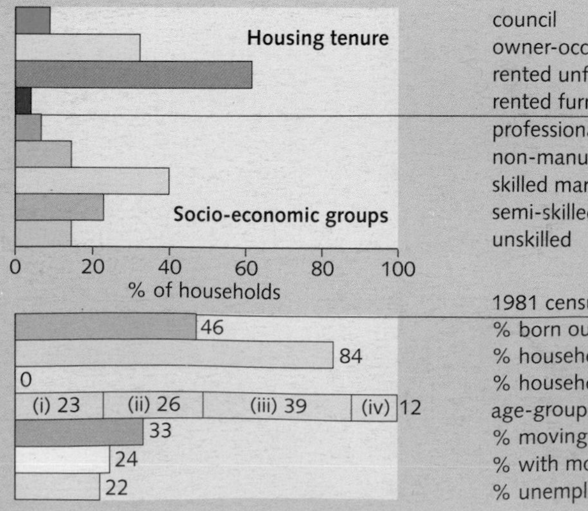

Housing tenure: council, owner-occupied, rented unfurnished, rented furnished

Socio-economic groups: professional managerial, non-manual, skilled manual, semi-skilled, unskilled

B Inner city redevelopment

0 20 40 60 80 100

% of households

1981 census figures	A Old inner city area	B Inner city redevelopment
% born outside the UK	46	3
% households with WC, hot water, bath	84	96
% households with garages	0	10
age-groups (i) 0–14 (ii) 15–34 (iii) 35–59 (iv) 60+	(i) 23 (ii) 26 (iii) 39 (iv) 12	(i) 30 (ii) 23 (iii) 32 (iv) 15
% moving into area in last 5 years	33	6
% with more than 1 person per room	24	9
% unemployed (1985)	22	28

C Suburbia

The rapid outward growth of cities began with the introduction of public transport, and accelerated with the popularity of the private car. This outward growth (also known as urban sprawl) led to the construction of numerous private, 'car-based' suburbs.

The houses built in the outer suburbs before the Second World War are characterised by their front and back gardens. Usually they have garages and are semi-detached with bay windows. The more recent estates have housing which differs in both style and type, but they remain well planned and spacious.

D Outer city council estate

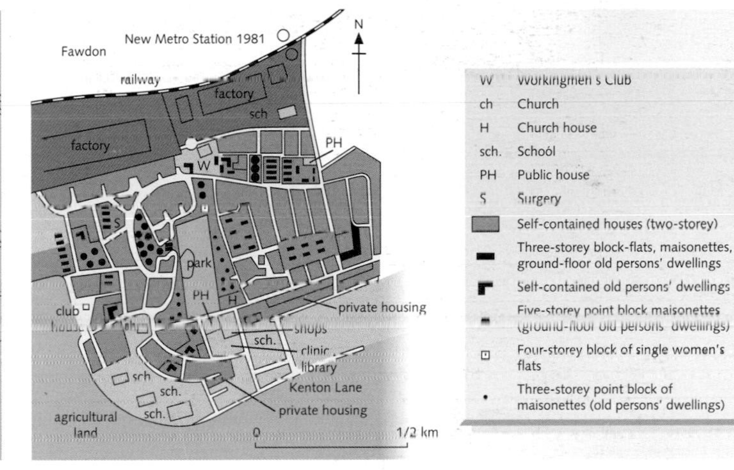

As local councils cleared the worst of the slums from their inner city areas in the 1950s and 1960s, many residents were rehoused on large council estates on the fringes of the city. Attempts were made to vary the type and size of accommodation:

- High rise tower blocks, often 10–12 storeys high.
- Low-rise tower blocks, usually 3–5 storeys high. These were built nearer the city boundaries, where there was more open space.
- Single-storey terraces with some gardens and car parking space.

C Suburbia / D Outer city council estate

Housing tenure

council
owner-occupied
rented unfurnished
rented furnished

Socio-economic groups

professional managerial
non-manual
skilled manual
semi-skilled
unskilled

0 20 40 60 80 100
% of households

1981 census figures	C Suburbia	D Outer city council estate
% born outside the UK	6	2
% households with WC, hot water, bath	100	100
% households with garages	85	5
age-groups (i) 0–14 (ii) 15–34 (iii) 35–59 (iv) 60+	(i) 29 (ii) 18 (iii) 46 (iv) 7	(i) 37 (ii) 19 (iii) 35 (iv) 9
% moving into area in last 5 years	15	4
% with more than 1 person per room	4	20
% unemployed (1985)	7	17

Inner cities – decline and decay

Figure 3.10 Land use in the traditional inner city

Most inner city areas developed along with industry in the nineteenth century. As industry grew, so too did the demand for workers. As an increasing number of people moved from rural areas to the towns for work, they needed low-cost houses in which to live. At that time, without either public or private transport, people also wanted to live as close as possible to their place of work. Despite difficult conditions, local inhabitants often created a strong community spirit. However, many of the early advantages of living and working in an inner city have long since become disadvantages (Figure 3.10).

Nineteenth-century houses were built as close together as possible creating a *high density*. Most were built in terraces (Figure 3.6) with some, especially in industrial cities, being 'back-to-back'. By the 1960s, when many of the houses lacked basic amenities, large-scale 'slum' clearances often led to the erection of high-rise flats (Figure 3.11). Unfortunately high-rise flats created as many social problems as they solved. Today it is still possible to find areas of older terraced housing or high-rise buildings, although many houses have been abandoned, boarded up and vandalised, while some of the high-rise flats have already been pulled down.

Inner cities were characterised by large factories built on land near to the city centre and adjacent to canals and railways. Many of these have been forced to close (Chapter 9) either due to a lack of space for expansion and modernisation, or due to the narrow, congested roads. Some factories, like the terraced housing, have been left empty. Others have been pulled down leaving large areas of derelict land – land ideal for the dumping of rubbish. Little has been done to turn these waste sites into recreational open spaces.

Figure 3.10 shows a disused canal and former railway sidings as well as the narrow, unplanned streets which are unsuitable for modern vehicles. The canal, and any river, is probably polluted although the air is likely to be cleaner following the closure of so many factories. Inner cities often lack modern services. Many of the traditional corner shops have been forced to close due to competition from city-centre shops, or demolished in slum clearance schemes. Schools are often in old buildings and there are insufficient places of entertainment or open space.

Figure 3.11 High-rise flats in Sheffield

The inner city, with its low-cost housing, has today become an area where people on low incomes, such as pensioners, single-parent families, first-time home buyers and the disabled, are forced to seek accommodation. It is also a place where ethnic groups, and especially recent arrivals, tend to concentrate (see Places – London below, and pages 74–75).

Over the years most cities have attempted to improve living conditions either, initially, by bulldozing large areas and building high-rise flats (*urban redevelopment*) or, later, by improving existing properties (*urban renewal*). However, although most inner city houses now have the basic amenities, the area still experiences cramped housing conditions together with high levels of poverty, unemployment and crime.

Ethnic concentrations in London

According to a report published in June 1997 by the London Research Centre, there are more than 37 ethnic groups of over 10 000 people living in London.

> *'The various groups tend to live in clusters and form ethnic "villages". However, there are no racial ghettos in London, unlike in the USA where almost everyone is one of an ethnic group, and where practically everyone belonging to that group lives [e.g. Chinatown, Japantown, Koreatown and Filipinotown in Los Angeles].*
>
> *In London, Brixton has the highest concentration of West Indians but black Caribbean people form only about one-third of its population – and less than 3 per cent of London's West Indians live there. London's ethnic minority population, currently about 1.6 million (out of Britain's total of 3 million) is growing slowly, as is its percentage of London's population as a whole – currently it stands at 23.4 per cent. This is not due to immigration but mainly because the birth rate of indigenous Britons is going down, and many choose to move out of London and live elsewhere.*
>
> *However, as the London Research Centre makes clear throughout its report, the term "ethnic" is becoming increasingly difficult to define as people's perception of their own identity – racial and social – changes. Soon the majority of "ethnic" Britons will have been born here.'*
>
> *London Evening Standard*, 3 June 1997

Figure 3.12 shows that there are areas within London where particular ethnic groups tend to concentrate. Even so, within these areas there are likely to be several smaller ethnic groups living fairly closely together in a multicultural society. This contrasts with North American cities where each ethnic group tends to be segregated from other ethnic groups.

Figure 3.12
Concentration of ethnic groups in the London boroughs, 1996

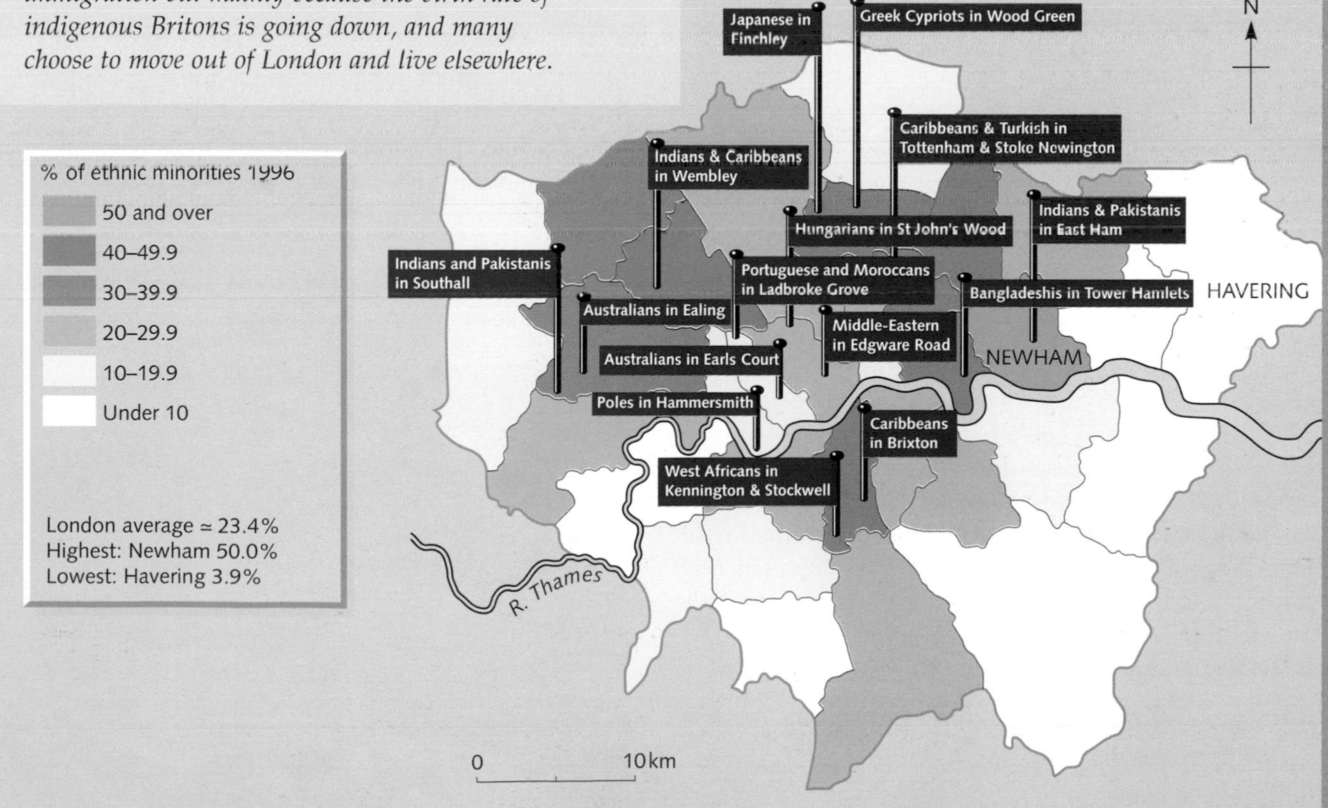

Places

Inner cities – redevelopment and regeneration

Urban development corporations (UDCs)

UDCs were first set up by the government in 1981 in an attempt to regenerate areas which had large amounts of derelict and unused land and buildings. Their aim was to encourage maximum private-sector investment and development. UDCs have powers to acquire, reclaim and service land and restore buildings to effective use. They promote new industrial, housing and community developments. The first two, the London Docklands Development Corporation (LDDC) and the Merseyside Development Corporation (MDC), came into existence in 1981. Since then 11 more UDCs have been created (Figure 8.23 and pages 130–131).

Places

London Docklands

During the nineteenth century and up to the early 1950s, London was the busiest port in the world. Because of a series of changes after that, many of them due to improvements in technology, the docks were virtually abandoned and became derelict. By 1981 larger ships could no longer reach the port of London, and containerisation did away with the need for large numbers of dockers. By that time the area had very few jobs, the docks had closed, over half the land was derelict, many of the nineteenth-century terraced houses needed urgent repair, transport was poorly developed, and there was a lack of basic services, leisure amenities and open space. The LDDC was set up to try to improve the economic, social and environmental conditions of the area.

Figure 3.13
London Docklands in 1997

Changes July 1981 – April 1996 (Figure 3.13)

Physical–environmental regeneration

- 728 hectares of derelict land reclaimed
- 160 000 trees planted
- 130 hectares of open space
- 17 conservation areas created

Economic regeneration

Improved transport links mean that central London can now be reached within ten minutes. The Docklands Light Railway (Figure 3.14) carries 320 000 passengers a week over its 29 km of track. It connects with Bank and Monument Underground stations and, since 1996, eastwards to Beckton and, by early 2000, southwards to Lewisham. The Jubilee Underground extension, due to open in March 1998, will give direct access to Waterloo and London Bridge mainline stations. The City Airport, built in the Royal Docks, handled over half a million passengers in 1995 (Figure 3.15). Over 135 km of new roads leading to, and within, Docklands have also been built, including a link with the M11.

Estimates claim that both employment and the number of businesses doubled between 1981 and 1996 (employment from 27 000 to 69 975 and businesses from 1000 to 2400), while unemployment has fallen (from 14.2 to 9.5 per cent). Financial and high-tech firms such as the Stock Exchange, Limehouse ITV studios and the Guardian and Daily Telegraph newspapers were attracted by the low rates of the UDC. By 1996, 76 per cent of office space in the prestigious Canary Wharf business complex, with its dominating 245 metre tower (Figure 3.16), had been let.

Figure 3.14
The Docklands Light Railway

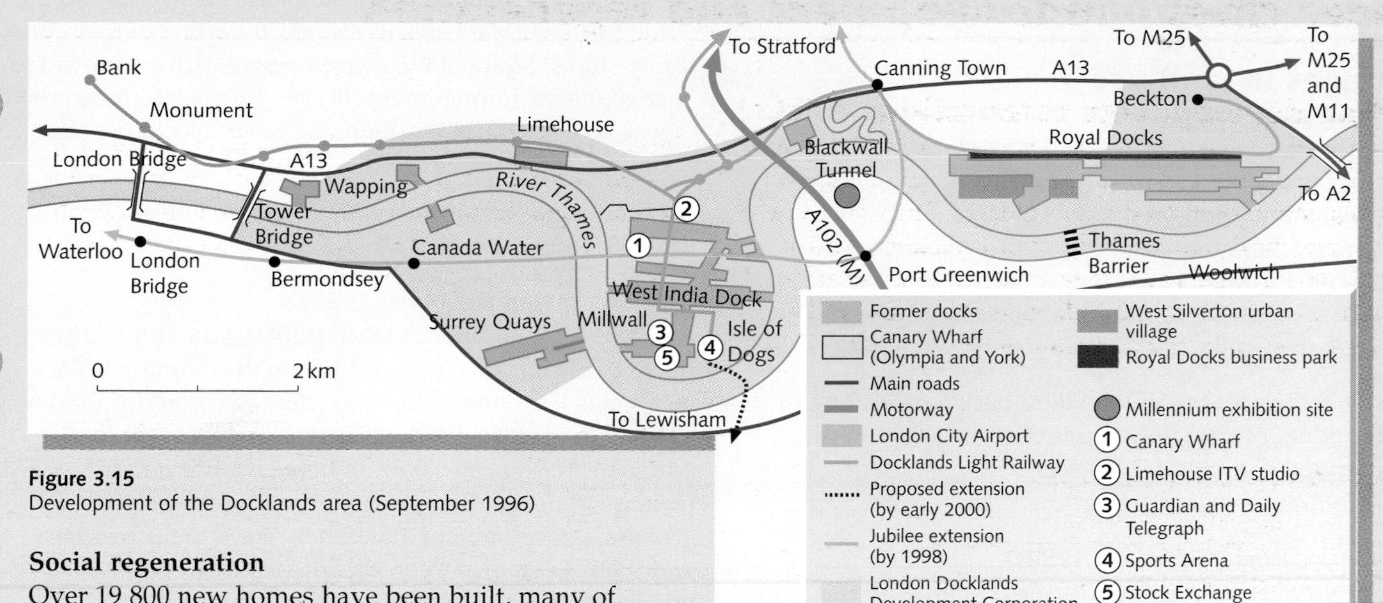

Figure 3.15
Development of the Docklands area (September 1996)

Social regeneration

Over 19 800 new homes have been built, many of them former warehouses converted into luxury flats, and 7900 existing local authority homes refurbished. More recently, Newham Council have concentrated on providing low-cost housing. The improvement in home provision has increased the resident population from 39 400 (1981) to 76 850 (1996) and the proportion of owner-occupied homes from 5 to 40 per cent. New shopping centres have been developed together with a post-16 college and a technology college, a national indoor sports centre, a marina for water sports, and several new parks. Almost £100 million has also been spent on health, education, training and community programmes. West Silverton urban village, one of the first of its type in Britain, is to be a self-contained community of 5000 people.

Which groups have been involved in the redevelopment of Docklands?

- Local **housing associations** have helped by gaining home improvement grants.
- The local **Newham Council** have built, when they had sufficient money, low-cost, affordable houses, and upgraded older properties. They have also tried to improve local services.
- The **LDDC** have been responsible for planning and redeveloping the Docklands.
- The **national government** created the Isle of Dogs enterprise zone with its reduced rates. This encouraged private investment and has given financial help towards improving transport systems.
- **Property developers** have been responsible for building the large office blocks, e.g. Canary Wharf (Olympia and York), and converting derelict warehouses into luxury flats.
- **Conservation groups** have supported tree planting and other schemes aimed at improving the quality of the environment.

Conflicting opinions

Some groups of people have benefited from these changes and so are in favour of the scheme. Others feel disadvantaged and so are against it. Local residents cannot afford the expensive flats built by speculative property developers, and there is still a shortage of low-cost housing. Jobs in the new high-tech industries are relatively few in number, and they demand skills not possessed by former dockers. The 'yuppie' newcomers rarely mix with the original 'Eastenders' who find that their close-knit community has been broken up. Property developers have not provided enough services, e.g. hospitals, or care for the elderly.

The LDDC is due to be wound down by 31 March 1998.

Figure 3.16
Canary Wharf, Isle of Dogs, London

The rural–urban fringe

Changes in land use

The nineteenth century

Before the nineteenth century, most towns in Britain were small in size and nucleated in shape (page 20). During the Industrial Revolution:

- factories and mills were erected on the nearest available land to the town centre
- low-cost housing was built as close as possible to these places of work.

Towns, therefore, began to extend outwards into the surrounding countryside. Continued urban growth surrounded the early factories and houses, leaving them today with an inner city location.

Inter-war period (1920s and 1930s)

The rapid outward growth of cities began with the introduction of public transport, and accelerated with the increased popularity of the private car. This outward growth, known as *urban sprawl*, led to the development of *suburbia* – that is, large estates consisting mainly of semi-detached houses (Figure 3.6) and corresponding with Burgess's zone of medium-cost housing (Figure 3.1). Suburbia had areas of open space and its own shopping parades, but rarely any industry.

1960s to the 1980s

Urban sprawl continued, with land on the urban fringe being used mainly for one of two types of housing:

- Private estates with *low-density*, high-quality housing (Figure 3.17). The most recent estates are likely to have large, detached houses with modern amenities both inside and out.
- Outer city council estates which were created mainly in the 1960s as local councils cleared the worst of their inner city slums. Many of the evicted residents were moved to edge-of-city, former *greenfield sites* where accommodation was provided in either high-rise tower blocks, low-rise flats or single-storey terraces (Figure 3.18).

Land on the edges of urban areas was also used, at this time, for industrial and trading estates.

Recent developments – the 1990s

An increasing number of land users see the *rural–urban fringe* as the ideal location for future development. This location is less congested, has easier access and provides cheaper land and a more attractive environment than places nearer the city centre. There is, however, an increase in the conflict between those who wish to see the economic development and extension of the urban area and those who wish to protect the rural environment which surrounds it (Figure 3.18).

Figure 3.17
A private estate in suburbia

Figure 3.18
Housing on outer city council estates

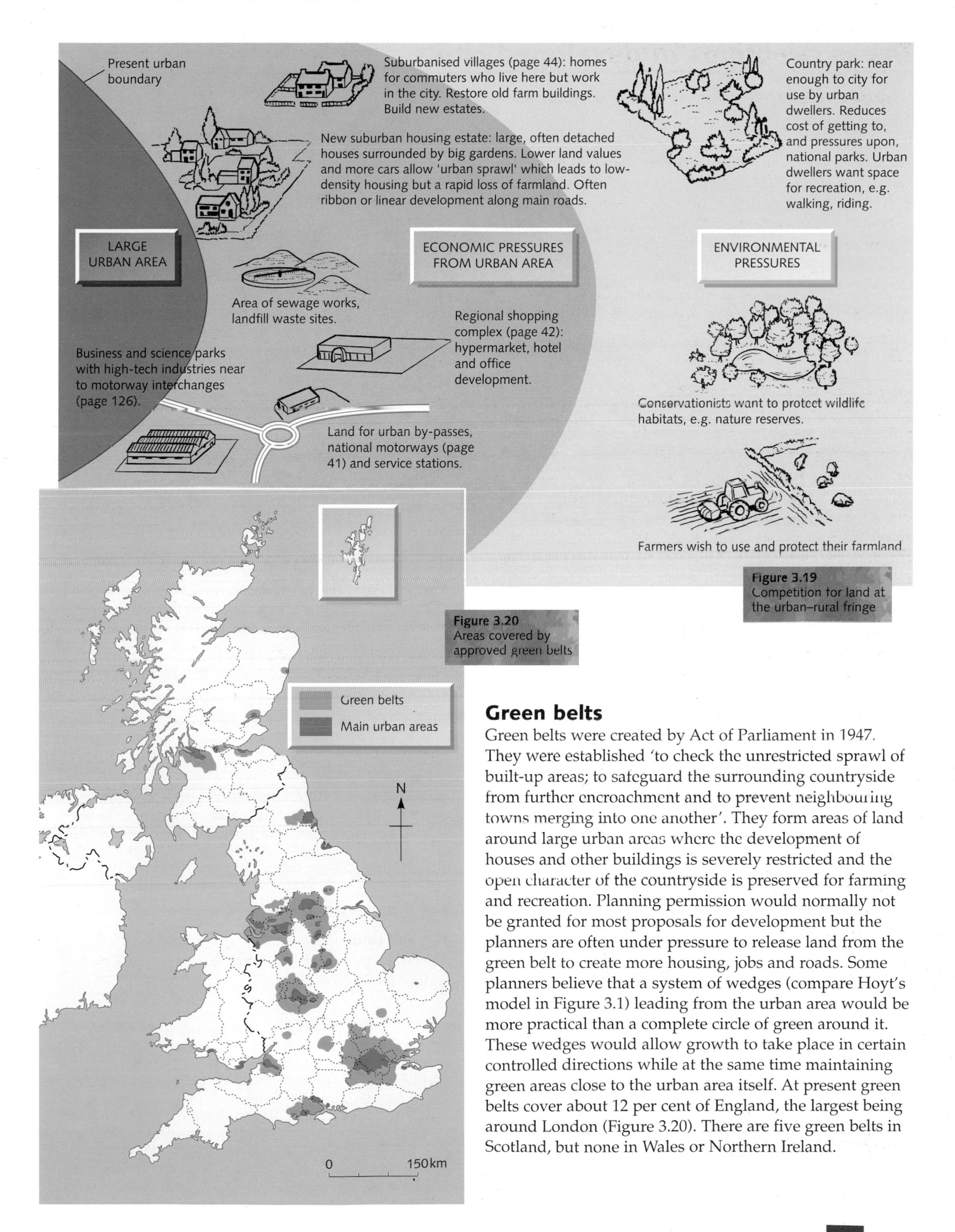

Figure 3.19 Competition for land at the urban–rural fringe

Figure 3.20 Areas covered by approved green belts

Green belts

Green belts were created by Act of Parliament in 1947. They were established 'to check the unrestricted sprawl of built-up areas; to safeguard the surrounding countryside from further encroachment and to prevent neighbouring towns merging into one another'. They form areas of land around large urban areas where the development of houses and other buildings is severely restricted and the open character of the countryside is preserved for farming and recreation. Planning permission would normally not be granted for most proposals for development but the planners are often under pressure to release land from the green belt to create more housing, jobs and roads. Some planners believe that a system of wedges (compare Hoyt's model in Figure 3.1) leading from the urban area would be more practical than a complete circle of green around it. These wedges would allow growth to take place in certain controlled directions while at the same time maintaining green areas close to the urban area itself. At present green belts cover about 12 per cent of England, the largest being around London (Figure 3.20). There are five green belts in Scotland, but none in Wales or Northern Ireland.

Urban problems in developed cities

Los Angeles

Figure 3.21 Downtown Los Angeles

What is your impression of Los Angeles? One that is frequently projected on our TV screens is that of film stars living in huge, luxurious, ranch-style homes which overlook the sea. Even to many Americans, Los Angeles is their 'city of dreams', with job opportunities, a hot sunny climate, sandy beaches and surf, and a chance to mingle with the stars. The city has grown rapidly, with downtown LA (the CBD) dominated by skyscrapers and dissected by freeways (urban motorways) (Figure 3.21). Yet, if this is the true image, why is the number of residents moving out of the city growing so rapidly?

Physical problems

Shortage of space The Los Angeles basin is surrounded by mountains and the sea. The city has grown outwards to fill the basin. By 1996, the Greater Los Angeles–Long Beach metropolitan area extended a distance of 130 km and had a population of 15 million. New houses have to be built on unsuitable land.

Figure 3.22
Smog over Los Angeles

Earthquakes The city lies on the San Fernando and Santa Monica Faults and adjacent to the notorious San Andreas Fault. An earthquake in 1994 (6.7 on the Richter scale – page 252) killed over 60 people, injured several thousand and caused buildings to collapse.

Water supply The city is built on the edge of a desert. Water has to be brought long distances from northern California and the Colorado River.

Floods and landslides Rain, when it does fall, is heavy. Runoff from the deforested and urbanised steep hills surrounding the basin (page 261), causes floods, landslides and mudflows.

Brush fires These occur after the long, hot dry summers. In both 1970 and 1993 many media stars saw their homes in Malibu destroyed.

Fog and smog Los Angeles often has coastal fog. This traps pollutants from traffic, power stations and industries to give smog (Figure 3.22).

Figure 3.23
Understanding the riots

> The riot was not a product of wanton thuggery. It was a symptom of sickness in the centre of our city. The predictable result of the idleness and despair rampant in densely populated areas where the conditions of life itself are marginal. Twice as many jobless were in south Los Angeles as in the rest of the city. Crowded schools produced dropouts. Bus services were inadequate, stifling efforts to find jobs outside the area and fuelling a sense of isolation from the rest of the city.
>
> *Los Angeles Times* 1992, p.10

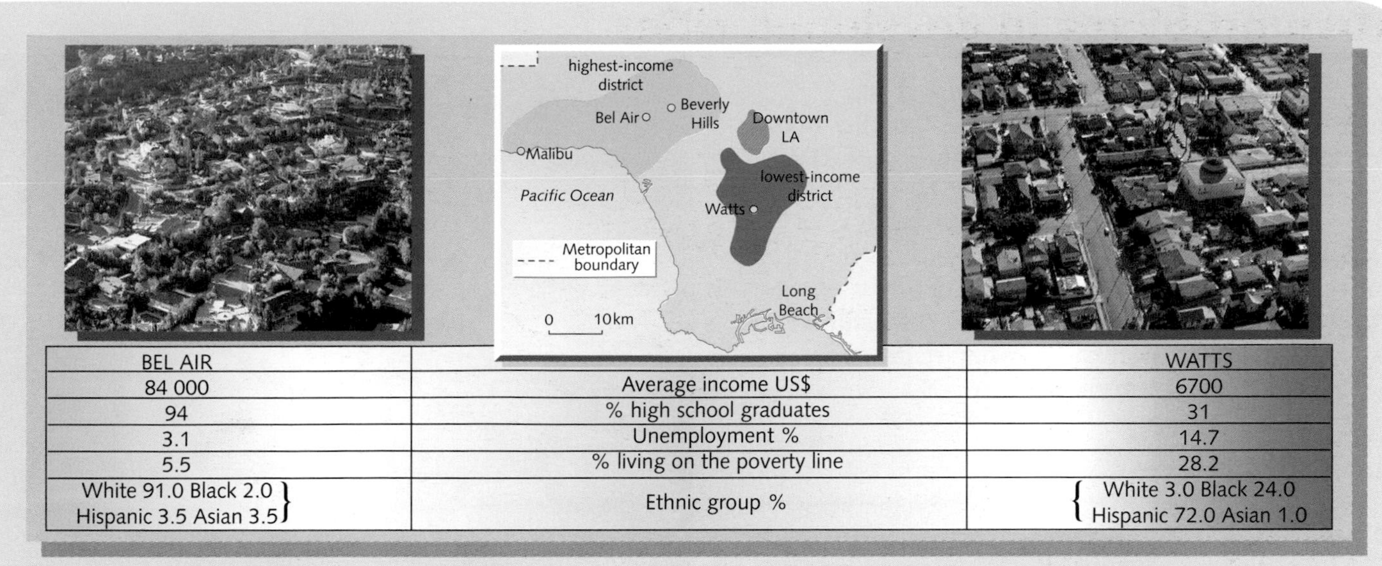

BEL AIR		WATTS
84 000	Average income US$	6700
94	% high school graduates	31
3.1	Unemployment %	14.7
5.5	% living on the poverty line	28.2
White 91.0 Black 2.0 Hispanic 3.5 Asian 3.5	Ethnic group %	White 3.0 Black 24.0 Hispanic 72.0 Asian 1.0

Figure 3.24 Inequalities in LA

Economic and social problems

Traffic The numerous freeways, created to give freedom to individual car owners, can no longer cope with the volume of traffic. People who live in suburban areas but work downtown face long, expensive journeys which add to the city's air pollution problem.

In-migration Although many immigrants come from other parts of the USA, 85 per cent arrive from Mexico (page 80) and the Asian Pacific Rim countries of Vietnam, China, Korea and the Philippines. Overseas migrants, who often have to take low-paid jobs, are forced to live in poorer districts such as Watts. Certainly people living in districts such as Watts have a much lower quality of life than those in Malibu or Bel Air (Figure 3.24).

High land values Competition for land and prime sites, whether for downtown offices or suburban housing, has made property expensive to buy or to rent.

Crime LA, like other large cities, has a high incidence of crime, violence, gang warfare and drugs. It is a major reason for out migration.

Racial tension The 1992 race riots led to 25 deaths, 572 injuries and over a thousand fires (Figure 3.23).

Edge cities

Many more wealthy inhabitants are moving out to the new 'edge cities' that now surround LA (Figure 3.25). Edge cities, often 50 km from the downtown area, have developed at nodal points (page 19) where major freeways intersect. They have a better environment (less crime, less pollution), cheaper housing and are nearer places of work. Offices and factories are relocating here as the cost of land is lower and, with improvements in technology (page 126), there is less need for a CBD site. Edge cities have their own shopping centres but, as yet, often lack places of entertainment.

Figure 3.25
An edge city close to Los Angeles

Transport in urban areas

Traffic in general and the car in particular create numerous problems in urban areas. Most families in the more developed countries of Western Europe, North America and Japan own their own car. The car gives people greater mobility and improves their access to places where they work, shop, are educated or find recreation. However, it is the widespread use of the car that is bringing traffic in most cities to a standstill (Figure 3.26) and which is a major cause of environmental, economic and social problems (Figure 3.27).

Commuting

Urban traffic problems are often at their worst during the early morning and late afternoons when commuters are travelling to and from work. A commuter is a person who lives in a smaller town or a village in the area surrounding a larger town or city, and who travels to that town or city for work. The term can also be used for residents who live in the suburbs of a large town or city. The increase in car ownership and the improvement in the road network means that more commuters can live further from their place of work. This has led to increasingly larger *commuter hinterlands* (the areas around large cities) where commuters live.

A more recent trend is a 'reversed' flow of commuters. This group tends to include either the less skilled and poorly paid or members of ethnic minorities, who live in low-cost inner city housing. As jobs in this part of the city have declined, these groups have to make long journeys to their place of work on newer edge-of-city industrial and retail parks.

Figure 3.26
Traffic in Los Angeles

	Yesterday's reading	Forecast
Nitrogen dioxide	**40 ppb**	**Very good**
Less than 50 = Very good, 50–99 = Good, 100–299 = Poor, 300+ = Very poor		
Ozone	**28 ppb**	**Very good**
Less than 50 = Very good, 51–89 = Good, 90–179 = Poor, 180+ = Very poor		
Sulphur	**5 ppb**	**Very good**
Less than 60 = Very good, 60–124 = Good, 125–399 = Poor, 400+ = Very poor		

Air quality is calculated in ppb (parts per billion)
London Evening Standard, 9 June 1997

Figure 3.28
Air quality in London, 9 June 1997

Figure 3.27
Traffic in urban areas

WHY HAS TRAFFIC IN URBAN AREAS INCREASED?	• GREATER AFFLUENCE AND INCREASED CAR OWNERSHIP. MANY FAMILIES EVEN HAVE TWO CARS (FIGURE 3.26). • PEOPLE COMMUTING TO WORK OR TRAVELLING TO CITY CENTRES FOR SHOPPING OR ENTERTAINMENT. • REDUCTION IN PUBLIC TRANSPORT AT EXPENSE OF PRIVATE CARS. • INCREASED ROAD FREIGHT, e.g. DELIVERY LORRIES.	
WHAT ARE THE DAMAGING EFFECTS OF INCREASED TRAFFIC IN URBAN AREAS?	• ENVIRONMENT	• AIR POLLUTION FROM VEHICLE EXHAUSTS – ESPECIALLY IN LARGE CITIES LIKE LOS ANGELES (FIGURE 3.22), MEXICO CITY AND TOKYO. SOME CITIES, e.g. LONDON, ISSUE DAILY AIR QUALITY FORECASTS/REPORTS (FIGURE 3.28). • NOISE POLLUTION FROM CARS, LORRIES AND BUSES. • VISUAL POLLUTION OF MOTORWAYS (FIGURE 3.21) AND CAR PARKS.
	• ECONOMY	• CONGESTION, ESPECIALLY AT PEAK TIMES (RUSH HOUR). IN CENTRAL LONDON, EVEN THOUGH MANY PEOPLE TRAVEL BY PUBLIC TRANSPORT, THE AVERAGE SPEED OF TRAFFIC IN 1996 WAS THE SAME AS IT WAS IN 1900, i.e. 20 KM/HR. • TIME WASTED SITTING IN TRAFFIC JAMS AND GRIDLOCK CONDITIONS. • COST OF BUILDING AND MAINTAINING ROADS. • COST OF PETROL/DIESEL AND THE USE OF A NON-RENEWABLE RESOURCE (OIL).
	• PEOPLE	• DANGER OF ACCIDENTS AND AN INCREASE IN STRESS TO BOTH DRIVERS AND PEDESTRIANS.
	• BUILDINGS	• DESTRUCTION OF PROPERTY FOR NEW OR WIDER ROADS AND CAR PARKS. • DAMAGE TO FOUNDATIONS CAUSED BY TRAFFIC VIBRATION.
HOW CAN TRANSPORT SYSTEMS IN URBAN AREAS BE MANAGED TO REDUCE THE DAMAGING EFFECTS OF INCREASED TRAFFIC?	• EXCLUDE, REDUCE OR ACCOMMODATE TRAFFIC BY SCHEMES SUCH AS TRAFFIC-FREE ZONES, PARK AND RIDE SCHEMES AND URBAN MOTORWAYS. • TRY TO REDUCE POLLUTION, ESPECIALLY FROM VEHICLE EXHAUSTS. • IMPROVE PUBLIC TRANSPORT, e.g. SUPERTRAMS IN SHEFFIELD, METROLINK IN MANCHESTER, METROS IN TYNE AND WEAR, SINGAPORE AND HONG KONG. THESE RAPID TRANSPORT SYSTEMS HAVE RESULTED FROM IMPROVED TECHNOLOGY (PAGE 41).	

BART – the San Francisco Bay Area Rapid Transit System

Like many other large cities, San Francisco receives thousands of commuters each weekday (Figure 3.29). During the 1960s an increasingly large percentage travelled by car, causing pollution (noise, fumes and visual), accidents and congestion (it could take an hour to cross Oakland Bridge at peak travel times). Increasing demands were made on the city authorities to construct more freeways into, and car parks within, the CBD. However, as this would not have reduced the congestion, especially at the bridges, it was decided instead to build a new transport system. Opened in 1974 and completed in 1978, the Bay Area Rapid Transit System (Figure 3.30) is a 120 km electric railway (with underwater, underground and elevated sections) designed to ease traffic congestion in the CBD. The underwater section was designed to withstand earthquakes by moving as the ground moved. It experienced no problems during the 1989 earthquake, whereas several car drivers were killed when part of the Bay Bridge, almost above this section, collapsed.

Figure 3.29 Downtown San Francisco

Figure 3.30 San Francisco Bay Area Rapid Transit System (BART)

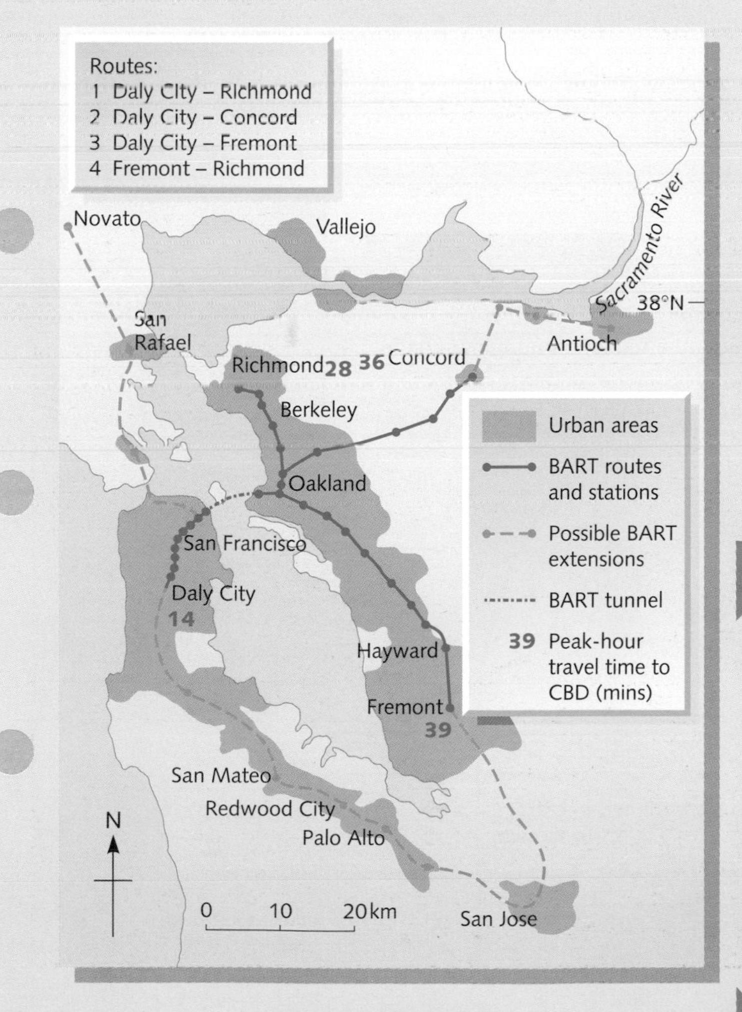

Advantages

- Electric and so pollution free (Figure 3.31).
- Fast conveyance of 350 000 commuters a day.
- Trains can travel up to 120 km/hr. Travel time at peak periods over the Bay between Oakland and San Francisco is 9 minutes (11 km) instead of over 40 minutes by road.
- Trains run every 1.5 minutes at peak times and every 20 minutes through the night.
- Modern carriages are noiseless, air-conditioned and carpeted.
- The whole system is fully automatic and computerised – drivers only take over in an emergency.
- Long platforms ensure rapid alighting and boarding.
- Lower fares than by bus to attract users.
- Cars left at suburban stations reduce CBD congestion.
- It has helped regenerate commercial life in downtown San Francisco.

Figure 3.31 BART train

Shopping in urban areas

Shops provide a service. As with settlements (Figure 2.12), it is possible to construct an urban shopping hierarchy. The *traditional hierarchy* is shown in Figure 3.32. At the base of the hierarchy is the corner shop (the village shop performs a similar function). It sells *low-order, convenience goods* which people are not prepared to travel far to purchase as they are needed regularly (perhaps daily). At the top of the hierarchy is the city centre. It provides *high-order goods* which people are prepared to travel much further to buy as they are more specialised and needed less regularly. Figure 3.33 shows how the sphere of influence (page 22) of a shop is related to its position in the hierarchy. Figure 3.34 shows the location of the various types of shops.

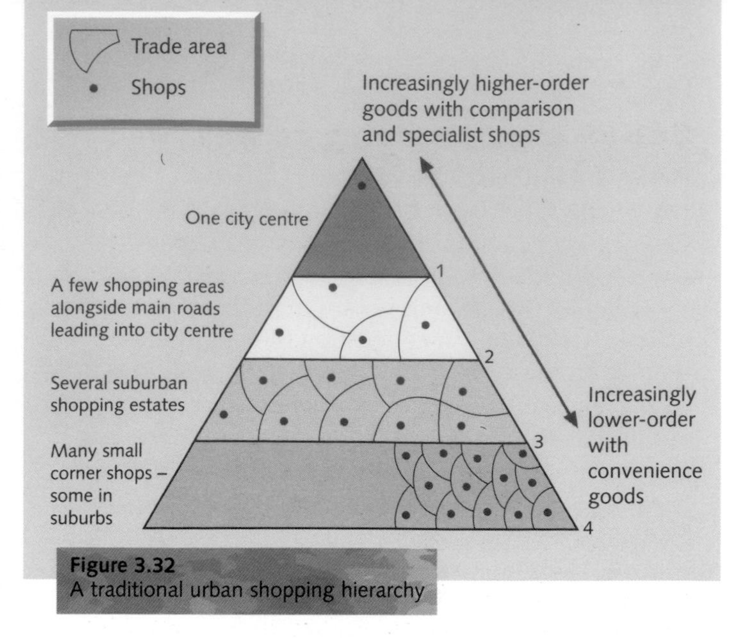

Figure 3.32
A traditional urban shopping hierarchy

Changing patterns and locations

Shopping areas, as with other land users in the urban area, are dynamic and are constantly changing. These changes may be due to shoppers having greater mobility, wanting somewhere to park the car, buying in bulk, preferring to do all their shopping under one roof and, perhaps because more women are now at work, late night shopping. Changes include:

- the development of, initially, hypermarkets and, more recently, regional shopping centres on edge-of-city sites (e.g. MetroCentre, page 43) and Meadowhall (Sheffield)
- the decline of many city centres as the major shopping area, mainly due to competition from the regional centres – although more recently some city centres have shown signs of recovery
- the closure of many corner shops due partly to being cleared in inner city redevelopment schemes or being priced out by supermarkets.

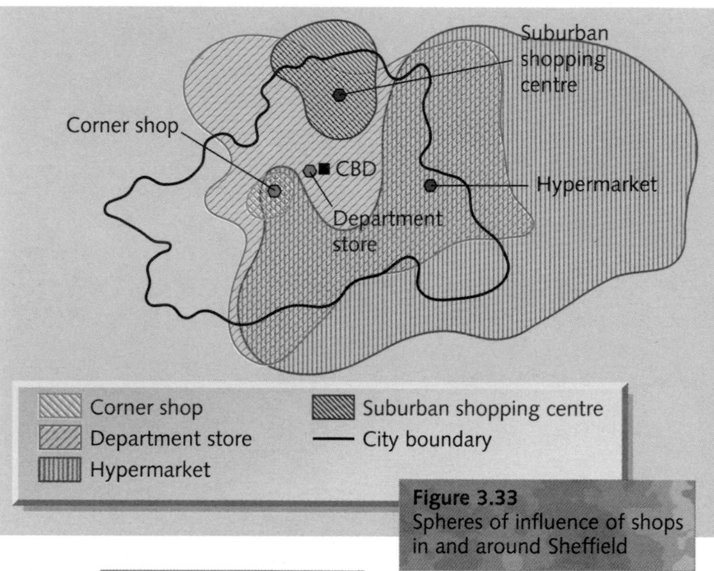

Figure 3.33
Spheres of influence of shops in and around Sheffield

Figure 3.34
Urban shopping hierarchy

Motorway or large urban by-pass
Several large suburban or district shopping centres
Regional shopping centre
Hypermarket
Ribbon development of shops along main roads
Transition zone with warehouse-type buildings, e.g. DIY furniture
Many small corner shops and local shopping centres
CBD: mainly precincts and under cover
Many small corner shops and local shopping centres
City boundary
Edge of CBD
T Transect line across city
Main roads out of city
0 10km

Frequency of visit	Weekly/ monthly	Two or three times a week	Corner shops daily	Monthly	Once or twice a week	Monthly	Impulse	Two or three times a week	Weekly/ monthly
Type of purchase	Bulk-buying	Mainly convenience, low-order	Convenience, low-order	Bulk DIY	Comparison, specialist, high-order, service	Bulk DIY	Convenience, low-order	Mainly convenience, low-order	Bulk-buying, specialist
Method of travel	Car, bus	Foot, car	Foot	Car, bus	Car Bus	Car, bus	Foot	Foot, car	Car, bus
Distance travelled (trade area or sphere of influence)	Many km	1–2 km	Under 1 km	Up to 5 km	All the urban area and smaller surrounding towns	Up to 5 km	Under 1 km	1–2 km	Many km

Out-of-town shopping centres

MetroCentre in Gateshead

Family shopping has evolved from the corner shop to the supermarket, and from the hypermarket to the 'out-of-town' shopping centre. Sir John Hall, whose brainchild is the MetroCentre, claims that since the 1960s shopping has evolved in three stages around central malls: the Arndale Centres of the 1960s, the Brent Cross and Eldon Square complexes of the late 1970s, and the MetroCentres of the 1980s. The main concept of the MetroCentre is to create a day out for the family, with the emphasis on family shopping and associated leisure activities.

The site for Gateshead's MetroCentre (Figure 3.35) was surveyed in 1980, and at that time received little interest from the city centre 'magnet' (or 'anchor') shops. However, by the time plans were published in 1983, retailing had changed and the success of out-of-town DIY shops led such retailing outlets as Marks & Spencer to reconsider their future policy. Indeed the MetroCentre was Marks & Spencer's first out-of-city location.

The scheme

There is free parking for 10 000 cars, with special facilities for the disabled driver, and new bus and rail stations for the non-motorist. Inside there are over 300 shops and 40 eating places.

Much attention has been paid to creating a pleasant shopping environment (Figure 3.37) – wide, tree-lined malls, air conditioning, one kilometre of glazed roof to let in natural light (supplemented by modern lighting in 'old world' lamps), numerous seats for relaxing, window-boxes, and escalators and lifts for the disabled. A market effect has been created by traders selling goods from decorative street barrows, and there is a wide variety of places to eat in. Leisure is a vital part of the scheme. There is a ten-screen cinema, a crèche for children, a space city for computer and space enthusiasts, a covered fantasy-land with all the attractions of the fair without the worries of the British climate, and a children's village with children's shops. A 150-room luxury hotel has been built as part of the complex.

Figure 3.35
The site of Gateshead's MetroCentre beside a dual carriageway (foreground), mainline railway and the River Tyne (behind)

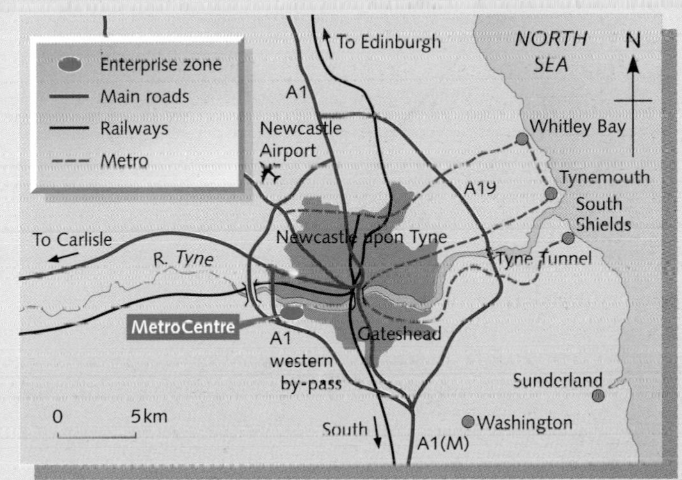

Figure 3.36
Location

Advantages of the site (Figure 3.35)

- It was in an enterprise zone (page 130) which initially allowed a relaxation in planning controls and exemption from rates.
- The area was previously marshland and relatively cheap to buy, and the 47 hectare site had possibilities for future expansion.
- It is adjacent to the western by-pass (2 km of frontage) which links with the North East's modern road network – essential for an out-of-town location (Figure 3.36).
- 1.3 million people live within 30 minutes' drive.
- It is adjacent to a main railway line, with its own railway station.

Figure 3.37
The emphasis inside the MetroCentre is on a pleasant, bright layout based on two-tiered malls

Movement out of cities

Suburbanised villages

In parts of the developed world there has been a reversal of the movement to large urban areas, with groups of people moving out into surrounding villages. This movement, a process known as *counterurbanisation*, has led to a change in the character of these settlements and to their being called *suburbanised* (because increasingly they adopt some of the characteristics of urban areas – Figure 3.38). They are also known as *commuter* or *dormitory* towns because many residents who live and sleep there travel to nearby towns and cities for work.

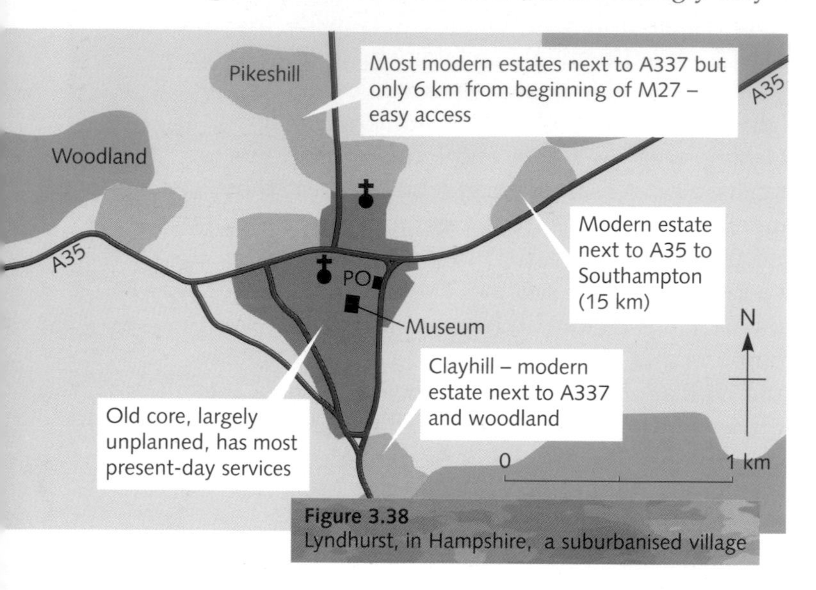

Figure 3.38
Lyndhurst, in Hampshire, a suburbanised village

Who moves into these villages?

- The more wealthy urban residents and those with improved family status. These groups have the money to afford the larger and often expensive houses, and the cost of travel to work, shops and amenities.
- Those wishing to move to a more attractive environment with less pollution and more open space.
- Elderly people who have retired and wish to live in a quieter environment.

How do the villages change?

- Newcomers begin to outnumber the original residents creating, on occasions, a social divide.
- House prices of existing properties increase, older buildings and barns are renovated and new estates are built on former farming land.
- An increase in the school population (primary) and more shops.
- More cars, causing congestion on narrow lanes and the need for wider roads.

Places

Braithwaite

Braithwaite is a small village in the English Lake District. In 1925 it consisted of a nucleated core of tightly grouped farms, outbuildings and terraced cottages along narrow lanes (coloured brown on Figure 3.39). The village green gave an open character to the western half. Most buildings dated from the eighteenth and nineteenth centuries. Employment was either in farming, the woollen, pencil and flour mills or the nearby mines. The community was self-contained and included a church, chapel, village hall, school and inn.

Seventy years later, the character of the village has changed mainly due to improved mobility and accessibility. The village, some 5 km from the tourist centre of Keswick, is next to the improved A66 which links West Cumbria with the M6. Although the old core remains, it is surrounded by modern houses and accommodation and services for tourists (coloured green on Figure 3.39). Only a handful of residents are employed locally, either in farming or tourism.

Figure 3.39
Braithwaite (Cumbria) in 1925 and today

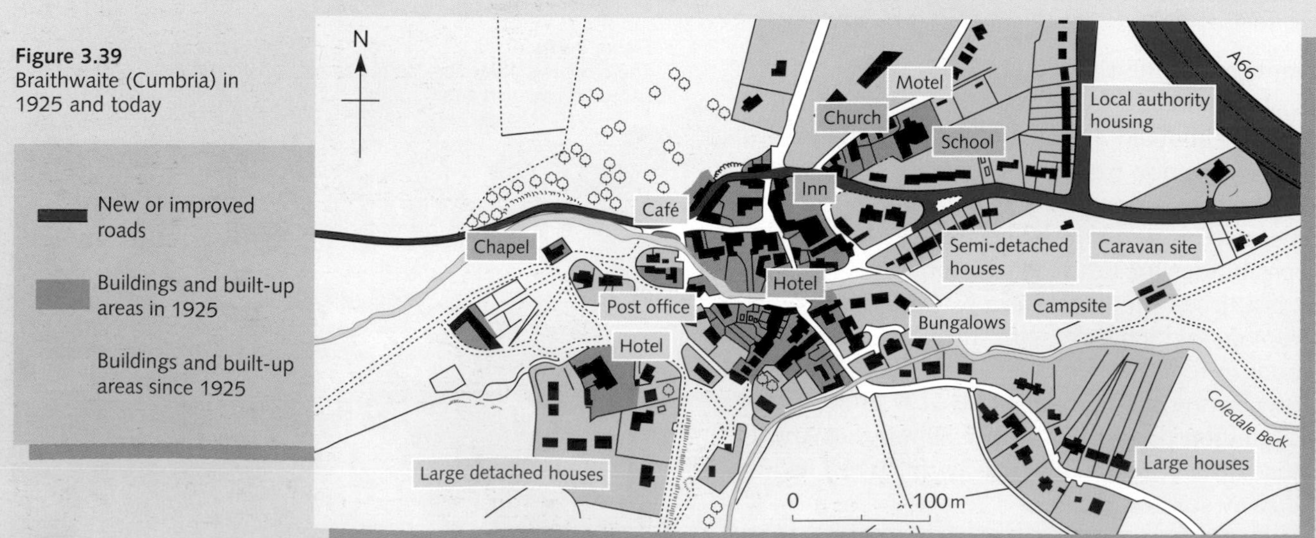

New towns

New towns have been built in several developed countries for a variety of reasons. These include:

- to take overspill from expanding conurbations and large cities or from inner city clearance schemes
- to attract new industries to areas of high unemployment
- to create a more pleasant environment and a better quality of life.

The main aim when planning a new town is to make it self-contained. New towns provide housing, jobs (to limit the time, distance and cost of journeys to work) and services (schools, hospitals, shops and leisure amenities), all in an attractive environment.

Lelystad (Netherlands)

The Flevoland Polder was reclaimed between 1950 and 1957 (Figure 3.40a). The development of Lelystad, chosen in 1948 as the polder's capital, began in 1960. The new town was expected to have a population of 100 000 by the year 2000. The town plan was to have an elongated central area surrounded by four districts (Figure 3.40b). Each district was to be divided into neighbourhoods of 5000 to 6000 inhabitants. The residential districts were to have a low housing density (usually fewer than 30 houses per hectare) with 90 per cent of the buildings being low-rise (Figure 3.41). Blocks of dwellings within the residential districts were grouped along narrow streets and around small squares which little or no traffic was allowed to enter (Figure 3.42). The residential districts are separated from one another by main roads which are both wide and straight. The roads were set at a lower level and are crossed by bridges only usable by pedestrians and cyclists. Lelystad prides itself on its 'green lungs'. Each residential district has large areas of open space and the town itself is surrounded on three sides by parks (the fourth is water). The town centre has a railway station (1988), town hall, provincial hall, shops, places of entertainment and a hotel-cum-congress centre. Growth has not been as rapid as first thought, mainly due to rival new towns being built on the polder. Lelystad has, in the late 1990s, a population of 60 000.

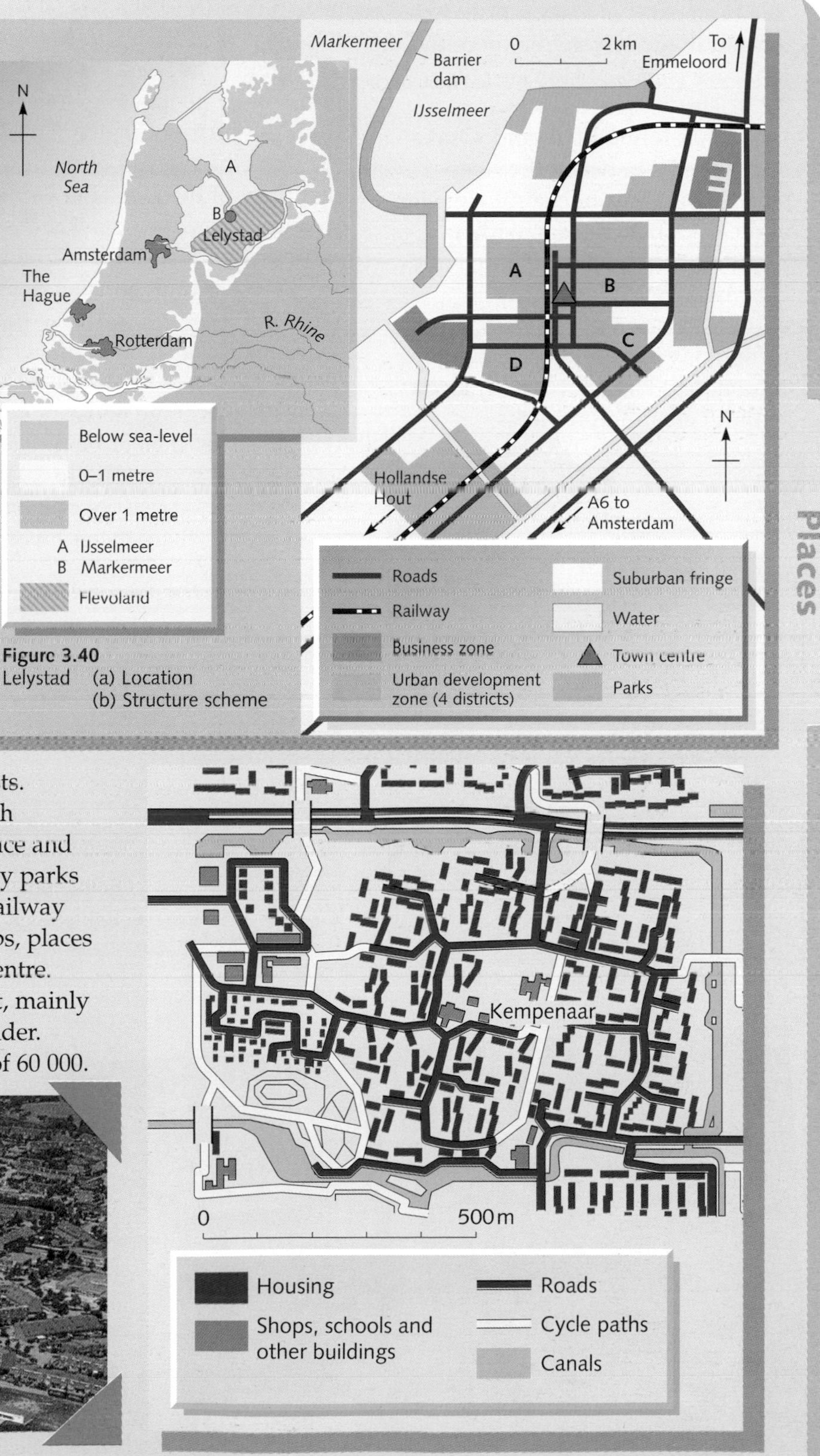

Figure 3.40
Lelystad (a) Location
(b) Structure scheme

Figure 3.42
A residential neighbourhood in Lelystad

Figure 3.41
Lelystad new town

CASE STUDY 3

Osaka–Kobe 2000

The twin cities of Osaka–Kobe are located in the Kansai region of the Japanese island of Honshu (Figure 3.43). This region, as elsewhere in Japan, has limited flat land. Osaka and Kobe lie between the steep-sided mountains of central Honshu and Osaka Bay. Kansai, with a population of over 9 million, has become a world leader in education, science, business, technology and industry . . . but this success has not been achieved without creating problems.

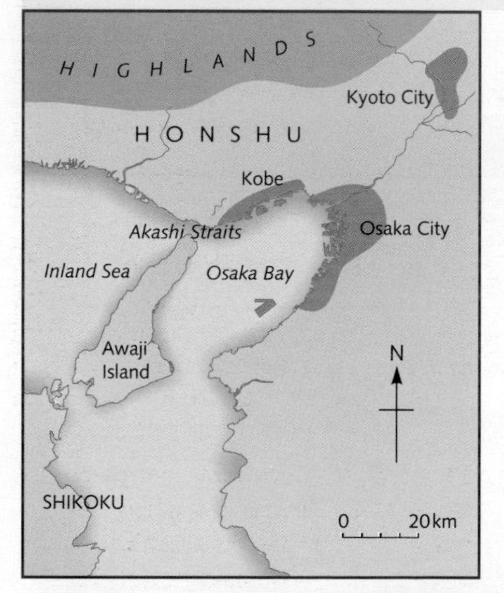

Figure 3.43 Location of Osaka–Kobe

Problems

Housing Osaka and Kobe have grown rapidly in the last 50 years (Osaka 2.60 million and Kobe 1.42 million in July 1996). As large numbers of people migrated here from the surrounding countryside, the two cities spread outwards so that most of the available flat land has now been used up. The resultant competition for land has led to a very high housing density with, in places, up to 10 000 people living within a square kilometre. The houses themselves are very small. The average house of 4.5 m^2 is eight times smaller than the average UK house (Figure 3.44). Many houses consist of only one main room, where the occupants have to cook, eat, work and sleep, and a small bathroom. There is no space for gardens and very little for parks or recreation (Figure 3.47).

Figure 3.44
Inside a Japanese home

High land values Osaka's city centre is the headquarters of many Japanese banks and corporations and is a major shopping and entertainment centre. The competition for prime sites has led to exceptionally high land prices and the building of high-rise office blocks (Figure 3.45).

Transport The unplanned growth of Osaka–Kobe and a rapid increase in road traffic has caused major transport problems. Roads are congested and often in a state of near gridlock. Emissions from vehicle exhausts cause severe air pollution. Movement westwards along the coast is made difficult by mountains extending down to the sea.

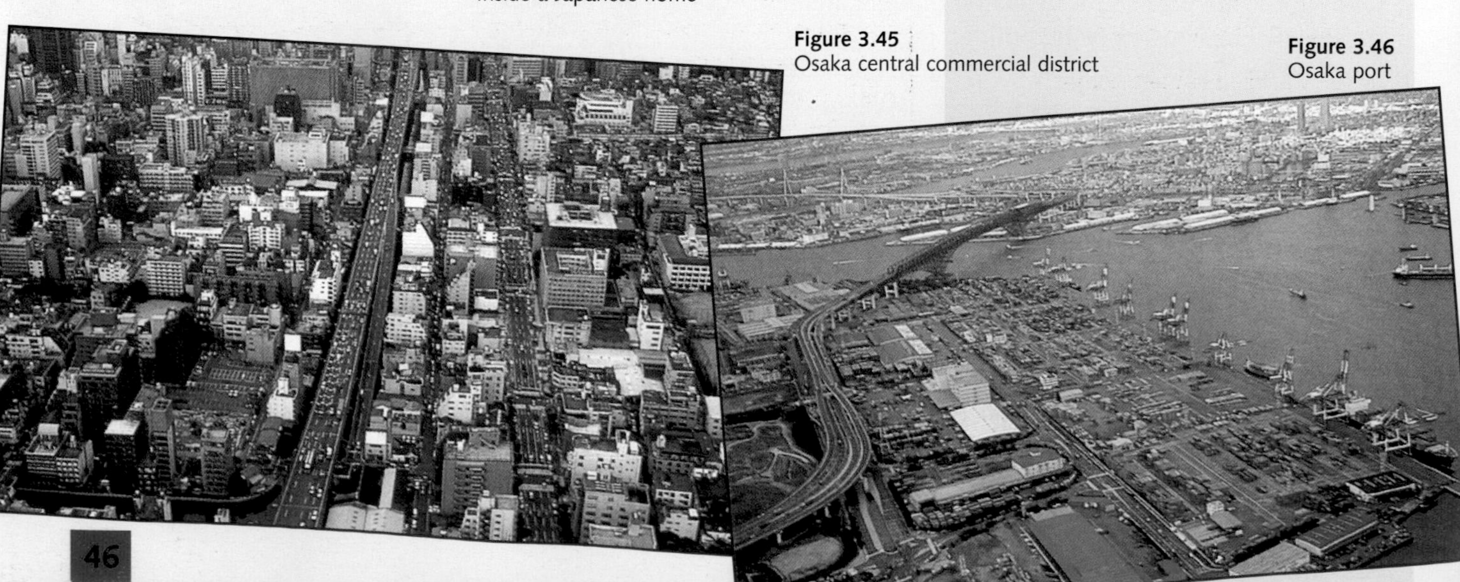

Figure 3.45
Osaka central commercial district

Figure 3.46
Osaka port

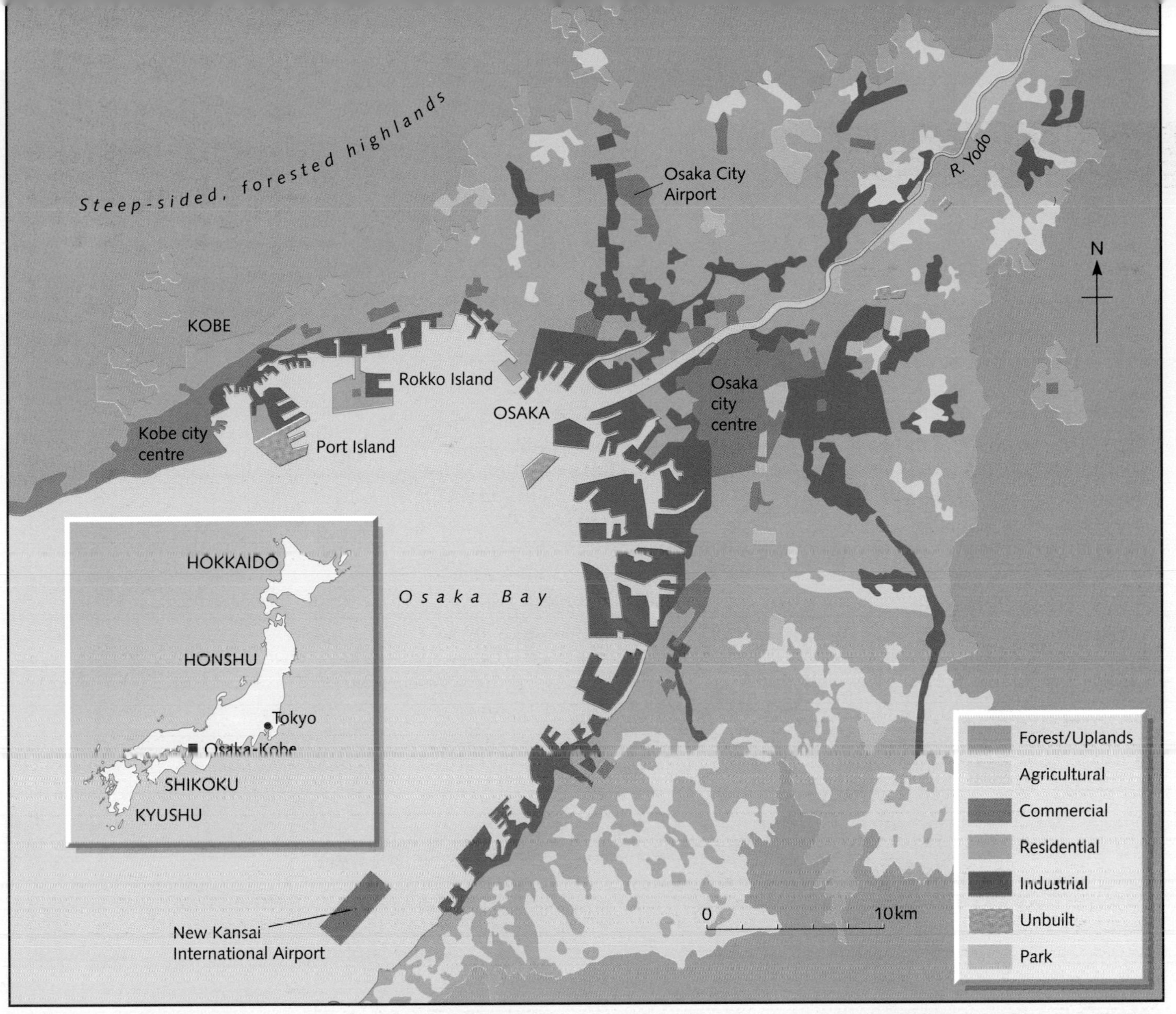

Figure 3.47 Land use in Osaka–Kobe

Port Up to 1400 ships a day enter the docks at Osaka (Figure 3.46). Many bring in oil and other raw materials. The port also exports large amounts of Japanese manufactured goods (page 184).

Industry Kansai is a major industrial region. Much of the industry is concentrated along Osaka Bay (Figure 3.47) where large companies manufacture incoming raw materials. As Japan has to import over 99 per cent of its oil, coal and iron ore, the port area has large oil refineries, steelworks and other industries processing imported raw materials. Once again, a major problem is a lack of space, for new docks, for enlarging the container port, for storing goods being imported or exported, and land needed by processing industries. Other industries, many of them small, are located alongside either the River Yodo or the outer city motorways. Even today 90 per cent of Japanese companies are small, often family concerns making parts for larger transnationals (page 132). Most of the components are used by car assembly or electronic and other high-tech companies.

Pollution The large volume of road traffic, smoke from some of the older heavy industries and rubbish incinerated at sea all cause serious air pollution. Rubbish dumped at sea and oil spilled from ships, especially oiltankers, pollutes the water of Osaka Bay.

Physical environment Japan is located at a destructive plate margin (see page 248). As the Pacific Plate is pushed under the Eurasian Plate and destroyed, it forms volcanoes and fold mountains. Osaka and Kobe are hemmed in by these mountains which, although not very high, are steep and forested. Destructive margins also experience severe earthquakes, the most recent in Japan destroying much of Kobe in January 1995 (page 252). Japan is also in the path of tropical storms known as typhoons. Typhoons, with heavy rain and hurricane-force winds, can affect Japan 30 times in an average year.

Some solutions

Housing Port Island (Figure 3.48) and Rokko Island have both been recently reclaimed from the sea in Osaka Bay (Figure 3.47). Much of Rokko Island is a new town with housing in the form of flats. Even before these flats were built there was a huge waiting list and the first occupants were those whose names were drawn from a hat. Compared with those in Osaka and Kobe, the houses are spacious (Figure 3.49), the majority having two large and one smaller room. The residential areas are near to schools, shopping centres, hospitals, parks and places of entertainment. Traffic is, where possible, segregated from the housing districts, while a new railway has been built to take commuters to Kobe. The main problem with living on Rokko Island is the break with Japanese tradition: there is an absence of temples and shrines, and women now have to go to work as two salaries are usually needed to pay for the expensive housing.

Port development Land has also been reclaimed from the sea for port extensions. Kobe's new container port, on Port Island, extends 7 km out to sea (Figure 3.48).

Industry Industrial development has taken place at four locations:

1 On land adjacent to Osaka Bay, previously used for the Nippon steelworks.
2 On land reclaimed from Osaka Bay by levelling inland areas behind Osaka, transporting the material and depositing it on the sea-bed.
3 By creating up to 11 'science cities' (some on the newly levelled sites) each with its high-tech industries and research centres. The centres are being financed by leading Japanese corporations in an attempt to maintain their country's global supremacy in this field.
4 Where space permits, alongside major motorways.

Figure 3.49
Inside a modern home on Rokko Island

Figure 3.48
Rokko Island, looking north to Kobe and the forested highlands

Figure 3.51 Kansai International Airport

Transport

- The Shinkansen (wrongly translated into English as 'the bullet train'), provides the most reliable rail service in the world (Figure 3.50a). One of three Shinkansen lines passes through Osaka and Kobe. The latest models can travel at speeds of up to 300 km/hr and can reach Tokyo, 553 km to the east, in about $2\frac{1}{2}$ hours. The trains carry 275 million people per year, run at 7 minute intervals, are computer controlled, arrive prompt to the second and have never been involved in a serious accident (Figure 3.50b). The route to the west passes through a series of tunnels cut through the coastal mountains.
- In an attempt to improve road traffic along the northern shore of the Inland Sea, a new road has been built linking Osaka and Kobe with the islands of Awaji and Shikoku (Figure 3.43). The Akashi Bridge, 4 km in length, was the world's largest when it was built. A proposed extension to the road to the southern island of Kyushu would create a 400 km by-pass of the urban-industrial centres on the mainland.
- Kansai International Airport, built on an artificial island in Osaka Bay, opened in 1994 (Figure 3.51). The airport has a main runway of 3500 metres, operates 24 hours a day and is connected to Osaka by a 13 km road and rail causeway. The terminal, the world's largest, can handle over 30 million passengers a year.

Figure 3.50 The Shinkansen

(a)

(b)

Pollution Japan, with money available, has made serious attempts to clean up some of the pollution caused during its rapid industrialisation. The Inland Sea, once a 'dead' sea, now has fish and oyster farms. The new high-tech industries are less polluting than the older, heavier industries. Nuclear energy, despite its risks, causes less air pollution than fossil fuels.

The environment Earthquakes and typhoons cannot be controlled but attempts can be made to reduce their effects. New buildings are designed to withstand earthquakes – as indeed many did in the 1995 Kobe 'quake (page 252). The Akashi Bridge is a fine example of how the Japanese use their technology. The foundations of the bridge have been sunk to a great depth under the sea-bed to resist earthquakes; the foundations and supports have been strengthened to combat the fast sea-currents; and the superstructure has been designed to withstand 130 km/hr typhoon winds.

QUESTIONS

1

(Pages 28 and 29)

a Complete the key to diagram A by adding each of the following statements to the correct letter:
- oldest, poorest-quality housing intermixed with industry
- large shops and office blocks
- newest and most expensive housing
- traditionally the centre of light manufacturing
- mainly interwar, medium-cost housing. (5)

b Diagram B shows land use zones and land values in a city in the developed world.
- i) Which of the five zones is the CBD? (1)
- ii) What do the letters CBD stand for? (1)
- iii) Name two kinds of land use found in the CBD. (1)
- iv) Give two reasons why there is little open space in zone B. (2)
- v) Which zone is most likely to have large houses with gardens? (1)

c
- i) Explain the meaning of *twilight zone*. (1)
- ii) What is the relationship between the value of land and its distance from a city centre? (1)
- iii) Explain why this relationship occurs. (2)

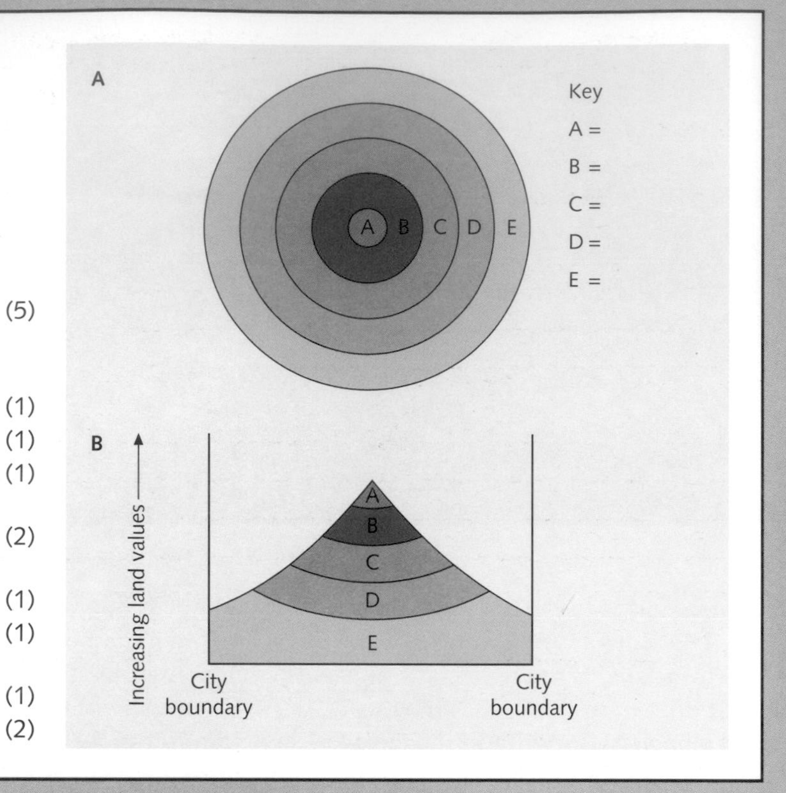

2

(Pages 30 and 31)

a
- i) For map 1, suggest the type of houses and age of houses. (2)
- ii) For map 2, suggest the type of houses and the age of houses. (2)
- iii) Which of the two areas has most:
  - gardens
  - car parking
  - places of work
  - open space
  - residents? (5)

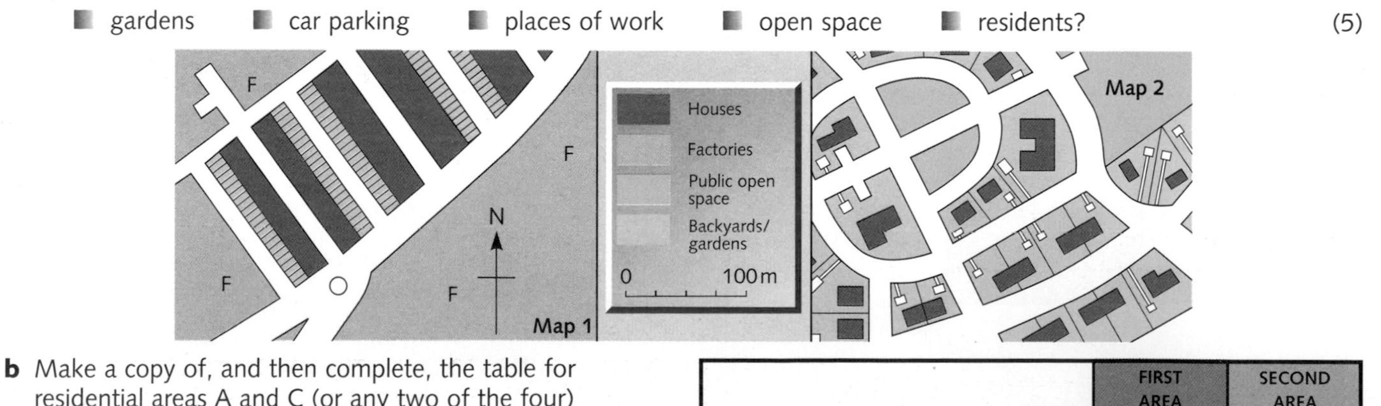

b Make a copy of, and then complete, the table for residential areas A and C (or any two of the four) described on pages 30 and 31. (8)

c For the two areas you described in part (b), give one reason for the differences in:
- type, design and age of housing
- road pattern
- land use
- socio-economic groups
- household amenities
- quality of the environment. (6)

	FIRST AREA	SECOND AREA
NAMED EXAMPLE		
LOCATION IN CITY		
TYPE, APPEARANCE, AGE OF HOUSING		
2 MAIN TYPES OF LAND USE		
MAIN TYPE OF TENURE		
2 MAIN TYPES OF SOCIO-ECONOMIC GROUP		
% BORN OUTSIDE THE UK		
% WITH WC, HOT WATER AND BATH		

3

(Pages 30 and 31)

a i) Which set of statistics below, A, B or C, is most likely to represent the area located at 1 on the map? Give three reasons for your answer. (1 + 3)

ii) Which set of statistics, A, B or C, is most likely to represent the area located at 3 on the map? Give three reasons for your answer. (1 + 3)

b i) Give two advantages of living in area 1. (2)

ii) Give two disadvantages of living in area 1. (2)

iii) Give two advantages of living in area 3. (2)

iv) Give two disadvantages of living in area 3. (2)

c How may changes in family wealth, family size or the age of the family affect the type of area where people might live? (3)

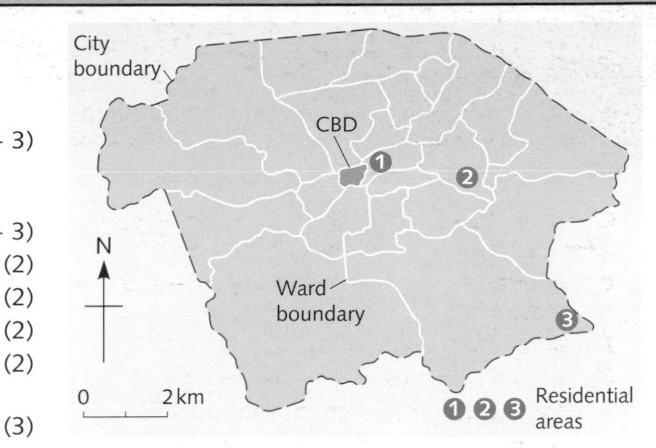

	HOUSING TENURE			HOUSING DENSITY		HOUSING QUALITY
RESIDENTIAL AREAS	% OWNER OCCUPIED	% OWNED BY THE COUNCIL	% RENTED	HOUSEHOLDS WITH OVER 1.5 PERSONS PER ROOM	HOUSEHOLDS WITH LESS THAN 0.5 PERSONS PER ROOM	% HOUSEHOLDS WHICH SHARE OR LACK A WC
SET OF STATISTICS A	98	0	2	1	84	0
SET OF STATISTICS B	24	34	42	19	35	19
SET OF STATISTICS C	62	14	24	8	14	10

4

(Pages 34 and 35)

a i) Give two reasons why the London docks lost their trade. (2)

ii) By 1981 London's Docklands faced severe social, economic and environmental problems. List six of these problems. (6)

b Since 1981 many changes have taken place in London's Docklands. Briefly describe how the area has undergone:

- environmental regeneration
- economic regeneration
- social regeneration (6)

c The advertisement here appeared in a national newspaper in June 1997.

i) On which street is Dunbar Wharf located? (1)

ii) What were the prices of the apartments for sale? (1)

iii) Give four advantages of living in Dunbar Court. (4)

d Name four groups that helped in the regeneration of the Docklands. For each group give one way in which they helped to improve the area. (4)

e How do you think each of the following will answer the question: 'Do you think the changes in London's Docklands have improved the area or made it worse?' (5)

a school-leaver ◆ a local shopkeeper ◆ a local retired couple ◆ a former docker ◆ a bank manager who lives in Dockland but works in central London

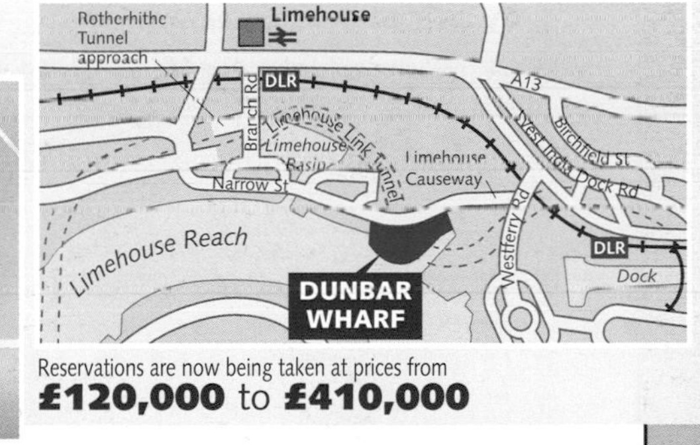

5

(Pages 38 and 39, 46–49)

a Describe, with the help of the diagram, five problems that have resulted from urban growth in a city that you have studied in an economically more developed country (e.g. Los Angeles, Osaka–Kobe). (5)

b How are these problems being tackled in the city you have named in part (a)? (5)

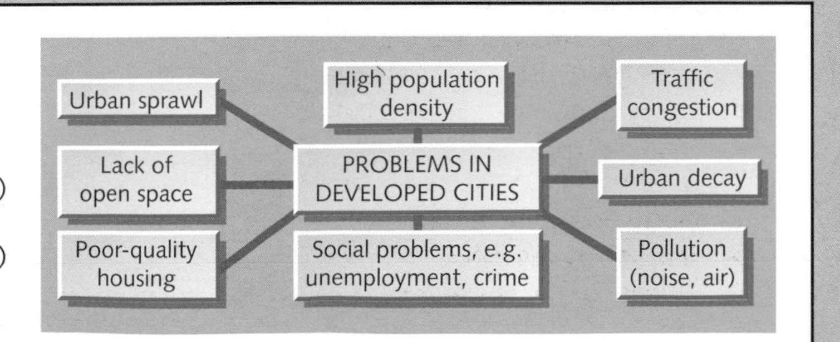

6

(Pages 40 and 41)

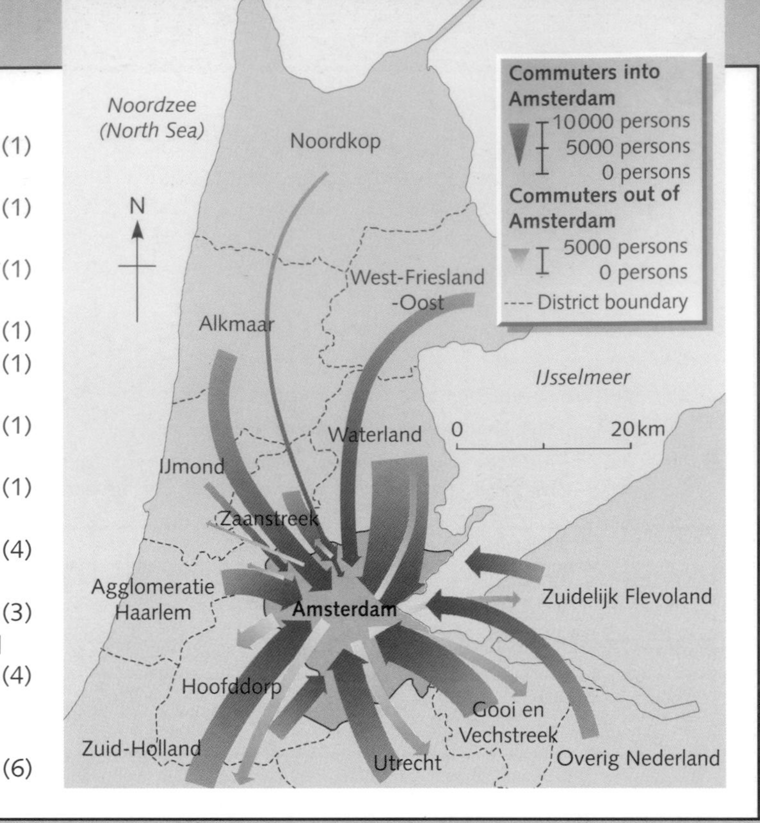

a i) What is a commuter? (1)
ii) Why do more people commute into Amsterdam than out of Amsterdam? (1)
iii) From which district do most of Amsterdam's commuters come? (1)
iv) Why do so few people commute from Noordkop to Amsterdam? (1)

b i) What is a commuter hinterland? (1)
ii) How many people commute from Zuid-Holland to Amsterdam? (1)
iii) How many people commute from Amsterdam to Zuid-Holland? (1)

c i) Give four reasons why city centres often experience traffic congestion. (4)
ii) Name three problems, other than congestion, created by traffic in city centres. (3)

d Describe two attempts that have been made in your local town or city to reduce the problems created by traffic. (4)

e With reference to a developed city outside the UK, describe the attempts made there to reduce its traffic problems. (6)

7

(Pages 42 and 43)

a i) What is meant by the following terms?
- comparison goods
- convenience goods
- specialist shops
- department store
- low-order and high-order goods
- bulk buying
- sphere of influence (7)

ii) Why has a department store a larger sphere of influence than a corner shop? (2)

b Study the diagram.
i) Describe the location of the corner shops. (1)
ii) Why are so many corner shops shown on the diagram? (1)
iii) How often might people shop here? (1)
iv) Give two reasons why corner shops have closed in recent years. (2)

c i) Describe the location of neighbourhood (suburban) shopping parades. (1)
ii) How far are people prepared to travel to these shops? (1)
iii) Give one advantage and one disadvantage of neighbourhood shopping parades. (2)

d i) What type of shopper uses shops along main roads in inner city areas? (1)
ii) What types of goods are sold here? (2)

e In the CBD of most British cities there is a large, modernised, undercover, air-conditioned, pedestrianised shopping centre.
i) Give three reasons why such shopping centres were built. (3)
ii) What are the advantages for shoppers and shopkeepers? (2)

f Many towns now have an edge-of-city hypermarket.

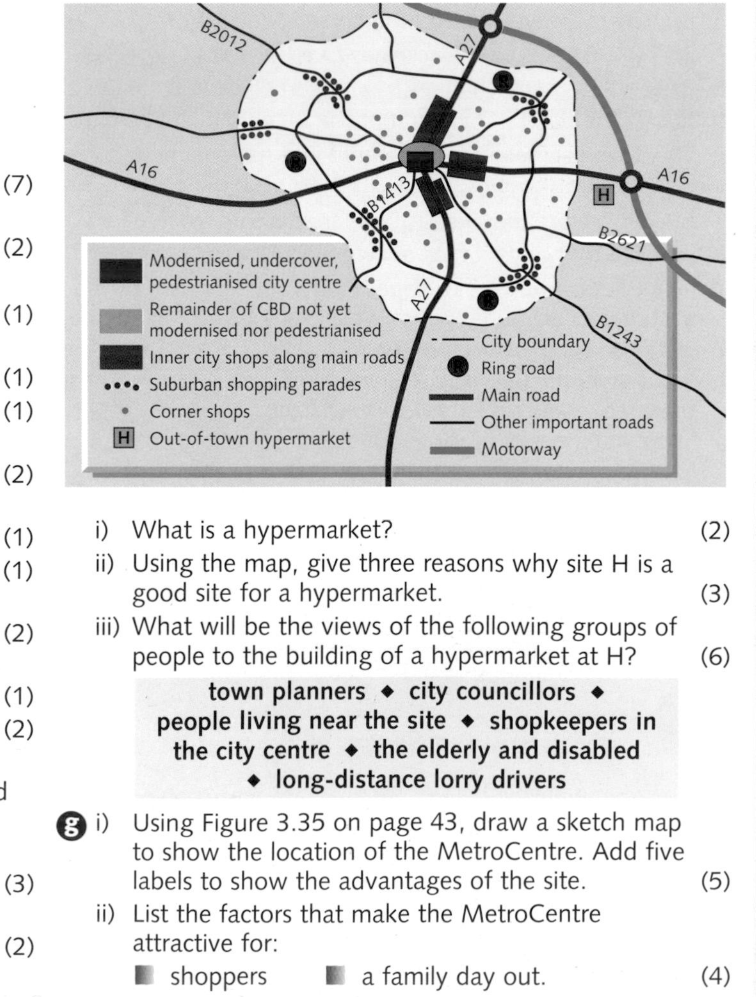

i) What is a hypermarket? (2)
ii) Using the map, give three reasons why site H is a good site for a hypermarket. (3)
iii) What will be the views of the following groups of people to the building of a hypermarket at H? (6)

town planners ◆ city councillors ◆ people living near the site ◆ shopkeepers in the city centre ◆ the elderly and disabled ◆ long-distance lorry drivers

g i) Using Figure 3.35 on page 43, draw a sketch map to show the location of the MetroCentre. Add five labels to show the advantages of the site. (5)
ii) List the factors that make the MetroCentre attractive for:
- shoppers
- a family day out. (4)

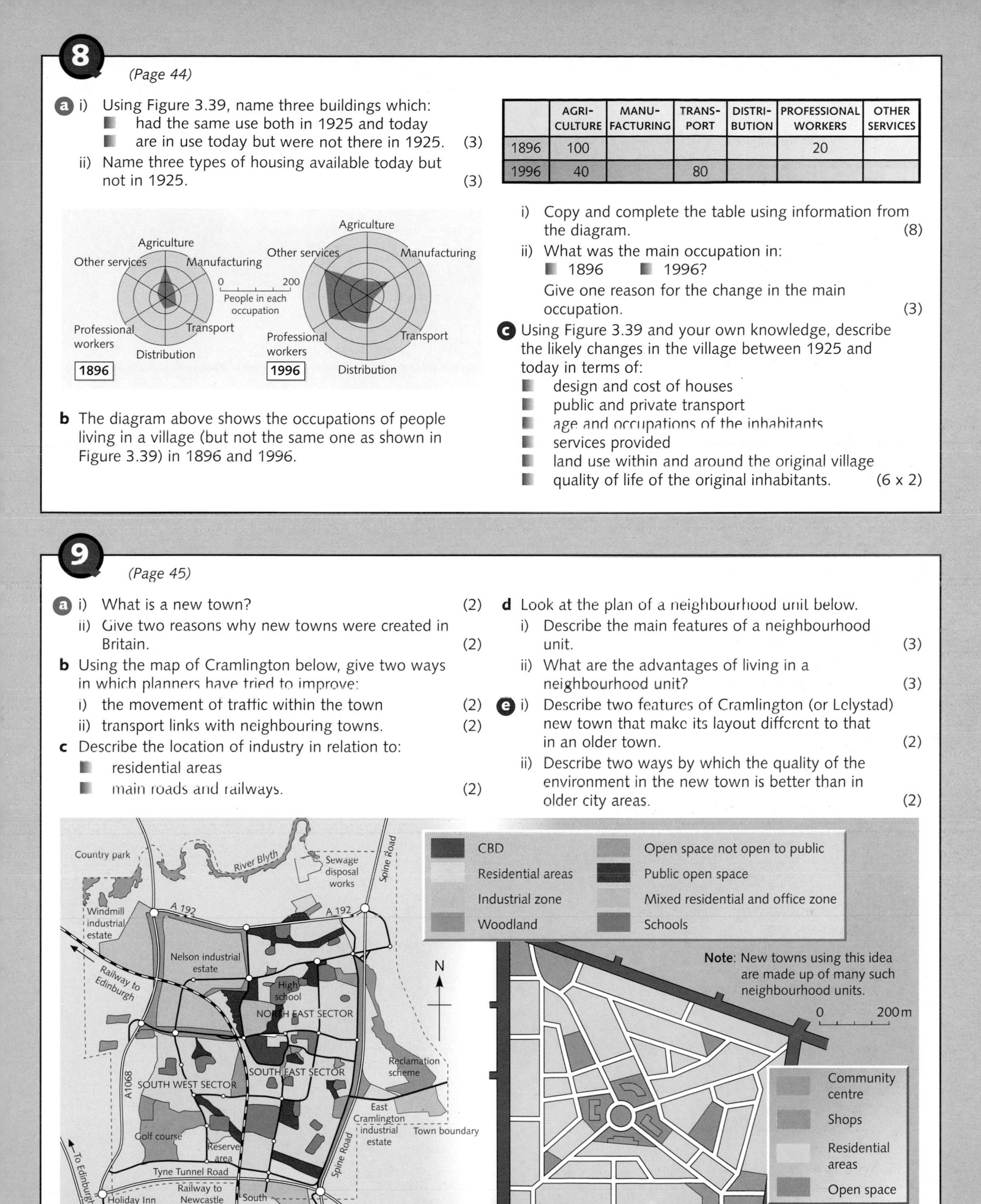

8

(Page 44)

a i) Using Figure 3.39, name three buildings which:
- had the same use both in 1925 and today
- are in use today but were not there in 1925. (3)

ii) Name three types of housing available today but not in 1925. (3)

b The diagram above shows the occupations of people living in a village (but not the same one as shown in Figure 3.39) in 1896 and 1996.

	AGRI-CULTURE	MANU-FACTURING	TRANS-PORT	DISTRI-BUTION	PROFESSIONAL WORKERS	OTHER SERVICES
1896	100				20	
1996	40		80			

i) Copy and complete the table using information from the diagram. (8)

ii) What was the main occupation in:
- 1896
- 1996?

Give one reason for the change in the main occupation. (3)

c Using Figure 3.39 and your own knowledge, describe the likely changes in the village between 1925 and today in terms of:
- design and cost of houses
- public and private transport
- age and occupations of the inhabitants
- services provided
- land use within and around the original village
- quality of life of the original inhabitants. (6 x 2)

9

(Page 45)

a i) What is a new town? (2)

ii) Give two reasons why new towns were created in Britain. (2)

b Using the map of Cramlington below, give two ways in which planners have tried to improve:

i) the movement of traffic within the town (2)

ii) transport links with neighbouring towns. (2)

c Describe the location of industry in relation to:
- residential areas
- main roads and railways. (2)

d Look at the plan of a neighbourhood unit below.

i) Describe the main features of a neighbourhood unit. (3)

ii) What are the advantages of living in a neighbourhood unit? (3)

e i) Describe two features of Cramlington (or Lelystad) new town that make its layout different to that in an older town. (2)

ii) Describe two ways by which the quality of the environment in the new town is better than in older city areas. (2)

4 URBANISATION IN DEVELOPING COUNTRIES

Growth of cities

Urbanisation, as defined at the beginning of Chapter 3, is the proportion of people living in towns and cities. Whereas in 1800 only 3 per cent of the world's population lived in urban areas, by 1950 this proportion had risen to 29 per cent and is predicted to exceed 50 per cent by the year 2006 (Figure 4.1). Although most developed countries have experienced urbanisation since the early 1800s, the process has only gathered pace in developing countries since the 1950s. Figure 4.2 shows the proportion of the total population of each country living in towns and cities in 1995. Notice that:

- the developed countries of North America, Western Europe and Oceania are usually those with the highest proportion
- the developing countries in Africa and South-east Asia often have the lowest proportion
- South America, the most economically advanced of the developing continents, is an exception (an *anomaly*), as several of its countries have an urban population exceeding 75 per cent.

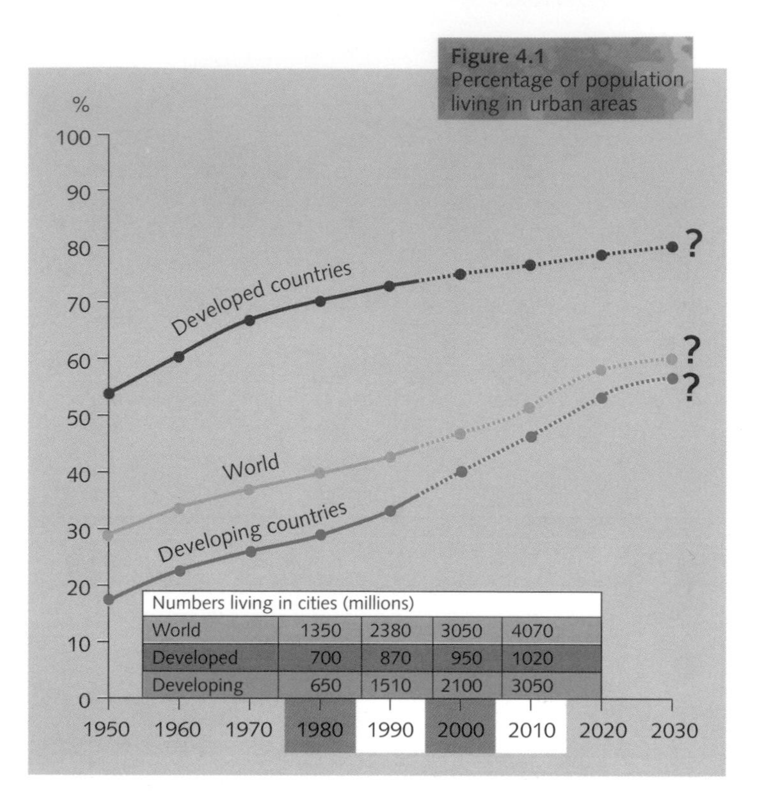

Numbers living in cities (millions)				
World	1350	2380	3050	4070
Developed	700	870	950	1020
Developing	650	1510	2100	3050

Figure 4.1 Percentage of population living in urban areas

There are two additional factors to consider:

1 The growth of very large cities with a population exceeding 1 million (Figure 4.3). In 1850 there were only two 'million' cities – London and Paris. This number increased to 70 by 1950 and to an estimated 290 in 1995. A new term, *megacity*, refers to places with a population in excess of 10 million.

2 The change in the distribution and location of large cities. Prior to 1950, most of the 'million' cities were in developed countries and in temperate latitudes north of the Equator (i.e. 40° to 50°N in Europe and North America). Since then there has been a dramatic increase in 'million' cities in developing countries, the majority of which lie within the tropics (Figure 4.4).

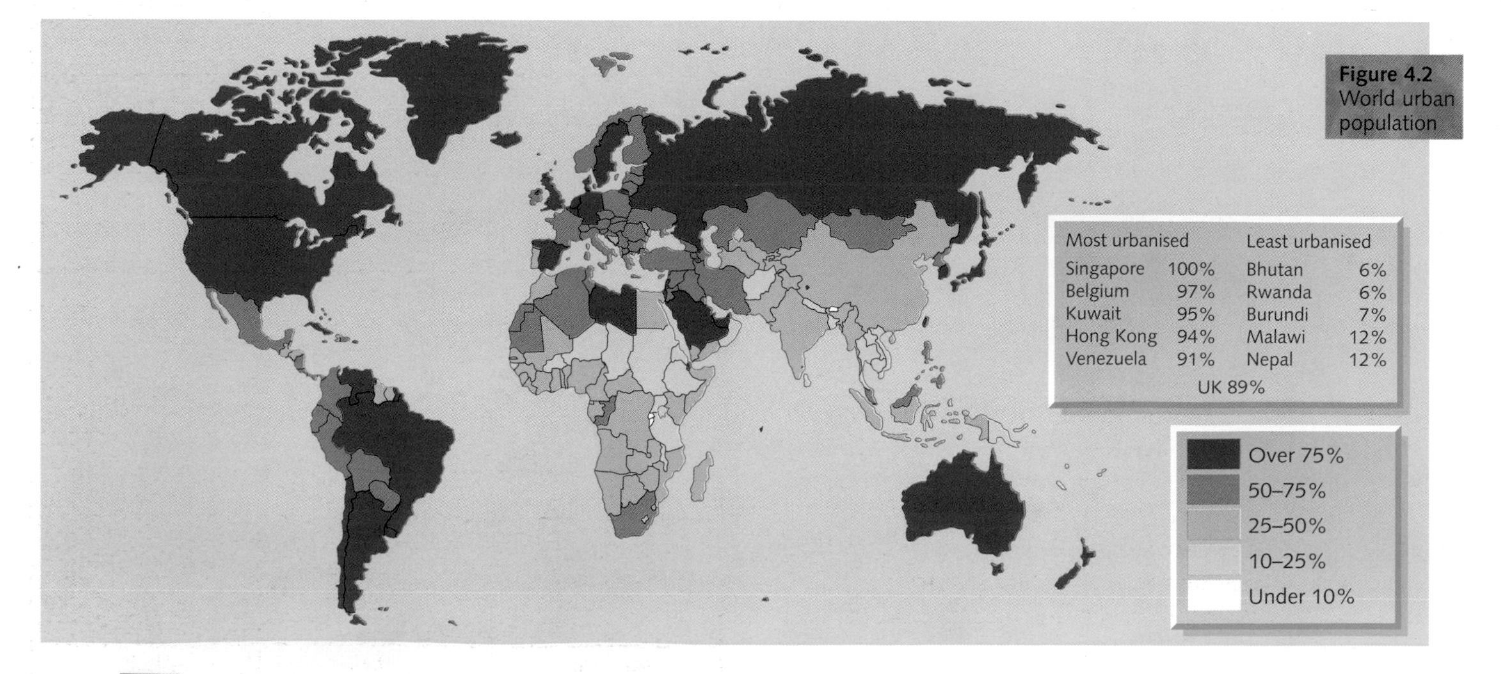

Most urbanised		Least urbanised	
Singapore	100%	Bhutan	6%
Belgium	97%	Rwanda	6%
Kuwait	95%	Burundi	7%
Hong Kong	94%	Malawi	12%
Venezuela	91%	Nepal	12%

UK 89%

Figure 4.2 World urban population

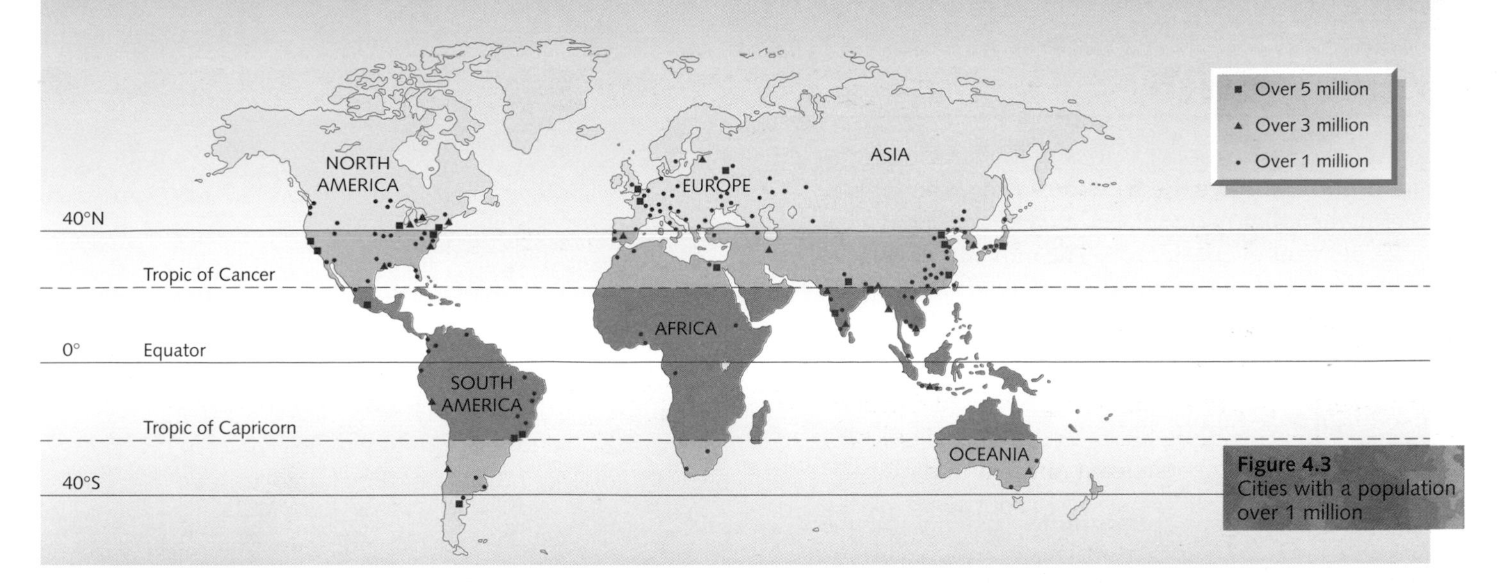

Figure 4.3 Cities with a population over 1 million

Figure 4.4 Growth and location of 'million' cities

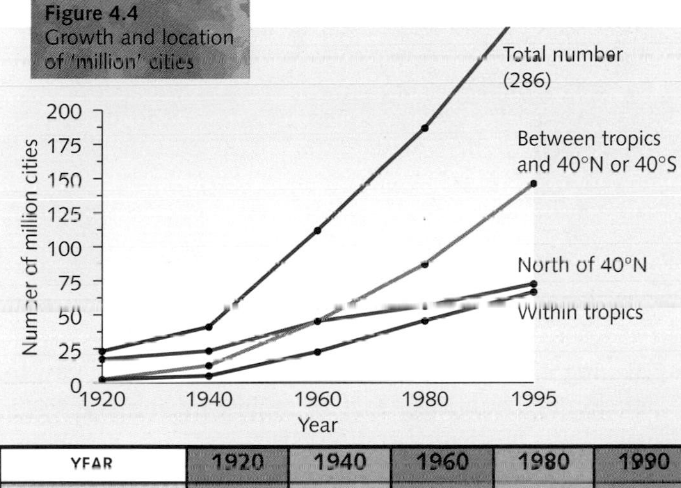

YEAR	1920	1940	1960	1980	1990
POPULATION IN MILLION CITIES (MILLIONS)	51.4	91.8	270.0	482.2	940.9
MEAN POPULATION OF MILLION CITIES (MILLIONS)	2.14	2.25	2.39	2.55	3.25
WORLD POPULATION (MILLIONS)	1800	2300	3100	4432	5292
% POPULATION IN MILLION CITIES	2.86	4.01	8.71	10.88	17.60

Figure 4.5a shows the rank order of the world's 10 largest cities over a period of years. In 1970 half of these cities were still in the industrialised, developed continents of North America and Europe. By 1985 there were no European cities in the top 10. Estimates suggest that by the year 2000 the two largest cities could be in Latin America, and 12 of the top 15 in developing countries in Latin America and Asia. Estimates may differ due to:

- variations between countries in their definition of the term 'urban' and in how they delimit the urban area (e.g. do they refer to Liverpool and Manchester or to Merseyside and Greater Manchester?)
- inaccuracies in collecting population data, e.g. birth, death and migration rates, and the assumption that these rates will remain constant over a period of time.

During the 1970s and 1980s the growth rates of most cities in developed countries slowed down. At that time the two fastest-growing urban areas were Mexico City and São Paulo – both in Latin America. At present, with these two places now growing less quickly, the fastest-growing cities are in Asia (Figure 4.5b). The population of Dhaka (Bangladesh) is believed to have doubled between 1985 and 1995, and that of other cities in the region will double between 1980 and 2000. Can you imagine all the inhabitants of a place such as Liverpool or Glasgow suddenly arriving in Dhaka in one year? Think of the problems this must pose for the newcomers, the existing inhabitants and the city authorities.

RANK ORDER	1970	1985	2000 (EST.) 1	2000 (EST.) 2
1	NEW YORK 16.5	TOKYO 23.0	TOKYO 28.0	MEXICO CITY 25.8
2	TOKYO 13.4	MEXICO CITY 18.7	SAO PAULO 22.6	SAO PAULO 24.0
3	LONDON 10.5	NEW YORK 18.2	BOMBAY 18.1	TOKYO 20.2
4	SHANGHAI 10.0	SAO PAULO 16.8	SHANGHAI 17.4	CALCUTTA 16.5
5	MEXICO CITY 8.6	SHANGHAI 13.3	NEW YORK 16.6	BOMBAY 16.0
6	LOS ANGELES 8.4	LOS ANGELES 12.8	MEXICO CITY 16.2	NEW YORK 15.8
7	BUENOS AIRES 8.4	BUENOS AIRES 11.6	BEIJING 14.4	SEOUL 13.8
8	PARIS 8.4	RIO DE JANEIRO 11.1	LAGOS 13.5	TEHRAN 13.6
9	SAO PAULO 7.1	CALCUTTA 9.2	JAKARTA 13.4	RIO DE JANEIRO 13.3
10	MOSCOW 7.1	BOMBAY 8.2	LOS ANGELES 13.2	SHANGHAI 13.3

(a) The largest cities (figures in millions)

Figure 4.5 The world's cities

1 North American/Oceania sources
2 Europe/Internet sources

LATIN AMERICA

EUROPE AND FORMER USSR

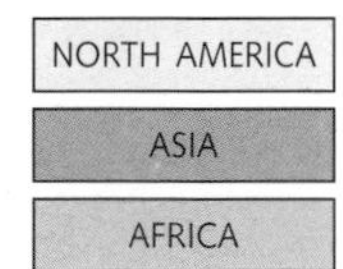

(b) The fastest-growing cities (1990s)

CITY	GROWTH RATE % PER YEAR	DOUBLING PERIOD (YEARS)
DHAKA	5.7	12.6
TEHRAN	4.1	17.5
KARACHI	4.0	17.8
DELHI	3.9	18.0
BANGKOK	3.8	18.5
NEW YORK	0.1	816.2
LONDON	0.1	1086.5

Urban growth

The movement of people to cities in the developing world began in the early twentieth century. Since then it has accelerated so that many places are expanding at a rate of over 25 per cent every decade. The movement from country areas to towns and cities is called *rural–urban migration*. Figure 4.6 is a quotation from the Brandt Report. It suggests that in developing countries, movement to the city is partly due to *rural push* and partly due to *urban pull* (Figure 4.7).

'Push' factors (why people leave the countryside):

- Lack of employment opportunities (usually the main factor).
- Pressure on the land, e.g. division of land among sons – each has too little to live on.
- Many families do not own any land.
- Overpopulation, resulting from high birth rates.
- Starvation, resulting from either too little output for the people of the area, or crop failure. Often, it may also be caused by a change in agriculture – from producing crops for local/family consumption to a system that produces cash/plantation crops for consumption in the developed world.
- Limited food production due to overgrazing, or to misuse of the land resulting in soil erosion or exhaustion (page 238).
- Mechanisation has caused a reduction in jobs available on the land together with, in many areas, reduced yields.
- Farming is hard work with long hours and little pay. In developing countries a lack of money means a lack of machinery, pesticides and fertiliser.

"The rush to the towns has created the same kind of misery as existed in the nineteenth-century cities of Europe and America. But industrialisation in those days was labour-intensive, so that the cities grew as the jobs expanded. .... The migration in today's developing world is often due to the lack of opportunity in the countryside – it is 'rural push' as much as 'urban pull'. The consequences of high birth rates and rapid migration are all too visible in many cities of the Third World, with abysmal living conditions and very high unemployment or underemployment. The strains on families, whose members are often separated, are very heavy. In São Paulo in Brazil, the population was growing at around 6–7 per cent annually in the late sixties and early seventies, in such appalling conditions that infant mortality was actually increasing. The fact that people still migrate to these cities only underlines the desperate situation which they have left behind."

Figure 4.6 Extract from the Brandt Report

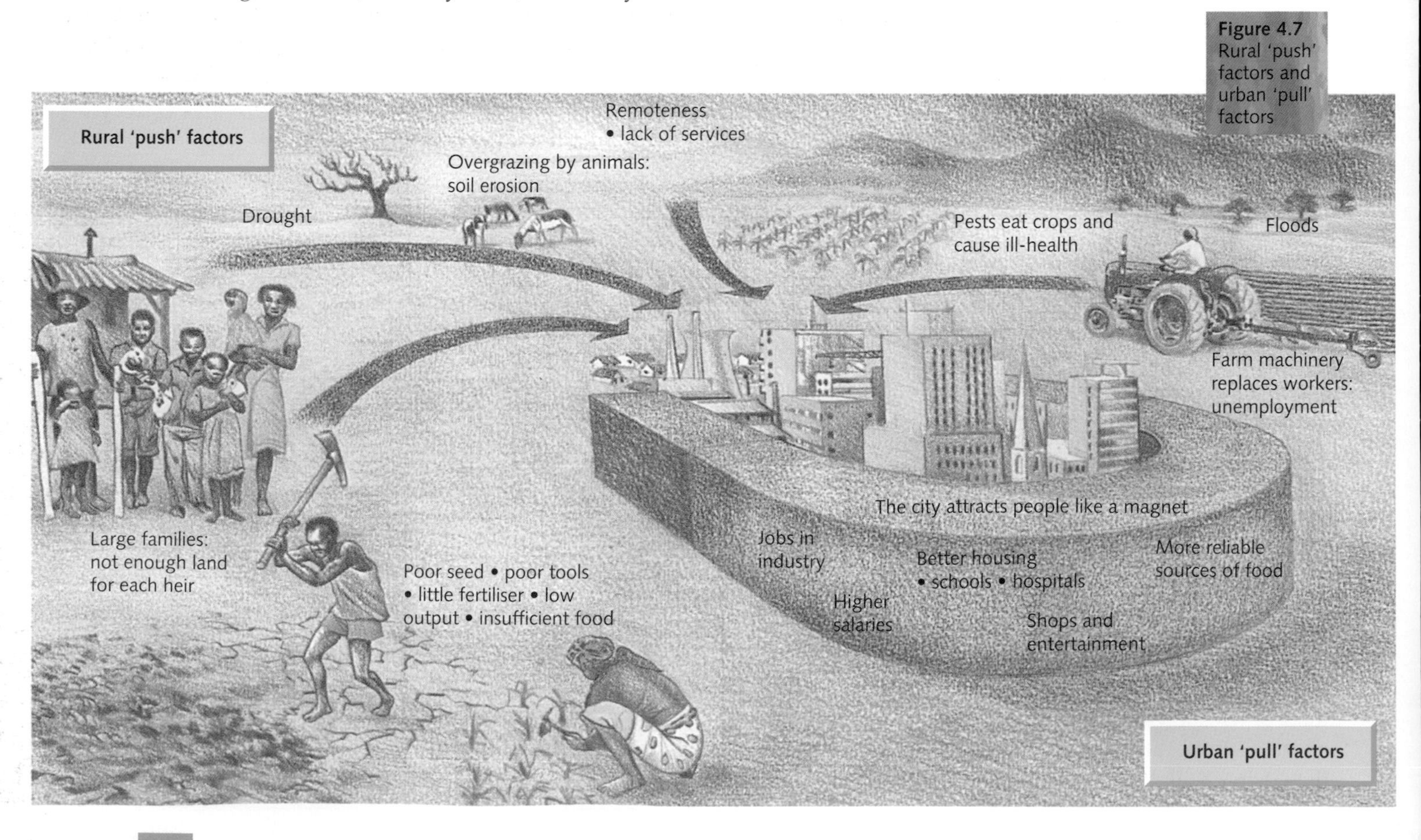

Figure 4.7 Rural 'push' factors and urban 'pull' factors

Figure 4.8
Central Hong Kong

- Natural disasters such as drought (Sahel countries), hurricanes (West Indies), floods (Bangladesh) and volcanic eruptions (Colombia) destroy villages and crops.
- Extreme physical conditions such as aridity, rugged mountains, cold, heat and dense vegetation.
- Local communities (Amazon Amerindians) forced to move.
- Lack of services (schools, hospitals).
- Lack of investment as money available to the government will be spent on urban areas.

'Pull' factors (why people move to the city):

- They are looking for better-paid jobs. Factory workers get about three times the wages of farm workers.
- They expect to be housed more comfortably and to have a higher quality of life.
- They have a better chance of services such as schools, medical treatment and entertainment.
- They are attracted to the 'bright lights'.
- More reliable sources of food.
- Religious and political activities can be carried on more safely in larger cities.

In Brazil the 'pull' factors, especially the opportunity for work, are usually much stronger than the 'push' factors.

Figure 4.7 presents a family's 'perceptions of the city'. This is what they think, expect or were led to believe the city is like (Figure 4.8). The reality is often very different (Figure 4.9). As many new arrivals to the city are unlikely to have much money, they will be unable to buy or rent a house, even if one was available. They will probably have to make a temporary shelter using cheap or waste materials. The gap between the rich and poor is far greater in developing countries than it is in ones that are more developed.

Figure 4.9
New arrivals to developing cities often encounter conditions like these in Ho Chi Minh City, Vietnam

Problems in residential areas

Just as in the developed world, in developing countries there are marked differences between residential areas of cities. However, there are three noticeable differences.

1 The gulf between the rich (who are relatively few in number) and the poor (many of whom are recent migrants from surrounding rural areas) is much greater. This means that the contrast between the well-off areas and the poorest areas is much greater in terms of quality and density of housing, the quality of the environment and the provision of amenities.
2 Most of the better-off areas are located near to the city centre while most of the poorer areas are found towards the edge of the city (Figure 4.10).
3 An increasing number of the inhabitants are forced to live in *shanty* or, using the UN term, *'informal' settlements*.

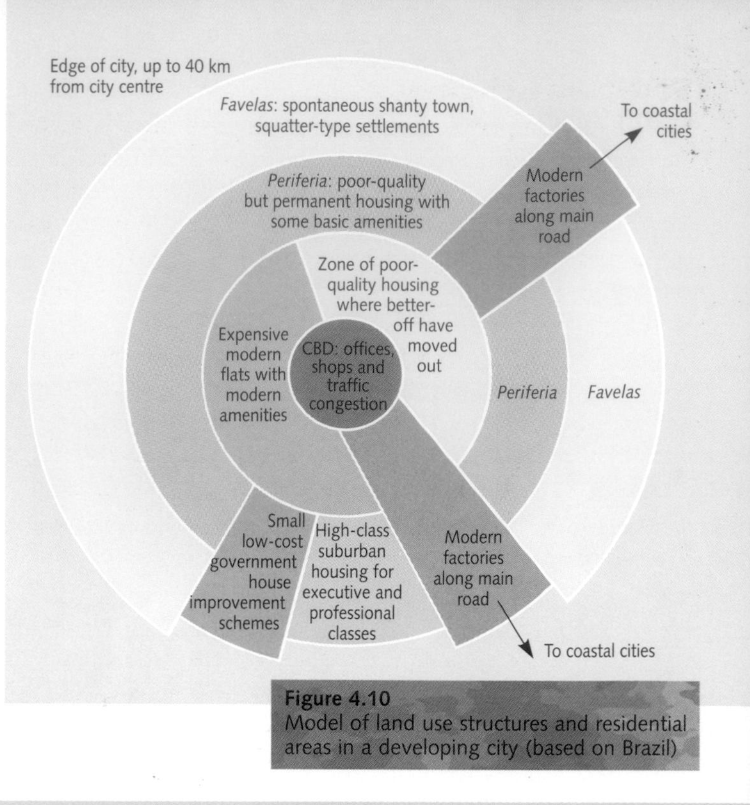

Figure 4.10
Model of land use structures and residential areas in a developing city (based on Brazil)

Places

São Paulo

São Paulo is the largest, richest and most industrialised city in Brazil. Its growth has resulted from a massive influx of immigrants, initially from Europe (after 1850) and more recently from rural Brazil, and a high birth rate. This rapid growth has been accompanied by increased social and economic segregation, with the poor being relegated to the city's periphery with its shanty settlements (*favelas*) and tenement slums (*corticos*). With the city's population predicted to grow from 7 million in 1970 to over 23 million by 2001, the authorities have an almost impossible task in providing sufficient housing and services.

Housing for the well-off

This group of people live in expensive housing ranging from elegant apartment complexes, each with its own social and recreational facilities, to Californian-style detached houses with large gardens and individual swimming pools (Figure 4.11). Family size is generally limited to two children, with housemaids. These properties, protected by security guards, are located near to the CBD where most of the residents work and shop (Figure 4.10). The children, who are healthy and well educated, are likely to find well-paid jobs in the future.

Figure 4.11 Californian-style housing for the rich in São Paulo

Housing for the poor (favelas)

Many of São Paulo's poor live in temporary accommodation. This is sometimes on vacant space next to modern factories or alongside main roads leading to the city, but is usually on the outskirts of the city. These people are 'squatters' and have no legal right to the land they occupy. The rapid growth of these spontaneous settlements, or *favelas* as they are known in Brazil, is common to cities in the developing world (Figures 4.12, 4.16 and 4.22). Favelas are found on land that has little economic value and which the well-off find unsuitable for development. Favelas often develop on steep hillsides which are liable to landslides, or on badly drained, unhealthy valley floors.

The housing is often a collection of primitive shacks made from any material available – wood, corrugated iron, cardboard or sacking. Some may only have one room in which the family has to live, eat and sleep. Others may have two rooms, one being used to sleep the family which is likely to consist of at least six children. Most houses lack such basic amenities as electricity, clean running water, toilets and main sewerage. Any empty space between houses will soon be occupied either by later migrants or, as there is no refuse collection, with rubbish. Favelas are overcrowded and have a high housing density.

Self-help schemes

Over 1 million people living in developing cities have no shelter of any kind, while over one-third live in squatter settlements – sites only vacant because they are subject to flooding, landslips or industrial pollution. Although most local authorities would probably prefer to remove shanty settlements from their cities, few can find the necessary resources that would be needed to provide alternative accommodation. As a result, shanty settlements become permanent. Self-help schemes seem, therefore, the only hope for the squatters to improve their homes (Figure 4.14). The most immediate needs of the poor are often simple: a plot on which to build, a small loan to improve or extend the house, cheap building materials, and basic services (see Places below). There is no need for advanced technology or expensive accommodation.

Figure 4.14
A community 'self-help' housing project in São Paulo – a 'slum of hope'

Figure 4.12
A favela in São Paulo – a 'slum of despair'

Community housing projects

Low-cost improvements (Figure 4.13)
Existing homes may be improved by rebuilding the houses with cheap and quick and easy-to-use breeze-blocks. A water tank on the roof collects rainwater and is connected to the water supply and, in turn, to an outside wash basin and an indoor bathroom/toilet. Electricity and mains sewerage are added. Most inhabitants of this type of housing, which is found in the peripheral parts of São Paulo, will have some type of employment enabling them to pay a low rent.

Self-help schemes (Figure 4.14)
Groups of people are encouraged to help build their own homes. Each group will do basic work such as digging ditches to take the water and sewage pipes. The local authority will then provide breeze-blocks and roofing tiles, and the group provide the labour. The money which this saves the authorities can be used to provide amenities such as electricity, a clean water supply, tarred roads and a community centre. The advantages of self-help schemes are that they can be done in stages, they can create a community spirit and, as the cost of building is relatively cheap, more houses can be provided.

Figure 4.13
A self-help housing scheme in São Paulo

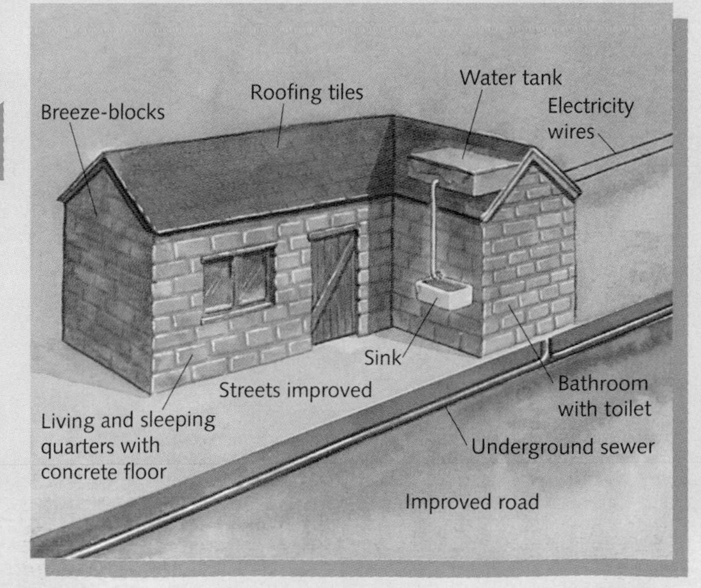

Problems in developing cities

Calcutta

Calcutta is one of the most notorious examples of the way in which problems are created when cities grow too quickly. In describing these problems, which in Calcutta are severe, it is easy to provide a negative stereotype of life in a developing city. However, despite the poverty and squalor, and even with the scale and range of the problems facing the city, and its lack of resources, the authorities have made real improvements in the urban environment.

Location and growth Calcutta is situated on flat, swampy land alongside the River Hooghly, one of numerous rivers that form the Ganges delta (Figure 4.15). It developed during colonial times as a major port and is now India's largest city. Its population of 7 million in 1970 had increased to 12 million by 1990 and is estimated to reach between 13 and 16 million by the year 2000.

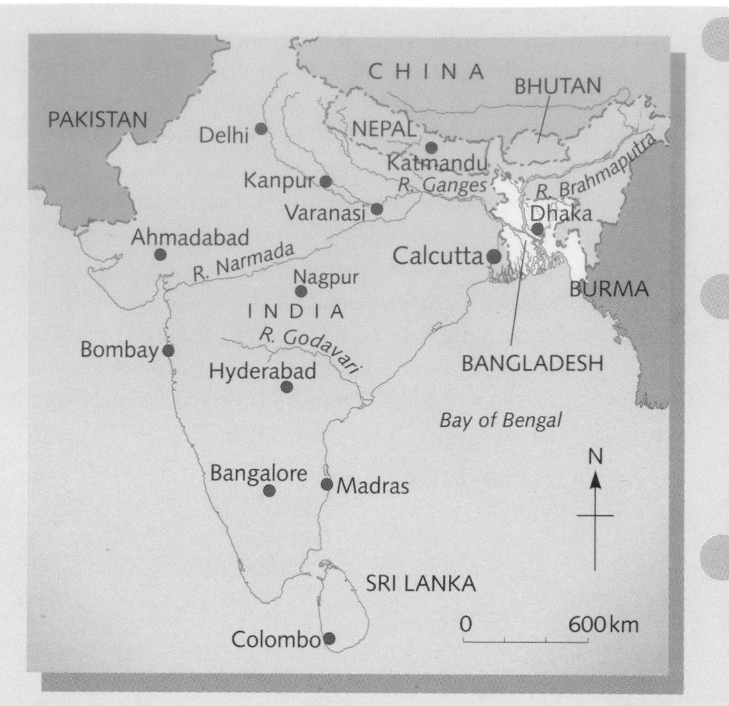

Figure 4.15 Location of Calcutta

Physical problems With the highest point of the city being only 10 metres above sea-level, there is a constant threat of flooding. Flooding can result from heavy rainfall associated with the annual monsoon (page 198) or be brought by tropical storms (typhoons – page 200). At these times, surplus storm water (often contaminated by sewage) has to be pumped away.

Housing Many families have no home other than the pavement. Reports suggest that over a quarter of a million people are forced to sleep in the open, covered only by bamboo, sacking, polythene or cardboard (Figure 4.16). Another 3 million residents live in *bustees* (Figure 4.17). Bustees are collections of houses built of non-permanent materials, such as wattle, with tiled roofs and mud floors – not the best materials to withstand the heavy monsoon rains. The houses, packed closely together, are separated by narrow alleys. Inside there is often only one room, and that is usually no larger than an average British bathroom. In this room the family, which can number up to eight, live, eat and sleep. Despite this overcrowding, the insides of the dwellings are clean and tidy. The houses belong to landlords who rent them out to bustee dwellers – the tenants are evicted if they cannot pay the rent. Estimates suggest a population density in some bustees of over 150 000 people per km^2.

Figure 4.16 Street dwellers in Calcutta

Figure 4.17 In the bustees

Figure 4.18 Problems with clean water

Water supply, sanitation and health Although three-quarters of Calcutta's population has access to piped water, it is not uncommon, especially in the bustees, for a single street tap to be used by 35 to 45 families (Figure 4.18). Sanitation is almost non-existent in the bustees and an estimated one-third of the city's population is without a toilet of any kind. Here human effluent runs down the narrow lanes. Elsewhere, the 100-year-old and more drains and sewage pipes tend to crack, spilling their contents onto the streets. Sewage often contaminates drinking water, especially after heavy rain. Rubbish, dumped in the streets, is rarely collected. A survey in 1991 recorded an infant mortality rate of 123 per 1000 live births. The major causes of death were cholera, typhoid, dysentery, tetanus and measles – diseases linked to poor sanitary conditions and overcrowding.

Figure 4.20 A stall in a Calcutta bustee

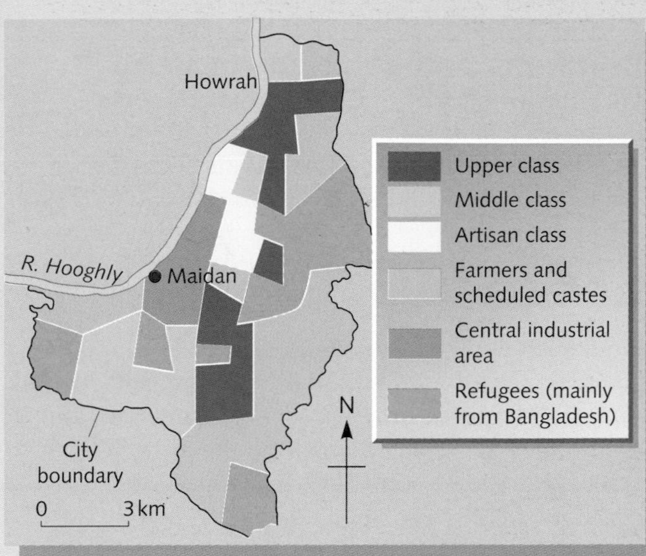

Figure 4.19 Residential areas

Segregation Figure 4.19 is a simplified map showing segregation between peoples of different caste, language, religion and occupations. The original Bengali-speaking Hindus live apart both from non-Bengalis and from the later Hindu refugees from Bangladesh.

Provision of services The provision of electricity, clean water, schools and hospitals, together with the collection of rubbish, all require considerable amounts of money – a resource in short supply to the authorities.

Transport Most people have either to walk or to use the overcrowded public transport system. The buses, many of which are old, are insufficient to carry everyone, and passengers can be seen hanging on to the outsides. Rickshaws are cheap but their numbers and slowness add to the congestion. Traffic noise, congestion and pollution continue despite the addition of a second bridge over the Hooghly and an underground system.

Employment Those with jobs tend to work in the informal sector (page 134) and use their home as a place of work. Often the fronts of houses are 'opened up' to allow the occupants to sell wood, food and clothes (Figure 4.20). Although few people are totally unemployed, many jobs only occupy a few hours a week and provide a very low income.

The Calcutta Metropolitan Development Authority This was set up in 1970. Since then it has attempted to make the bustees more habitable by paving some of the alleys, digging extra drains and providing more water taps and public toilets. Despite financial support from many voluntary agencies, a lack of money and high birth and immigration rates have combined to slow down progress in providing better housing, services, transport and more jobs.

Problems and solutions in Cairo

Location and growth Cairo is located on the edge of the desert where the River Nile leaves its valley and enters the delta (Figure 4.21). Although the Giza pyramids were built nearby, it was not until the tenth century that Cairo became a permanent settlement. Today it is the largest city in Africa with a population estimated to be over 12 million.

Housing Unlike most other cities in developing countries, Cairo has few squatter settlements. Instead, most buildings are brick-built and permanent. Many newcomers to the city live within the old medieval core (Figure 4.22). Some find accommodation in overcrowded, two-roomed apartments within tall flats. Others 'camp' in any empty space in roof-top slums on tops of offices and other buildings (the desert climate allows roofs to be flat). Later an extra storey may be added to the building, often illegally, making the temporary homes permanent. Another estimated 2 million people (the authorities have no real idea) live in Cairo's 'City of the Dead' (Figure 4.23). This is a huge Muslim cemetery where people find shelter, often inside the tombs themselves. The tombs are dry and clean but lack electricity, sewerage and running water – services not needed by the dead! The 'city' now has its own market and workshop industries.

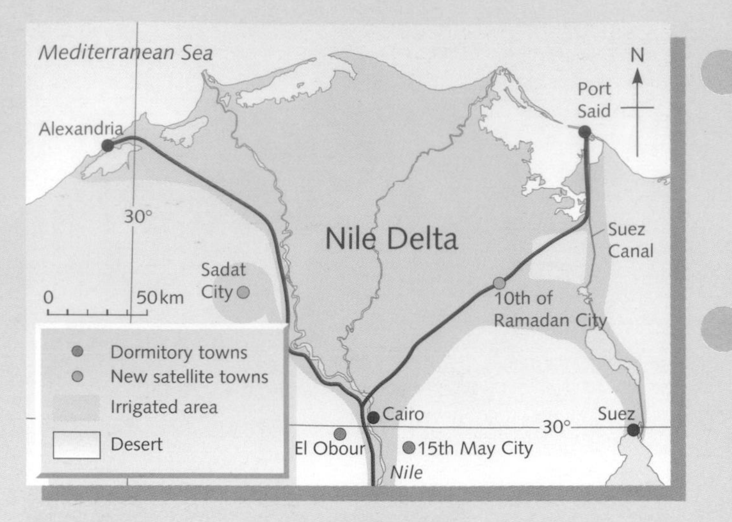

Figure 4.21 Location of Cairo

Other problems The authorities in Cairo, in common with other large and rapidly growing cities, are faced with numerous problems other than just housing. Some of these are summarised in Figure 4.24.

Figure 4.22 Central Cairo

Figure 4.23 In the City of the Dead

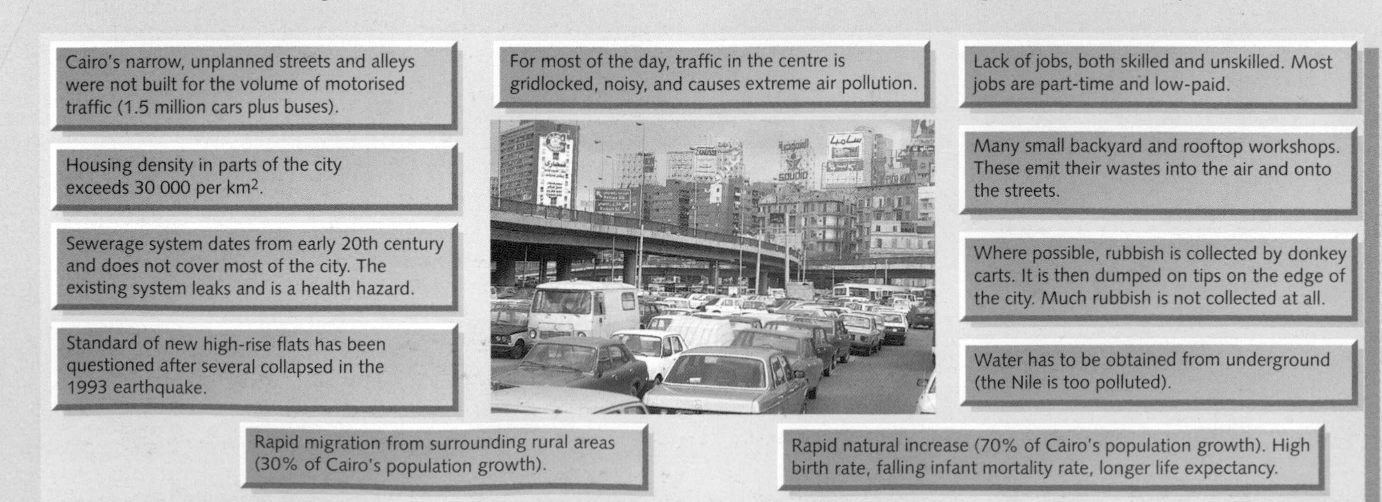

Figure 4.24 Urban problems in Cairo

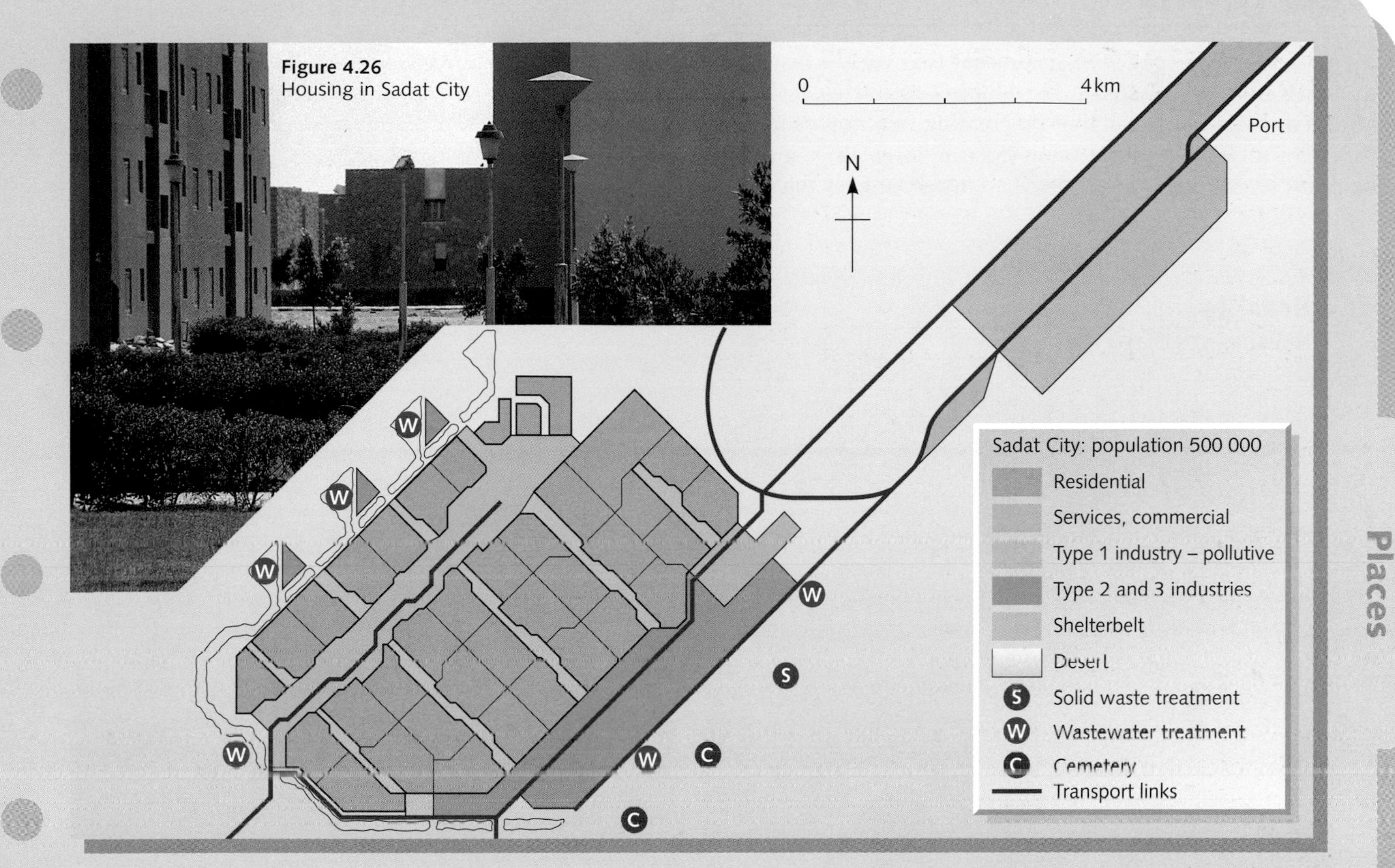

Figure 4.26 Housing in Sadat City

Figure 4.25 Structure plan, Sadat City

Attempts by the authorities to solve these problems

- Implementing an aid-funded scheme to repair and extend the old sewerage system.
- Constructing an outer ring road, new urban motorways and a metro line (this already carries over a million people a day).
- Organising refuse-collection by licensing donkey carts.
- Erecting numerous high-rise apartment blocks, mainly in the suburbs.
- Extending the area served with electricity (electricity comes from the High Dam at Aswan – page 276).
- Building new dormitory towns which are linked to Cairo (El Obour and 15th of May City) and new 'satellite' towns which are self-contained (Sadat City and 10th of Ramadan City) (Figure 4.21).

The satellite new town of Sadat City

Sadat City was planned as a self-contained, independent new town with the main aim of encouraging people and industry to move away from Cairo. It is located halfway along the desert road between Cairo and Alexandria, far enough from them to discourage daily commuting, but close enough to benefit from the proximity of Cairo and its international airport and from the port of Alexandria. The plan, which began to be implemented in the late 1970s, is shown in Figure 4.25. The population is expected to reach 400 000 by 2001.

The master plan consisted of a central spine flanked by neighbourhoods of 4000 to 6000 people at one end and industry at the other. Housing consists of three- or four-storey blocks of flats built from either prefabricated concrete panels or sandstone blocks (Figure 4.26). The houses, with an average size of 140 m^2, are built on a self-help basis. Residents are encouraged to buy their own homes, but only those with reliable, well-paid jobs in the formal sector (page 134) can afford the loans. Each neighbourhood has its own primary school and a nursery and is linked, by footpaths, to five other neighbourhoods. At the centre of the six neighbourhoods are services which include secondary education, health care and social services as well as facilities for recreation, cultural and religious activities. The water supply comes from underground.

The mixed economic base consists of industry supplemented by several central government offices and a university research and training centre. The informal sector is also being encouraged, as is the use of public transport.

New towns

As already stated (page 55), at present the world's fastest-growing cities are located in developing countries. It would appear, therefore, that by creating new towns in those countries, the pressure on existing large cities might be reduced. Unfortunately, these are the countries that have the fewest resources available to plan and develop self-contained new towns. Brazil is an exception, where Brasilia, described as 'a planners' dream', was designed as the nation's new capital.

Places

Brasilia

Brasilia is perhaps the most famous example of an elaborately planned new town. For decades, concern had been expressed at the relative richness and overpopulation of the south-east of Brazil (São Paulo and Rio de Janeiro) compared with the rest of the country. In 1952 Congress approved, by only three votes, the move of the capital from Rio de Janeiro (where most politicians preferred to live) inland to Brasilia to try to open up the more central parts of the country. Building began in 1957.

Figure 4.28
A view of the superblocks

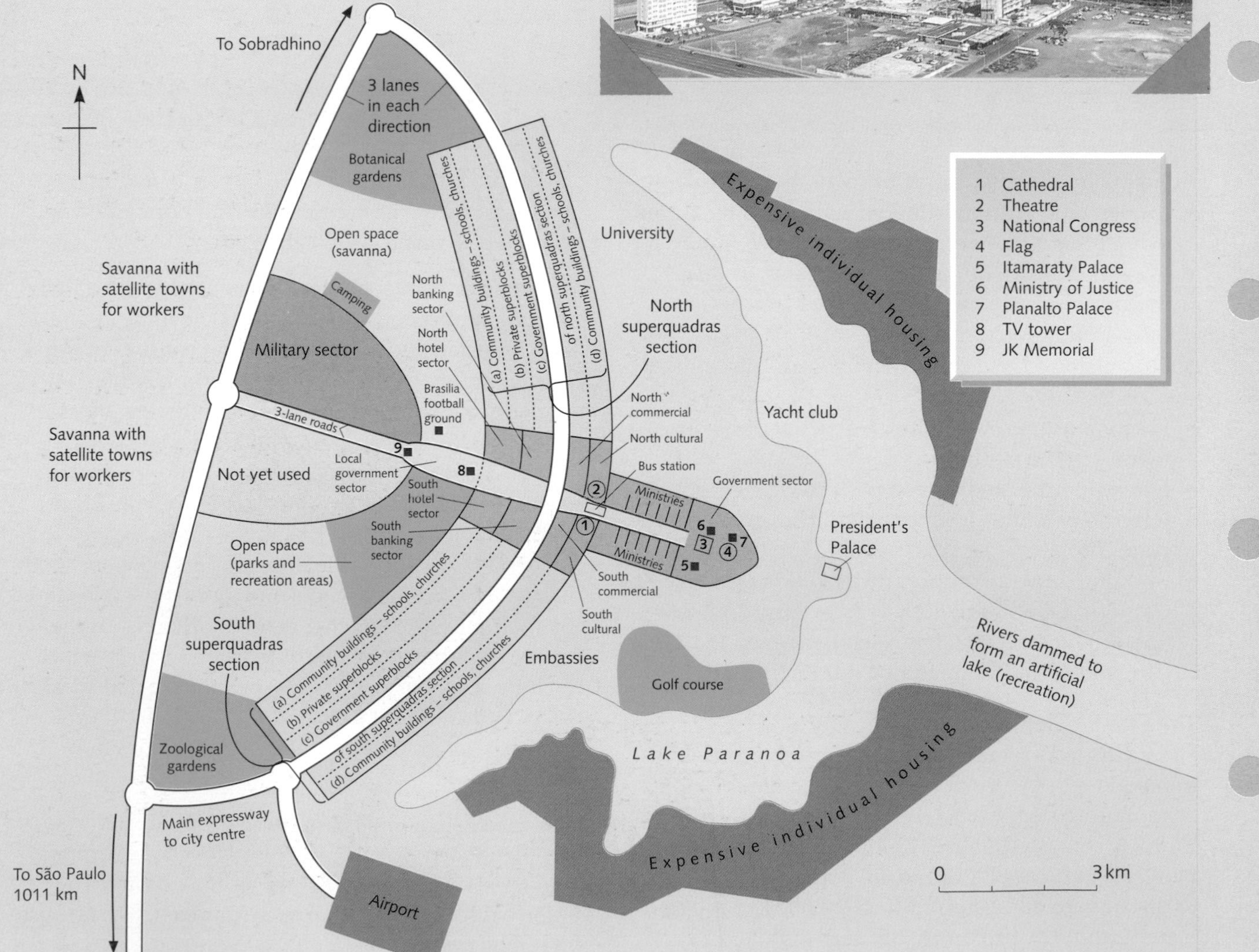

Figure 4.27
Plan of Brasilia

Figure 4.29 Inside one of the superblocks, Brasilia

Brasilia was built in the shape of an aeroplane. The two 'wings', each 6.5 kilometres long, were used for housing (superblocks); the 'fuselage', 9 kilometres long, has been divided into sections for local government, hotels, commerce, culture and national government (Figure 4.27). Each section is paired along the length of the fuselage. Many of the cultural and civic buildings are noted for their futuristic architecture.

Brasilia was built for the motorist. The idea was to have a road network enabling people to get to work and shops without being slowed down by traffic lights, and in safety. The result is that there are great distances between places (often too far to walk), so it is dangerous for those who did want to walk (as they had to cross three-lane expressways), and there were an increasing number of accidents. Traffic lights eventually had to be introduced.

Several local rivers have been dammed to form a large artificial lake (mainly used for recreation). Around this lake are the very expensive individual houses and foreign embassies. By 1985 the city had already reached 1 million inhabitants – a figure initially planned for the year 2000 – and 1.7 million by 1995. Brasilia is a bureaucratic city with little industry.

The superblocks (superquadras)

The wings of the 'aeroplane' are used for housing and are divided into superblocks (*superquadras*), each of which is meant to be a complete unit in itself (Figure 4.30), similar to a neighbourhood unit in a British new town. Each superblock is approached by a slip road, and at each entrance is a post box, public telephone and a newspaper kiosk. Inside the superblock are nine to eleven apartment blocks, each ten storeys high, and housing about 2500 people. The apartments are luxurious, unlike most British high-rise flats, and are well looked after. The apartments shown in Figures 4.28 and 4.29 belong to the Bank of Brazil and are occupied by its employees. Almost all the other apartments are inhabited by local or national government workers, all of whom have well-paid jobs. The apartments' rear windows are covered by a grid for privacy, and the fronts have shutters which can be opened or closed. The whole area is very clean and well maintained. Many flowering trees and shrubs have been planted (Brasilia has 30 000 new trees planted each year). The inhabitants like living in these apartments as it gives them security (although Brasilia's crime rate is well below the national average), the environment is attractive, services are nearby, and those working in the town centre can easily travel home at lunchtime (unlike workers in other Brazilian cities). Each superblock has its own kindergarten, to save the children crossing roads. Each pair of superblocks has a play area and a self-contained shopping area. On the edge of the housing area are community buildings with a Catholic church, a senior school and a community centre for each set of four superblocks.

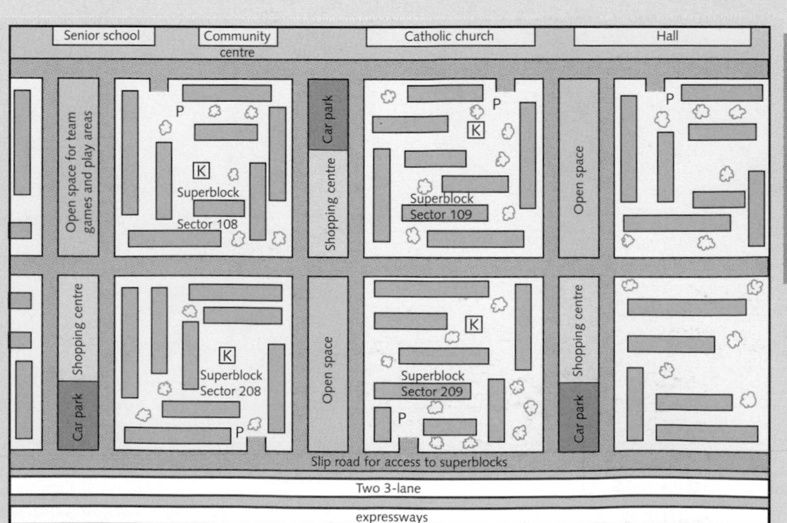

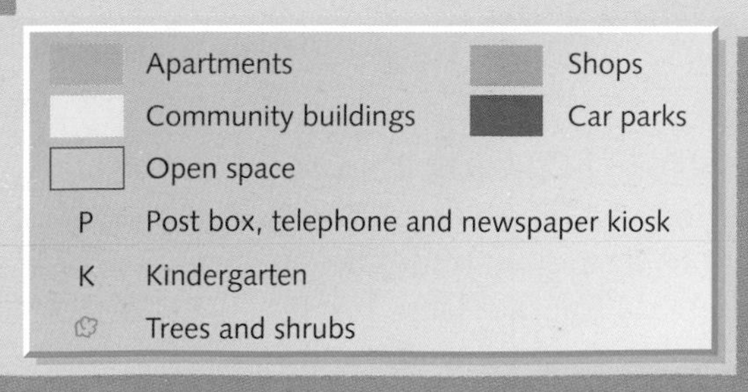

Figure 4.30
Plan of part of the south superquadras (residential) section

CASE STUDY 4

Rio de Janeiro 2000

Location and growth

Rio de Janeiro is situated around the huge natural harbour of Guanabara Bay in south-east Brazil (Figure 4.31). Even though it has been replaced by Brasilia as the country's capital and by São Paulo as the largest city and the centre of industry and commerce, Rio is still one of the world's megacities (Figure 4.5). Estimates suggest that by the year 2000, 6 million people will live in the main conurbation and 12 million in the metropolitan region (Figure 4.32). Like most cities, there are two sides to Rio:

- the beaches of Copacabana (Figure 8.35) and Ipanema backed by luxury housing
- the problems of rapid urban growth including favelas, traffic and crime.

Problems within Rio de Janeiro

Housing Apart from an estimated half-million homeless street dwellers, over 1 million people live in favelas and another 1 million in poor-quality local authority housing (Figure 4.32). A *favela* is a wildflower that grew on the steep hillsides (*morros*) which surrounded early Rio de Janeiro. Today, these same morros are covered in favelas (informal shanty settlements). Of over 600 favelas, the largest are Roçinha and Morro de Alemao (see Figure 4.31, bottom left), each with a population of 100 000. The official definition of a favela is a residential area of 60 or more families living in accommodation that lacks basic services (no running water, sewerage or electricity), and who have no legal right to the land on which they live. The houses are constructed from any materials available – wood, corrugated iron, broken bricks and tiles (Figure 4.33). Most favelas in Rio are built on hillsides considered too steep for normal houses. The most favoured sites are at the foot of the slope near to the main roads and water supply – although these places may receive sewage running downhill in open drains from houses built uphill. People living near the hilltops have to carry everything thing they need – including water in cans several times a day. When it rains on the steep slopes, flash floods and mudslides (page 235) can carry away the flimsy houses – over 200 people died as a result of storms in 1988 (Figure 4.34). Several attempts have been made to clear favelas (once before the visit of the Pope). In some cases the authorities built new homes (*conjuntos habitaçionais*) to replace those they had cleared (although often they are little better than the shacks they had cleared), or the evicted residents, with nowhere else to go, returned and rebuilt their homes. The local authority now accepts the presence of favelas and is working with resident associations to improve conditions there (page 68).

Figure 4.31
Rio de Janeiro and Sugar Loaf Mountain

Figure 4.32 Land use in Rio de Janeiro

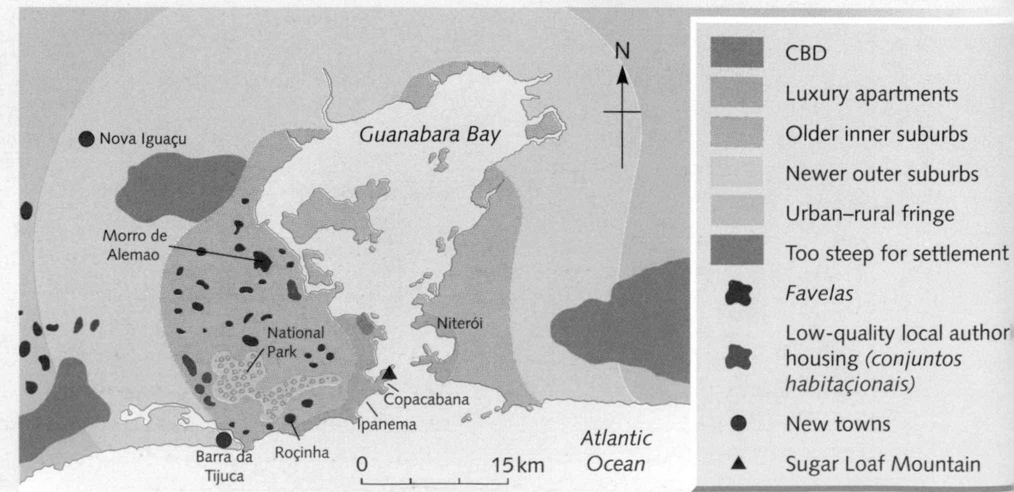

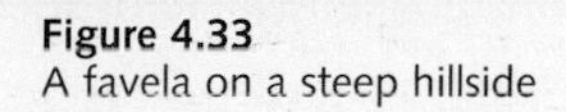

Figure 4.33
A favela on a steep hillside

Crime Many residents and visitors see crime as one of Rio's main problems. The favelas are perceived – certainly by non-residents – to be areas associated with organised crime, violence and drug trafficking (residents, on the other hand, claim that crime has decreased and the community spirit has increased). As a city, Rio has gained worldwide notoriety for its use of drugs, especially cocaine. Many of the well-off Rio residents are moving out of the city to places such as Barra da Tijuca (page 69) which they see as a safer environment for their family. Tourists to Rio's famous beaches of Copacabana and Ipanema (Figure 8.35) are warned not to take valuables with them or to wear jewellery or watches.

Traffic Although the mountains add to Rio's attractiveness, they also hem the city in. One effect of this is to channel traffic along a limited number of routes. For much of the day there is severe congestion, pollution and (even through the night) noise.

Pollution An industrial haze, intensified by traffic fumes, often hangs over much of Guanabara Bay. Along the coast, the beaches and sea are also polluted. A city the size of Rio produces huge amounts of waste and rubbish. In favelas the rubbish is unlikely to be collected and its presence, together with possibly polluted water supplies and the sewage in open drains, cause health hazards (there was, for example, an outbreak of cholera in 1992).

Figure 4.34
Landslide on a Rio hillside

Despite these problems, living in a favela has its positive side. There is often a strong community spirit, a rich street life and plenty of samba music.

CASE STUDY 4 continued

Attempts to solve some of Rio's problems

Self-help housing schemes – Roçinha

The residents of Roçinha (Figure 4.32) have slowly transformed the favela into a small city. Most of the original temporary wooden buildings have been upgraded to brick and tile and many have been extended to utilise every square centimetre of space (Figure 4.35). Many residents have lived here since the favela developed in the 1950s, and have set up their own shops and small industries – the so-called 'informal sector' (page 134), and created their own places of entertainment. The authorities, having now accepted the existence of favelas such as Roçinha and working with local resident associations, have added electricity (satellite TV has reached here too!), paved and lit some of the steep streets, and added more water pipes. However, improvements are restricted by the high density of housing and the steepness of the hillsides.

The local authority Favela Bairro project

The city authorities have recently set aside £200 million to improve living conditions in 60 of the 600 favelas within their boundary. Initially 16 mid-size favelas have been chosen because:

- being smaller, the problems are easier and cheaper to tackle than in larger favelas such as Roçinha and Morro de Alemao
- they had fewer drug, crime and unemployment problems
- in some cases they were more visible from the well-off areas and to tourists to the city.

The city authorities claim that they wish to transform the favelas socially and culturally and to integrate them as part of the city. The plan includes:

- replacing buildings that were either made of wood or built on dangerous slopes, with brick-built housing – the new houses are much larger (5 x 4 metres) and have a yard of equal size
- widening selected streets, so that emergency services and waste-collection vehicles can gain access
- laying street pavements and concrete paths
- laying pipes for water and cables for electricity
- improving sanitation, adding health facilities and providing sports areas
- using labour from within the favela so that residents can develop and use new skills (Figure 4.36).

In return the residents have to pay taxes to the authorities.

Figure 4.35
Improvements to a house in Roçinha

Figure 4.36
A self-help scheme in Roçinha

Figure 4.37
View of Barra da Tijuca

The new town of Barra da Tijuca

In an attempt to find more space and a safer place to live, many of Rio's more wealthy inhabitants have begun to move out of the city – the process of counterurbanisation (page 44). Counterurbanisation has led to a population loss in central Rio, a situation normally only associated with cities in the developed world.

The nearest available flat land to central Rio is 20 km along the coast, but because the mountains between here and Rio extend down to the sea, it was largely inaccessible. In 1970, when the area was still virtually uninhabited, a four-lane motorway was cut through tunnels under the mountains and on stilts built over the sea. By 1995, the new town of Barra da Tijuca already had a population of 130 000.

Barra da Tijuca is the 'new Rio'. It is not a suburb but a self-contained city built along the coastal motorway (Figure 4.37). It has 5 km of shops (the largest complex in South America), schools, hospitals, offices and places of entertainment. Barra has a culture based around the beach (20 km of sand and surf), shopping malls, expensive restaurants and leisure centres. The new residents live in spacious and luxurious accommodation. Three-quarters of the accommodation is in high-rise apartments which vary in height between 10 and 30 storeys (Figure 4.38). The apartments have been built in blocks, are protected by security guards and contain every possible modern amenity and gadget. The remainder of the accommodation is single- and double-storey detached houses. Although usually both adult members of the family work through choice, often they have to seek high-paid jobs to pay for their expensive accommodation. Despite each family having its own car, there is an efficient local bus service linking the apartments, hypermarkets and leisure amenities.

However, Barra already has its own new favelas. The rich new residents still need housekeepers, cooks, cleaners and gardeners, and these people need somewhere to live.

Figure 4.38
High-rise apartment blocks, Barra da Tijuca

Figure 4.39
Inside an apartment block, Barra da Tijuca

QUESTIONS

1

(Pages 54 and 55)

a The map below shows predicted urban population changes 1980–2000.
  i) Estimate the urban population of Europe in 1980 and in 2000. (2)
  ii) Which two regions are expected to show the greatest urban population growth between 1980 and 2000? (2)
  iii) State two differences in the urban population growth between North America and Latin America. (2)

b What evidence does the map show that urbanisation is faster in poorer parts of the world? (3)

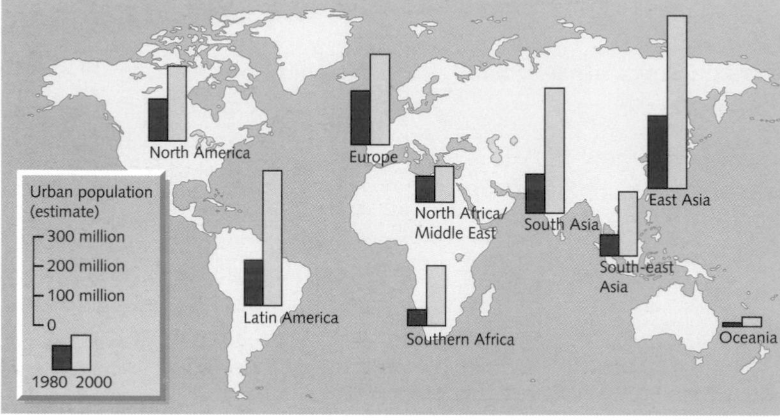

c i) What is meant by the term a *'million city'*? (1)
  ii) How many 'million cities' were there in:
  - 1920
  - 1995? (2)

  iii) Name the two fastest-growing cities in the 1970s. (1)
  iv) Name the three fastest-growing cities in the 1990s. (1)
  v) Using Figure 4.5, complete the table below to show the changing location of the world's largest cities. (2 x 3)

d Using Figure 4.3 and an atlas, try to name as many as possible of the:
  i) 34 cities with a population of over 5 million in 1995
  ii) 32 cities with a population between 3 and 5 million in 1995.

(For each, 1 mark for every 5 cities named)

		1970	1985	2000 (EST.)
DEVELOPED CONTINENTS	EUROPE			
	NORTH AMERICA			
DEVELOPING CONTINENTS	LATIN AMERICA			
	ASIA			

2

(Pages 56 and 57)

A feature of cities in the developing world is that they are growing very rapidly, partly due to people moving to them from surrounding rural areas.

a i) What is this movement from the countryside called? (1)
  ii) Give three reasons why people may wish to move into a big city from the surrounding countryside (urban 'pull' factors). (3)
  iii) Give three reasons why people may have to move away from the countryside (rural 'push' factors). (3)

b Name four problems likely to occur in urban areas when large numbers of people move into them. (4)

3

(Pages 58–61)

The diagram on the right is an incomplete model of a city in the developing world.

a Match up the following with letters A to F on the model:
- Modern, luxury high-rise flats
- Squatters who have built shanty towns (favelas)
- A large shopping centre with tall office blocks
- A suburban luxury estate for professional workers
- Modern factories built alongside main roads leading out of the city
- An area where some houses have had piped water and electricity added, yet are still of poor quality (6)

b Explain how the urban land use pattern has developed. (3)

c Give two ways in which this model showing urban land use in a developing city differs from the model (Figure 3.1) for a developed city. (2)

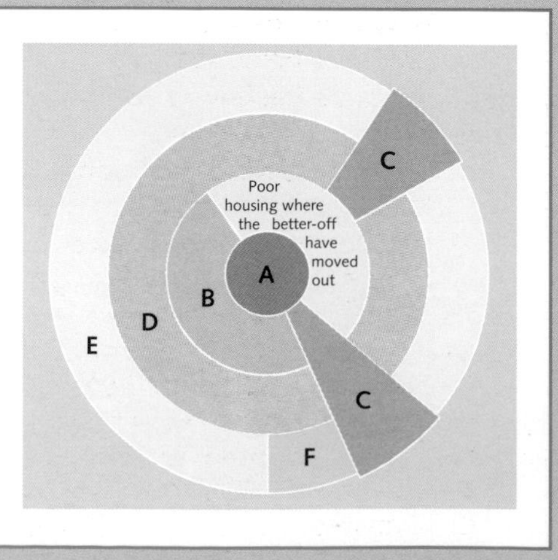

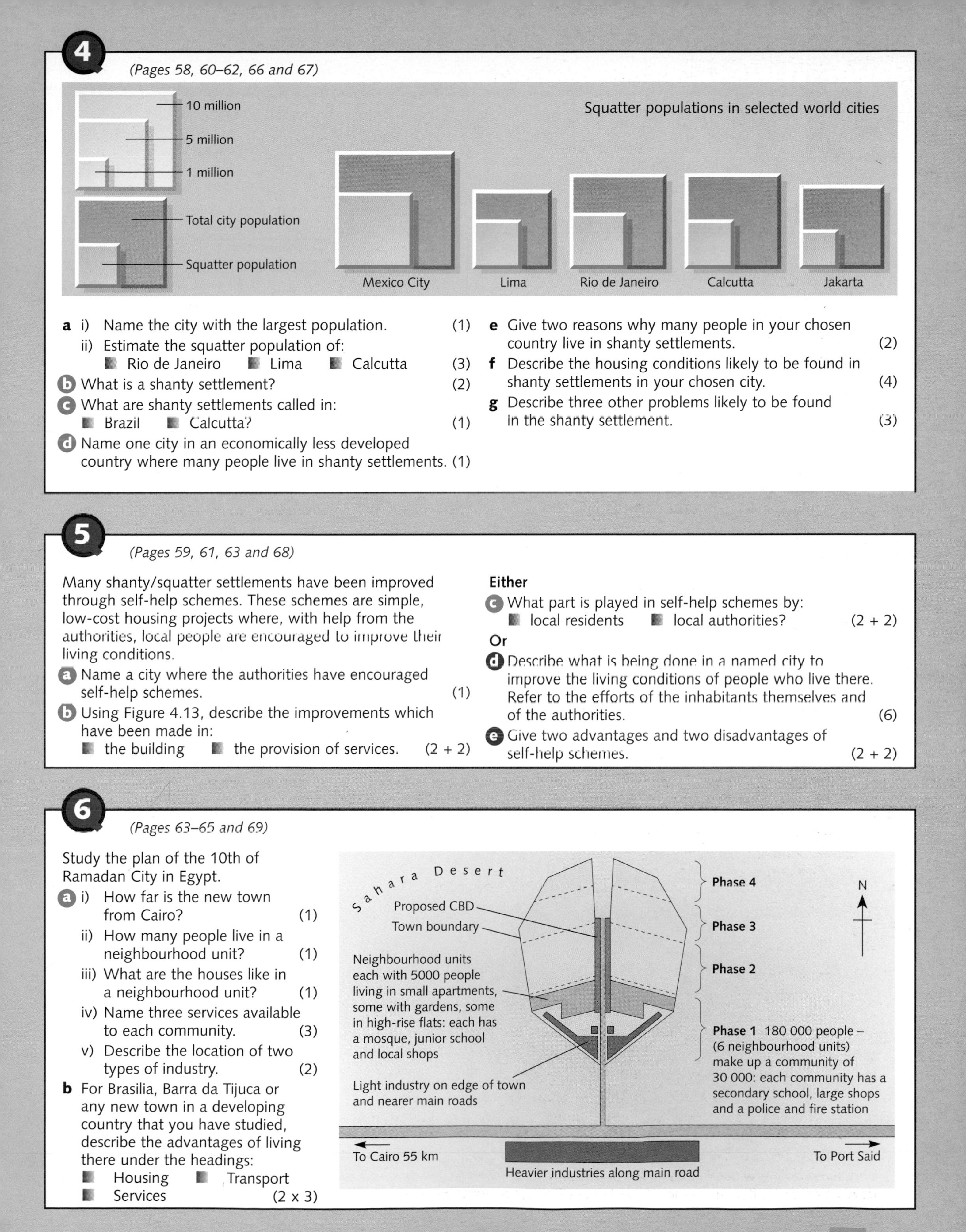

4

(Pages 58, 60–62, 66 and 67)

a i) Name the city with the largest population. (1)

ii) Estimate the squatter population of:

- Rio de Janeiro
- Lima
- Calcutta (3)

b What is a shanty settlement? (2)

c What are shanty settlements called in:

- Brazil
- Calcutta? (1)

d Name one city in an economically less developed country where many people live in shanty settlements. (1)

e Give two reasons why many people in your chosen country live in shanty settlements. (2)

f Describe the housing conditions likely to be found in shanty settlements in your chosen city. (4)

g Describe three other problems likely to be found in the shanty settlement. (3)

5

(Pages 59, 61, 63 and 68)

Many shanty/squatter settlements have been improved through self-help schemes. These schemes are simple, low-cost housing projects where, with help from the authorities, local people are encouraged to improve their living conditions.

a Name a city where the authorities have encouraged self-help schemes. (1)

b Using Figure 4.13, describe the improvements which have been made in:

- the building
- the provision of services. (2 + 2)

Either

c What part is played in self-help schemes by:

- local residents
- local authorities? (2 + 2)

Or

d Describe what is being done in a named city to improve the living conditions of people who live there. Refer to the efforts of the inhabitants themselves and of the authorities. (6)

e Give two advantages and two disadvantages of self-help schemes. (2 + 2)

6

(Pages 63–65 and 69)

Study the plan of the 10th of Ramadan City in Egypt.

a i) How far is the new town from Cairo? (1)

ii) How many people live in a neighbourhood unit? (1)

iii) What are the houses like in a neighbourhood unit? (1)

iv) Name three services available to each community. (3)

v) Describe the location of two types of industry. (2)

b For Brasilia, Barra da Tijuca or any new town in a developing country that you have studied, describe the advantages of living there under the headings:

- Housing
- Transport
- Services (2 x 3)

MIGRATION

What is migration?

Migration is a movement and in human terms usually means a change of home. However, as seen in Figure 5.1, it can be applied to temporary, seasonal and daily movements as well as to permanent changes both between countries and within a country.

Permanent international migration is the movement of people between countries. *Emigrants* are people who leave a country; *immigrants* are those who arrive in a country. The *migration balance* is the difference between the numbers of emigrants and immigrants. Countries with a *net migration loss* lose more people through emigration than they gain by immigration, and depending on the balance between their birth and death rates (page 6) they may have a declining population. Countries with a *net migration gain* receive more people by immigration than they lose through emigration, and so will have an overall population increase (assuming birth and death rates are evenly balanced).

Figure 5.1
Types of migration

	EXTERNAL (INTERNATIONAL)	BETWEEN COUNTRIES
PERMANENT	I) VOLUNTARY	WEST INDIANS TO BRITAIN
	II) FORCED (REFUGEES)	NEGRO SLAVES TO AMERICA, KURDS, RWANDANS
	INTERNAL	WITHIN A COUNTRY
	I) RURAL DEPOPULATION	MOST DEVELOPING COUNTRIES
	II) URBAN DEPOPULATION	BRITISH CONURBATIONS
	III) REGIONAL	NORTH-WEST TO SOUTH-EAST OF BRITAIN
SEMI-PERMANENT	FOR SEVERAL YEARS	MIGRANT WORKERS (TURKS INTO GERMANY)
SEASONAL	FOR SEVERAL MONTHS OR SEVERAL WEEKS	MEXICAN HARVESTERS IN CALIFORNIA, HOLIDAY-MAKERS, UNIVERSITY STUDENTS
DAILY	COMMUTERS	SOUTH-EAST ENGLAND

International migration can be divided into two types – voluntary and forced (Figure 5.2).

Figure 5.2
Voluntary and forced migration

Voluntary migration is the free movement of migrants looking for an improved quality of life and personal freedom. For example:

- Employment – either to find a job, to earn a higher salary or to avoid paying tax
- Pioneers developing new areas
- Trade and economic expansion
- Territorial expansion
- Better climate, especially on retirement
- Social amenities such as hospitals, schools and entertainment
- To be with friends and relatives

Forced migration is when the migrant has no personal choice but has to move due to natural disaster or to economic or social imposition. For example:

- Religious and/or political persecution
- Wars, creating large numbers of refugees
- Forced labour as slaves or prisoners of war
- Racial discrimination
- Lack of food due to famine
- Natural disasters caused by floods, drought, earthquakes, volcanic eruptions or hurricanes
- Overpopulation, when the number of people living in an area exceeds the resources available to them

Figure 5.3
Major international migrations since 1945

Foreign born as a % of total population
- Over 7.5%
- 3.0–7.5%
- 1.5–3.0%
- Less than 1.5%
- No data

Voluntary
1 East Europeans to Germany
2 Europeans to North America
3 Jews to Israel
4 Irish and Commonwealth to UK
5 Europeans to Australia
6 North Africans to France
7 Chinese to Japan and Korea
8 Mexicans to North America
9 Central Americans and West Indians to North America
10 Bantu workers to South Africa
11 South Asian workers to the Gulf
12 Egyptian workers to Libya and the Gulf
13 Hong Kong Chinese and South-east Asians to western USA and Canada

Forced
A Palestinian refugees
B Indian and Pakistani refugees
C Bangladeshi and Pakistani refugees
D Vietnamese and Cambodian refugees
E Afghan refugees
F Mozambique refugees
G Yugoslav refugees
H Rwanda-Burundi refugees
I Somali, Ethiopian, Sudanese refugees
J Liberian and Sierra Leonian refugees

Refugees

Refugees are people who have been forced to leave their home country for fear of persecution for reasons of race, religion, politics, internal strife (civil war) or due to environmental disaster. They move to other countries hoping to find help and asylum. Refugees do not include displaced persons, who are people who have been forced to move within their own country. By the end of 1992, 600 000 citizens of the former Yugoslavia had become refugees seeking sanctuary in other European countries, while over 2 million had become displaced persons mainly due to 'ethnic cleansing'.

The United Nations (UN) suggested that, at the beginning of 1996, there were almost 15 million refugees in the world (Figure 5.4). However, as most refugees are illegal immigrants, the UN admits that this figure could be very inaccurate. More than half of the world's refugees are children and most of the adults are women. Over 80 per cent of refugees are in developing countries – countries that are least able to help. Refugees live in extreme poverty, lacking food, shelter, clothing, education and medical care. They have no citizenship, few (if any) rights, virtually no prospects, and are unlikely to return to their homeland (Figure 5.5).

The present problematic refugee situation began over 50 years ago in war-torn Europe, although many of those refugees were later assimilated by their host country. It was the Palestinian Arab refugee camps, set up after the creation of the state of Israel in 1948, which first showed that the problem had become permanent and apparently insoluble, and 1.75 million people still live in camps in this part of the Middle East. Trends in the last decade confirm that while the number of refugee movements has continued to increase, most movements are still between developing countries. Those movements include:

- 6 million Afghans forced by war to leave for neighbouring Pakistan and Iran
- 1.5 million Ethiopians, Sudanese and Somalis driven from their homes in East Africa by drought, famine and civil war
- inhabitants from the southern African countries of Angola and Mozambique; from the central African states of Rwanda, Burundi and the Democratic Republic of the Congo; and from the former European country of Yugoslavia – all due to civil wars (Figure 5.6).

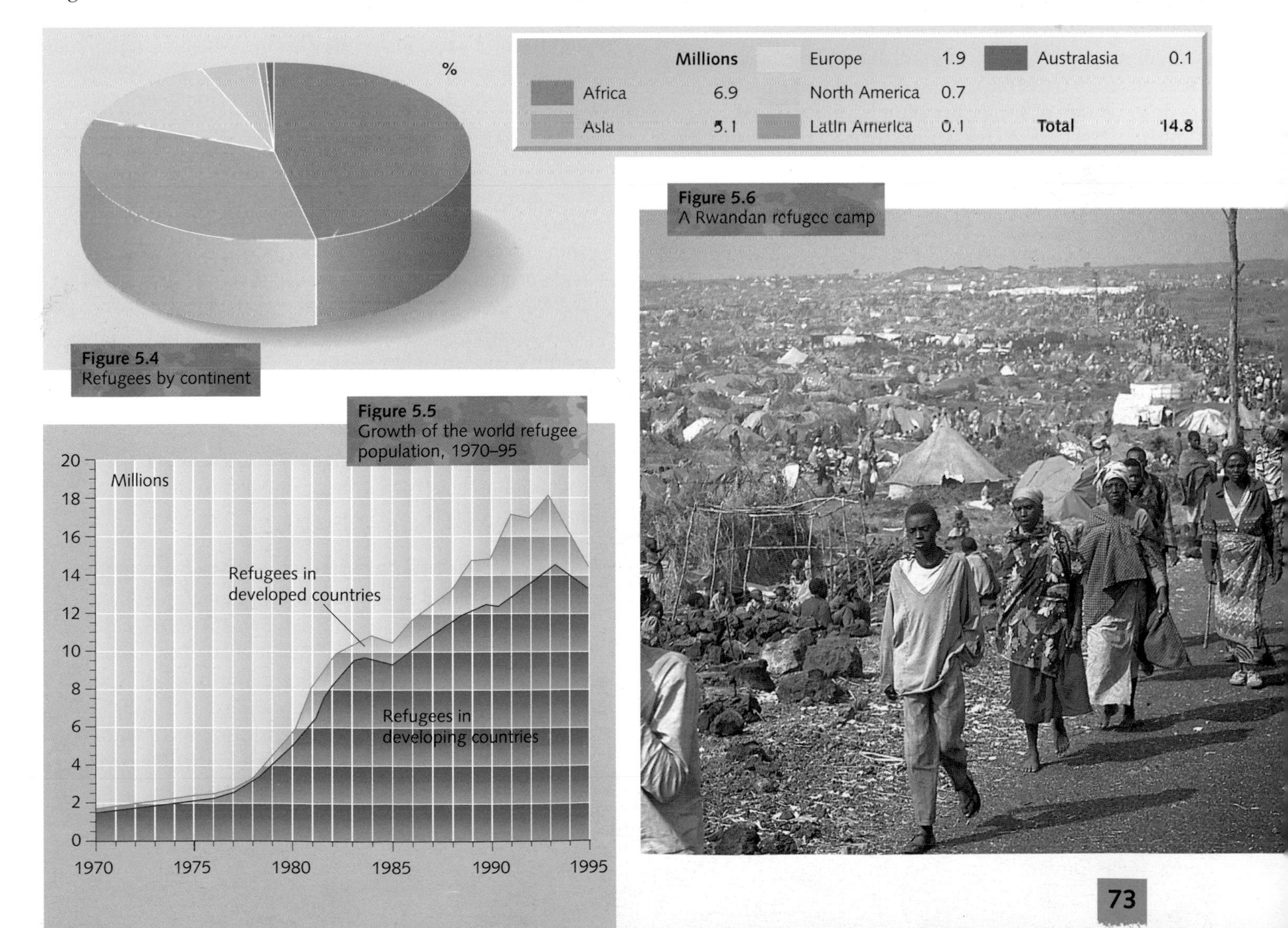

Figure 5.4
Refugees by continent

Figure 5.5
Growth of the world refugee population, 1970–95

Figure 5.6
A Rwandan refugee camp

Migration into the UK

	1991 CENSUS (%)	MID-1995 ESTIMATE (%)	% BORN IN UK (1991 CENSUS)	% LIVING IN CONURBATIONS (1991)
WHITE	94.5	94.2	96	32
ETHNIC MINORITIES	5.5	5.8	46	78
BLACK (AFRICAN AND CARIBBEAN)		1.5	55	86
INDIAN		1.5	36	79
PAKISTANI		1.0	44	72
BANGLADESHI		0.4	45	80
CHINESE		0.4	51	59
OTHERS		1.0	47	

Figure 5.7
Ethnic minorities in the UK

The UK has experienced waves of immigrants, giving it a society that has always been one of mixed races and cultures. The majority of UK residents are descended from immigrants – Romans, Vikings, Angles, Saxons and Normans. The Irish have settled in Britain for several centuries, while many other Europeans migrated here both during and after the Second World War. Usually the number of immigrants was small enough to be easily assimilated into the existing population. However, the immediate post-war years saw a much larger influx (around 3 million), many of whom had different colours of skin. It is this biological difference, rather than a social or cultural difference, that has led to racial tensions in parts of the UK.

A small proportion of immigrants have come from the *Old Commonwealth* (Australia, New Zealand and Canada) and are descendants of earlier British migrants to those countries. The largest proportion are *New Commonwealth* immigrants from former British colonies in the Indian subcontinent (India, Pakistan and Bangladesh), Africa (Nigeria) and the West Indies (Jamaica). This group (Figure 5.7), who are mainly non-white, migrated here because:

- Britain had a labour shortage after the Second World War. Many West Indians came here partly due to the 'push' factor of overcrowding in their own islands, but mainly due to the 'pull' of jobs in Britain (page 56). The British government actively encouraged people to apply for specific jobs, for example with London Transport. Unfortunately, many immigrants were underskilled and were forced to take poorly-paid jobs (see page 78).
- Groups of Asians, including Hindus, Muslims and Sikhs, found themselves as religious or political refugees following the division of India. They came to Britain in the 1950s.

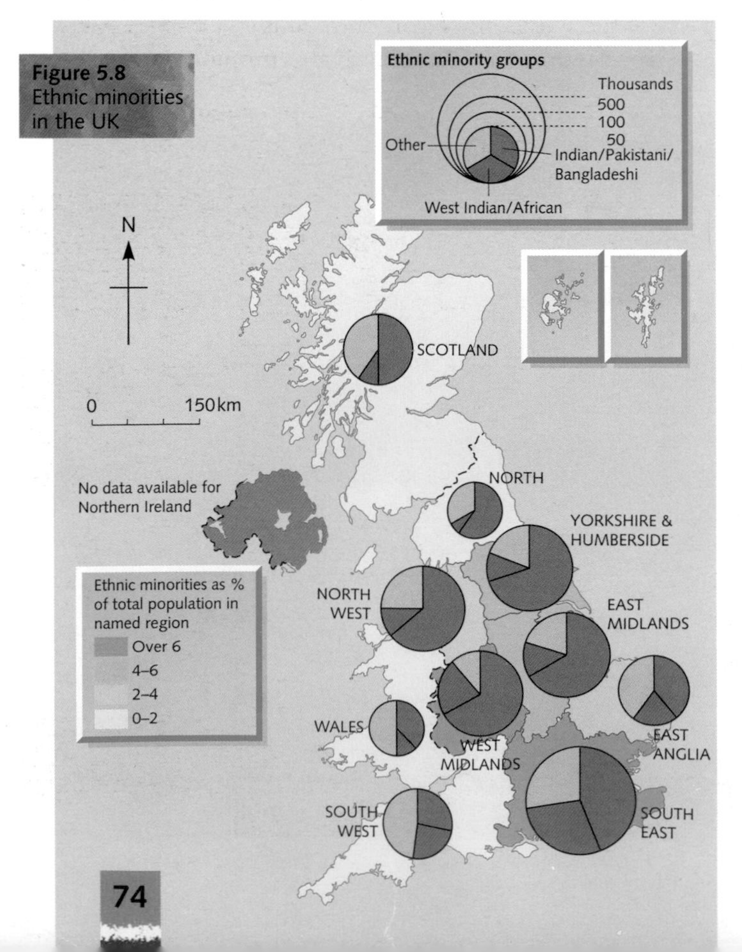

Figure 5.8
Ethnic minorities in the UK

Successive British governments have, since then, tried to restrict and control the number of non-white immigrants unless they:

- were dependants of relatives already living in the UK
- had specific jobs, especially those involving skills that were in short supply in Britain, e.g. doctors
- were British passport-holders evicted from their home country, e.g. Ugandan Asians
- could prove themselves to be genuine refugees, e.g. Bosnian Muslims.

Ethnic minorities only make up 5.8 per cent of the UK population, with an increasing number, almost half, having been born in Britain.

Uneven concentrations of ethnic groups

Immigrants avoided areas that had high unemployment (Scotland and Northern Ireland) and went to large cities and conurbations (not small towns) where there were better job opportunities. The greatest concentrations of ethnic groups are in London (page 33), the West and East Midlands and West Yorkshire (Figure 5.8). There has also been a tendency for one ethnic group to concentrate in a particular area, e.g. Pakistanis in West Yorkshire and West Indians in Birmingham.

Problems facing immigrants to Britain

The present-day issue in the UK is more about the problems experienced by existing ethnic minorities than about the actual number of new immigrants.

The majority of residents in Britain who have come directly as immigrants from the New Commonwealth, or who are descended from them, not only live in conurbations, but also tend to group together with members of their own ethnic group in inner city areas (Figures 3.12 and 5.9). The segregation of various ethnic groups in British cities could have resulted from differences in wealth, colour, religion, education and the quality of the environment. Many activities in these communities are positive, such as the Notting Hill Carnival. However, we are usually presented with a negative description of ethnic-based activities, perhaps because it is considered to be more media-worthy.

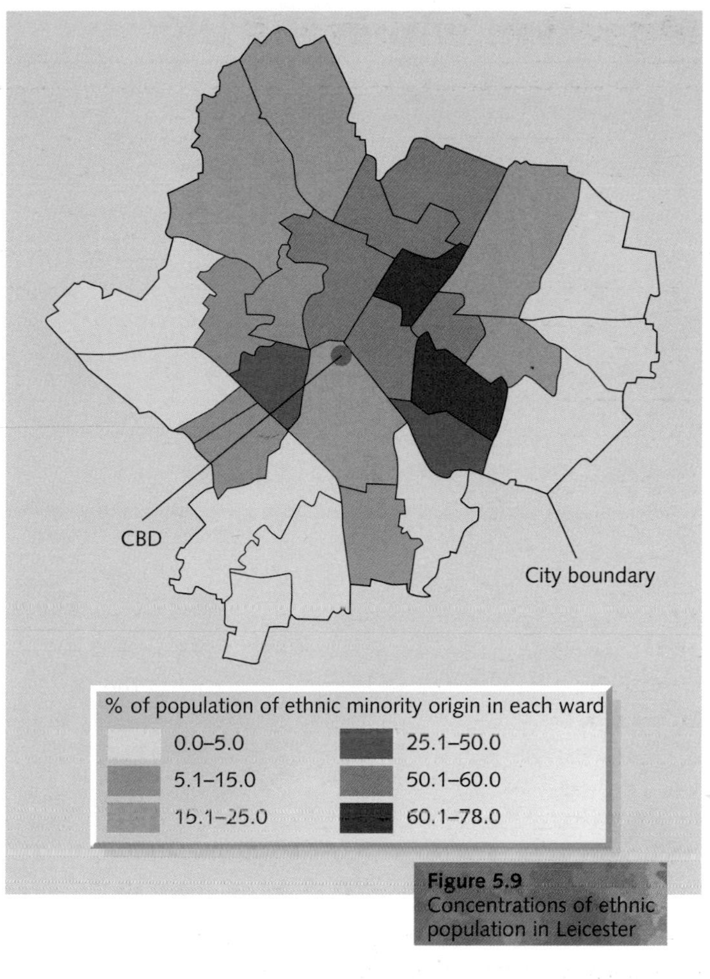

Figure 5.9 Concentrations of ethnic population in Leicester

The major problems confronting immigrants generally result from difficulties with the English language, cultural differences, and racial prejudices. Many members of ethnic minorities have to live in overcrowded, poor-quality housing. Overcrowding becomes worse in some groups that have a higher birth rate and larger families (Figure 5.10). Unemployment often exceeds 70 per cent, and much of it is long-term. Fewer education opportunities and, often, lower expectations, mean the inhabitants develop few skills. A lack of money on the part of the inhabitants and of the various levels of government means that services that are provided are inadequate. While the authorities speak of the high rate of crime in these areas – violence, drugs and muggings – residents complain of police harassment. The resultant lack of trust leads to further tension. These problems were highlighted during the 1981 and 1985 inner city riots.

Although ghettos like those in New York have not developed in Britain, ethnic groups that have a similar religion, language, diet, social organisation and culture tend to concentrate together, e.g. Jamaicans in Brixton, Anguillans in Slough, Sikhs in Southall and Bengalis in East London (Figure 3.12). Yet, as history has shown in many parts of the world, these concentrations can lead to fear, prejudice and jealousy among rival communities.

The Scarman Report, following the 1981 riots in English cities, identified four main problems (Figure 5.11).

GROUP	% OF ENGLAND'S POPULATION	% UNDER 16	% OVER 65	% LIVING IN CONURBATIONS	AVERAGE FAMILY SIZE	% PROFESSIONAL/ MANAGERIAL	% SEMI AND UNSKILLED	UNEMPLOYED 16–29 YEARS OLD	WITHOUT BASIC WC/HOT WATER/BATH
WHITE	94.2	19.5	19.3	32	3.1	40	18	15	6
BLACK AND ASIAN	5.8	32.8	4.2	78	3.5 (BLACK) 5.9 (ASIAN)	14	37	25	28

Figure 5.10 Inequality in England and Wales

Housing Most blacks and Asians (80 per cent) live in overcrowded buildings which they are able to rent or buy cheaply because the dwellings are substandard or in undesirable areas. There is often a reluctance to sell better-quality housing to non-whites.

Education Blacks and Asians often experience difficulties with the English language, which puts them at a disadvantage. Also, the schools they attend are often old and lack resources.

Jobs These are difficult to find because fewer skills have been acquired, and much industry has moved out of inner city areas. Those with jobs are poorly paid and so cannot afford good housing – a vicious circle. Despite legislation there is still considerable bias against non-white job applicants.

Discrimination This was regarded by Scarman to be (and still is) a major obstacle to assimilation.

Figure 5.11 Summary of parts of the Scarman Report

Migration within the UK

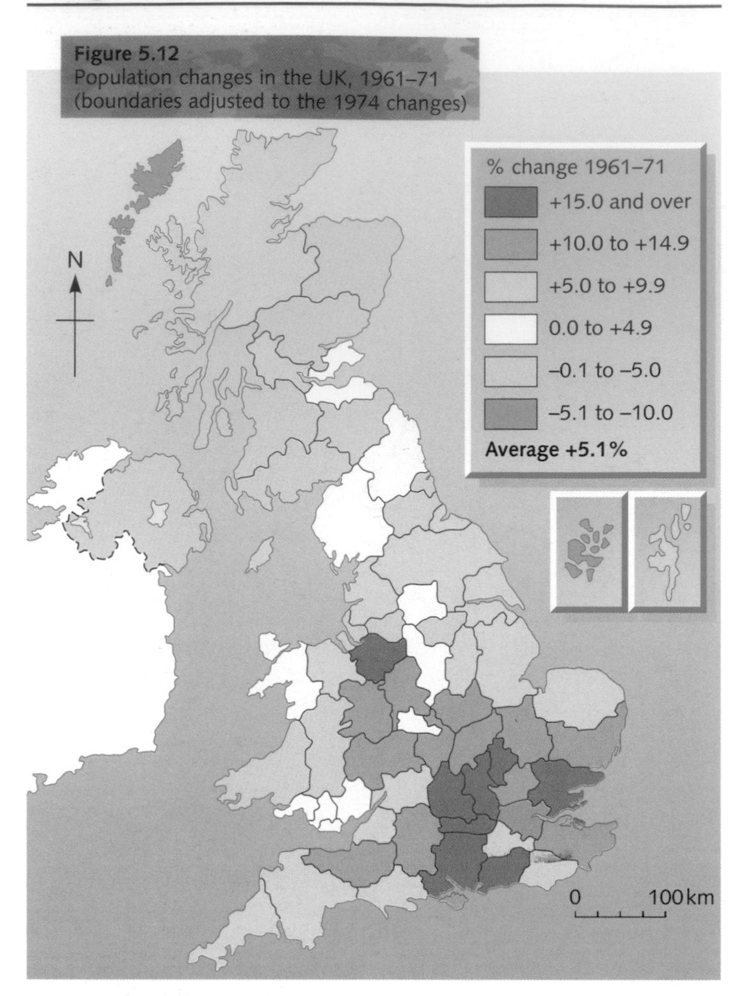

Figure 5.12
Population changes in the UK, 1961–71 (boundaries adjusted to the 1974 changes)

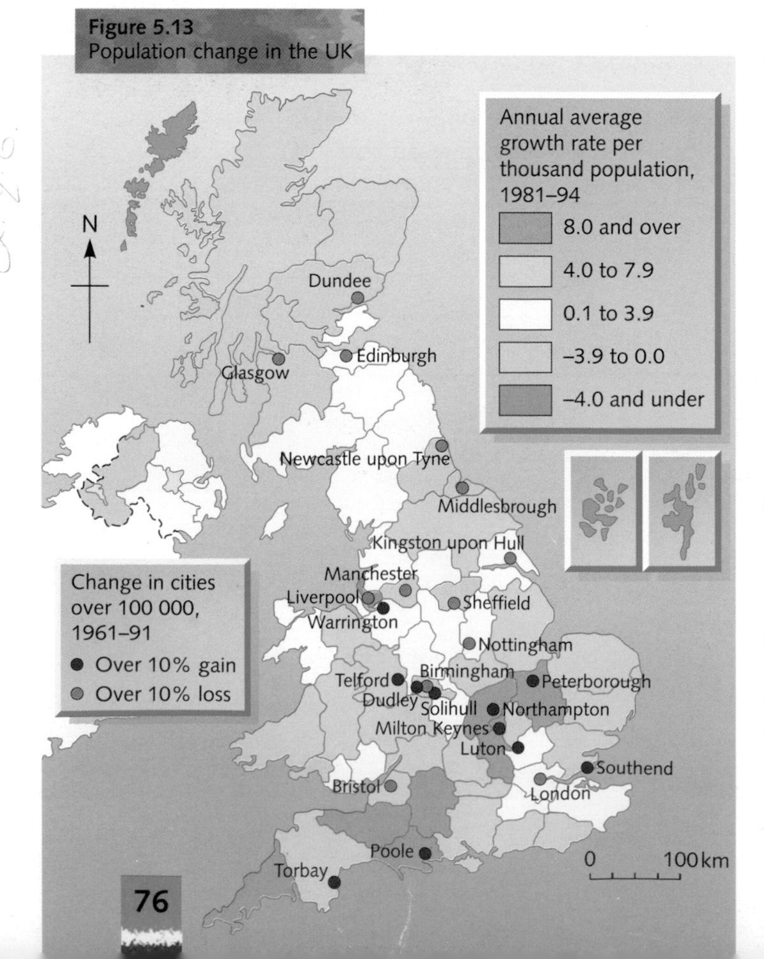

Figure 5.13
Population change in the UK

Rural-to-urban migration

During the Industrial Revolution of the nineteenth century, many people in Britain were either:

- forced to leave the countryside (*'push'* factors causing *rural depopulation*), or
- attracted to the growing towns (*'pull'* factors leading to *urbanisation*).

Workers were needed in the rapidly growing:

- textile towns of Yorkshire and Lancashire
- coalfield towns in South Wales, north-east England and central Scotland where, initially, iron and, later, steel was produced
- ports of London, Liverpool and Glasgow.

Regional migration

Between about 1930 and 1980 there was a steady drift of people from the north and west of Britain to the south-east of England. Figure 5.12 shows that former industrial areas in north-east England, Lancashire, Yorkshire and central Scotland together with the more rural parts of Northern Ireland, Scotland and central Wales, all lost population. People left these areas because of:

- older, poorer-quality housing
- exhaustion of minerals (coal/iron ore)
- decline of older industries (ships/steel/textiles)
- poorly paid, mainly manual, jobs
- poorer transport links/accessibility
- many polluted former industrial environments
- decline of older ports (Glasgow/Liverpool)
- fewer cultural amenities and social/sporting events
- colder, wetter climate.

At the same time places in the south-east increased their population. People moved here because:

- newer, better-quality housing
- growth of newer/lighter/footloose industries
- better-paid, more skilled jobs
- better transport links (motorways/rail/airports)
- fewer polluted industrial environments
- growing links with Europe
- better services (shops/schools/hospitals)
- many cultural amenities, more social/sporting events
- warmer, drier climate (holidays/retirement).

Counterurbanisation

Since the 1960s in Inner London, and increasingly since the 1980s in the rest of Britain, there has been a movement away from the conurbations and larger cities (Figure 5.13). Most of this movement, known as *counterurbanisation*, has been to new towns (page 45), dormitory/overspill towns and suburbanised villages (page 44). Although most of these places are in southern England, rural areas in the north and west have also gained in population, reversing the trend of the nineteenth century.

Those people who are leaving the cities (Figure 5.14) do so for economic, social and environmental reasons.

Employment As industry has declined in inner city areas, it has relocated on edge-of-city sites or in smaller rural towns. People move for promotion, for better-paid jobs or simply to find a job.

Housing When people become more affluent they are likely to move from the high-density small terraced houses and high-rise flats of the inner city to larger, modern houses with better indoor amenities, garages and gardens.

Changing family status People move as a result of an increase in family wealth or family size.

Environmental factors These include moving away from the noise, air and visual pollution created by increasing traffic and declining industry in large urban areas to quieter, less polluted environments with more open space.

Social factors People may move out of cities due to prejudice against neighbours and ethnic groups, an above-average local crime and vandalism rate, or poorer educational facilities.

The 1991 census showed that fewer people, when changing home, are making long-distance moves (compare the keys in Figures 5.12 and 5.13). Most moves are short, with over two-thirds being under 10 km and within the same county (Figure 5.15).

Who moves out?

- Those with higher incomes now capable of buying their own homes in suburbia
- Those with higher skills and qualifications – especially moving to new towns
- Parents with a young family wishing for gardens, open space and larger houses

Figure 5.14
Moving out of large cities

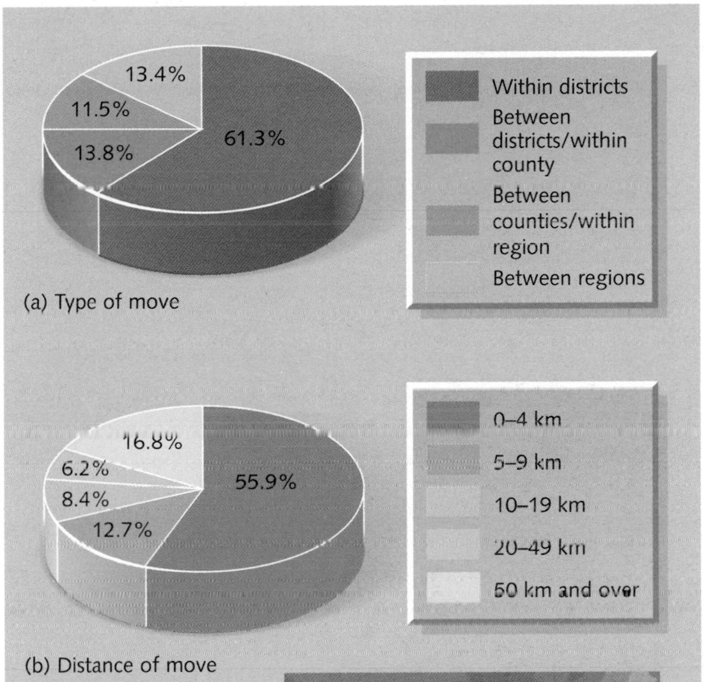

(a) Type of move

Within districts 61.3%; Between districts/within county 13.8%; Between counties/within region 11.5%; Between regions 13.4%

(b) Distance of move

0–4 km 55.9%; 5–9 km 12.7%; 10–19 km 8.4%; 20–49 km 6.2%; 50 km and over 16.8%

Figure 5.15
A classification of migration in the UK

Cycle of change in London

Figure 5.16 shows the outward movement of people from London. For several decades it has been the inner city areas, often adjacent to the River Thames, that have suffered the greatest loss. Until the 1970s people moved to the suburbs, but since the 1980s even these outer areas have lost population. The only anomaly (that is, an instance that does not fit the usual pattern) is 'The City', where there have been new housing developments, as for example, at the Barbican.

Recently, however, there has been a reversal in movement, with wealthy people returning to central London. Older, derelict property has been bought relatively cheaply and refurbished and transformed into expensive housing and flats, as in Islington, along the banks of the Thames and in the former Docklands (page 35).

Figure 5.16
Population movement in Greater London, 1971–91

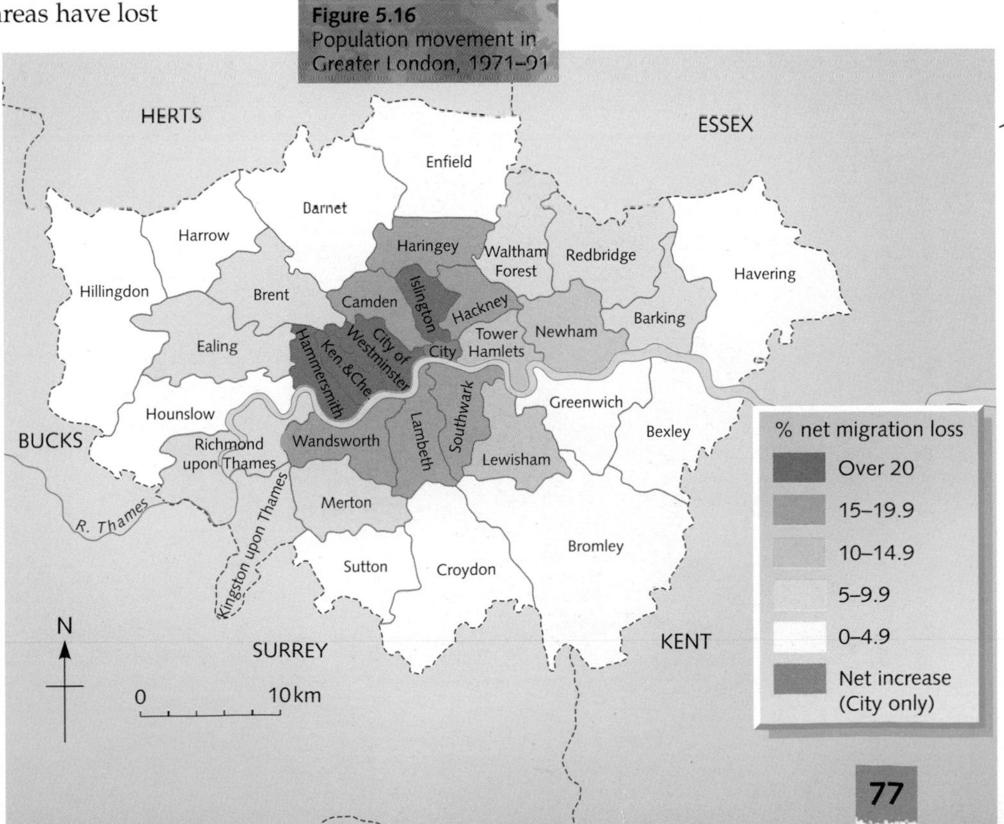

Migrant workers

In countries where there is a low standard of living and a shortage of jobs, groups of people will migrate to nearby, wealthier countries hoping to find work. Many migrant workers, the majority of whom are likely to be men, may move with the intention of:

- returning home after several years, having 'made their fortune' (e.g. Turks into Germany, North Africans into France, Egyptians and Jordanians into Saudi Arabia)
- bringing their family to join them a later date (e.g. Indians into the UK)
- working seasonally at harvest times (e.g. Mexicans into California).

Places

Turks into Germany

Migrants into West Germany, 1945–89

Like other West European countries, West Germany needed rebuilding when the Second World War ended in 1945. There were many more job vacancies than workers available and so extra labour was needed. As West Germany became increasingly wealthy, it attracted workers from the poorer parts of southern Europe and the Middle East. At the same time, several of those countries in southern Europe and the Middle East had too many workers for the number of jobs available, and those who had jobs usually received very low wages. The result was a movement of people from poorer countries like Turkey to richer countries like West Germany.

As many of the early Turkish immigrants came from farming villages (Figure 5.17), they initially found agricultural jobs in West Germany. Soon, however, they turned to relatively better-paid jobs in factories and the construction industry. These were jobs which the Germans themselves did not want as they were often dirty, unskilled, poorly paid and demanded long and unsociable hours. By 1989, West Germany had 4.5 million *Gastarbeiter* or 'guest workers', accounting for 7.4 per cent of its total workforce. Of these, 29 per cent had come from Turkey (Figure 5.18).

Figure 5.17 A Turkish village

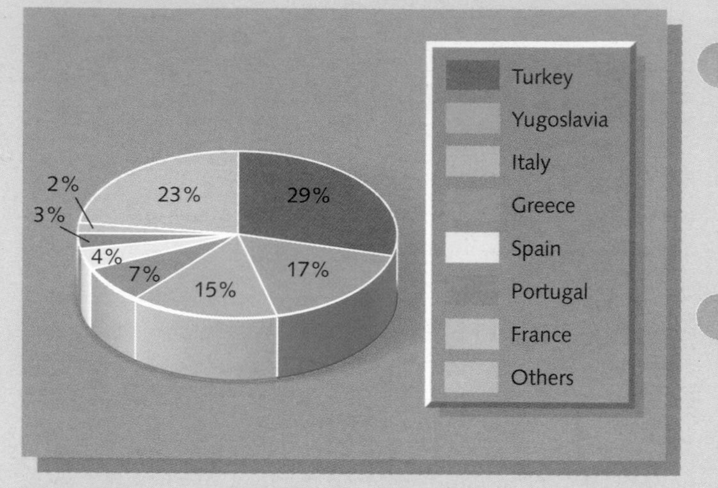

Figure 5.18
Source of immigrant workers arriving in West Germany in 1988

Figure 5.19 shows the imbalance between males and females and between different age-groups of Turkish migrants into West Germany – a population structure (pyramid) typical of other countries receiving large numbers of migrant workers. The movement of migrant workers from Turkey (the 'loser' country) to Germany (the 'receiving' country) has benefits and creates problems to both countries. Some of these advantages and disadvantages are listed in Figure 5.20.

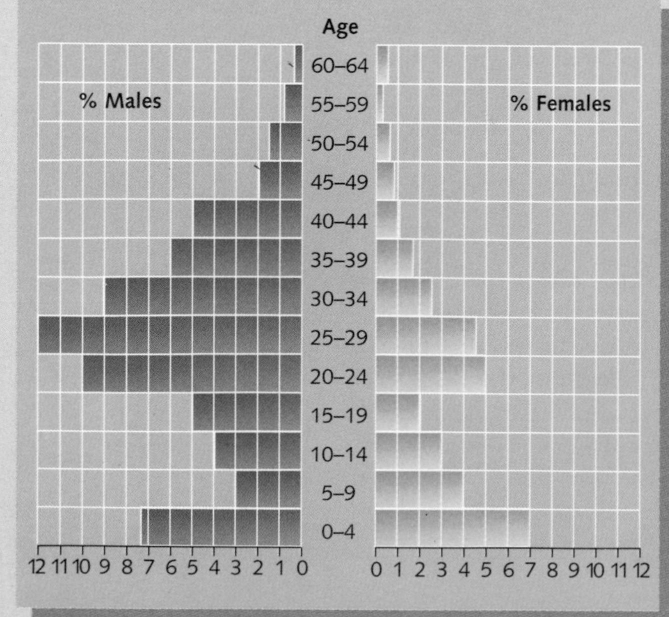

Figure 5.19
Age–sex structure for Turks in the former West Germany

ADVANTAGES	DISADVANTAGES
LOSING COUNTRY (e.g. TURKEY)	
• REDUCES PRESSURE ON JOBS AND RESOURCES, e.g. FOOD	• LOSES PEOPLE IN WORKING-AGE GROUP
• LOSES PEOPLE OF CHILD-BEARING AGE, CAUSING DECLINE IN BIRTH RATE	• LOSES PEOPLE MOST LIKELY TO HAVE SOME EDUCATION AND SKILLS
• MIGRANTS DEVELOP NEW SKILLS WHICH THEY MAY TAKE BACK TO TURKEY	• MAINLY MALES LEAVE, CAUSING A DIVISION OF FAMILIES
• MONEY EARNED IN GERMANY SENT BACK TO TURKEY	• LEFT WITH AN ELDERLY POPULATION AND SO A HIGH DEATH RATE
RECEIVING COUNTRY (e.g. GERMANY)	
• OVERCOMES LABOUR SHORTAGE	• PRESSURE ON JOBS BUT MOST LIKELY TO BE THE FIRST UNEMPLOYED IN A RECESSION
• PREPARED TO DO DIRTY, UNSKILLED JOBS	• LOW-QUALITY, OVERCROWDED HOUSING LACKING IN BASIC AMENITIES (INNER CITY SLUMS – *BIDONVILLES* IN FRANCE, *FAVELAS* IN SOUTH AMERICA)
• PREPARED TO WORK LONG HOURS FOR A LOW SALARY (LONDON UNDERGROUND)	• ETHNIC GROUPS TEND NOT TO INTEGRATE
• CULTURAL ADVANTAGES AND LINKS (e.g. NOTTING HILL CARNIVAL)	• RACIAL TENSION
• SOME HIGHLY SKILLED MIGRANTS (e.g. PAKISTANI DOCTORS)	• LIMITED SKILLED/EDUCATED GROUP
• IN A DEVELOPING COUNTRY THESE MIGRANTS COULD INCREASE THE NUMBER OF SKILLED WORKERS	• LACK OF OPPORTUNITIES TO PRACTISE THEIR OWN RELIGION, CULTURE, ETC.
	• LANGUAGE DIFFICULTIES
	• OFTEN LESS HEALTHY

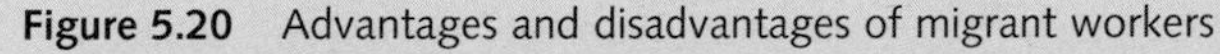

Figure 5.20 Advantages and disadvantages of migrant workers

Figure 5.21 A Turkish worker In Germany

Figure 5.22
Immigrants to Germany
(a) Source of immigrant workers arriving in the former West Germany in 1989
(b) Non-German population living in Germany in 1994

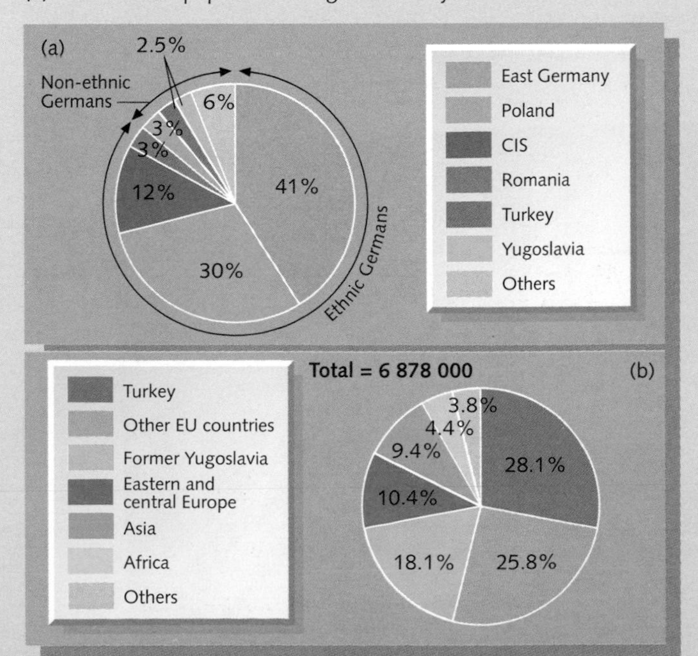

West Germany was more liberal towards immigrants than other West European countries, but it did impose a ban on the recruitment of foreign workers after 1973 – although Turks still arrived to reunite their families or to seek political asylum. In 1980 new laws reduced the right of asylum and grants were offered to Turks wishing to return home. Very few Turks took advantage of this offer and even fewer have taken out German citizenship. The Turks have their own centres in most large German cities where they speak their own language and have their own culture (food, dress and entertainment – Figure 5.21). While this is causing increasing resentment among some Germans, without the Turks, transport, hospital and electricity services would probably come to a halt.

Migrants into Germany since 1989

Large numbers of German-speaking people began leaving Eastern Europe in mid-1989. Their initial route to West Germany was via Hungary and Austria. Later, following the dismantling of the Berlin Wall, they migrated directly from East Germany (Figure 5.22). Many immigrants were relatively unskilled and were prepared to accept the types of jobs previously taken by Turkish and other 'guest workers'. The situation was made worse in the early 1990s by large numbers of former Yugoslavs seeking asylum from the civil war in their home country, and by a world economic recession. Although both Germans and Turks had found full employment in the 1950s and 1960s, by the mid-1990s nearly 10 per cent of Germans and 30 per cent of Turks were unemployed. This led to the German government trying to restrict the immigration of non-German ethnic groups, and to an increase in racist attacks on those groups.

CASE STUDY 5

Immigrants into California 2000

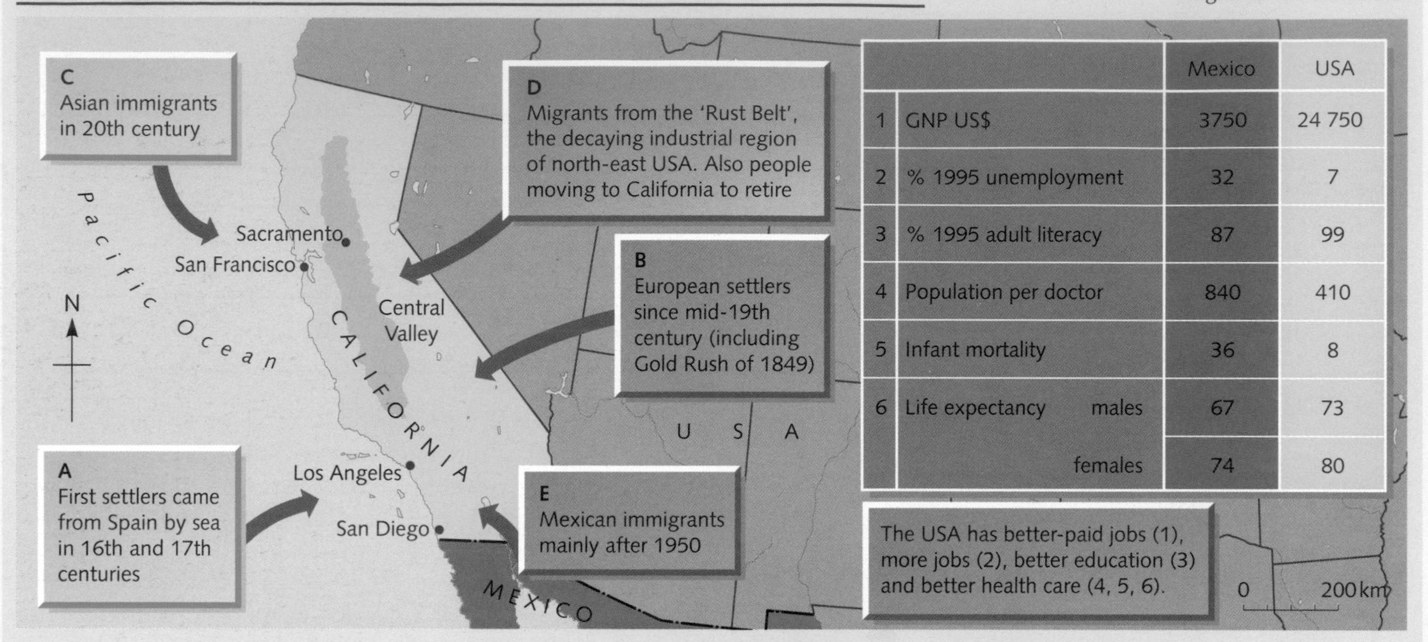

			Mexico	USA
1	GNP US$		3750	24 750
2	% 1995 unemployment		32	7
3	% 1995 adult literacy		87	99
4	Population per doctor		840	410
5	Infant mortality		36	8
6	Life expectancy	males	67	73
		females	74	80

Figure 5.23 Migrants into California

California was, until the mid-nineteenth century, sparsely populated mainly by native Americans. The first migrants did not arrive in any large numbers until the mid-nineteenth century (Figure 5.23). Since then, successive waves have helped increase the resident population to 32 million:

- **Nineteenth century:** from Western Europe – external, voluntary, permanent (Figure 5.1).
- **Early twentieth century:**
  - eastern and southern Europe
  - eastern Asia

  both external, mainly voluntary/some forced, permanent.
- **1930s:** American Mid-west (the 'Dust Bowl') – internal, partly forced, permanent.
- **Since 1950:**
  - north and east of the USA – internal, voluntary, permanent
  - eastern Asia – external, voluntary and forced, permanent
  - Mexicans (Hispanics) – external, voluntary, initially seasonal/increasingly permanent.

Mexican workers

Hispanics are people from Spanish-speaking countries in Latin America. In recent years, many Hispanics have migrated northwards to form the fastest-growing ethnic minority in the USA. Compared with its rich northern neighbour, Mexico has a relatively low standard of living, insufficient jobs and poorer education and health provision. As a result many Mexicans, especially those living in villages and working on the land (Figure 5.24), migrate into the USA, although often on a temporary basis. At one time only the men migrated, returning when they had earned enough money. Now they often stay permanently, sometimes taking their families with them (some villages in Mexico have lost half their population), sometimes deserting their family.

Figure 5.24 A Mexican village

Migrants – illegal and legal

Estimates suggest that between 1 and 2 million Mexicans try each year to cross into the USA, many illegally. Although the USA has set up elaborate border security controls using horses, helicopters and other advanced detective equipment, it is possible that nearly a million migrants manage to slip through each year. As one border guard claimed, 'We catch somebody one night, return him to Mexico the next day, for him to try again that night and the next, until he is finally successful'. The number of migrants who arrive legally with their 'Green Card' issued by the Department of Immigration, rises and falls with the seasonal availability of jobs.

Jobs taken by migrant workers

Although Mexican workers are viewed as a drain upon America's social security and welfare system, they are essential to the nation's economy. The migrants take the harder, dirtier, seasonal, more monotonous, more dangerous, less skilled and less well-paid jobs. For example:

- Large numbers find seasonal work on large agricultural estates, mainly in California's Central Valley, at harvest times (Figure 5.25). Despite the low pay by American standards, some migrants can earn more during three or four months in the USA than in a full year back in Mexico.
- Many others obtain employment in large urban areas, mainly Los Angeles, either in the construction industry or in hotels and restaurants.

Mexicans in Los Angeles

Many Mexicans perceive Los Angeles as the 'City of Opportunity'. The reality is very different (Figure 5.26). Most of the immigrants are young, have little money and few qualifications. Unless they have a Green Card, they cannot work legally and are forced to take very low-paid jobs, often in the informal sector (page 134). Some have to move around to avoid detection while others find language a problem when seeking a permanent job. Many of those who do get jobs find their income is insufficient to obtain decent accommodation and are forced to live in ghettos in the poorest districts. In one such district, that of Watts (Figure 5.27), Hispanics make up nearly three-quarters of the total population.

Figure 5.25 Mexican migrant workers packing vegetables in California

Figure 5.26 Poor-quality Hispanic housing in Watts, Los Angeles

Figure 5.27 Rapid increase in the Hispanic population of Los Angeles and the district of Watts

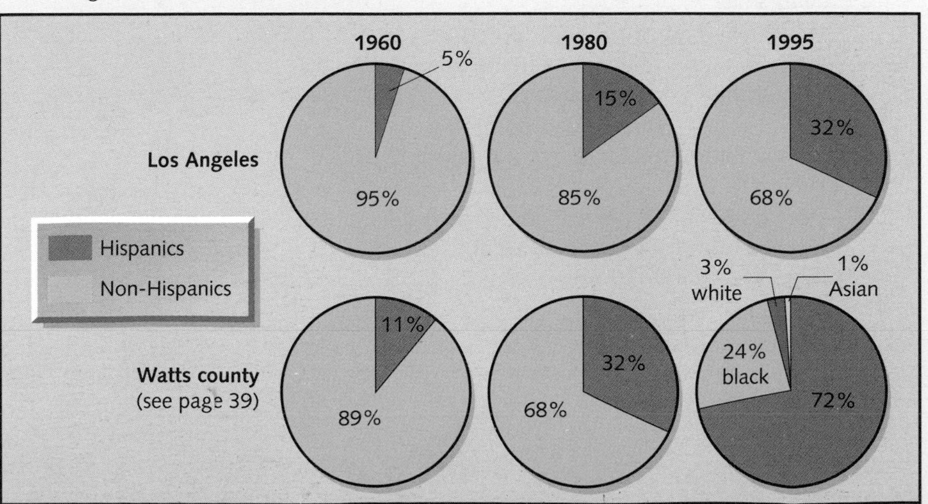

QUESTIONS

1

(Page 72)

a What is the difference between the following types of migration:
- voluntary and forced
- permanent and temporary? (2)

b Give a possible reason for each of the following examples of voluntary migration:
- Bantu into South Africa
- British doctors to the USA
- Mexicans into California
- Pop groups to America
- Development of British colonies
- West Indians to the UK
- Elderly, wealthy Americans to Florida (7)

c Give a possible reason for each of the following examples of forced migration:
- Africans into the USA
- Rwandans into the Democratic Republic of the Congo
- Jews from Nazi Germany
- Pilgrim Fathers to New England
- Palestinian Arabs from Israel
- Afghans from Afghanistan (6)

2

(Page 73)

a What is a refugee? (1)

b According to the pie-graph:

i) Which *country* has produced the most refugees? (1)

ii) Which *continent* has produced the most refugees? (1)

iii) What proportion of world refugees have come from Asia and Africa? (1)

iv) Which European country has produced the most refugees? (1)

c Give reasons why so many refugees have recently fled from any *named* country that you have studied. (3)

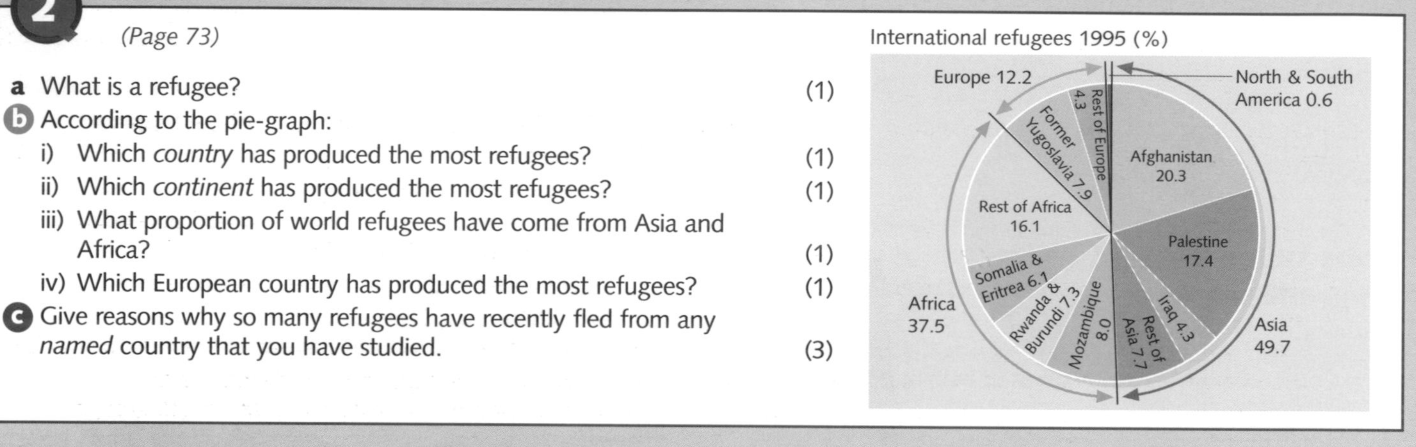

3

(Pages 74 and 75)

a i) What is meant by the term *New Commonwealth immigrant*? (2)

ii) Name two New Commonwealth countries. (2)

iii) Why did New Commonwealth immigrants come to Britain? (3)

iv) Why did most immigrants settle in large cities? (3)

v) Why did most immigrants settle in inner city areas? (3)

b Four wards have been labelled A, B, C and D on the map.

i) Which ward has the greatest number of people living in it? (1)

ii) Which ward has the highest proportion of New Commonwealth immigrants living in it? Why is this? (2)

iii) Which ward has the lowest proportion of New Commonwealth immigrants living in it? Why is this? (2)

iv) Why does ward B have an above-average number of New Commonwealth immigrants for an edge-of-city ward? (1)

c What are the problems any immigrant might face when he or she arrives in a new country, as far as the following are concerned?
- accommodation
- employment
- language
- culture
- prejudice (5)

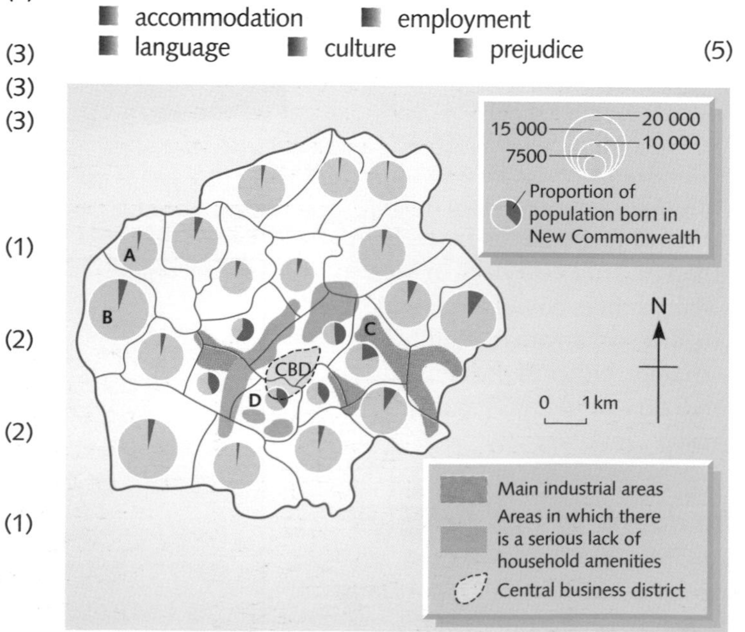

4

(Pages 76 and 77)

a The graph shows regional changes in population in the UK, 1984–91.

i) Which three regions lost population between 1981 and 1994? (1)

ii) Describe the location of these three regions. (1)

iii) Give three reasons why these regions lost population. (3)

b i) Which three regions had the largest gain in population? (1)

ii) Describe the location of these three regions. (1)

iii) Give three reasons why these regions gained in population. (3)

c Using Figure 5.13:

i) Name three cities that showed an increase in population. (3)

ii) Name three cities that showed a decrease in population. (3)

d Why are people moving away from large cities and conurbations? (4)

% change
12
10
8
6
4
2
0
–2
GAIN
LOSS
Scotland
North
Yorks & Humberside
North West
East Midlands
East Anglia
West Midlands
Wales
South West
Rest of South East
Greater London
Northern Ireland

5

(Pages 78 and 79)

The map shows the migration of workers into West Germany between 1970 and 1989.

a **Either**

i) Put in rank order (the highest first) the five countries from which most migrant workers came. (5)

ii) Suggest two reasons why workers might migrate from one part of Europe to another. (2)

iii) Why have some countries, such as Germany, encouraged inward migration of workers in the past? (2)

iv) What problems might develop in a country that has encouraged inward migration? (2)

Or

i) Describe the pattern of migrant workers to West Germany as shown on the map. (3)

ii) What are the advantages and disadvantages for countries such as Germany, that have encouraged large numbers of migrants to work there? (6)

b Describe and suggest reasons for the types of jobs available to foreign workers on their arrival in Germany. (3)

c State three problems that might arise in countries from which large numbers of migrants leave to work elsewhere. (3)

d Why did the West German government try to stop the inflow of migrant workers in the 1980s? (2)

e How has the unification of Germany affected the Turkish community? (2)

f With reference to Figure 5.19:

i) What percentage of the Turkish migrant population is aged between 15 and 19? (1)

ii) Which age-group has the largest number of immigrants? (1)

iii) Why are there more male immigrants than females? (2)

g What are the advantages and disadvantages to Germany of having immigrants with those age-groups and the sex ratio shown in Figure 5.19? (4)

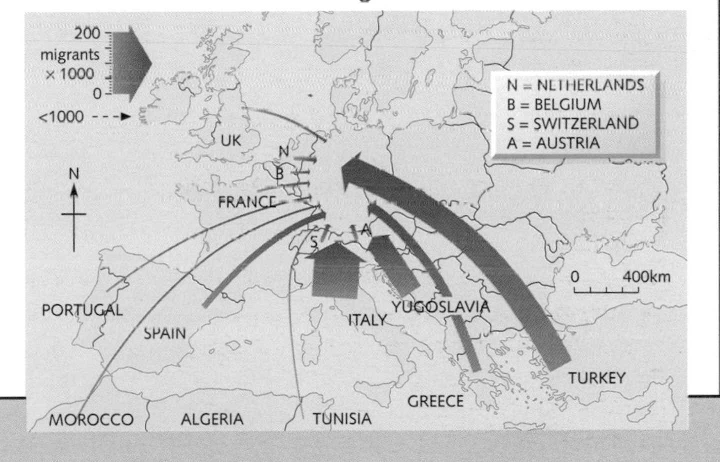

6

(Pages 80 and 81)

a Name a state in the USA that attracts Mexican migrant workers. (1)

b Give two reasons why Mexican workers try to find work in the USA. (2)

c Why do Mexicans try to enter the USA as 'illegal immigrants'? (1)

d Give two reasons why Americans try to restrict Mexican migrant labour entering the USA. (2)

e Give one reason why Americans need seasonal Mexican labour. (1)

f Many Mexicans move to Los Angeles. Describe what sort of accommodation and jobs they are likely to find once they arrive. (4)

Farming systems and types

Farming is an industry and operates like other industries. It is a *system* with *inputs* into the farm, *processes* which take place on the farm and *outputs* from the farm (Figure 6.1).

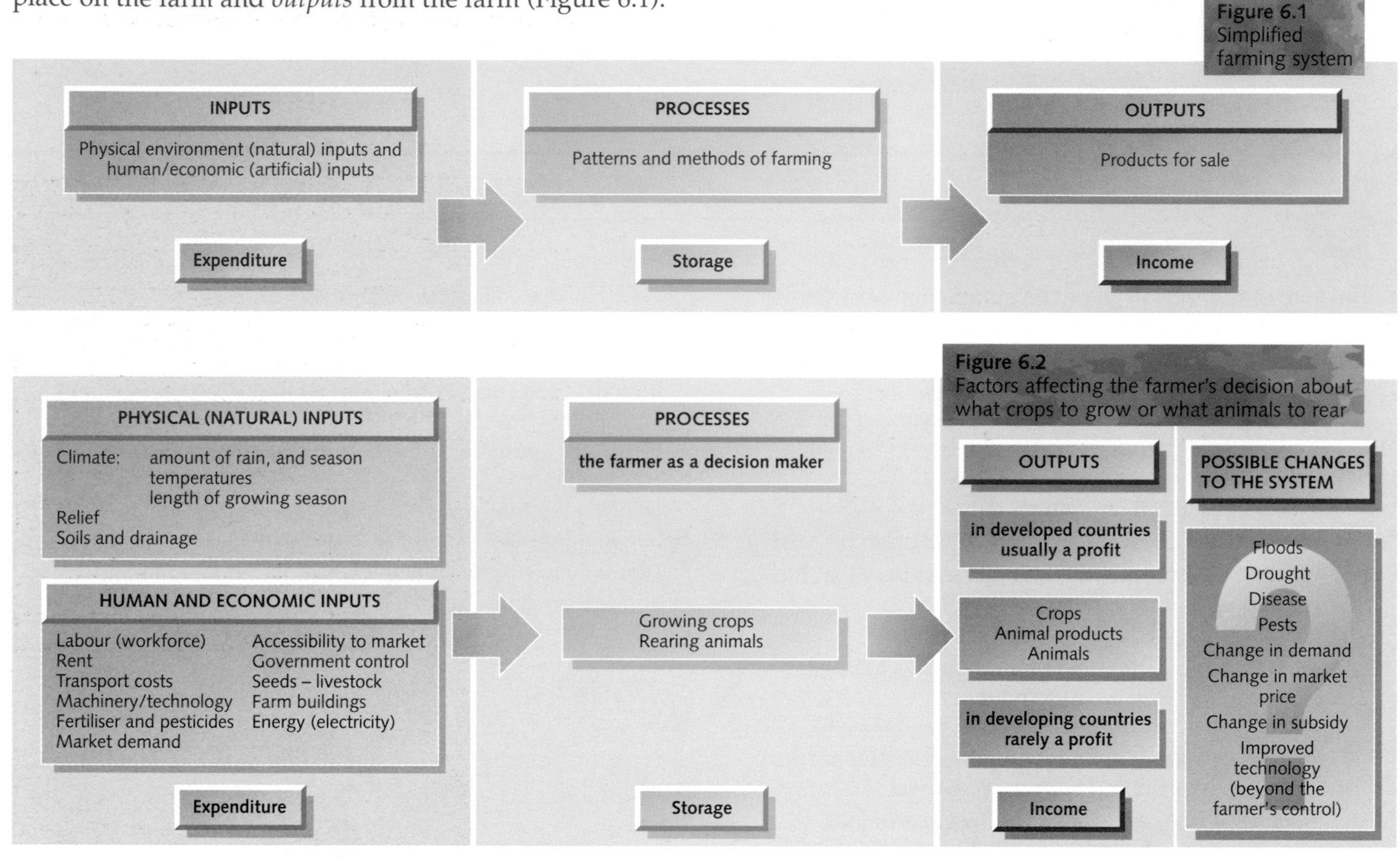

Figure 6.1 Simplified farming system

Figure 6.2 Factors affecting the farmer's decision about what crops to grow or what animals to rear

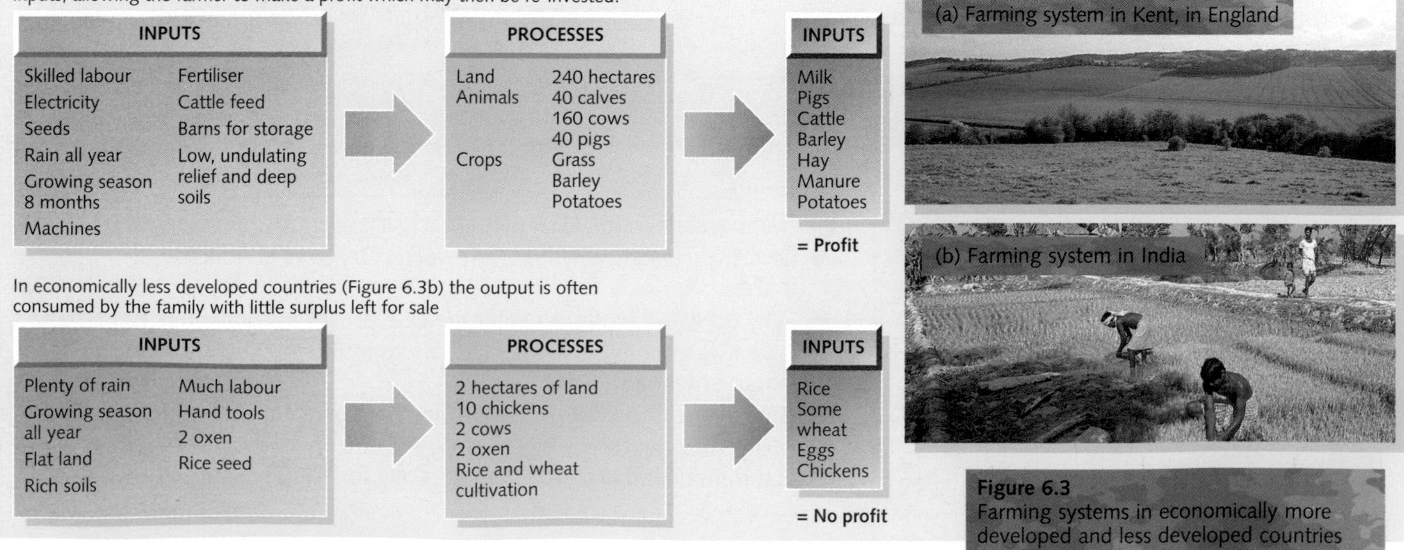

Figure 6.3 Farming systems in economically more developed and less developed countries

The farmer as a decision maker

Each individual farmer's decision on what crops to grow or animals to rear, and which methods to use to maximise outputs, depends on an understanding of the most favourable physical and economic conditions for the farm (Figure 6.2). Sometimes the farmer may have several choices and so the decision may depend upon individual likes and expertise. On other occasions the choice may be limited by extreme physical conditions or economic and political pressures.

Classification of types of farming

The classification shown in Figure 6.4 is based upon the following criteria:

Specialisation This includes *arable* (the growing of crops), *pastoral* (the rearing of animals) and *mixed* (both crops and animals) farming.

Economic status *Commercial* farming is the growing of crops or rearing of animals for sale (i.e. outputs exceed inputs). *Subsistence* farming is when just sufficient food is provided for the farmer's own family (i.e. outputs may be the same or less than the inputs and so the family may struggle for survival).

Intensity of land use This depends upon the ratio between land, labour and capital (money). *Extensive* farming is where the farm size is very large in comparison with either the amount of money spent on it (Amazon Basin) or the numbers working there (American Prairies). *Intensive* farming is when the farm size is small in comparison with either the numbers working there (Ganges Delta) or the amount of money spent on it (Denmark).

Land tenure *Shifting* (and *nomadic*) cultivation is where farmers move from one area to another. *Sedentary* is where farming and settlement is permanent.

Remember that the map in Figure 6.4 is simplified. It only shows the generalised world location of the main types of farming. It does not show local variations, transitions between the main farming types nor if several types occur within the same area.

	TYPE OF FARMING	NAMED EXAMPLE
1	NOMADIC HUNTING AND COLLECTING	AUSTRALIAN ABORIGINES
2	NOMADIC HERDING	MAASAI IN KENYA, SAHEL COUNTRIES
3	SHIFTING CULTIVATION	AMERINDIANS OF AMAZON BASIN
4	INTENSIVE SUBSISTENCE AGRICULTURE	RICE IN THE GANGES DELTA
5	PLANTATION AGRICULTURE	SUGAR CANE IN BRAZIL
6	LIVESTOCK RANCHING (COMMERCIAL PASTORAL)	BEEF ON THE PAMPAS
7	CEREAL CULTIVATION (COMMERCIAL GRAIN)	CANADIAN PRAIRIES, RUSSIAN STEPPES
8	MIXED FARMING	NETHERLANDS, DENMARK
9	'MEDITERRANEAN' AGRICULTURE	SOUTHERN ITALY, SOUTHERN SPAIN
10	IRRIGATION	NILE VALLEY, CALIFORNIA
11	UNSUITABLE FOR AGRICULTURE	SAHARA DESERT

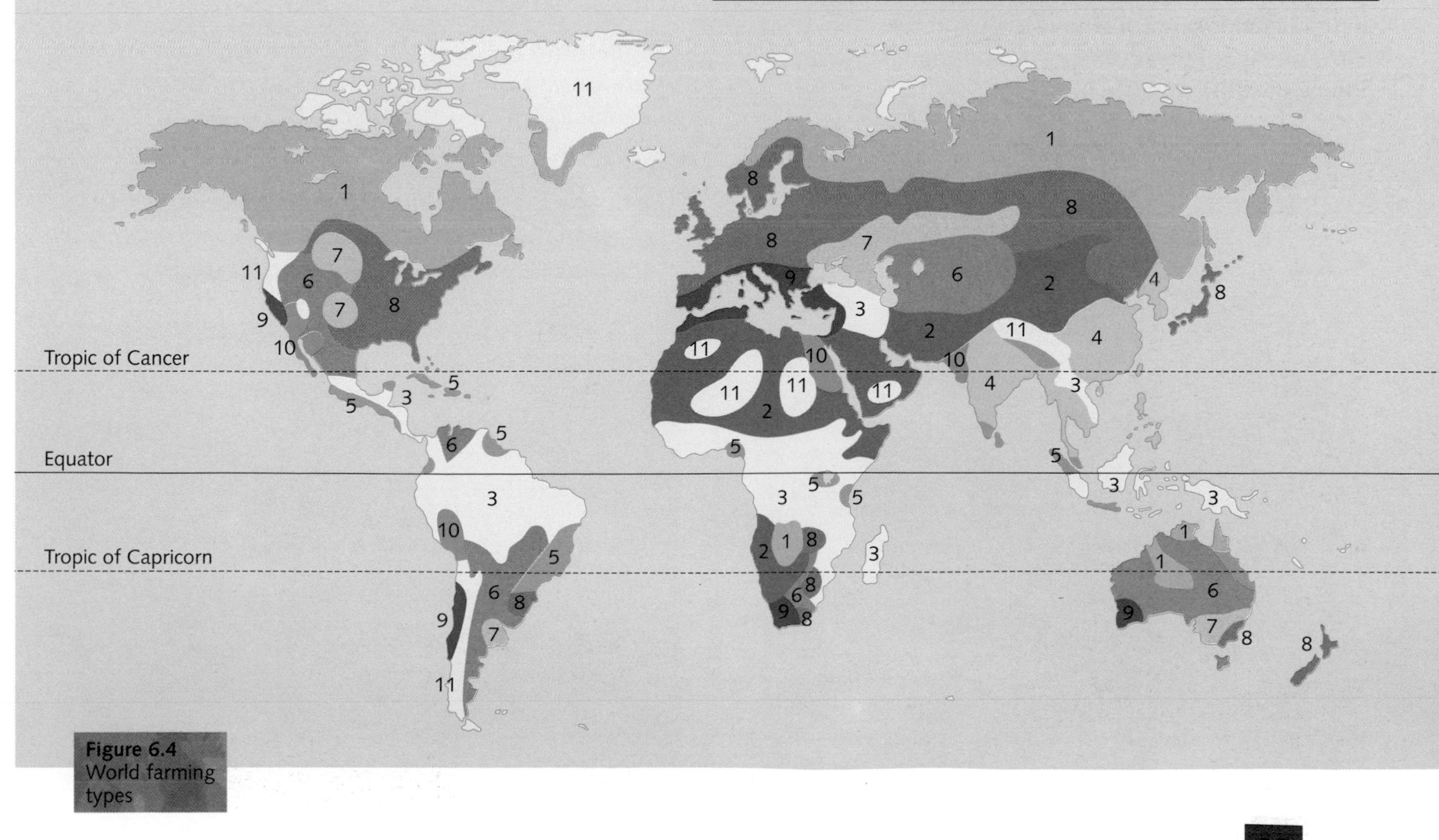

Figure 6.4 World farming types

Farming in the EU

Factors affecting farming in the EU

Farmers' decisions on which crops to grow or animals to rear and what methods to use to produce the outputs depend upon an understanding of the most favourable physical and economic conditions of their locations. A simplified map to show the major areas of specialisation is shown in Figure 6.5.

Physical inputs

Relief Usually the flatter the land, the larger and more efficient is the farm (Figure 6.6). Output tends to decline as land gets steeper and higher (Figure 6.7).

Soils The deeper and richer the soil, the more intensive the farming and the higher the output, e.g. limon of the Paris Basin, alluvium of the Po Valley and estuarine deposits in the Netherlands. Soils ideally should be reasonably well drained.

Rainfall Those areas in the more northern and western parts of the EU with adequate and reliable rainfall throughout the year tend to produce good grass and so rear animals. The drier areas with less reliable rainfall in the south and east are more suited to arable farming.

Temperatures In the north the length of the growing season is limited, whereas in the warmer, sunnier south, cereals and fruit ripen more readily. Aspect is an important local factor.

Human inputs

Government aid Farmers rely on grants for new stock and machinery, and subsidies to guarantee a fixed price. Farm prices are fixed under the Common Agricultural Policy (CAP).

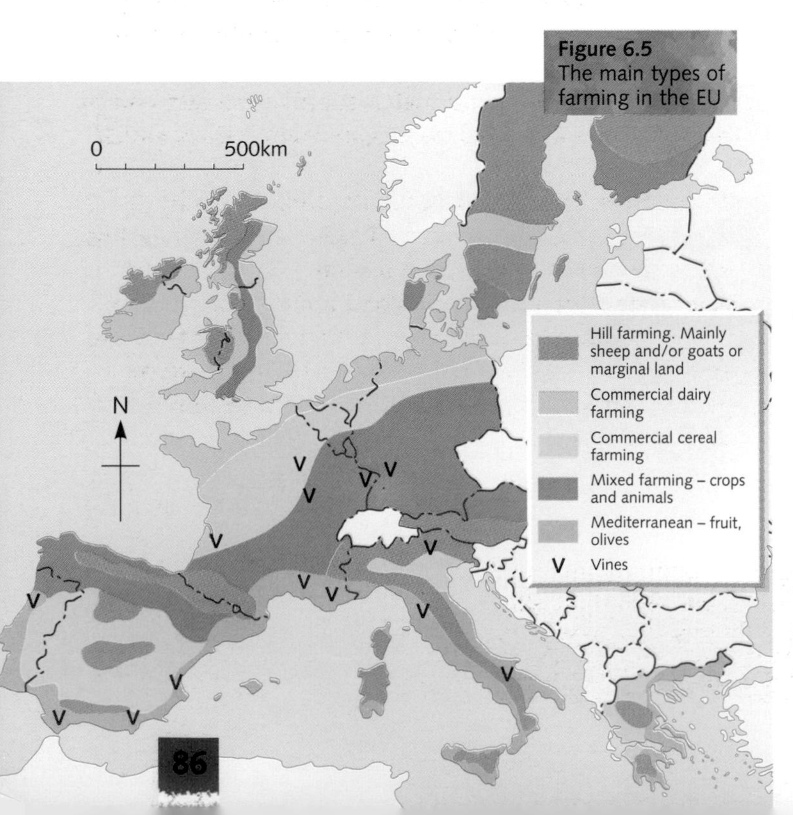

Figure 6.5 The main types of farming in the EU

Figure 6.6 Flat farming landscape – Dutch polders

Fertiliser These have increased in variety and effectiveness, raising output, especially in the more affluent farming regions.

Mechanisation The introduction of many new labour-saving machines has increased output, but has led to a sharp decline in the numbers employed in agriculture (Figure 6.8).

Improvement in varieties Output has also increased due to better strains of seed and better-quality animals.

Marketing Perishable goods need to be grown near markets for freshness, and bulky crops in a similar location to minimise transport costs.

Size of farm Apart from an area around the southern parts of the North Sea, the size of most farms in Europe is very small (Figure 6.8). Attempts are being made to amalgamate farms but while this may increase their efficiency it also increases rural depopulation. Notice how farm sizes decrease in the peripheral areas such as Portugal and Greece.

Competition for land Traditional farming areas are under threat from urban growth and recreational demands.

Variable inputs

The farmer is vulnerable to government policies, to changes in market prices and demands, and especially to changes in the weather (flood, drought, frost).

Figure 6.7 Mountainous farming landscape – southern Greece

Common Agricultural Policy (CAP)

The basic aims of the CAP when it was set up in 1962 were to:

- create a single market in which agricultural products could move freely
- make the European Community (EC, as it then was) more self-sufficient by giving preference to EC produce and restricting imports from elsewhere
- give financial support to EC farmers which included guaranteed prices (*subsidies*), and therefore a guaranteed market, for unlimited production
- increase the average field size, farm size and farmers' income.

These aims replaced all existing national policies, although they often led to conflict between member states. Trying to get a balanced interpretation of how successful the CAP has been is difficult as all member states have both 'pro-marketeers' and 'anti-marketeers'. Figure 6.9 gives some of the arguments used by the rival 'pro' and 'anti' groups.

Figure 6.8 Agriculture in the EU

	% OF ECONOMICALLY ACTIVE POPULATION IN AGRICULTURE		% OF TOTAL LAND USED FOR AGRICULTURE		AVERAGE SIZE OF FARMS (HECTARES)		% GNP FROM AGRICULTURE	
	1960	1995	1980	1995	1980	1995	1980	1995
AUSTRIA	24	7	43	41	9	14	5	2
BELGIUM	8	3	52	47	15	24	2	2
DENMARK	16	6	63	64	25	38	5	5
FINLAND	36	9	10	8	10	16	9	5
FRANCE	22	6	59	55	25	26	4	3
GERMANY	14	3	53	49	15	28	2	1
GREECE	55	23	56	70	4	13	18	17
REP. OF IRELAND	36	14	69	58	22	36	11	8
ITALY	30	9	58	56	7	18	6	3
LUXEMBOURG	15	3	76	49	28	34	2	1
NETHERLANDS	10	5	57	46	16	26	4	4
PORTUGAL	44	17	55	43	5	12	8	6
SPAIN	42	11	71	50	15	26	7	5
SWEDEN	14	3	10	7	11	23	4	2
UK	4	2	77	65	69	94	2	2

1970s and 1980s – increasing concerns over the CAP

- 70 per cent of the EC's budget was spent supporting farming when farming only provided 5 per cent of the EC's total income.
- As farmers were encouraged to produce as much as possible and as improved technology increased their output, large surpluses were created – the so-called cereal, butter and beef 'mountains' and the wine and milk 'lakes'.
- Imports were subject to duties to make them less competitive with EC prices. This handicapped the economically less developed countries.
- Although EC farms became larger (Figure 6.8) and more efficient, only the most prosperous farmers benefited – often at the expense both of farmers on the periphery and of the environment (pages 98–99).

1992 – agricultural reform

Five aims were defined. These were:

1. To increase the EU's agricultural competitiveness by concentrating on quality rather than quantity and in training young farmers.
2. To stabilise markets and match supply with demand by reducing subsidies and quotas on commodities which had a surplus.
3. To ensure a fair standard of living for farmers by providing income support and early retirement to those in less favoured and marginal areas.
4. To maintain jobs on the land and reduce migration to the towns by introducing alternative forms of land use.
5. To protect and enhance the natural environment by paying farmers to 'set aside' land or change the use of their land.

Achievements

Achieved a larger measure of self-sufficiency. This reduces the costs and unreliability of imports.

Created higher yields due to input of capital for machinery and fertiliser.

In NW Europe the average farm size has increased almost to the recommended level.

Amalgamation of fields – in parts of France the number of fields has been reduced to one-eighth of the 1950 total.

Production has changed according to demands, e.g. less wheat and potatoes and more sugar beet and animal products.

Subsidies to hill farmers have reduced rural depopulation.

Poorer farmers gain an opportunity to receive a second income by working in nearby factories ('5 o'clock farmers') or from tourism.

Higher income for farmers.

Subsidies have reduced the risk of even higher unemployment in such rural areas as the Mezzogiorno.

Reduced reliance on crops imported from developing countries which themselves have a food shortage.

A surplus one year can offset a possible crop failure in another year.

Problems

An increase in food prices, especially in the net importing EU countries of Germany and the UK.

Creation of food surpluses – the so-called 'mountains and lakes'.

Selling of surplus products at reduced prices to East European countries (causes both political and economic opposition).

Increased gap between the favoured 'core' agriculture regions and the periphery.

Peripheral farm units still very small and often uneconomic.

High costs of subsidies. 'Industrial' countries such as the UK object to 70% of the EU budget being spent on agriculture.

'Five o'clock farmers' spend insufficient time on their farms. In France 15%, and in Germany 30% of farmers have a second income.

Destruction of hedges to create larger fields destroys wildlife and increases the risk of soil erosion.

By reducing imports from developing countries the latter's main source of income is lost thus increasing the trade gap between the two areas.

Figure 6.9 A balance sheet showing some of the achievements and some of the problems still to be faced by the EU's Common Agricultural Policy

Farming in Denmark

Although Denmark today has a high agricultural output , its land was not always fertile. Soils are sandy in the west and clay in the east (Figure 6.10). Since the nineteenth century much money and effort has gone into improving these soils. Modern farms in the west specialise in rearing animals while those in the east concentrate on growing cereals.

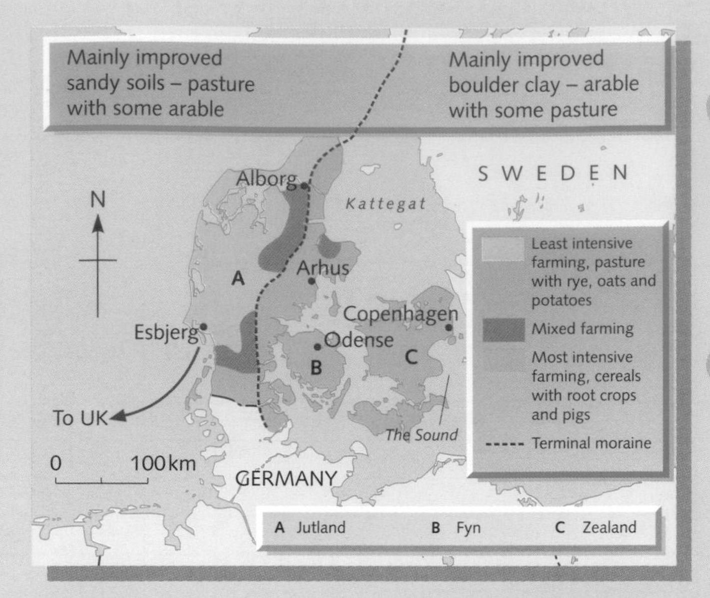

Figure 6.10 Farming types in Denmark

Danish farming before 1960

The combination of relief (low-lying and relatively flat), soils and climate (winter frosts, warm sunny summers and rainfall below average) meant that Denmark was ideally suited to the growing of cereals. Denmark was a major producer of wheat and barley until, in the 1870s, the building of railways across Canada opened up the Prairies. Denmark could not compete with the flood of cheap wheat but realised that there was a demand for dairy products in the rapidly industrialising countries of Britain and Germany.

Danish farming since 1960

Denmark's climate has always been more suitable for the growing of cereals than for grass. The reliance upon one type of farming makes farmers more vulnerable to adverse climatic conditions, to changes in market demand and price, and to disease and pests. In mixed farming, on the other hand, farmers can earn their income from several sources and alternatives. Figure 6.11 shows a modern farming landscape in Denmark. Dairy cattle remain important (they also provide the skimmed milk for the pigs), but in winter they are kept indoors. Pigs and poultry are kept indoors all year.

Selective livestock breeding produced:
- the Danish red cow which thrives on relatively poor pasture and gives a high butter fat content;
- the Landrace pig whose long back gives top-quality bacon (Figure 6.15).

Imported Friesian cattle give high milk yields.

Figure 6.11 Danish agricultural landscape

Many farmers have adopted an eight-year rotation system which may be: wheat – root crops – barley – sugar beet – barley – grass – mixed barley and oats – sugar beet.

The cereals are often cut green for silage, while root crops and sugar beet are not only valuable crops in their own right, but help to replace nitrogen in the soil previously used by the cereal crops.

Figure 6.12 The benefits of the co-operative system

Co-operatives
Most Danish farms are owner-occupied, but the farmers work together to try to get the maximum benefit from buying in bulk and selling on a collective basis. Co-operatives can help individual farmers by:

Bulk buying Co-operatives buy such items as cattle feed, fertiliser and seed in bulk, so reducing the cost to farmers.

Market products 200 or 300 farmers may, among them, own a dairy. The milk is sent to this dairy and the dairy does the processing and selling. The profits are shared according to the amount of milk that each farmer supplies. The dairy will also provide transport, which is a more efficient method than each individual having to do so.

Finance The co-operatives have their own banks which allow cheaper loans for machinery and buildings.

Quality Each farmer is expected to produce goods of the highest quality. The co-operatives help to ensure that goods, such as those branded Lurpak, are equated with top quality.

Back-up services These include a veterinary service, free advice on new methods of farming and quality control, research into new techniques, colleges to train young people and to retrain practising farmers, and links with associated agricultural industries (e.g. bacon curing, brewing, milling, the production of butter and cheese, and the manufacture of farm machinery).

	1960	1982	1995
LAND USE (% OF FARMLAND)			
UNDER CEREALS	47	75	91
UNDER ROOTS	18	8	3
UNDER GRASS	32	9	6
ANIMALS			
DAIRY COWS (1000s)	1438	1063	658
PIGS (1000s)	6147	9348	10 870
MECHANISATION			
NUMBER OF HORSES	171	42	–
NUMBER OF TRACTORS	111 300	176 300	155 000
NUMBER OF FARMS	196 076	107 500	60 600
% OF POPULATION IN AGRICULTURE	16.0	6.1	6.0
AVERAGE SIZE OF FARM (HECTARES)	15.8	26.7	38.2

Figure 6.13 Changes in Danish agriculture, 1960–95

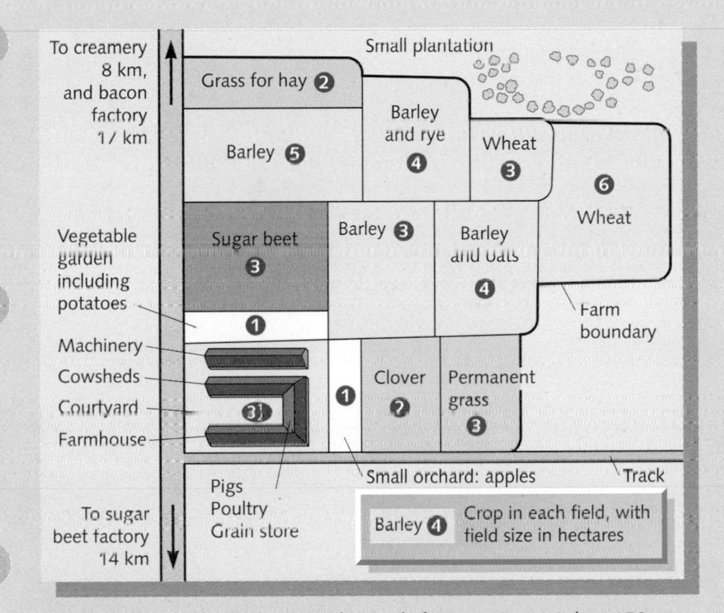

Figure 6.14 Layout of a typical Danish farm on Fyn – it has 150 cows (120 Red Danish, 30 Friesian), 240 Landrace pigs and 220 hens

Figure 6.15 Livestock breeding produces top-quality pigs

Recent changes in Danish farming

- An increase in the average size of Danish farms and a decrease in the number of smaller farms (Figure 6.13).
- An increase in mechanisation.
- A shift of cattle and milk production from the east to the west (as well as fewer cows overall), and an increase in cereals in the east.
- A decrease in the agricultural workforce and a pronounced movement to the towns for jobs.
- Difficulty in getting labour for the twice-daily milking, leading to an increase in pig farming and cereal growing. In 1980 only 2 per cent of school leavers went into farming.

Since joining the EU

- EU membership has strengthened traditional markets with the UK and Germany, and opened up new ones. 26 per cent of Denmark's exports are agricultural products, and 66 per cent of these go to the EU.
- The EU has provided the subsidies necessary to guarantee fixed prices for farm produce.
- CAP has encouraged an increase in cereal production, especially barley. Barley was also needed to meet the increased demands of the Danish brewing industry.
- CAP has recently cut milk quotas. This has reduced dairy farming and butter production.

A typical Danish farm on Fyn

Figure 6.14 gives the layout of a typical farm on the island of Fyn. All farms, even modern ones, are built around the traditional courtyard. Notice the dwelling house on one side of the yard facing the cowsheds. The third side is usually occupied by accommodation for pigs and poultry and facilities to store grain.

Despite its emphasis on dairy produce, most of Denmark is arable land (Figure 6.10). The land is divided into large fields, most of which are bordered by wire fences not hedges. Wheat, barley, oats and rye, together with the tops of the sugar beet, are all grown as fodder for the animals, for Danish domestic consumption, or use in industry (e.g. milling, brewing). Only a small proportion of the farm is actually under grass mainly because the grass is not of good enough quality. It is still fairly common to see the Danish red cow tethered, or at least limited to small areas of grazing ground. These animals are usually kept indoors from late September to late April and are fed on fodder crops grown on the farm. Most farms are not large enough to store the large quantities of winter fodder needed. When cereals are harvested they are sent to the co-operatives for storage until they are needed. Root crops and clover are also grown for fodder, although sugar is also obtained from the beet.

Farming in southern Italy (the Mezzogiorno)

The Mezzogiorno, which means *the land of the midday sun*, lies south of Rome and includes the two islands of Sicily and Sardinia (Figure 6.16). It contains the poorest parts of the EU, with Basilicata described as 'the most disadvantaged of the 160 regions in the EU' (EU Report). Figure 6.17 gives some of the causes of poverty and high unemployment, and the reasons why so many people have emigrated. To this list can be added the lack of mineral and energy resources, industry, commerce, services and skilled labour. The resultant low standard of living meant that between 1950 and 1975, 4.5 million people emigrated to the north of Italy, to the USA and as 'guestworkers' to Germany and Switzerland (page 78).

The following two accounts describe what life was like in Basilicata. The first describes the village of Aliano in the late 1930s, the second, farming in the 1960s.

1 *'The village itself was merely a group of scattered white houses at the summit of the hill [Figure 6.18].*

'The houses were nearly all of one room, with no windows, drawing their light from the door. The one room served as kitchen, bedroom, and usually as quarters for the barnyard animals. On one side was the stove; sticks brought in every day from the fields served as fuel. The walls and ceiling were blackened with smoke. The room was almost entirely filled with an enormous bed; in it slept the whole family, father, mother and children. The smaller children slept in reed cradles hung from the ceiling above the bed, while under the bed slept the animals.

'The second aspect of the trouble is economic, the dilemma of poverty. The land has been gradually impoverished; the forests have been cut down, the rivers have been reduced to mountain streams that often run dry, and livestock has become scarce. Instead of cultivating trees and pasture lands there has been an unfortunate attempt to raise wheat in soil that does not favour it. There is no capital, no industry, no savings, no schools; emigration is no longer possible, taxes are unduly heavy, and malaria is everywhere. All this is in a large part due to the ill-advised intentions and efforts of the State, a State in which the peasants cannot feel they have a share, and which has brought them only poverty and deserts.'

Carlo Levi, *Christ stopped at Eboli*

2 *'To go to the fields is almost a reflex, conditioned by the absolute lack of any other work. You go, even when you might not have to. The donkey needs fodder: you cut it by hand, sometimes just along the verges, and shove it into the sacks that hang from the saddle. While you're about it, you pull up wild greens for a salad – a little sorrel, dandelions, whatever there is. You weed your patch of wheat, you loosen the dirt around the beans, tie up a vine. You look over to see how your neighbour's crops are coming. Not really that you hoped yesterday's hailstorm had beaten them down, but there would be some justice if ... You collect a few twigs for kindling. And finally at some invisible cue of light not yet changed but about to change, the long walk back to town. Five miles, ten – a long way with nothing to think about except how to get a bit more land, a job – how to feed your children. Your children. Will they live like this?'*

Ann Cornelisen, *Women of the Shadows*

REGION	BIRTH RATE	INCOME PER HEAD (AV. = 100)	UNEMPLOYMENT %	NET MIGRATION (% PER YEAR) 1980	1994
1 NORTH-WEST	7.6	115	8.9	+0.8	+4.4
2 LOMBARDY	8.7	130	6.1	+0.7	+4.0
3 NORTH-EAST	9.0	115	6.3	–0.2	+3.9
4 EMILIA-ROMAGNA	7.2	124	6.6	+0.1	+7.2
5 CENTRE	7.8	104	7.9	–0.1	+5.0
6 LATIUM	10.0	115	11.0	+0.6	+2.9
7 ABRUZZI	10.1	87	11.3	–1.3	+4.4
8 CAMPANIA	14.6	68	23.1	–1.0	+0.4
9 SOUTH	12.6	67	17.5	–1.8	–1.0
10 SICILY	13.8	68	21.9	–1.4	+1.6
11 SARDINIA	9.8	75	20.3	–1.1	+1.4

The North
60% of population
80% of GNP

The Mezzogiorno
40% of population
20% of GNP

Milan 3
Turin 1 2 Venice
Genoa 4
5
Rome 6 7
11 Naples 8 9 Basilicata
10
0 200km
N

Figure 6.16 Population and wealth distribution in Italy, 1994

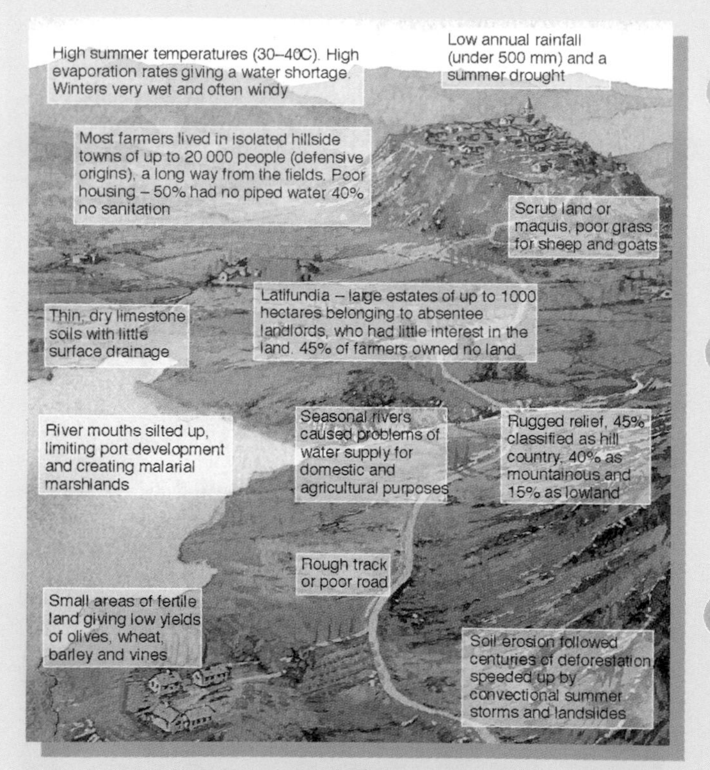

Figure 6.17 The Mezzogiorno in 1950

Figure 6.18 A hilltop town in Italy

Farming in the Mezzogiorno today

Although half the jobs in farming have been lost in the last 30 years, agriculture still employs 25 per cent of the working population. The government, through a scheme called Cassa per Il Mezzogiorno, tried to improve agriculture as well as services and industry in the region. Although several coastal areas benefited from a few large-scale developments, the many mountainous areas of the Mezzogiorno saw little improvement. In 1984 the Comunità Montana scheme was introduced with the emphasis on smaller, more practical developments.

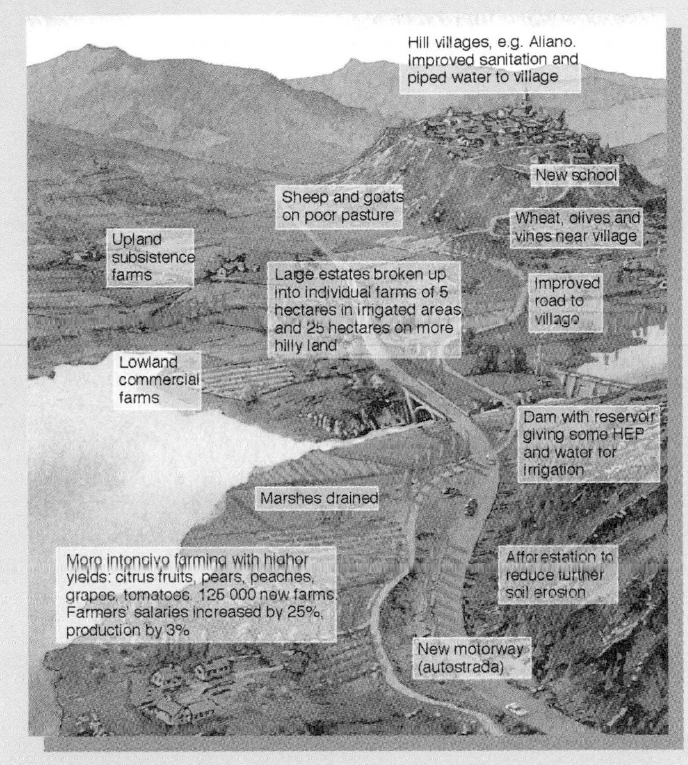

Figure 6.19 The Mezzogiorno in the mid-1990s

Figure 6.20 Vines growing on the hillside in southern Italy

Upland subsistence farms

These have hardly changed (Figure 6.19). Wheat is the main crop but yields are low largely due to the soils which are stony, shallow, dry and easily washed away. Olives and vines are still important as they can tolerate the high temperatures, the summer drought and the thin soils (Figure 6.20). These two crops need a lot of cheap hand labour, especially at harvest time. Much of the higher, poorer areas are left as grazing land for sheep and goats. Most farmers can, through hard work, feed their own families but there is rarely any surplus for sale.

Lowland commercial farms

These are found either where the large estates have been broken up into smaller units or where the malarial marshes have been drained (Figure 6.19). New motorways have been opened and large dams have been built across several rivers. The water stored in the resultant reservoirs is used for both domestic purposes and irrigation. Pipes carry water to coastal farms where large sprinkler systems enable soft fruit (peaches and pears) and citrus fruit (oranges and lemons) to grow commercially. Citrus fruits in particular flourish in the hot summers (their thick skins limit moisture loss), mild winters, alluvial river delta soils and good drainage. Flowers, tomatoes and vines are also grown (Figure 6.20).

Problems

Although wealthier than in the 1950s, most of the Mezzogiorno remains poor by EU standards.

- Farming tends to be done by the older generation and attracts few young people.
- Increased mechanisation has led to increased unemployment (Figure 6.16). There are few alternative jobs and an increasing number of abandoned fields and farms.
- Although many former emigrants have returned home wealthy, the area is receiving numerous immigrants from poorer countries in Africa and the Middle East.
- The ground is unstable with soil erosion, landslides and earthquakes (parts of Aliano and other settlements remain uninhabited after the 1980 'quake).
- This is a peripheral area in the EU, a long way from the main markets, and has received little real help from the CAP.

Commercial farming in Japan

Only 17 per cent of Japan is classed as flat land – the remainder is highland dissected by steep-sided valleys. With so little flat land there is great competition for space for housing, industry, communications and farming. Even away from the large cities many areas are half urban and half rural, with numerous small fields being surrounded by large villages. The average Japanese farm is only 1.3 hectares, the equivalent of two football pitches. It has to be used *intensively* and by all members of the farmer's family. As Japan has become a major industrial country, farming too has undergone a rapid change.

Rice

Rice gives very high yields wherever the land is flat enough for it to be grown. Southern Japan has hot and very wet summers and mild winters. The soils are rich either from volcanic deposits or alluvium brought down from the mountains by fast-flowing rivers. No land is wasted, and the lower hillsides are terraced (Figure 6.22). Small machines are used to plough, plant and harvest the crop (Figures 6.23 and 6.25) and these have now replaced most of the older back-breaking jobs. Machinery is also important as most farms are run by an increasingly older workforce, since their children prefer better-paid, less demanding jobs in the new high-tech factories in nearby cities (Figure 6.21). Japanese farmers and their wives may work ten hours a day in their fields in order to get two or, in the extreme south, three crops a year from their tiny farms. Vegetables and young rice plants are grown under vinyl sheets or in vinyl greenhouses. The climate is warm enough for crops such as barley, wheat and soya beans to be grown in the winter months. The increase in mechanisation and the use of fertiliser (Japan uses more fertiliser per hectare than any other country) led to rice overproduction. However, this surplus has disappeared as the number of farms and farmers has decreased and the consumption of Western-type food has increased. Since 1993 Japan has had to import rice, mainly from Thailand. Imported rice is four times cheaper than that grown at home.

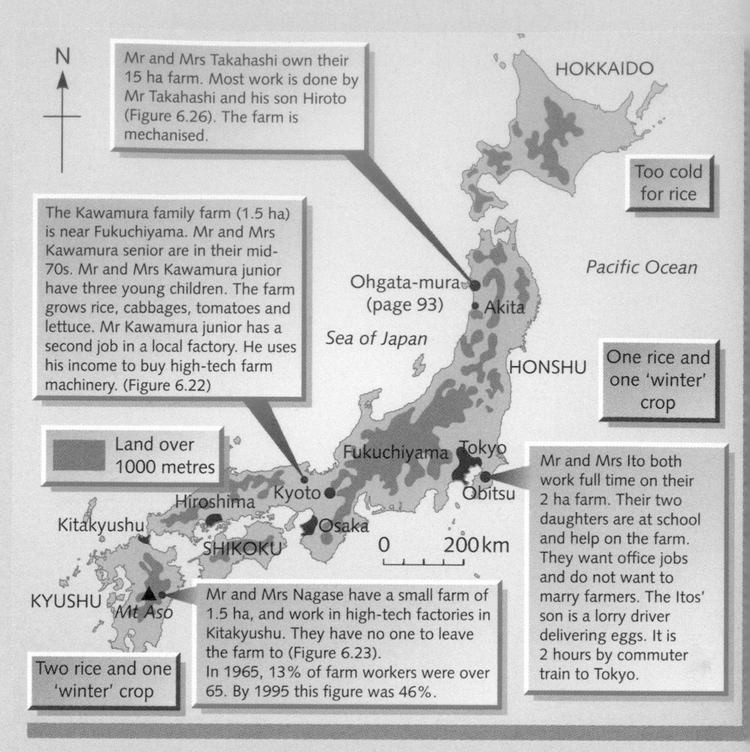

Figure 6.21 Location of four farming families in Japan

Figure 6.22
Farming landscape – Fukuchiyama Valley

Figure 6.23
Mr and Mrs Nagase

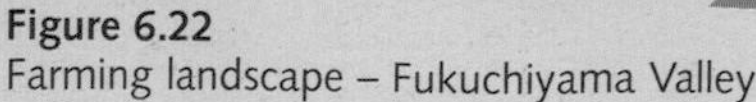

Places 2000

Northern Honshu

Lake Hachiro-gatu, although only 4 metres deep, was the second largest in Japan. Its floor was covered in rich alluvium. Reclamation, with guidance from the Dutch, began in 1964 and the first pioneer farmers arrived in 1966. Today, a pumping station pumps excess rainwater off the land to stop flooding, while an irrigation system can return the water if the land becomes too dry. The land was divided into fields of 1.25 hectares as this size made it easier for large-scale machinery. Each farm is made up of 12 fields (15 hectares), and there are 560 farms. Rice is grown on a much more commercial scale here than anywhere else in Japan. The landscape is very similar to that of the Dutch polders (Figures 6.6). The new town of Ohgata-mura, in which all 3400 inhabitants live, has an agricultural college and every modern amenity.

Figure 6.24
Ohgata-mura

Mr and Mrs Takahashi were amongst the first arrivals. Hiroto, their son (Figure 6.26), has pleased his parents by marrying (very important to Japanese that the family name is continued), and by staying to help run the farm.

Hiroto's year

April Rice planted in nurseries under vinyl cover. Fields ploughed by cultivators and fertiliser added.

May Paddies (fields) flooded by irrigation. Ploughed again with a rotivator with a paddle wheel. Rice planted by machine (Figure 6.25) to avoid back-breaking work. Any plant out of line has to be replanted by hand.

June Heavy rains begin. Keeps paddies flooded. Herbicides sprayed. Machines remove weeds.

July Continuous weeding.

August Ripening period. Excess water drained away.

September (late) Combine cuts five rows of rice at a time (Figure 6.26). Rice is taken to silos to be dried and stored.

October Next year's rice fields (the majority) left fallow – winters are too cold for crop growth. In a rotation, some fields are planted with barley which is harvested the next June and replaced with soya beans which are ready by the end of October.

Figure 6.25
Planting rice

Figure 6.26
Harvesting rice with a combine, Hiroto can work alone

Subsistence rice farming in the Lower Ganges Valley

The River Ganges flows south-eastwards from the Himalayas (Figure 6.27). The *alluvium* (silt) which it carries from the mountains has been deposited over many centuries to form, east of New Delhi, a flat plain (Figure 6.28) and, where it enters the Bay of Bengal, a large delta (page 265). The plain and delta is one of the most densely populated parts of the world (Figure 1.1 and Question 1 page 14). Many of the people who live here are *subsistence* farmers, growing mainly rice on an intensive scale. Despite their exceptionally hard work, many farmers can only produce enough food for their own family and, perhaps, village. Rice, with a high nutritional value, can form over three-quarters of the total local diet and is a *sustainable* form of farming.

Figure 6.28 Aerial view of the Lower Ganges Valley

Figure 6.30 Rice cultivation on the floodplain of the River Ganges

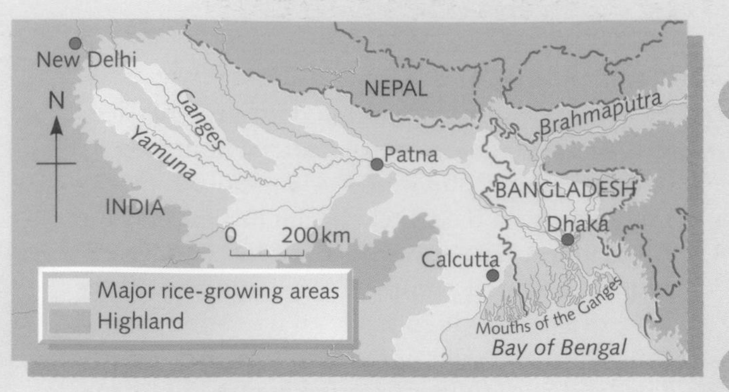

Figure 6.27
Rice-growing areas in the Lower Ganges Valley

Physical inputs

Wet padi, a variety of rice, needs a rich soil. It is grown in silt which is deposited annually by the Ganges and its tributaries during the monsoon floods. This part of India and Bangladesh has high temperatures, over 21°C, throughout the year (Figure 6.29), and the continuous growing season allows two crops to be grown annually on the same piece of land. Rice, initially grown in nurseries, is transplanted as soon as the monsoon rains flood the padi-fields. During the dry season, when there is often insufficient water for rice, either vegetables or a cereal crop is grown.

Human inputs

Rice growing is labour-intensive. Much manual effort is needed to construct the bunds (embankments) around the padi-fields, to build (where needed) irrigation canals, to prepare the fields, and to plant, weed and harvest the crop. Many farms, especially nearer the delta, are very small. They may only measure 1 hectare (the size of a football pitch) and be divided into 12 or 15 plots. The smallness of the farms and the poverty of the people means that hand-labour has to be used rather than machines (Figure 6.30). Water buffalo (oxen) provide manure and are used in preparing the padi-fields.

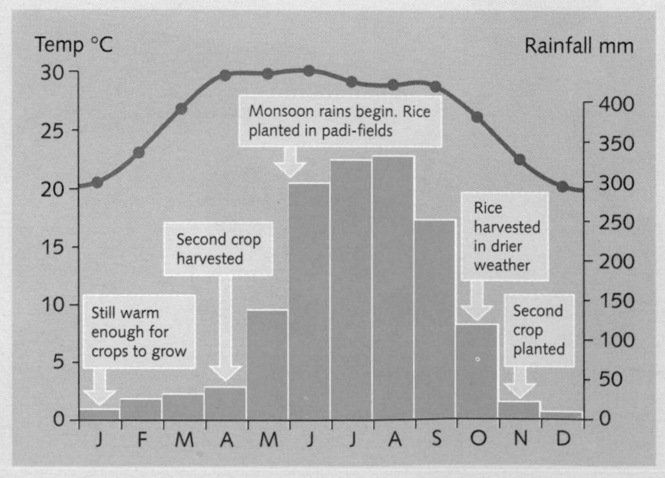

Figure 6.29 Climate graph for Calcutta

The Green Revolution

The Green Revolution, in its narrowest sense, refers to the introduction of high-yielding varieties (HYVs) of cereals to economically less developed countries. Its beginnings were in Mexico where, financed by money mainly from the USA, scientists developed new varieties (hybrids) of wheat and maize. Encouraged by this success, a research centre was set up in the Philippines which, in the early 1960s, evolved an improved strain of rice. The immediate effect was to increase rice yields sixfold at its first harvest. However, despite the improvements in food supplies in many parts of South-east Asia, the Green Revolution created a number of problems. Some of the successes and failures of the Green Revolution are summarised in Figure 6.31.

Figure 6.31 Successes and failures of the Green Revolution

Successes

- HYVs have increased food production. For example, India, which used to experience food shortages until the 1960s, became self-sufficient in cereals.
- The increase in yields led to a fall in food prices.
- Faster-growing varieties allow an extra crop to be grown each year.
- Yields are more reliable as many new varieties are more disease-resistant.
- Higher yields allow other crops, notably vegetables, to be grown, adding variety to the local diet.
- HYVs allow the production of some commercial crops.
- HYVs are not so tall as traditional varieties, enabling them to withstand wind and rain.
- Many of the more well-off farmers who could afford seed, fertiliser and tractors, have become richer.

Failures

- HYVs need large amounts of fertiliser and pesticides which increase costs, encourage the growth of weeds and can harm water supplies (page 96).
- HYVs need a more reliable and controlled supply of water. They are more vulnerable to drought and to waterlogging. Irrigation, where used, increases costs and can cause salinisation (Figure 6.37).
- HYVs are more susceptible to attacks by pests and diseases.
- Many of the poorer farmers who do not own the land they farm and cannot afford to buy seed, fertiliser and tractors, have become much poorer.
- Mechanisation has increased rural unemployment and migration to the towns.
- Farming has become less sustainable.

Farming and the environment

The use of chemicals

- *Pesticide* is defined as all chemicals applied to crops to control pests, diseases and weeds (Figure 6.32). The United Nations claimed in the 1960s that one-third of the world's crops were lost each year due to pests, diseases and weeds. Scientists have estimated that without pesticides, yields of cereal crops could be reduced by 25 per cent after one year and by 45 per cent after three years. Unfortunately, pesticides can also affect wildlife.
- *Fertiliser* is a mineral compound containing one or more of the six main elements (nutrients) needed for successful plant growth. The average soil does not contain sufficient nutrients, especially nitrogen, phosphorus and potassium, to provide either a healthy crop or a high yield. There is increasing concern where nitrate is washed (*leached*) through the soil either into:
  - rivers where, being a fertiliser, it causes rapid growth of algae and other plants, which use up oxygen (the process of eutrophication) leaving insufficient for fish life (Figure 6.33)
  - underground water supplies needed for domestic use.
- *Phosphate*, released from farm slurry (animal manure) and untreated human sewage, can also pollute water supplies (e.g. the Norfolk Broads). Slurry can be 100 times more polluting than household sewage.

Figure 6.33 Algae forming on a river during hot, sunny weather

Figure 6.32 Spraying pesticide

The removal of vegetation

The removal of the vegetation cover, whether trees or grass, can increase the risk of soil erosion by either running water or the wind (Figure 6.34). Soil erosion (page 238) is usually limited where tree crops are grown and greatest where the land has been ploughed.

Figure 6.34 Soil erosion in eastern England

Figure 6.35
The less acceptable face of modern farming – the removal of hedges to create a prairie-like landscape

The removal of hedgerows

Between 1945 and 1975, 25 per cent of hedgerows in England and Wales disappeared. Despite increased opposition by conservation groups, a further 35 per cent was lost between 1984 and 1993. Farmers began removing hedgerows and trees to create larger fields (Figure 6.35). Hedgerows are costly and time-consuming to maintain, take up space and money which could be used for crops, and limit the size of field machinery. Trees get in the way of mechanised hedge-trimmers. In contrast, Figure 6.36 shows the advantages of well-maintained hedgerows.

Loss of wildlife habitats

The intensification of agriculture in Western Europe has led to the loss of important wildlife habitats such as wetlands (page 216) and moorlands.

Irrigation

Irrigation, the artificial watering of the land, needs careful, expensive management to avoid *salinisation* (Figure 6.37). If water, which contains salts, is channelled onto the land it must also be drained away. If not, the soil can become increasingly saline. Where the water table rises, it brings the dissolved salts nearer to the surface and they affect the roots of plants. Where water remains on the surface, it will evaporate to leave a crust of salt. In both cases, as most plants are intolerant of salt and die, crops can no longer be grown. Estimates suggest that nearly half of the irrigated land in places like Pakistan and Egypt and one-third of the irrigated land in California is affected by salinisation.

Figure 6.36
The advantages of hedgerows

Figure 6.37
Salinisation

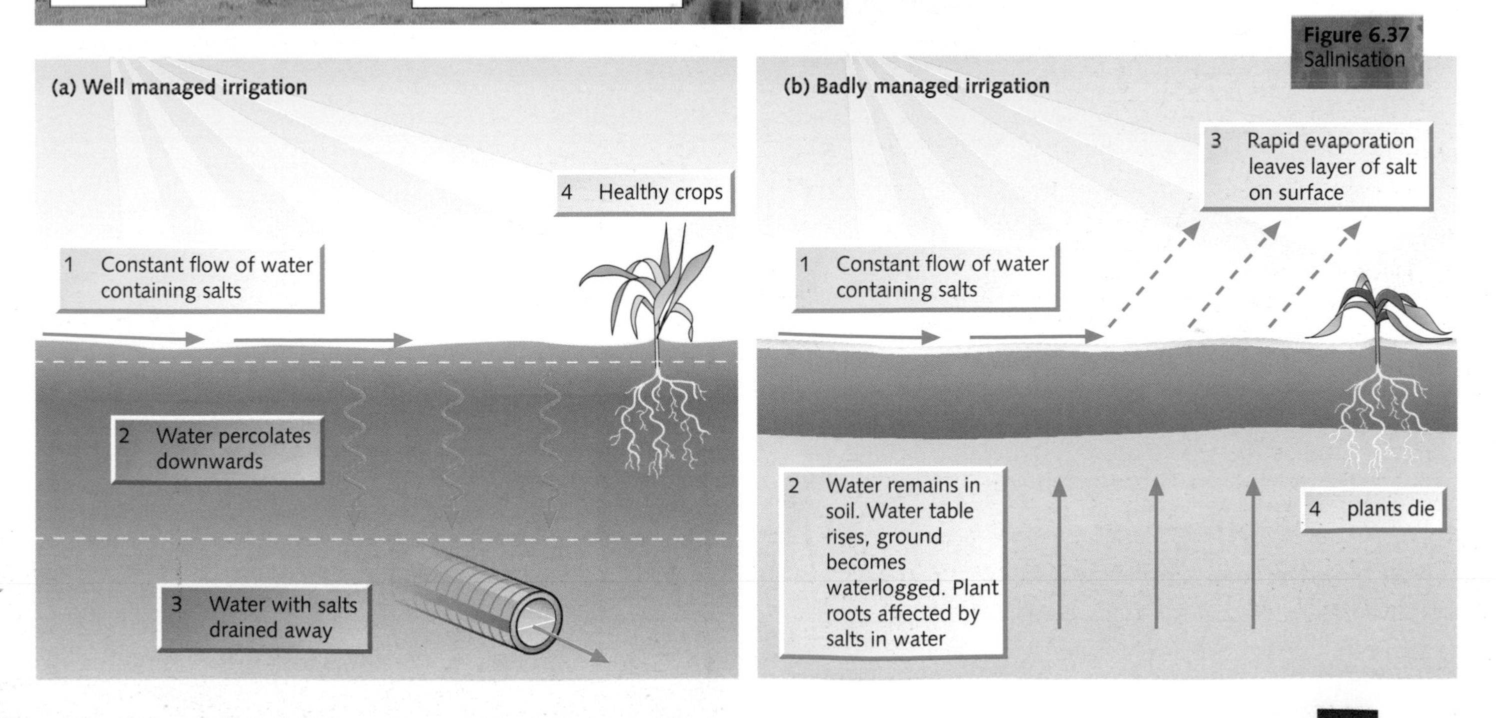

Food supply and malnutrition

The steady increase in global food production since the 1960s has led (with the exception of Africa where there has been a 10 per cent decrease) to a fall in the *proportion* of underfed people (Figure 6.38). At the same time, however, increases in world population and world poverty have meant a sharp increase in the *number* of people suffering from *malnutrition*. Malnutrition is caused by deficiencies in diet, either in amount (quantity) or type (quality). Until the 1970s, it was believed that malnutrition resulted from the population growing more rapidly than food supplies. Today it is attributed to poverty, as large numbers of the world's population are unable to afford to buy an adequate diet. While malnutrition results in starvation only under extreme conditions, it does reduce people's capacity to work and their resistance to disease. In children, it can retard mental and physical development, and cause illness.

Dietary energy supply (DES)

DES is the number of calories per capita (i.e. per person) available each day in a country. Like GNP (page 170) it does not take into account differences between individuals or between areas within a country. Between 1970 and 1990 there was an increase in available food supplies per capita in every developing region except sub-Saharan Africa (Figure 6.39). It has been estimated that in most developing countries, especially those within the tropics, a person consuming less than 2350 calories per day is likely to experience chronic malnutrition (Figure 6.40). In 1990, 20 per cent of people living in these countries were suffering from chronic malnutrition. Their numbers have increased from 435 million in 1975 to 600 million in 1990. This increase is mainly due to human factors (e.g. civil wars, political instability and international debt) rather than to physical causes (e.g. natural disaster such as drought).

Figure 6.40 shows that people living in developed countries need more calories per day than those in developing countries. This is partly because there is a greater proportion of adults in developed countries (children have smaller needs) and partly because most developed countries are in cooler latitudes (where more energy is needed for body heating).

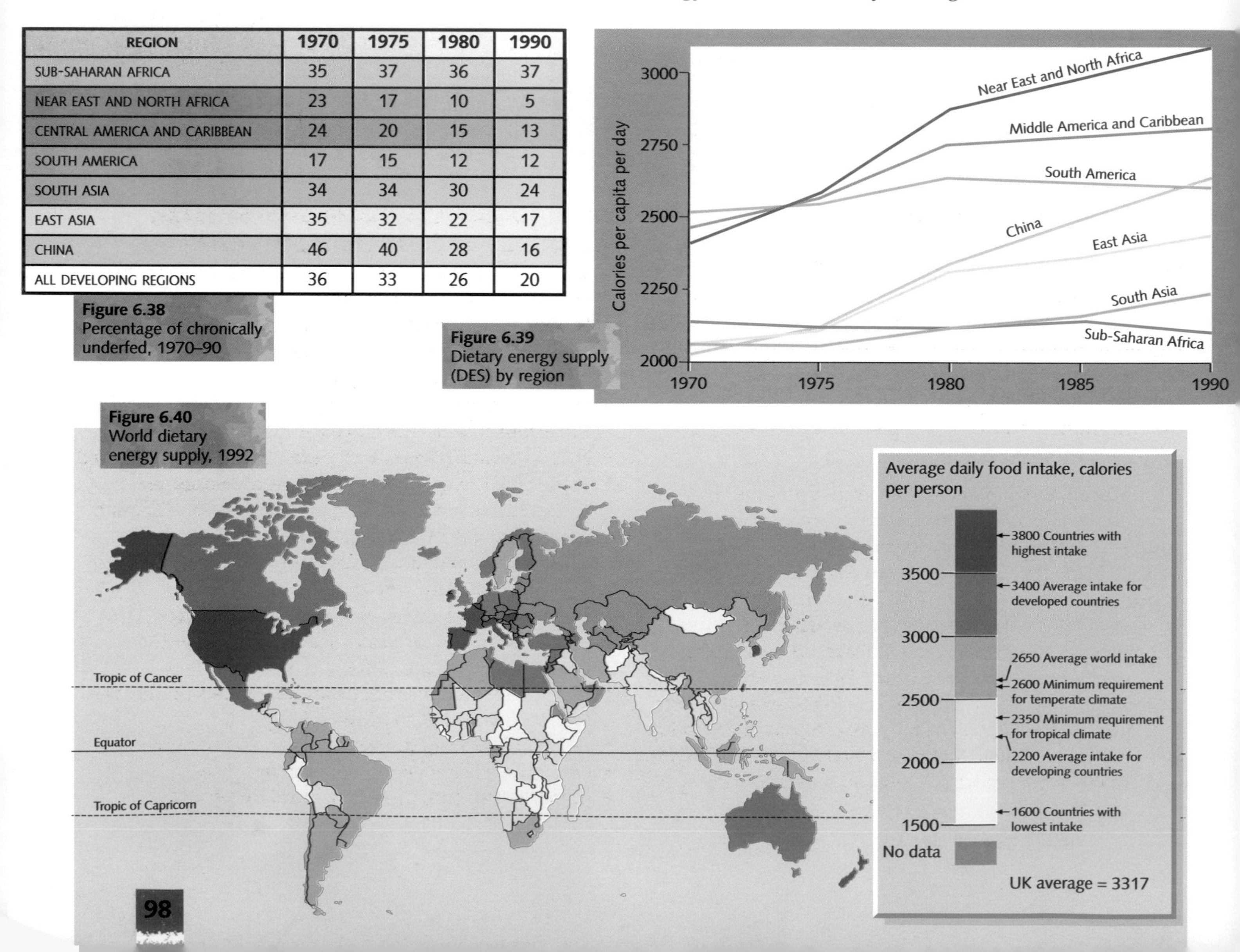

REGION	1970	1975	1980	1990
SUB-SAHARAN AFRICA	35	37	36	37
NEAR EAST AND NORTH AFRICA	23	17	10	5
CENTRAL AMERICA AND CARIBBEAN	24	20	15	13
SOUTH AMERICA	17	15	12	12
SOUTH ASIA	34	34	30	24
EAST ASIA	35	32	22	17
CHINA	46	40	28	16
ALL DEVELOPING REGIONS	36	33	26	20

Figure 6.38
Percentage of chronically underfed, 1970–90

Figure 6.39
Dietary energy supply (DES) by region

Figure 6.40
World dietary energy supply, 1992

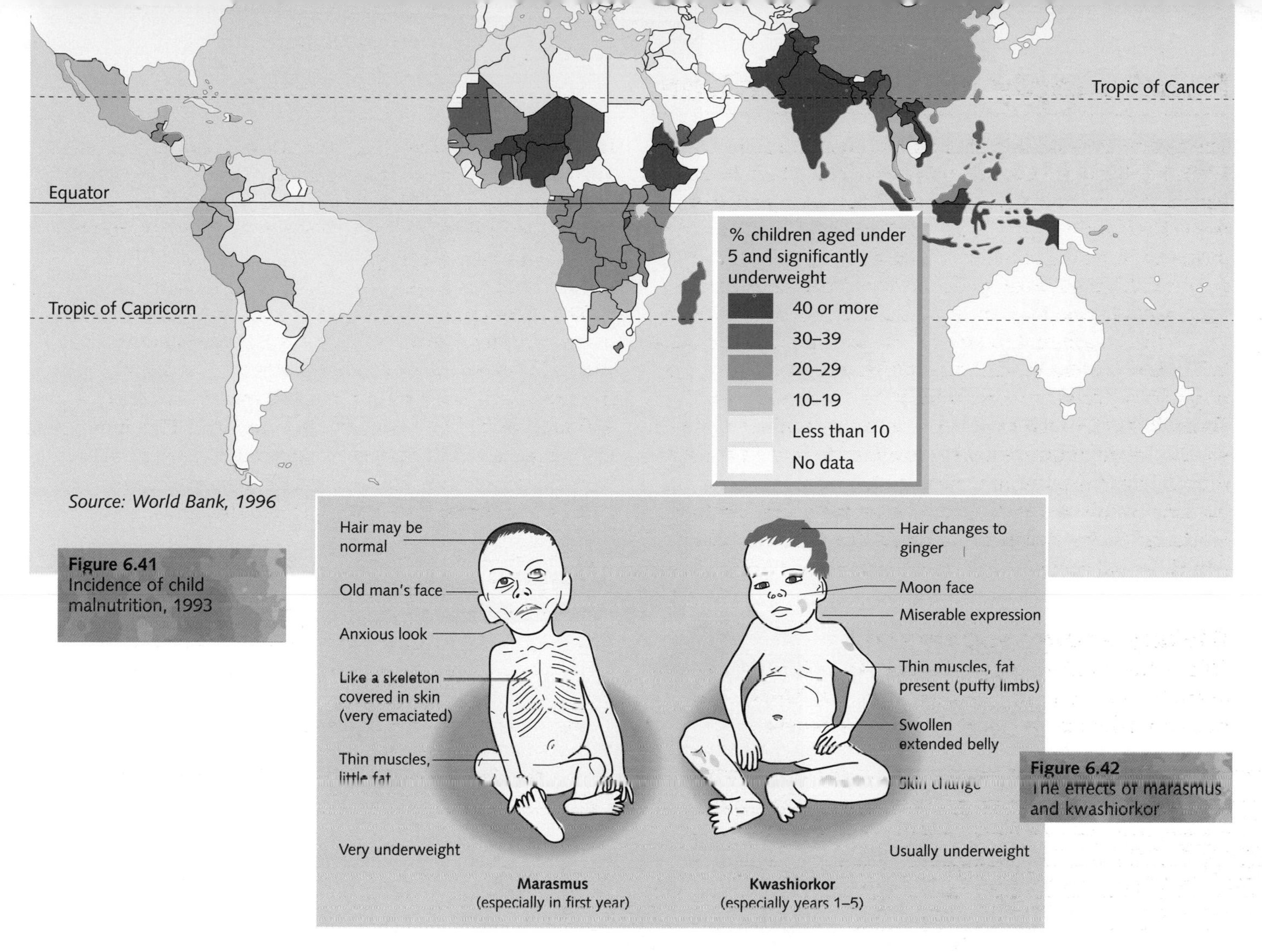

Figure 6.41 Incidence of child malnutrition, 1993

Figure 6.42 The effects of marasmus and kwashiorkor

Malnutrition in children

Children under the age of 5 are particularly susceptible to malnutrition. In 1990, 35 per cent of children in this age-group in the developing world were considered to be underweight (Figure 6.41). The percentage is, surprisingly, highest in South-east Asia where, despite improvements in food supply per capita since 1970 (Figure 6.38), 78 per cent of the developing world's underweight children live. In contrast, Africa has only 15 per cent of the total underweight children.

Children fall ill either because their diet contains too few proteins, which are particularly important during early stages of growth, or too few calories. The two major protein deficiency diseases are *marasmus* and *kwashiorkor* (Figure 6.42). Marasmus is most common in children in their first year of life. Kwashiorkor results from a predominance of cereals (e.g. rice) and a deficiency of protein (e.g. milk, eggs and meat). Two of several diseases resulting from a lack of vitamins are beri-beri and rickets. Beri-beri, due to a lack of vitamin B, can lead to a wasting and paralysis of limbs. Rickets, caused by a deficiency in vitamin D, causes deformities in bones, legs and the spine.

Why do many parts of sub-Saharan Africa suffer from malnutrition?

- The high birth rate and falling death rate means there are many more people to be fed.
- Few farmers have the money to buy high-yielding seeds, fertiliser, pesticides or machinery, or to implement irrigation schemes.
- When food is scarce, neither governments nor people can afford to buy high-priced surplus from overseas.
- During colonial times, European companies established commercial crops for their own profit instead of encouraging subsistence crops for local use. Although now independent, some governments still give tax concessions to overseas transnationals, allowing them to continue to grow these crops.
- The soil has been overused in the past and few nutrients remain. In places soil erosion has led to desertification (page 241).
- Many areas, receive small and unreliable amounts of rainfall (pages 208 and 242).
- Pests and diseases destroy crops and stored grain.
- Often there is not enough protein in the diet.
- In many countries there is political instability.

CASE STUDY 6

Farming in Brazil 2000

Figure 6.43
A clearing in the Amazon rainforest

Shifting cultivation

Shifting cultivation is a form of subsistence farming (page 101), and is a traditional form of agriculture found in many areas of the tropical rainforest. It tends now to be found in only the most inaccessible and least 'exploited' areas. The Amerindians use stone axes and machetes to fell about one hectare of forest. Any undergrowth has to be cleared immediately to prevent it growing rapidly in the hot, wet climate. After a time the felled trees, having been given time to dry, are burned. This burning helps to provide nutrients for the soil as the ash is spread over the ground as a fertiliser. This is also known as 'slash and burn'.

Within the clearing is built the tribal home or *maloca* (Figures 6.43 and 6.44). This usually consists of tree trunks lashed together with lianas, and thatched with leaves. Nearer the river the houses are built on stilts, as the water level can rise by 15 metres after the rainy season.

Figure 6.44
A *maloca*, or Indian house, built from materials of the tropical rainforests

The clearings, or gardens, are called *chagras*. Here the women grow virtually all of the tribe's carbohydrate needs. The main crop is manioc, the 'bread of the tropics', which is crushed to produce a flour called *cassava* (Figure 6.45). It can also provide sugar and a local beer. Other crops are yams (though these need a richer soil), beans and pumpkins. The men supplement this diet by hunting, mainly for tapirs and monkeys, fishing and collecting fruit. The blowpipe and bow and arrow are still used.

Figure 6.45
Cassava flour being prepared from manioc

Unfortunately the balance between plants and soil is very delicate. Once the canopy of trees has been removed, the heavy rains associated with afternoon storms can hit the bare soil. This not only causes soil erosion, but it leaches any minerals in the soil downwards. As the source of humus – the trees – has been

Figure 6.47
Raimondo Jose cutting open Brazil nuts

Figure 6.46
Raimondo Jose's clearing

removed and as there is a lack of fertiliser and animal manure, the soil rapidly loses its fertility (Figure 12.16). Within four or five years yields decline, and the tribe will 'shift' to another part of the forest to begin the cycle all over again.

Shifting cultivation needs a high labour input and large areas of land to provide enough food for a few people. Although it is a wasteful method of farming, it causes less harm to the environment than permanent agriculture would.

Recently the Amerindians of the Amazon rainforest have been forced to move further into the forest or to live on reservations. Large numbers have died, mainly because they lack immunity to 'Western' diseases, and those who survive have the difficult choice of either trying to live in increasingly difficult surroundings, or joining the other homeless and jobless in favelas in the larger urban areas.

Subsistence farming

Few of Brazil's farmers own the land on which they work, as most farms in the country are large in size and run by large organisations. In an attempt to open up the Amazon Basin to settlers, the Brazilian government has constructed 12 000 km of roads through the tropical rainforest since the 1960s (Figure 12.28). Alongside these roads are areas where land has been given to former landless farmers. One such farmer is Raimundo Jose. He moved to the state of Para, near the mouth of the Amazon, from the dry north-east of Brazil in 1987. He cleared just enough land for a small family house which he built out of wood and thatch (Figure 6.46). Although he is a subsistence farmer, his method is sustainable as he relies on collecting two tree crops – carob fruit and Brazil nuts (Figure 6.47). To him, unlike the big developers who clear large tracts of forest (page 222), the trees are essential for his survival. Although he earns very little, he is satisfied with his new quality of life.

Although other small farms were created in the region, many have failed mainly due to a lack of inputs into the system. The farmers have very little capital, no machinery, no fertiliser and limited access to a market. The soil, which lacks nutrients, is quickly leached (page 237) even if the protective tree cover is only partially, and not completely, removed.

Commercial farming – plantations

Plantations were developed in tropical parts of the world in the eighteenth and nineteenth centuries mainly by European and North American merchants. The natural forest was cleared and a single crop (usually a bush or tree) was planted in rows (Figure 6.48). This so-called 'cash crop' was grown for export, and was not used or consumed locally. Plantations needed a high capital investment to clear, drain and irrigate the land, to build estate roads, schools and hospitals, and to bridge the several years before the first crop could be harvested. Much manual labour was also needed. The managers were usually European while labourers were either recruited locally or brought in from other countries. Because they were willing to accept lower wages, this workforce secured a greater profit for the recruiting companies. The almost continuous growing season meant that the crop could be harvested virtually throughout the year. Today most plantations are still owned by transnational companies (page 132), with their headquarters in a more economically developed country.

Figure 6.50 Ripening coffee cherries

Figure 6.48 A coffee fazenda

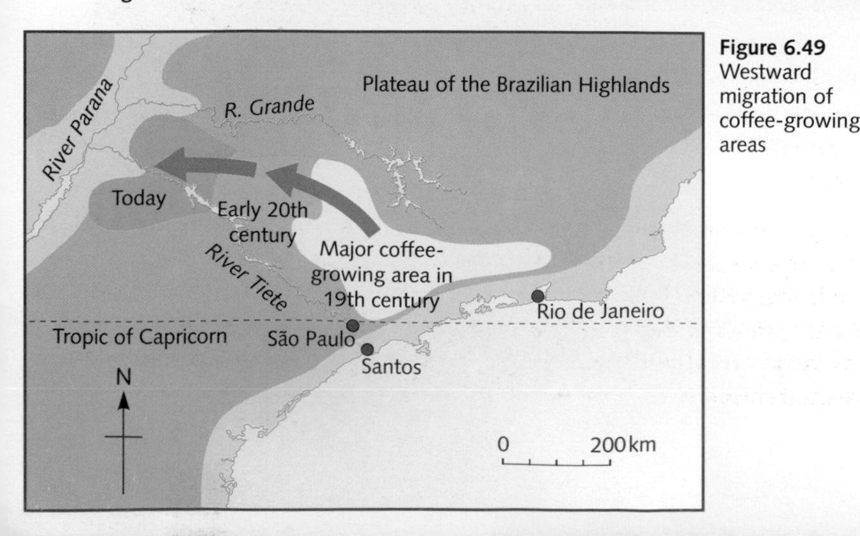

Figure 6.49 Westward migration of coffee-growing areas

Coffee plantations (fazendas)

Ideal conditions include:

- Gently rolling ground or valley sides at altitudes up to 1700 metres (Figure 6.48). Valleys which may become waterlogged or act as frost hollows are unfavourable (frost is coffee's worst enemy).
- A deep red soil called *terra rossa*.

The major producing states are Parana, São Paulo and Minas Gerais (Figure 6.49). The tree begins to yield after three years, reaches a maximum between 10 and 15 years and dies after 40. When harvested, the red cherries (Figure 6.50), as the ripe coffee is called, are stripped from the branches and cut into halves to expose two green 'beans' which are left out in the sun in huge drying yards. They are raked frequently, and large tarpaulins are kept nearby for protection against any rain.

Changes in coffee production

European (mainly Italian) immigrants in the 1870s developed new coffee fazendas. This led to an increase in coffee production. By 1906 Brazil produced 1.2 million tonnes of coffee when world demand was only 0.7 million, and coffee accounted for 70 per cent of Brazil's exports – two major problems! Meanwhile soils on the early plantations had become

Figure 6.51
The São João estate

Figure 6.52
The São João mill

exhausted and so new fazendas were developed westward in drier but less frosty areas. More immigrants in the 1960s again led to overproduction, but output was soon drastically cut by:

- the government offering incentives for other crops to be grown
- the spread of disease in coffee trees, and
- several killing frosts.

While Brazil had produced 43 million bags of coffee and 49 per cent of the world's total in 1960, it only produced 20 million bags and 21 per cent of the world's total in 1995.

Sugar cane – the São João estate

Sugar cane is now Brazil's most important crop and Brazil, with 5 per cent of the total, has become the world's leading producer. It is grown on large estates, such as São João, in the rich farming state of São Paulo.

The São João estate covers 50 000 hectares (50 km^2) and is 300 times larger than an average farm in the UK (Figure 6.51). In the centre of the estate is the sugar mill and other buildings (Figure 6.52). The estate is almost a self-contained community with a church, clinic, free housing for some of the workers and, being Brazil, a football pitch. The estate employs 2000 people.

Sugar cane needs plenty of moisture, high temperatures and a fertile, well-drained soil. Until fairly recently, most of the cane was cut by hand and the labourers were guaranteed a full-time job. Now two-thirds of the cane is harvested by machine with over half of the cutters on a day-to-day contract (Figure 6.53). The cutters work eight hours a day and six days a week during the eight months of harvest (for the rest of the year they are likely to be unemployed as there is little alternative work in the area). Wages depend upon the amount of cane cut. Many workers commute by estate bus from nearby towns (outward commuting), in some instances 60 km a day, leaving home at 5 o'clock in the morning and returning at 7 o'clock at night. Others are seasonal workers who travel 2000 km from the north-east of Brazil.

During harvest time many trucks, each carrying 60 tonnes of cane, arrive at the mill each day. The mill, which works 24 hours a day at this time of year, produces 4 million tonnes of cane a year. Of this, 40 per cent is turned into sugar (half of which is exported) and 60 per cent into alcohol which is used instead of petrol by most Brazilian cars. The waste from the cane is either returned to the fields as a fertiliser or used in the mill's power station to produce electricity for the estate and nearby towns.

Figure 6.53
Harvesting sugar cane

QUESTIONS

1

(Page 84)

Study these two farming systems. One is a subsistence farm in India, the other is a dairy farm in south-west England.

a i) What is the size of Farm A? (1)

ii) How many animals are kept on Farm A? (1)

iii) Which two places might use the outputs from Farm A? (2)

b The boxes labelled 'Inputs' and 'Outputs' for Farm A have been left empty. From the following list, fill in the two boxes to show four likely inputs and four likely outputs: barley, fertiliser, hay, labour, machinery, fattened pigs, manure, milk. (8)

c Explain the differences between the two farms under the following headings: Land, Labour, Machinery, Animals, Crops. (5)

d Is Farm A or Farm B the one found in India? (1)

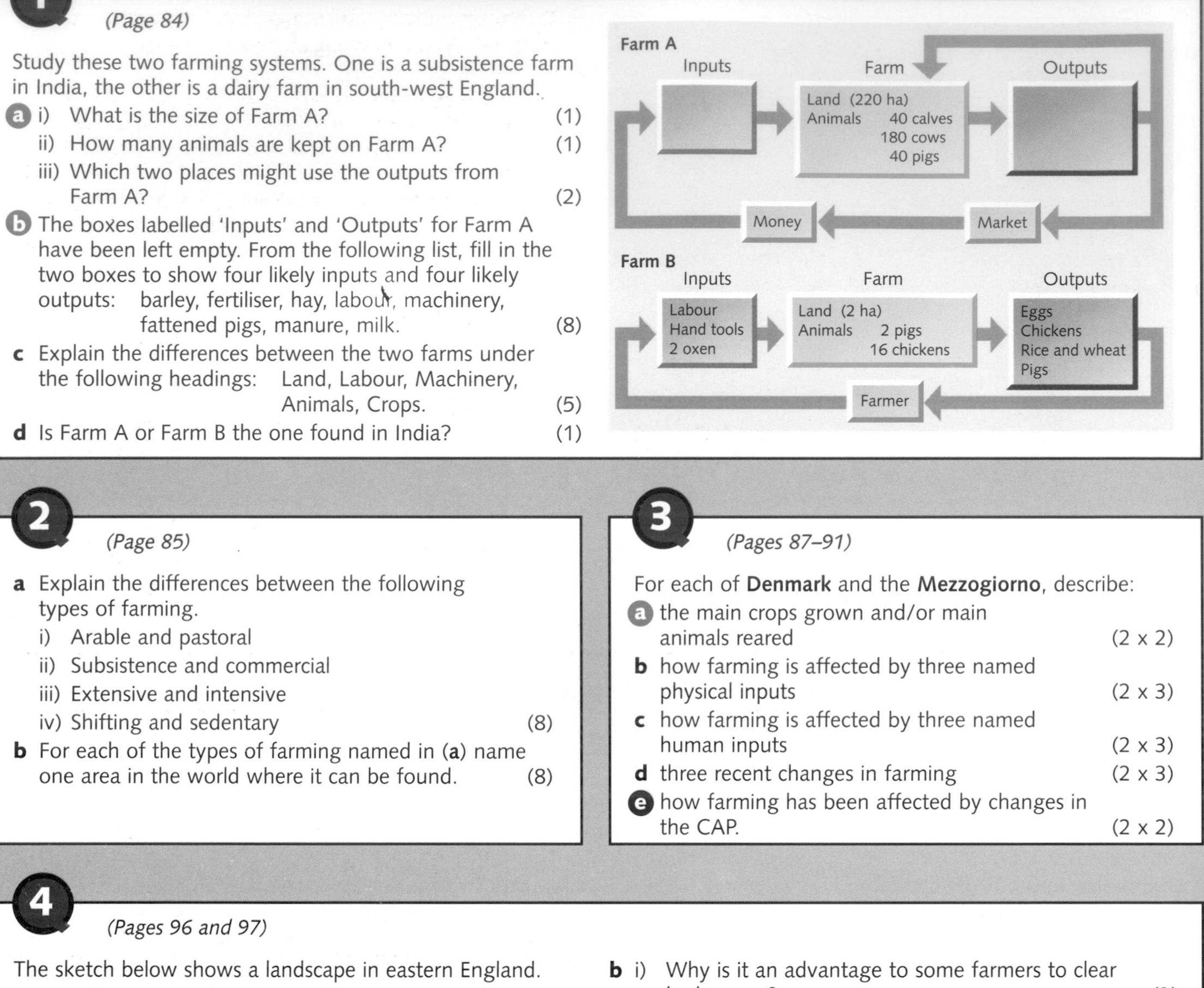

2

(Page 85)

a Explain the differences between the following types of farming.

i) Arable and pastoral

ii) Subsistence and commercial

iii) Extensive and intensive

iv) Shifting and sedentary (8)

b For each of the types of farming named in (**a**) name one area in the world where it can be found. (8)

3

(Pages 87–91)

For each of **Denmark** and the **Mezzogiorno**, describe:

a the main crops grown and/or main animals reared (2 x 2)

b how farming is affected by three named physical inputs (2 x 3)

c how farming is affected by three named human inputs (2 x 3)

d three recent changes in farming (2 x 3)

e how farming has been affected by changes in the CAP. (2 x 2)

4

(Pages 96 and 97)

The sketch below shows a landscape in eastern England.

a i) What is one advantage and one disadvantage to the farmer who uses nitrate fertiliser? (2)

ii) Describe how nitrate fertiliser may reach rivers. (1)

iii) How does an increase in nitrate in rivers affect:

- fish
- humans? (2)

b i) Why is it an advantage to some farmers to clear hedgerows? (3)

ii) Give three ways by which this removal can harm the environment. (3)

c Describe two other aspects of modern farming which it is claimed might harm the environment. (2)

1 Crops use up nutrients in the soil

2 Farmer adds nitrate fertiliser to increase crop yields. It is expensive to buy

3 Rain washes out (leaches) nitrates → Some enters rivers

4 Cattle: their manure becomes high-nitrogen slurry, which runs into rivers. Slurry is 200 times more polluting than domestic sewage

5 Nitrates in rivers encourage growth of algae and larger plants. These use up oxygen

6 Fish die due to lack of oxygen

7 Plants limit light getting to river bed

Water used for domestic supply affects human health

5

(Pages 92–95)

The photo below was taken at harvest time from a train as it passed through the southern Japanese island of Kyushu.

a i) Describe the size and shape of the fields. (2)
ii) What crop is growing in the fields? (1)
iii) How can you tell from the photo that it was harvest time? (1)
iv) Describe two of the buildings in the photo. (2)
v) For what do you think they are used? (2)

b Ohgata-mura is towards the north of Japan. Describe the differences between Ohgata-mura and Kyushu using the following headings:
- Relief of the land
- Farm size
- Field size
- Machinery used
- Summer crops
- Winter crops (6)

c Describe the farmer's year in Ohgata-mura. (3)

d Describe farming in the Lower Ganges Valley under the headings:
- Physical inputs
- Human inputs. (2 x 3)

e State, briefly, the meaning of the term 'Green Revolution'. (1)

f The two farmers shown here lived in the same village in India in the 1960s. Describe why, in the next few years, Farmer A became very wealthy while Farmer B became even poorer. (4)

g Write a report to show the advantages and disadvantages of the Green Revolution. What is your overall conclusion? (8)

Farmer A

I'm quite wealthy. I have enough rice to feed my family and sell some as well. I've also got some savings.

Farmer B

I'm very poor, have no savings and have only enough food to feed my family.

6

(Pages 98 and 99)

a What is *malnutrition* and what effect can it have on children? (2)

b i) What is meant by the term *dietary energy intake* (*DES*)? (1)
ii) How many calories does an average person need each day to avoid chronic malnutrition? (1)
iii) Which continent suffers most from chronic malnutrition? (1)
iv) Which region has most underweight children aged under 5? (1)

c Describe two child illnesses caused by malnutrition. (6)

d Give six reasons why people in several parts of sub-Saharan Africa suffer from malnutrition. (6)

7

(Pages 100–103)

Several tribes in the Amazon rainforest still practise shifting cultivation.

a Copy and complete the flow chart by putting the correct letter in the appropriate box.
A – Harvest B – Weed C – Spread ash as fertiliser D – Move to a new area
E – Sow crops including manioc F – Burn remaining vegetation G – Select a suitable area
H – Fell trees with stone axes I – Soil soon loses fertility (7)

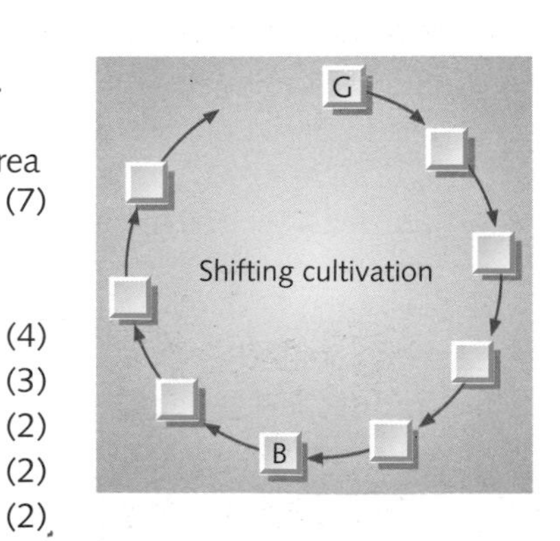

b Describe how clearing the forest may cause:
i) the soil to lose its fertility
ii) the land to suffer from soil erosion. (4)

c Describe the method of farming practised by Raimundo Jose. (3)

d What is a plantation? (2)

e i) What are the ideal growing conditions for coffee? (2)
ii) How is coffee grown and harvested? (2)

f Describe farming on the São João sugar cane estate using the following headings:
- Size of estate
- Buildings on the estate
- Growing conditions
- Harvesting
- Uses of sugar cane
- Workforce (labour). (12)

7 ENERGY RESOURCES

What are resources?

Resources can be defined as features of the environment which are needed and used by people. The term usually refers to *natural resources* which occur in the air, in water or on the land. These resources include raw materials (minerals and fuels), climate, vegetation and soils. Sometimes the term is widened (though not in this chapter) to include *human resources* such as labour, skills, machinery and capital (Figure 7.1).

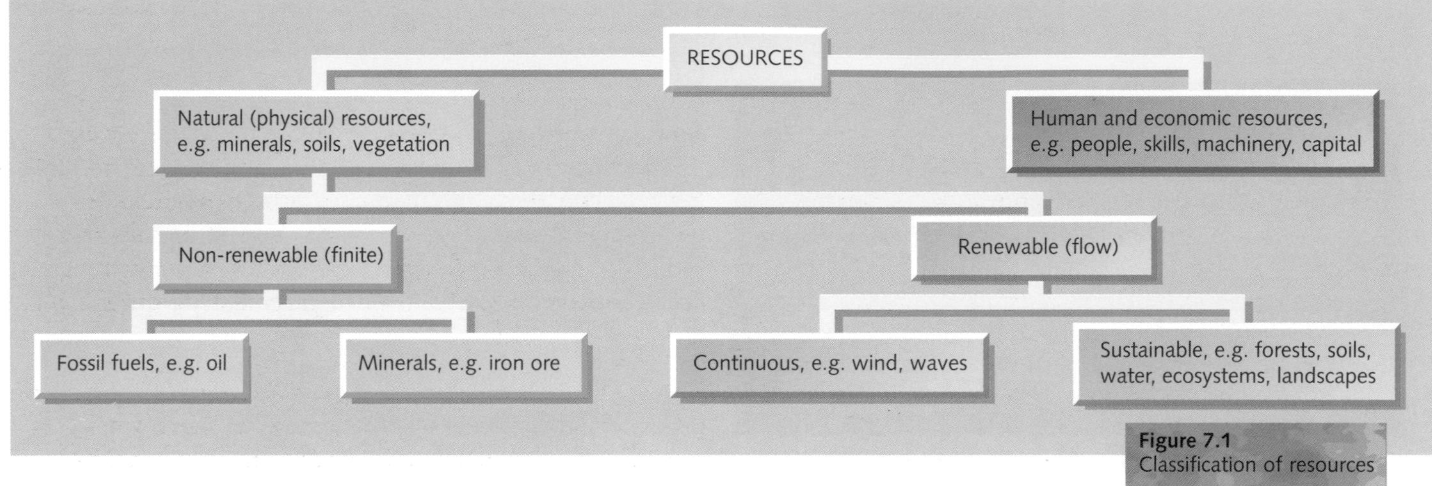

Figure 7.1
Classification of resources

Natural resources are commonly subdivided into two groups:

1. *Non-renewable resources* are said to be finite or non-sustainable as their exploitation and use will eventually lead to their exhaustion, e.g. fossil fuels and minerals.
2. *Renewable resources* can either be:
   - *a flow of nature* so that, being continuous, they can be used over and over again, e.g. solar and geothermal energy, wind and water power
   - *sustainable*, which means they are renewable and self-generating if left to nature, e.g. clean water, trees, fish, wildlife, soils, ecosystems and landscapes. However, if these sustainable resources are used carelessly or are over-used by people, then either:
     - their value may be reduced (i.e. degraded), as when soils lose their fertility and are eroded, water supplies are polluted or trees die due to acid rain, or
     - their existence is threatened, as with overfishing, deforestation or the draining of wetlands.

Managing resources

The demand for, and use of, the world's resources continues to grow at an increasingly rapid rate. This is mainly due to:

- *economic development* as more countries try to develop industrially and economically
- *population growth* as the world's population continues to increase.

The combined effects of economic development and population growth means that there is a growing need to manage the Earth's resources (Figure 7.2). Management might best be achieved through a range of approaches. These include conservation (wildlife, scenery), recycling (glass, waste paper), greater efficiency in existing resource use (home insulation), developing renewable resources (wind, waves and tidal power), controlling pollution (reducing emissions from vehicles and power stations) and adopting forms of sustainable development (page 175).

Figure 7.2
This . . .

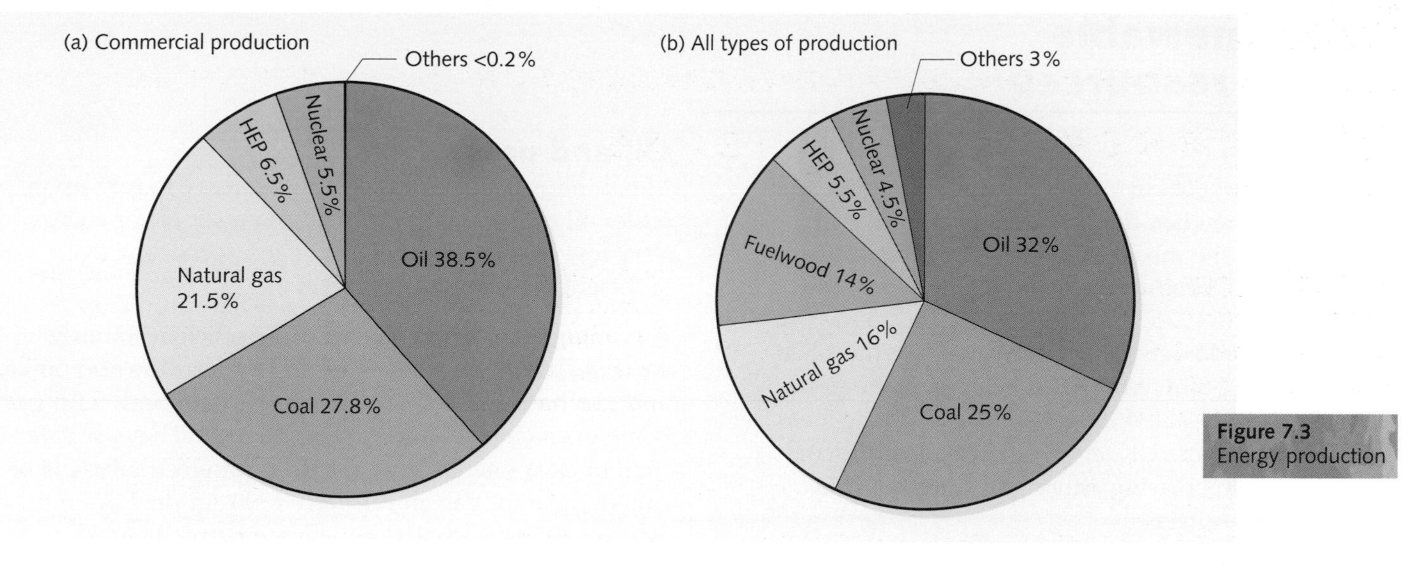

Figure 7.3 Energy production

Energy resources

The sun is the primary source of the Earth's energy. Without energy, nothing can live and no work can be done. Green plants convert energy into a form that can be used by people.

Non-renewable resources

In the mid-1990s, coal, oil and natural gas accounted for 87.8 per cent of the world's commercially produced energy (Figure 7.3a). These are forms of stored solar energy produced by photosynthesis in plants over thousands of years. As these three types of energy, referred to as *fossil fuels*, take so long to form and be replaced, they are regarded as non-renewable. Each year the world (in reality mainly the developed countries), consumes an amount of fossil fuel that took nature some 1 million years to provide – a rate far in excess of their replacement. Fossil fuels have, in the past, been relatively easy to obtain and cheap to use, but they have become major polluters of the environment.

Two other non-renewable sources of energy should be noted. Nuclear energy uses uranium and so is not a fossil fuel. Fuelwood is a non-commercial source of energy. If included with the other forms of energy, it provides an estimated 14 per cent of the world's and 35 per cent of developing countries' energy requirements (Figure 7.3b).

Renewable (alternative) resources

Renewable resources of energy, which are mainly forces of nature which can be used over and over again, are considered to be sustainable. At present only running water (hydro-electricity) is a significant source of renewable energy on a global scale (Figure 7.3a). Other sources, often more important on a local scale, are the sun (solar), the wind, vegetation waste (biomass) and heat from the Earth (geothermal). As yet, economic and technical problems tend to restrict the conversion of these sources of energy on a large scale. However, in time the world is likely to have to look to these and other sources of energy (e.g. waves and tides) as the supply of fossil fuels becomes exhausted.

. . . or this?

Non-renewable energy resources

Coal

The fortunes of coal in Western Europe have declined since the 1960s when the cheaper price of oil and the exhaustion of the most easily obtainable coal meant a decline in both production and the workforce. Rising oil prices, after the 1974 Middle East War, led to the exploitation of lower-cost coal resources in the USA, Australia and South Africa. The recessions of the early 1980s and 1990s sealed the fate of most mining communities in the UK and Belgium. Globally, coal production is increasing although little is used by developing countries.

Advantages Reserves are likely to last for over 300 years. Improved technology has increased the output per worker, allowed deeper mining with fewer workers, and made conversion to electricity more efficient. Coal is used for electricity, heating and making coke.

Disadvantages The most easily accessible deposits have been used up and production costs have risen. With increased competition from other types of energy, it means that whereas in 1980, 70 per cent of the UK's electricity was generated using coal, by 1998 the figure is expected to be less than 50 per cent. The burning of coal causes air pollution and, by releasing carbon dioxide, contributes to global warming (page 204). Deep mining can be dangerous while opencast mining temporarily harms the environment (Figure 7.4). Coal is heavy and bulky to transport.

Figure 7.4
Opencast coal mining

Oil and natural gas

Many industrialised countries have come to rely upon either oil or natural gas as their main source of energy. Very few of them, the UK being an exception, have sufficient reserves of their own.

Advantages Oil and gas are more efficient to burn, easier to transport and distribute (by pipeline and tanker) and less harmful to the environment than coal, with gas being even cheaper and cleaner than oil. They are safer than nuclear energy. Both are used for electricity (the so-called 'dash for gas' in the early 1990s in the UK) and heating. Oil is the basis of the huge petrochemical industry.

Disadvantages Reserves may only last another 50–70 years. New fields are increasingly difficult to discover and exploit (Figure 7.5). Terminals and refineries take up much space and there is the danger of spillage (oil), leaks (gas), explosions and fire. The burning of gas and oil releases, respectively, nitrogen oxide and sulphur dioxide which contribute to acid rain (page 202). Both oil and gas are subject to sudden international price changes and are vulnerable to political, economic and military pressures.

Figure 7.5
A marine oil platform

For

- Several independent experts predict that without nuclear power, Britain will face an energy gap by the year 2000, and that this will mean fewer jobs and a lower standard of living. Demand for electricity is predicted to increase by 30 per cent between 1990 and 2000.
- Only very limited raw materials are needed, e.g. 50 tonnes of uranium per year compared with 540 tonnes of coal per hour needed for coal-fired stations.
- Oil and natural gas could be exhausted by the year 2030. Coal is difficult to obtain and dirty to use.
- Numerous safeguards make the risks of any accident minimal.
- Nuclear waste is limited and can be stored underground.
- Nearly all the money spent in Britain on energy research has been on nuclear power.
- Nuclear energy schemes have the support of large firms and government departments.
- Nuclear power is believed to contribute less than conventional fuels to the greenhouse effect and acid rain.

Against

- It is not clear how safe it is. So far there have been no serious accidents in Britain, as, for example, at Chernobyl in the Ukraine, but there have been several leaks at Sellafield.
- There is a large conservationist lobby which claims that one accident may kill many, and ruin an area of ground for hundreds of years. The Irish Sea is increasingly contaminated.
- Many people think that Britain should concentrate on using renewable forms of energy rather than those that are non-renewable.
- Nuclear power cannot be used for two of industry's major demands, heating and transport, as costs are too high.
- There is less demand for energy by industry as declining industries (such as steel) used more energy than those that are replacing them (such as micro-electronics).
- Potential health risks. The high incidence of leukaemia around Sellafield and Dounreay has been linked to the proximity of the power station.
- Nuclear waste can remain radioactive for many years. There are problems with reprocessing and then storing nuclear waste.
- The cost of decommissioning old power stations, the first of which closed in 1989 at Berkely, is extremely high.

Figure 7.6
The nuclear energy debate

Nuclear energy

Few debates have generated such strong emotions as that on the merits of nuclear power (Figure 7.6). Several older industrialised countries (e.g. France and Belgium) and many newly industrialised countries (e.g. Japan, South Korea and Taiwan) which lack sufficient fossil fuels of their own are turning increasingly to nuclear power, despite grave fears over its safety (Figure 7.7).

Figure 7.7
Nuclear reactors at Wakasa Bay, Japan

Fuelwood

In Africa, trees have been called the 'staff of life' due to their vital role in preserving the environment and in providing rural communities with their basic needs (shelter, food, fuel and shade). Yet here, as in many rural areas in other developing regions, trees are being removed at an ever faster rate. Collecting fuelwood is a time-consuming job for women and children (Figure 7.8). Each day they may have to walk many kilometres to find enough wood upon which to cook their meals (women also have to do the farming and look after large families).

As Africa's population grows, and with it the demand for wood, then a cycle of environmental deprivation is created (Figure 7.9).

Figure 7.8
Collecting wood for fuel, Kenya

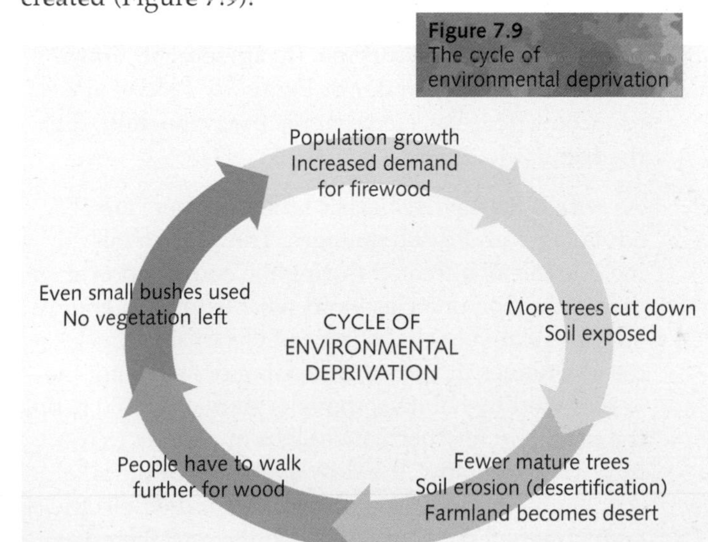

Figure 7.9
The cycle of environmental deprivation

Renewable energy resources

Hydro-electric power (HEP)

Hydro-electricity generates the highest proportion of the renewable types of energy. Although accounting for only 6.5 per cent of the world's total commercial energy (Figure 7.3a), in many countries it accounts for over 80 per cent, e.g. Paraguay 100 per cent, Norway 96 per cent and Brazil 86 per cent. It is, as the examples suggest, important to both developed and developing countries – providing they have a constant supply of fast-flowing water. Figure 7.10 shows some of the factors ideally needed for the location of a hydro-electric power station. Hydro-electricity can be generated at a natural waterfall (e.g. Niagara Falls), by building a dam across a valley (e.g. Aswan on the Nile (page 276) and at Itaipù – Figure 7.12) or where water flows rapidly down a hillside (e.g. Norway).

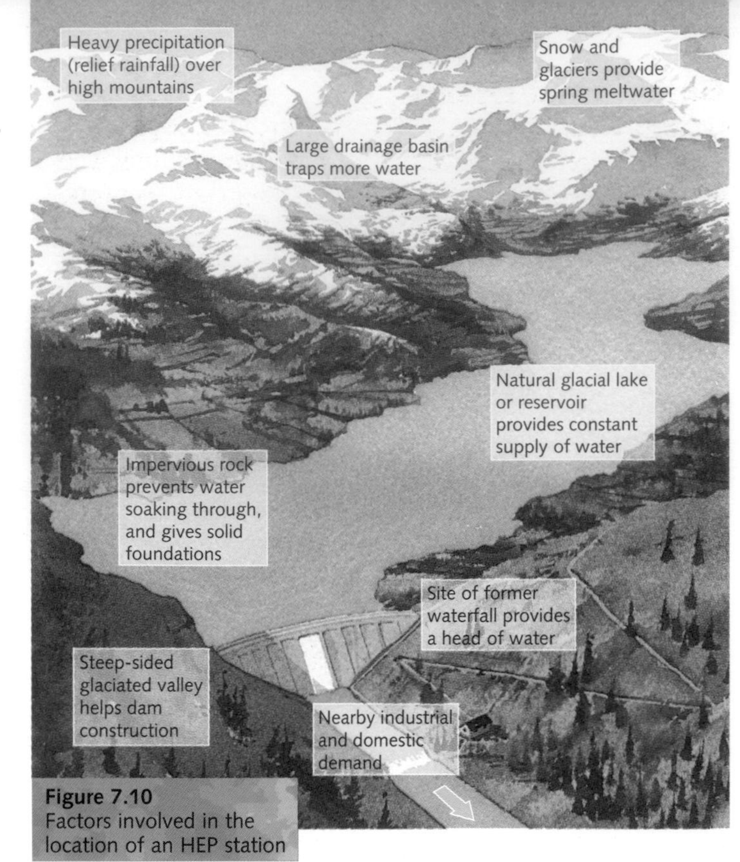

Figure 7.10 Factors involved in the location of an HEP station

Advantages Hydro-electric power is renewable, is often produced in highland areas where the population is sparse, is a relatively cheap form of electricity and creates only limited pollution. Dams, where built to store water, reduce the risks of flooding and water shortages.

Disadvantages Dams are very expensive to build, large areas of farmland and wildlife habitats may have to be flooded forcing people and animals to move, unsightly pylons can cause visual pollution, and there is always the possibility of the dam collapsing. Silt, previously spread over farmland, will be deposited in the lake. Quite recently (1966) it has been shown that if an area is flooded, the decaying vegetation can release methane and carbon dioxide – two greenhouse gases.

Itaipù, Brazil

In 1982 a dam was completed across the River Parana at Itaipù, on the border of Brazil and Paraguay (Figure 7.11). The lake behind the dam is 180 km long and 5 km wide (Figure 7.12). The 18 turbines fitted at Itaipù make it the world's largest HEP scheme. Although the venture was a joint one between Paraguay and Brazil, Paraguay receives its annual needs from just one turbine. By agreement, Brazil purchases the remainder of Paraguay's share at a reduced price. The electricity is then transmitted to the São Paulo area.

As with other hydro-electric schemes, there are advantages and disadvantages. The advantages include the jobs created during the construction stage, the production of a clean and reliable form of energy, and the relatively cheap price of electricity. The disadvantages include the flooding of farmland, the loss of wildlife habitats, the relocation of 42 000 people, the relatively few permanent jobs in relation to the huge cost and, now, the slow silting-up of the lake. Although Paraguay and São Paulo get their electricity and Paraguay gets much-needed income, there has been little benefit to the local residents of Itaipù.

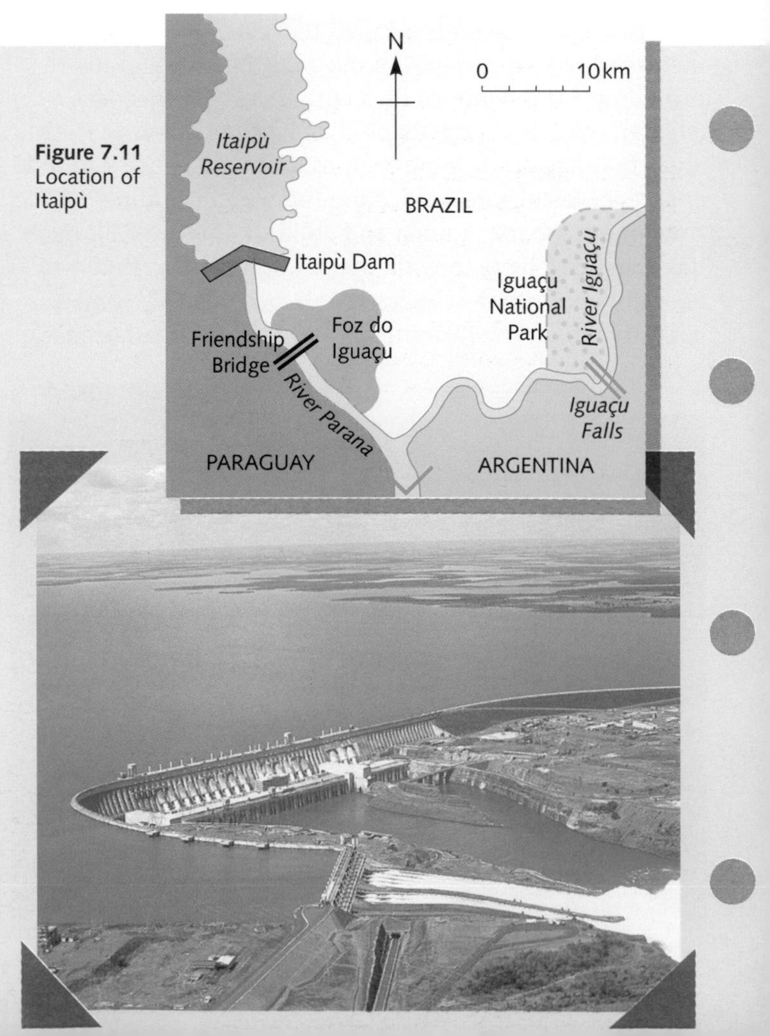

Figure 7.11 Location of Itaipù

Figure 7.12 The Itaipù dam in Brazil is 190 metres high

Geothermal energy

Geothermal means 'heat from the Earth'. Heat is stored in magma or in rocks beneath the Earth's surface. In volcanic areas, such as New Zealand, Iceland, Japan and central America, the heated rocks are near to the surface. In a natural geothermal system, water, usually originating as rain, seeps downwards through cracks and cavities. Where it comes into contact with heated rock, it is warmed and rises to the surface where it is ejected as geysers, hot springs or steam (Figure 7.13). In some areas of known heated rock, cold water is artificially pumped downwards through boreholes before returning to the surface, usually as steam (Figure 7.14). The advantages of geothermal energy include being renewable, providing a constant supply and being relatively pollution free. The disadvantages include the high cost of construction and maintenance to produce useful energy, its limitation to volcanic areas, the threat to power stations from eruptions and earthquakes, and the emission of sulphuric gases.

Figure 7.13 Geothermal activity – geysers, hot springs and steam – at Rotorua, New Zealand

Figure 7.14 How geothermal energy is produced

New Zealand

The Taupo volcanic area in New Zealand's North Island contains several geothermal fields (Figure 7.15). Some of these fields are used to generate electricity (e.g. Wairakei and Ohaaki), some are tourist attractions (e.g. Rotorua and Wairakei), some are used for domestic heating (e.g. Rotorua and Taupo) and some are used for industry (e.g. the pulp and paper mill at Kawerau).

Wairakei, commissioned in 1958, is the second oldest geothermal power station in the world. A power station was proposed following two dry years when there was insufficient water to generate hydro-electricity. Wairakei was chosen because it was in the centre of a geothermal field and was next to the Waikato river which could provide water for cooling. Approximately 50 production wells have been drilled in the area (see Figure 7.16 bottom left). On reaching the surface the steam is piped 2 km to the power station. Cold water from the river is then sprayed into condensers to help condense the steam (it takes 40 kg of water to condense 1 kg of steam). At present, Wairakei produces 4 per cent of New Zealand's electricity.

Places

Figure 7.15 The Taupo volcanic zone

Figure 7.16 Wairakei

Wind

Although windmills were used in Britain during the Middle Ages to grind corn, few attempts were made until recently to generate this form of energy. Now, apart from hydro-electricity, it is the only renewable option being developed commercially in the UK. Wind turbines, to be at their most efficient, need to be in areas with high and regular wind speeds. Such sites are usually found on exposed coasts or in upland areas in the more remote parts of western Britain (Figure 7.17). As 30 metre high turbines are expensive to build and maintain, it is an advantage to group a minimum of 25 machines together to form a 'wind farm' (Figure 7.18). Britain's first wind farm was opened in 1991 near Camelford in Cornwall. The farm, on moorland 250 metres above sea-level and where average wind speeds are 27 km/hr, generates enough electricity for 3000 homes. By 1996, when there were almost 60 wind farms either working or planned, the wind contributed about 0.2 per cent to the UK's total energy generation. Although the first large-scale wind farms were built in California, by 1996 Europe produced one-third more energy from wind power than the USA. At present, Denmark relies more on wind power than any other country.

As with other forms of energy, wind power has its advantages and disadvantages (Figure 7.18).

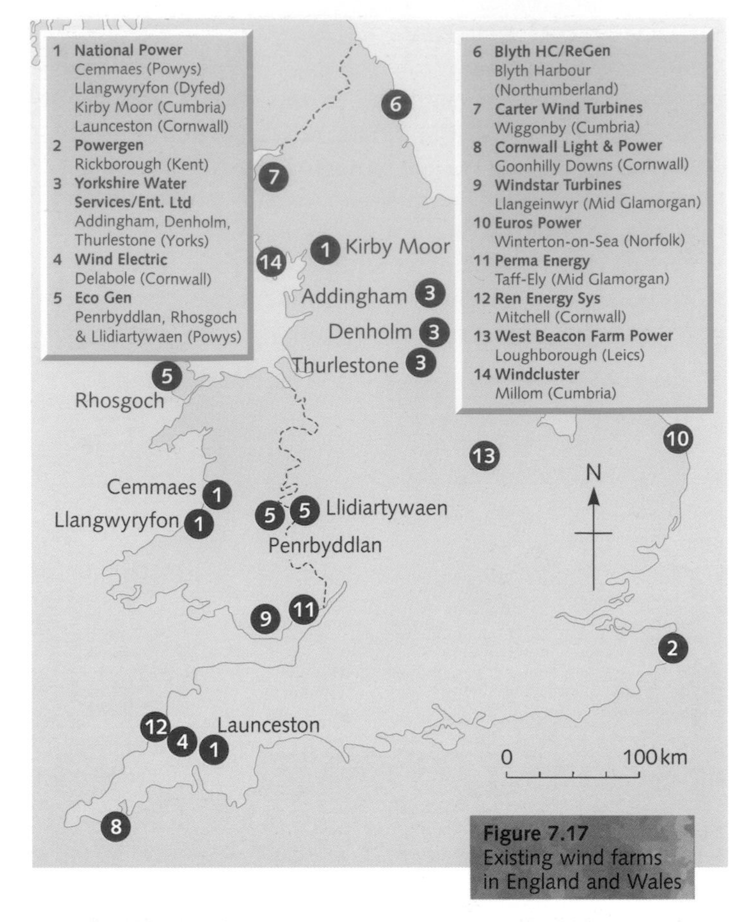

Figure 7.17 Existing wind farms in England and Wales

Figure 7.18 Wind farm at Camelford, Cornwall

Advantages

- Wind turbines do not cause air pollution and will reduce the use of fossil fuels.
- Winds are much stronger in winter, which coincides with the peak demand for electricity.
- After the initial expense of building a wind farm, the production of electricity from this renewable source is relatively cheap.
- Wind farms provide a source of income for farmers and may attract small industries to rural parts of Britain where job opportunities are at present limited.
- Wind could generate 10 per cent of the UK's total electricity.

Disadvantages

- Wind does not blow all the time. At present electricity generated during storms cannot be stored for use during calm periods.
- Groups of 30 metre tall turbines spoil the scenic attraction of the countryside.
- 7000 turbines are needed to produce the same amount of electricity as one nuclear power station. 100 000 wind farms may be needed if Britain is to generate 20 per cent of its total energy supply from the wind.
- Wind farms are noisy and can interrupt radio and TV reception for people living nearby.

Figure 7.19 A solar energy station

Figure 7.20 Using wave power, Norway

Figure 7.21 The Rance tidal barrage in France

Figure 7.22 Biogas – fermentation of animal dung

Solar energy

The amount of solar energy reaching the Earth far exceeds the total of all other forms of energy. It can be used to generate electricity directly by using solar panels (Figure 7.19) and photo-voltaic cells. Solar energy has the advantages of being a safe, pollution-free, efficient and limitless supply. Unfortunately the construction of solar 'stations' is expensive and requires further advances in technology. For Britain, the solar option is further hindered by the weather. Britain receives less sunlight than most places on Earth. Winter, when demand is highest, is the time when the UK gets most cloud, experiences shorter hours of daylight and, when it does shine, the sun's angle is low in the sky reducing its effectiveness (Figure 11.2). Whereas solar power is not expected to provide Britain with more than 1 per cent of its energy requirements, its potential is far greater in developing countries, many of which are located in warmer, sunnier latitudes. In developing countries such as India, solar energy can be used in more isolated, rural areas where mains electricity is too expensive to install.

Waves, tidal, biogas and biomass energy

Waves Waves, especially during storms, have exceptionally high energy levels. However, it is proving a problem to design machinery that can withstand the power of the waves and also convert their energy into electricity – the only existing station is in Norway (Figure 7.20).

Tidal Like the Rance in Brittany (Figure 7.21), many British rivers have a large tidal range which could generate electricity. However, schemes such as the proposed Severn Barrage would be very expensive to build, would destroy important wildlife habitats and could disrupt local shipping.

Biogas Fermenting animal dung gives off methane gas which can be used in developing countries instead of fuelwood (Figure 7.22). However, while this is a cheap source of energy, it means dung can no longer be used as a fertiliser. Also, methane is a greenhouse gas.

Biomass Crops like sugar cane (page 103), cassava and maize contain starch which, if it is fermented, can produce a type of alcohol. In Brazil, 'alcool' is used to operate cars (Figure 7.23).

Figure 7.23 Biomass energy

World energy

Figure 7.24 shows how much energy each country produced per capita (person) in 1992. It also shows the main areas where fossil fuels are exploited (remember, fossil fuels account for 87.8 per cent of the world's energy – Figure 7.3a). Figures 7.24 and 7.25 show how well endowed for energy are the USA and Russia as well as (for its size) is the UK. Figure 7.26 shows the amount of energy consumed per capita in each country. Notice, with the exception of oil-rich Saudi Arabia, how the largest consumers correspond closely with the more industrialised, economically developed countries.

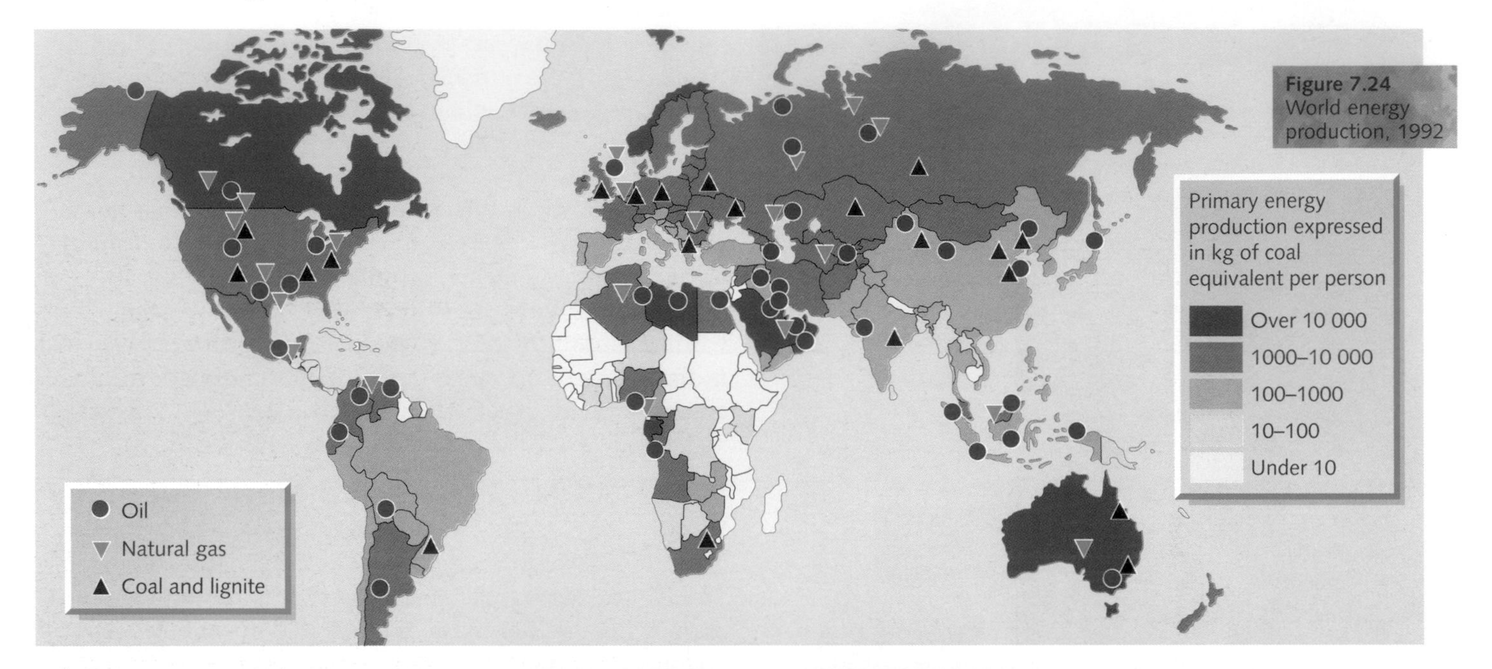

Figure 7.24 World energy production, 1992

OIL 1994		NATURAL GAS 1993		COAL (BITUMINOUS) 1993		COAL (LIGNITE) 1993		NUCLEAR POWER		HEP	
SAUDI ARABIA	13.2	CANADA	28.2	CHINA	36.0	USA	23.7	USA	31.0	CANADA	12.8
USA	12.6	NIGERIA	9.0	USA	17.6	GERMANY	17.5	FRANCE	16.3	USA	12.2
RUSSIA	9.9	KAZAKHSTAN	8.3	INDIA	7.9	RUSSIA	9.1	JAPAN	11.8	FORMER USSR	10.4
IRAN	5.7	UZBEKISTAN	8.0	RUSSIA	6.3	CHINA	7.4	FORMER USSR	7.9	BRAZIL	10.3
MEXICO	4.9	RUSSIA	7.4	AUSTRALIA SOUTH AFRICA	5.8	POLAND	5.4	GERMANY	6.9	CHINA	6.9
(UK 9TH	3.9)	(UK 6TH	3.0)					(UK 6TH	3.9)		
				(UK 10TH	2.7)						

Figure 7.25 Major world producers (percentage figures)

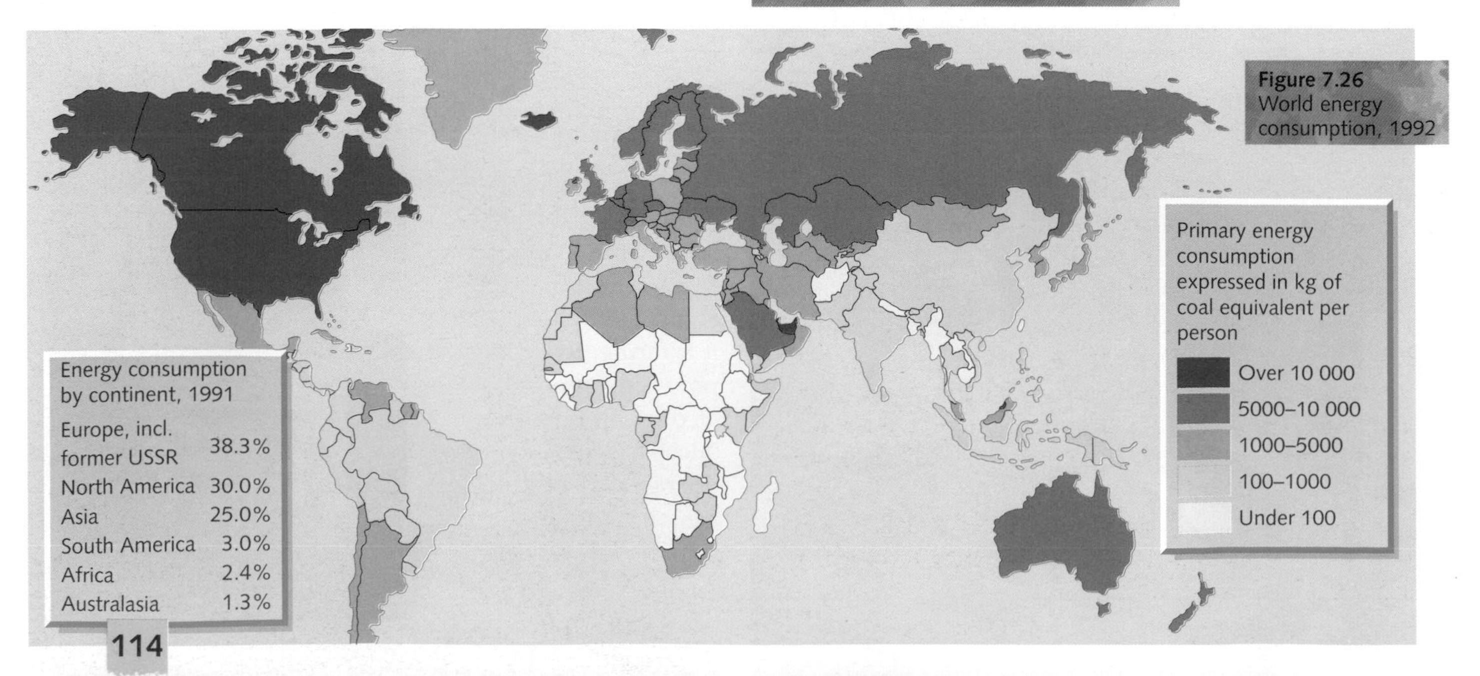

Figure 7.26 World energy consumption, 1992

	1960	1970	1980	1990	1994
COAL	76.5	50.7	35.2	30.6	23.2
OIL	22.6	42.7	39.7	34.7	35.7
NATURAL GAS	0.1	2.7	20.1	25.6	31.2
NUCLEAR	0.2	3.3	4.5	8.0	9.4
HYDRO-ELECTRIC	0.6	0.6	0.6	0.5	0.5

Figure 7.27
Changes in the sources of British energy (%), 1960–94

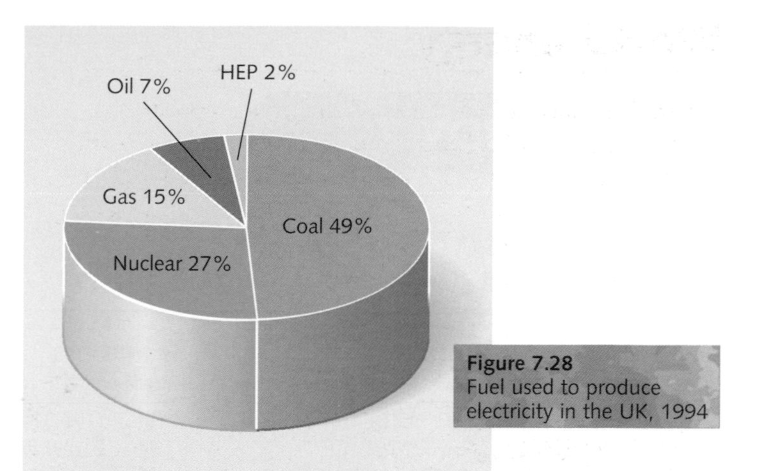

Figure 7.28
Fuel used to produce electricity in the UK, 1994

Electricity in the UK

In 1960 virtually all of Britain's energy came from just two sources – coal and oil. Figure 7.27 shows that, in the following 35 years, coal declined significantly as other sources became increasingly more important. The table does not show the steady increase in the use of energy until the 1980s and the slower increase since then.

Figure 7.28 shows the types of energy at present used to produce Britain's electricity. Electricity is supplied through the National Grid – a technically sophisticated system of power stations, electricity storage schemes and transmission lines. Most of Britain's electricity is still produced by heat in thermal power stations (Figure 7.29). Heat is obtained by burning one of the three non-renewable fossil fuels (coal, oil and, increasingly, gas). The fuel heats water to produce steam, as does the release of energy in nuclear power stations. The steam then drives a turbine which in turn operates a generator which produces the electricity.

Figure 7.29
Drax power station, Thorpe Marsh near Doncaster

Figure 7.30 shows the location of Britain's major electricity generating power stations in the late 1990s.

- **Coal-fired** power stations are ideally located on or near coalfields, large centres of population and older industrial areas. As large amounts of water are needed to cool the steam to condense it back into water, many of the more recent coal-fired stations were built either along main inland rivers (e.g. the Trent) or on coastal estuaries (e.g. the Thames).
- **Oil-fired** power stations are located on deep, sheltered coastal estuaries which can accommodate large tankers. Estuaries also provide the large amounts of water needed for cooling and, ideally, are away from large centres of population (e.g. Milford Haven). Present EU fuel policies make it unlikely that any more oil-fired stations will be built.
- **Gas-fired** power stations have become the fashion in the 1990s, partly because of their lower costs and partly because they produce a smaller amount of greenhouse gases. The increase in their number has mainly been at the expense of coal-fired stations and, consequently, the British coal-mining industry.
- Most **nuclear** power stations are located on coasts and estuaries where there is water for cooling, plenty of cheap and easily reclaimable land and away from large centres of population.
- **Hydro-electric** power stations are located in the more remote and upland areas of Britain which have large and reliable amounts of rain.

Figure 7.30
Britain's main electricity generating stations

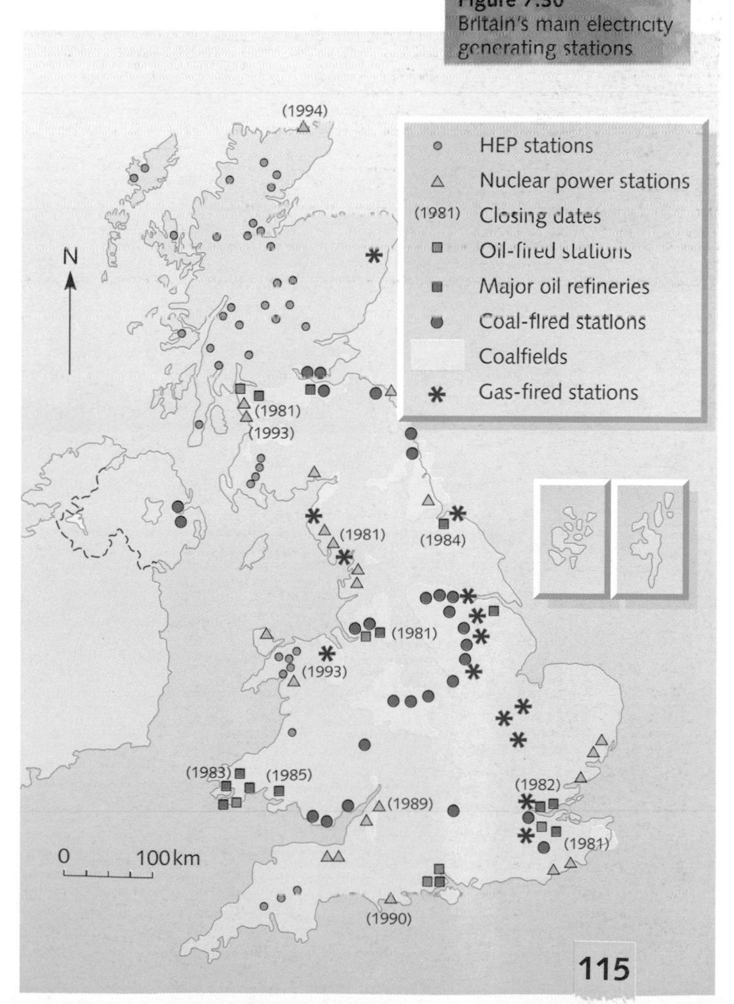

CASE STUDY 7

Energy and the environment – oil in Alaska

Oil exploration in the harsh arctic climate of northern Alaska began in the 1950s. In 1968, two years after exploration began there, North America's largest oilfield was discovered at Prudhoe Bay (Figure 7.31). The field contains one-third of the USA's known oil reserves and 12 per cent of its known gas reserves. However, before the oil could be used, a route and a method of transport had to be decided by which the oil could be moved south. Two routes were suggested.

1 By giant oiltankers through the Arctic Ocean and around Alaska. This route was rejected as being too dangerous (the northern ocean is ice-covered for most of the year), and not economically practicable, despite two trial runs by a supertanker.

2 By pipeline, 1242 km in length, southwards across Alaska to the ice-free port of Valdez. This route faced such enormous physical difficulties and environmental opposition that it took until November 1973 before the USA government finally gave the go-ahead for the pipeline to be constructed, and it was June 1977 before the first crude oil was pumped along it.

Physical problems and environmental concerns

Many of the physical problems facing the oil companies and the concerns expressed by numerous environmental groups in 1968 are summarised in Figure 7.32.

Figure 7.31
The route of the Alaska oil pipeline

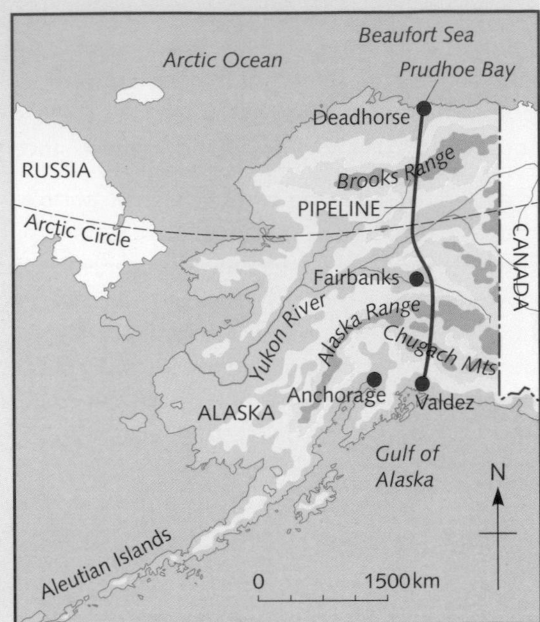

Figure 7.32
The challenge of building the Alaska pipeline

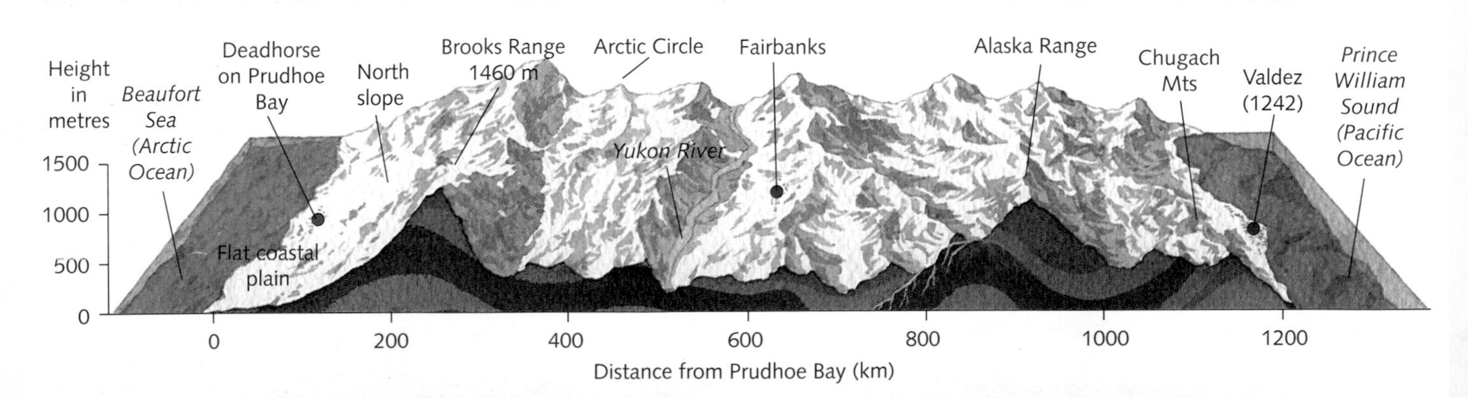

PHYSICAL PROBLEMS	
ISOLATION	• NO ROAD IN 1968
CLIMATE	• TEMPERATURES OF −50°C IN WINTER. PROBLEMS OF KEEPING OIL FLOWING, AND HUMAN WORKING CONDITIONS
	• HEAVY SNOWFALL: 1–2 M AT VALDEZ
	• FOG, GALES, ICEBERGS IN PRINCE WILLIAM SOUND AND GULF OF ALASKA
SOIL	• PERMAFROST OVER MUCH OF ALASKA MEANT PIPELINE COULD NOT GO UNDERGROUND
	• THREE MONTHS WHEN SURFACE THAWS IN SUMMER, GROUND BECOMES UNSTABLE AND PIPELINE COULD MOVE
	• BUILDINGS WOULD RAISE TEMPERATURE AND THAW THE PERMAFROST
RELIEF	• PIPELINE HAD TO GO OVER THREE MOUNTAIN RANGES
	• SOUTHERN AREA IS A MAJOR EARTHQUAKE ZONE – TSUNAMIS ON COAST
	• OVER 350 RIVERS TO CROSS (THOUGH MANY ARE FROZEN IN WINTER)

ENVIRONMENTAL CONCERNS	
VEGETATION	• TUNDRA ECOSYSTEM VERY FRAGILE – CONCERNS THAT ONCE DESTROYED IT COULD TAKE DECADES TO REGENERATE
WILDLIFE	• CROSSES CARIBOU MIGRATION ROUTE
	• NEAR CARIBOU BREEDING GROUNDS
	• MIGHT AFFECT HABITATS OF BEARS, WOLVES AND MOOSE
WILDERNESS	• CROSSES, OR IS ADJACENT TO, NATIONAL PARKS AND WILDLIFE REFUGES
LOCAL PEOPLE	• MAY BE FORCED TO SELL THEIR LAND AND TO MOVE
OIL SPILLS	• MAJOR THREAT

Some attempted solutions

Figures 7.33 to 7.36, taken in mid-1996, show some of the attempted solutions.

Figure 7.33 The Dalton Highway and Alaska pipeline north of the Yukon River

Figure 7.34 Winter at Prudhoe Bay

Figure 7.35 Approaching the Brooks Range

Figure 7.36 In the Atigun Pass

- *Figure 7.33* The first task, completed even before the government's go-ahead, was to extend the existing road, from just north of Fairbanks to Prudhoe Bay – the Dalton Highway. It was built on gravel pads to allow drainage and to limit the formation of ice. The pipeline was built on stilts 3 metres high to avoid melting the permafrost, and raised to 6 metres on caribou migration routes, so that animals can pass below the pipe. A total of 12 pumping stations (not shown) means that, in the event of a break, the flow of oil can be stopped. Notice how, after 20 years, the vegetation near the pipeline has still not fully recovered.
- *Figure 7.34* The pipeline supports are sunk into the permafrost for stability. The pipe itself, which has a diameter of 1.22 metres, rests on 'sliding shoes'. These can move horizontally by one or two metres in the event of an earthquake. The oil is pumped through at a temperature of about 80°C to prevent it freezing. Notice the tundra vegetation and the flat coastal plain.
- *Figure 7.35* shows the pipeline as it approaches the Brooks Range. The first snow of winter had just fallen – on 3 August!
- *Figure 7.36* The Atigun Pass, in the Brooks Range, is the highest point that the pipeline reaches (1460 m). It has recently been sunk underground in river gravels (an environmental improvement) and encased in concrete (to prevent the oil from freezing and the ground from thawing).

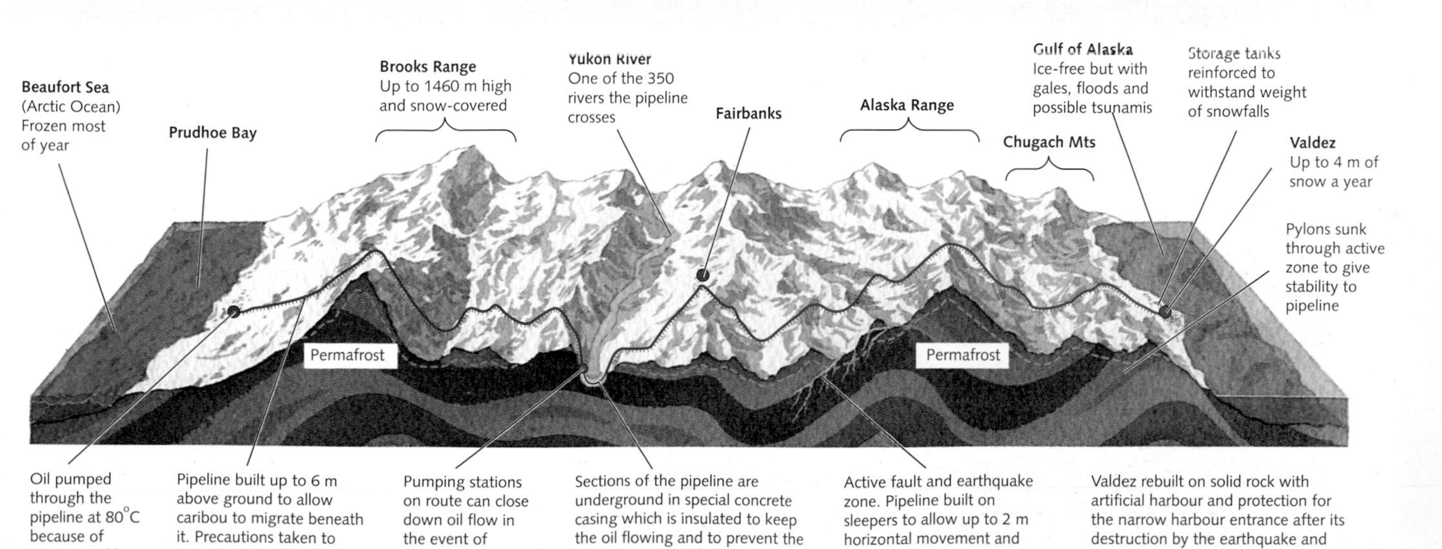

Figure 7.37 Cross-section of the route of the pipeline across Alaska

CASE STUDY 7 continued

Figure 7.38 Valdez oil terminal

Figure 7.39 The Valdez Arm

Up to 1990 there had been some 300 reports of relatively minor oil leaks. The oil companies have, so far, been able to act quickly to seal these leaks. However, their efforts tend to be in restoring the flow (economic gain) rather than in cleaning up the spillage (environmental loss) – although efforts to clean up often *increase* the damage to the ecosystem.

The Exxon Valdez oil spill, 1989

Valdez is the ice-free port at the end of the Alaskan pipeline (Figure 7.38). The port was re-sited after the original one was destroyed by an earthquake in 1964. To reach the open sea from Valdez, oiltankers have no option but to follow a potentially dangerous route which takes them through Prince William Sound (Figure 7.39) and then through a narrow gap between two islands (Figure 7.40).

When the long-feared disaster occurred on 24 March 1989, it was not due to any of the physical dangers shown on Figure 7.40 but to human negligence and incompetence. Shortly after midnight the supertanker *Exxon Valdez*, 40 km out of Valdez and carrying 50 million tonnes of crude oil, ran aground in near-perfect weather conditions on Bligh Reef. It was first assumed that the ship had veered off course to avoid an iceberg, but it was later discovered that the captain was drunk and no senior officer was on the bridge at the time. The Alyeska Pipeline Company, which was supposed to react to such an emergency within five hours, did nothing for twelve hours . . . and then took the wrong advice and continued to do nothing. By 2 April, and after several days of bad weather, the oil covered 2600 km^2 and the slick extended 900 km from the wreck. At this stage, the Exxon Oil company, which owned the tanker but was under no legal contract to do anything, took over the cleaning-up operations (Figure 7.41). Exxon worked until the end of the year, at an eventual cost of $600 million, before pulling out and leaving 60 per cent of the spilled oil still in Prince William Sound or along 1700 km of coastline. Even in 1997, traces can be found in remote inlets, and the ecosystem has still to

Figure 7.40 The *Exxon Valdez* disaster

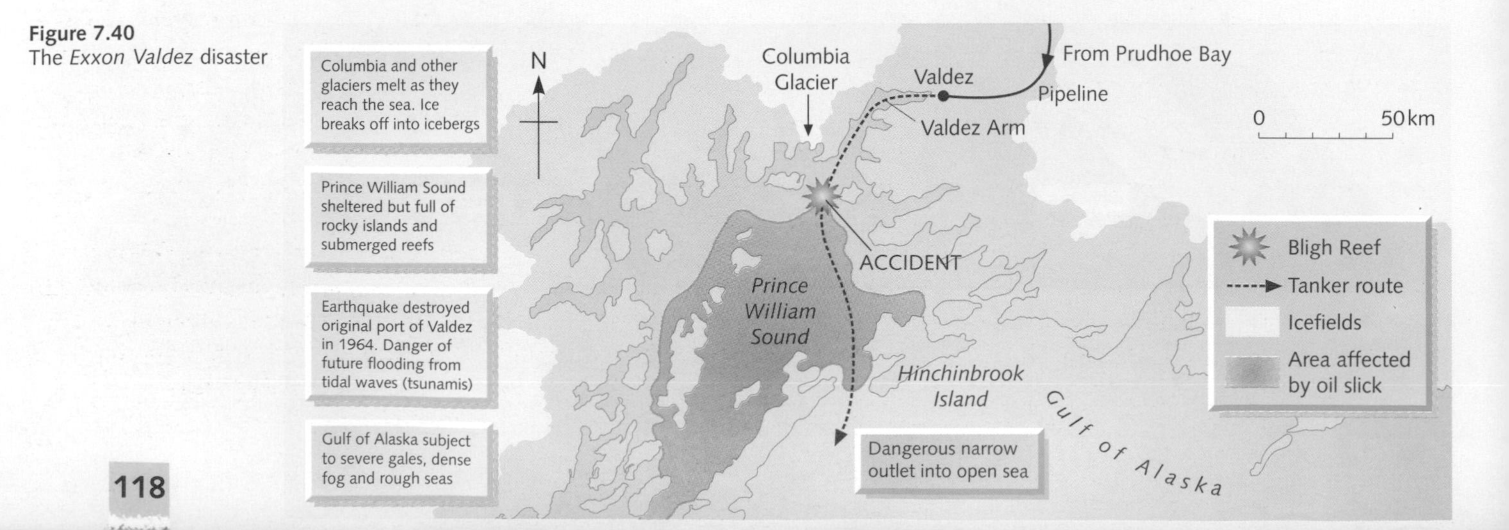

fully recover. The effect of the spillage on wildlife and the marine ecosystem was enormous (Figure 7.42). The bodies of 35 000 seabirds and 3000 sea-otters were recovered (thousands more must have perished, while the bodies of dead seals sink). Likewise the local economy, which relied heavily upon fishing, was badly hit as salmon hatcheries were polluted and deep-sea fishing ruined.

Figure 7.41 Clean-up operations

Today and the future

For the last two decades oil has accounted for 85 per cent of Alaska's income. During that time the state has invested £20 billion of this income into a communal fund that pays every person in the state, including children, an annual dividend (sometimes the equivalent of over £1000 per person). However, oil production is beginning to decline rapidly. In 1996, when it was 45 per cent down on its 1987 peak, the oil companies sought permission to explore new areas. One such area was the Arctic National Wildlife Refuge, a wilderness area on the North Slope where caribou, snowgeese and other wildlife are protected. As always there are groups of people in favour of and against the proposal. Their arguments include the following:

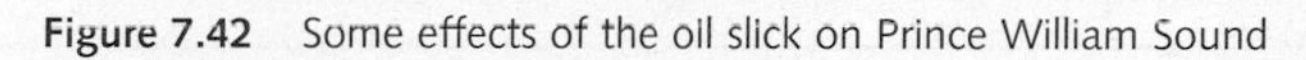

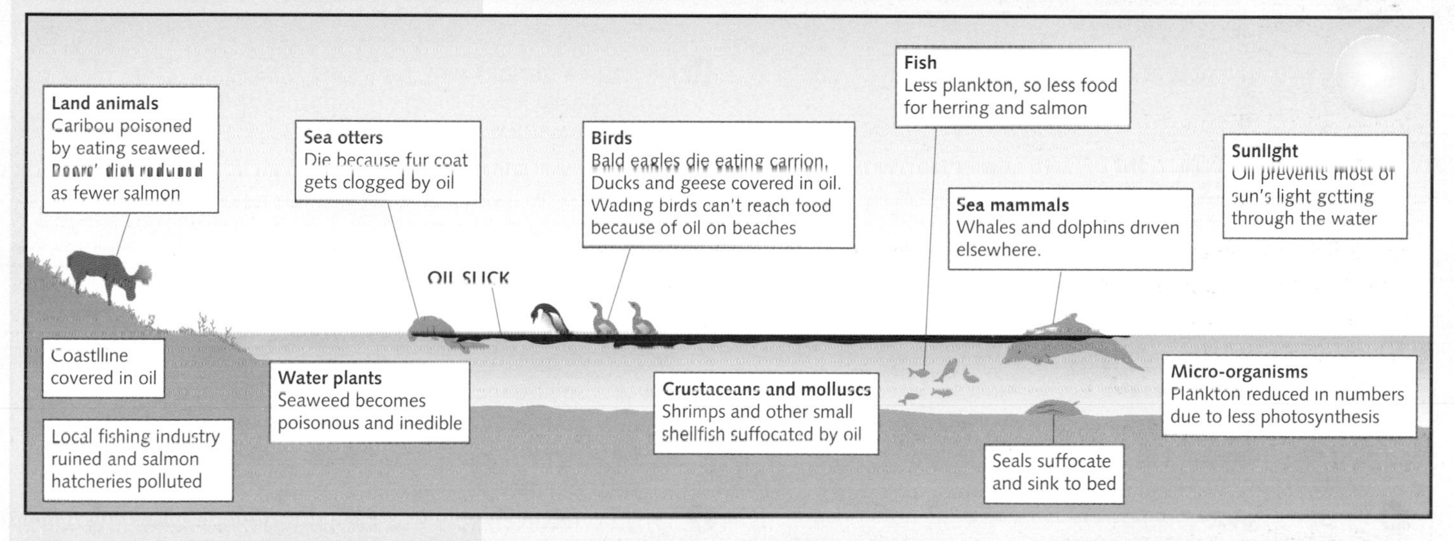

Figure 7.42 Some effects of the oil slick on Prince William Sound

For (mainly economic and social)

- Most Alaskans see oil as their main source of jobs and wealth.
- The income from oil has been used within Alaska to build schools, hospitals and public buildings, and to improve water supplies and roads.
- Many American statesmen appreciate the importance of Alaskan oil, which still provides nearly one-third of the USA's total oil reserves and one-eighth of its total natural gas reserves.
- Alaskans see little point in protecting 'a herd of flea-ridden caribou seen by only a few hundred humans a year'.

Against (mainly environmental)

- Environmental groups, mainly outside Alaska, want the fragile Arctic National Wildlife Refuge to be protected from all forms of development.
- Some native American groups claim they would prefer to live in their traditional way rather than benefit from oil income.
- There are real fears of a major oil spillage from either a pipeline breakage or another tanker accident.
- A criminal enquiry (1997) is investigating charges that one oil company has been illegally dumping toxic waste into its northern Alaskan field.

QUESTIONS

1

(Pages 106 and 107)

a What is the difference between:
- i) a renewable and a non-renewable resource? (1)
- ii) a finite resource and a sustainable resource? (1)

b i) What is a fossil fuel? (1)
- ii) Give two disadvantages of burning fossil fuels. (2)

c Complete the table by placing the following types of energy into the correct column. Some types of energy may fit into more than one column.

◆ biogas ◆ biomass ◆ coal ◆ fuelwood ◆ geothermal ◆ hydro-electric ◆ natural gas ◆ nuclear ◆ oil ◆ solar ◆ tidal ◆ waves ◆ wind (6)

Non-renewable	Fossil fuels	Renewable

2

(Pages 108 and 109)

a Give two advantages and two disadvantages of using each of the following types of energy:
- coal
- oil and natural gas
- nuclear (3 x 4)

b Give three reasons why fuelwood is used more in economically less developed countries than the types of energy named in (**a**). (3)

3

(Pages 110, 111 and 113)

a What is hydro-electric power? (1)

b Draw an annotated (labelled) sketch to show what makes the ideal site for a hydro-electric power station. (4)

c For Itaipù (Brazil) or any hydro-electric power scheme you have studied, list three advantages and three disadvantages of the scheme. (6)

d What is geothermal power? (1)

e Describe how geothermal energy is harnessed. (3)

f Why is geothermal power important in a country like Iceland or New Zealand? (3)

g What is solar energy? (1)

h i) List three advantages of solar energy. (3)
- ii) Give three reasons why Britain uses relatively little solar energy. (3)

i Why do relatively few tropical countries use solar energy? (2)

4

(Page 112)

Read this newspaper article.

a i) The farmer suggests that wind farms ruin the landscape. Describe three ways it might be argued they do this. (3)
- ii) Which group of people do you think might agree with the farmer? (1)
- iii) Which group of people, other than the Electricity Board, might disagree with the farmer? (1)

b Do you think wind farms are a good way to produce electricity? Give reasons for your answer. (4)

Mid-Wales Farmer Faces Blight from Wind Farm!

A farmer told our reporter today that plans to build a huge wind farm on Mynydd Carn should be scrapped. Mr Hywel Davies and his wife Mair, of Ty Coch farm, say they would be able to see the 60 metre tall turbines from every window of their house. 'They ruin the landscape,' said Mrs Davies. 'We don't want them here in mid-Wales, and lots of other people agree with us.' An Electricity Board official defended the proposals for the wind farm. 'We cannot keep using fossil fuels forever – the local farmer might not see the sense of a wind farm, but plenty of others do!'

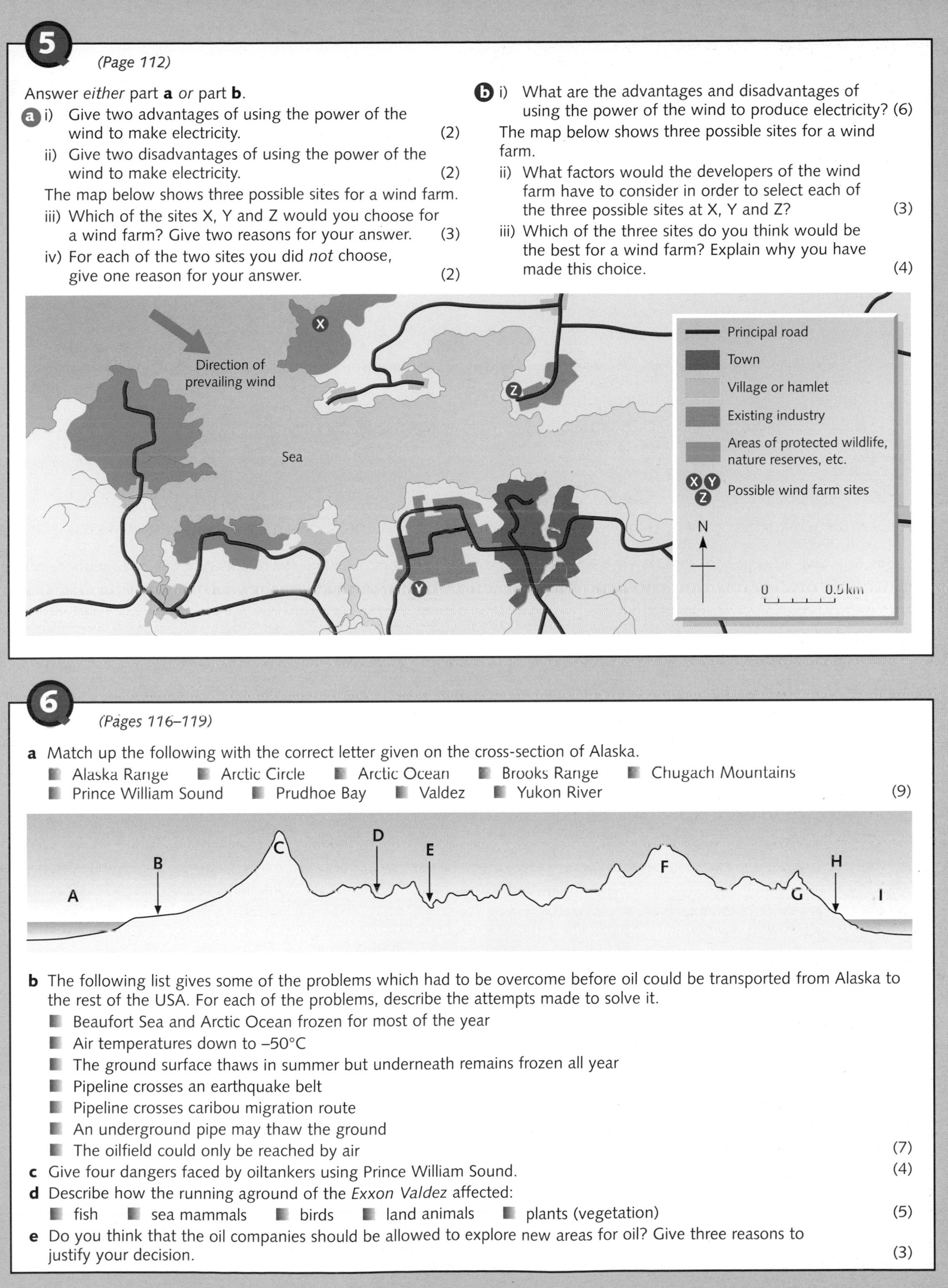

5

(Page 112)

Answer *either* part **a** *or* part **b**.

a i) Give two advantages of using the power of the wind to make electricity. (2)

ii) Give two disadvantages of using the power of the wind to make electricity. (2)

The map below shows three possible sites for a wind farm.

iii) Which of the sites X, Y and Z would you choose for a wind farm? Give two reasons for your answer. (3)

iv) For each of the two sites you did *not* choose, give one reason for your answer. (2)

b i) What are the advantages and disadvantages of using the power of the wind to produce electricity? (6)

The map below shows three possible sites for a wind farm.

ii) What factors would the developers of the wind farm have to consider in order to select each of the three possible sites at X, Y and Z? (3)

iii) Which of the three sites do you think would be the best for a wind farm? Explain why you have made this choice. (4)

6

(Pages 116–119)

a Match up the following with the correct letter given on the cross-section of Alaska.

- Alaska Range
- Arctic Circle
- Arctic Ocean
- Brooks Range
- Chugach Mountains
- Prince William Sound
- Prudhoe Bay
- Valdez
- Yukon River

(9)

b The following list gives some of the problems which had to be overcome before oil could be transported from Alaska to the rest of the USA. For each of the problems, describe the attempts made to solve it.

- Beaufort Sea and Arctic Ocean frozen for most of the year
- Air temperatures down to –50°C
- The ground surface thaws in summer but underneath remains frozen all year
- Pipeline crosses an earthquake belt
- Pipeline crosses caribou migration route
- An underground pipe may thaw the ground
- The oilfield could only be reached by air

(7)

c Give four dangers faced by oiltankers using Prince William Sound. (4)

d Describe how the running aground of the *Exxon Valdez* affected:

- fish
- sea mammals
- birds
- land animals
- plants (vegetation)

(5)

e Do you think that the oil companies should be allowed to explore new areas for oil? Give three reasons to justify your decision. (3)

8 INDUSTRY

The industrial system and industrial location

Industry as a whole, or a factory as an individual unit, can be regarded as a system. At its simplest, there are *inputs* into a factory (or industry), *processes* that take place in the factory, and *outputs* from the factory (Figure 8.1). For a firm to be profitable and to remain in business, the value of its outputs must be greater than the cost of its inputs. Some of the profit should then be re-invested, e.g. in modernising the factory and introducing new technology.

Before building a factory, the manufacturer should consider the major elements in the system diagram. It is unlikely that all the factors listed in Figure 8.2 will be available at one particular site and so a decision must be made as to which site is likely to provide the best location. This will be where the cost of raw materials, energy, labour, land and transport is lowest, and where there is a large market for the product (Figure 8.3).

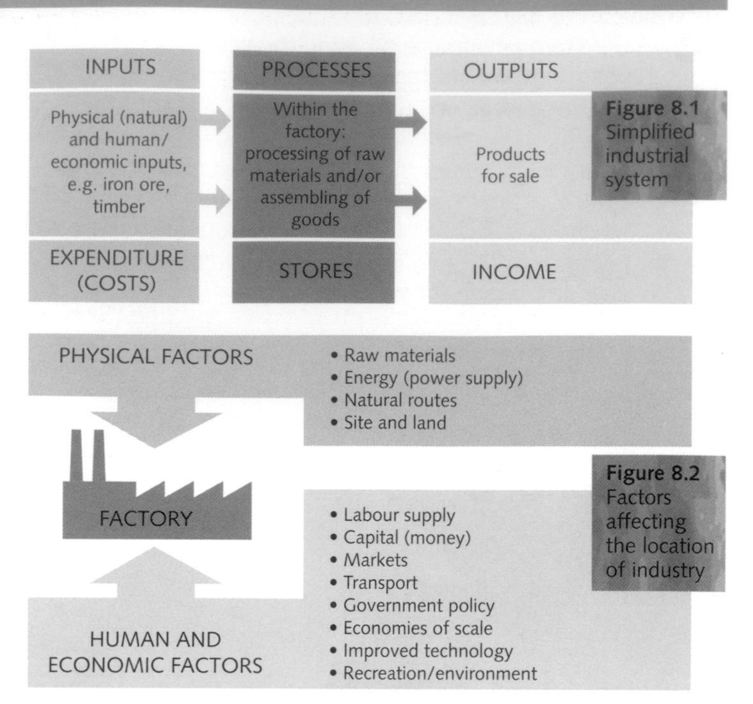

Figure 8.1 Simplified industrial system

Figure 8.2 Factors affecting the location of industry

PHYSICAL FACTORS

Raw materials The bulkier and heavier these are to transport, the nearer the factory should be located to the raw materials. This was even more important in times when transport was less developed.

Power – energy This is needed to work the machines in the factory. Early industry needed to be sited near to fast-flowing rivers or coal reserves, but today electricity can be transported long distances.

Natural routes River valleys and flat areas were essential in the days before the railway, car or lorry.

Site and land Although early industry did not at first take up much space, it did need flat land. As the size of plant increased (e.g. steelworks), more land was needed. Ideally such sites should be on low-quality farmland where the cost of purchase is lower. Last century many sites were in today's 'inner city' areas whereas now they tend to be on edge-of-city 'greenfield' locations.

- In the nineteenth century it was physical factors such as the source of raw materials (e.g. iron ore) and sources of energy (e.g. coal) which determined industrial locations.

HUMAN AND ECONOMIC FACTORS

Labour This includes both quantity (large numbers in nineteenth-century factories) and quality (as some areas demand special skills as technology develops).

Capital Early industry depended on wealthy entrepreneurs. Now banks and governments may provide the money.

Markets The size and location of markets have become more important than the source of raw materials.

Transport Costs increase when items moved are bulky, fragile, heavy or perishable.

Economies of scale Small units may become unprofitable and so merge with, or are taken over by, other firms.

Government policies As governments tend to control most wealth, they can influence industrial location.

Improved technology Examples are facsimile (fax) machines and electronic mail.

Leisure facilities Both within the town and the surrounding countryside, leisure activities are becoming more desirable.

- By the late twentieth century, the three main factors deciding industrial location were more likely to be the nearness to a large market, the availability of skilled labour, and government.

Figure 8.3 Factors affecting the location of industry in the UK

Figure 8.4 Shipyard on the Clyde in 1935

Figure 8.5 Modern industry – Woodlands, Bristol

Location of industry in the UK

Figure 8.6 shows the location and distribution of Britain's traditional heavy industries. Most of these industries were established in the nineteenth century. Their growth was based on the use of coal, the development of technology to process local and imported raw materials, the creativity of the people and the ability to export manufactured goods. Consequently the major industrial areas were either on Britain's coalfields or in coastal ports located on deep-water estuaries.

The location, distribution and type of Britain's present-day manufacturing industry has changed considerably (Figure 8.7). For a variety of reasons (Figure 8.8) coal mines began to close in the 1920s, textile mills in the 1960s, shipyards in the 1970s and steelworks in the 1980s. Modern replacement industries, many of which are high-tech and connected with electronics (page 126), employ fewer people and are often located well away from the traditional manufacturing areas. They are said to be '*footloose*' as, not being tied to raw materials, they have a relatively free choice of location.

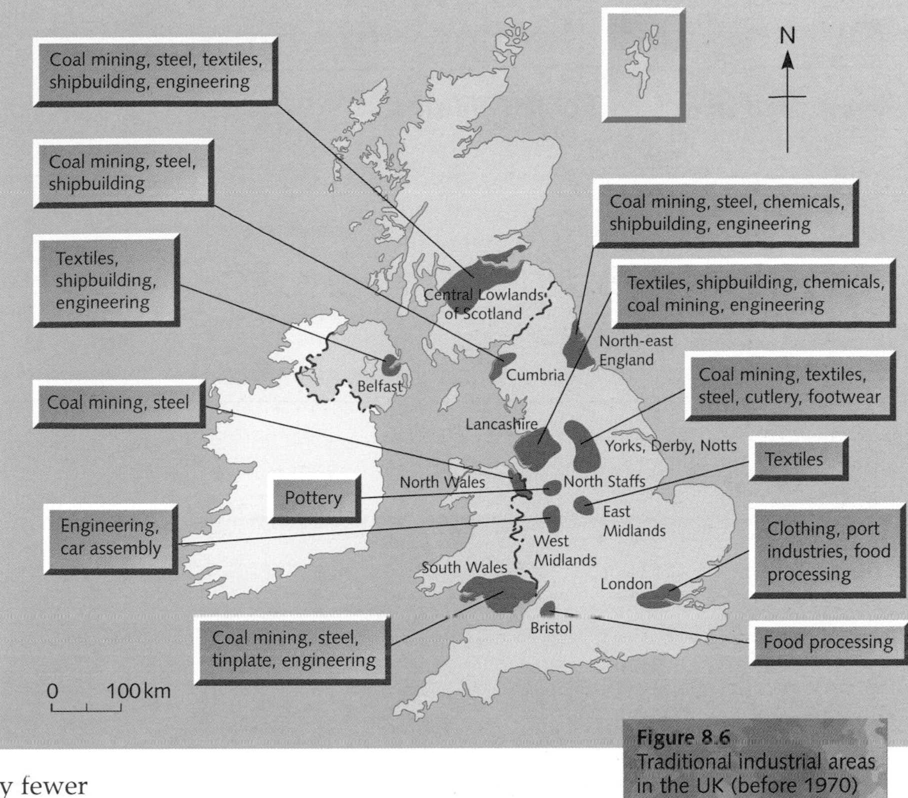

Figure 8.6 Traditional industrial areas in the UK (before 1970)

These newer industries have:

- either opted for a more pleasant working environment near to large markets and major transport routes, or
- especially in the case of foreign companies, been tempted by government policies to locate in former industrial areas which often had higher levels of unemployment

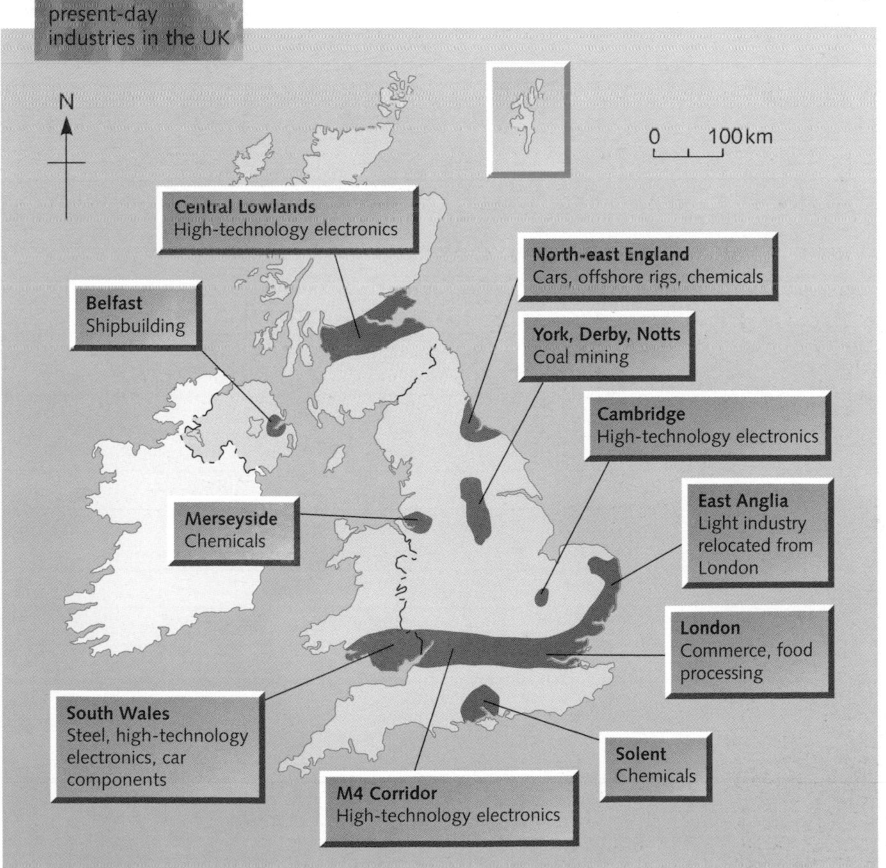

Figure 8.7 Location of present-day industries in the UK

Figure 8.8 Reasons for job losses

- Exhaustion of resources
- Introduction of new machinery or new methods needing fewer workers (automation)
- Fall in demand for product
- Site needed for other uses
- Large-scale redevelopment of inner city areas
- Closure due to high costs of production – high wages
- Closure due to high costs of production – old, inefficient methods or difficult conditions
- Rationalisation of programme of a larger company
- Competition from overseas
- Lack of money for investment
- Competition from rival products
- Political decisions which deny government financial assistance to ailing companies

Decline and changing location

Iron and steel in South Wales

In 1860 there were 35 ironworks in the valleys of South Wales (Figure 8.9). During the day, at that time, the sky was blackened by the smoke from chimneys. At night it turned red from the glare of the many furnaces. The shuddering noise of forge hammers lasted 24 hours a day. Whole villages were totally dependent upon the local ironworks. Their inhabitants lived in small terraced houses, built parallel to the railway along the valley floor in a linear pattern.

South Wales had the ideal location for iron making (Figure 8.10a). Coal and 'blackband' iron ore were often found together on valley sides. Limestone was quarried nearby. Fast-flowing rivers, the result of heavy rainfall draining down steep mountainsides, provided power to turn the early water wheels. The valleys themselves led to coastal ports where iron products, and surplus coal, were exported to many parts of the world. The iron industry became centred on places like Ebbw Vale and Merthyr Tydfil. In 1856 an improvement in the method of iron smelting meant that it became economic to manufacture steel rather than the previously brittle iron. After 1860, steelworks slowly began to replace the iron foundries.

Figure 8.9
Location of ironworks in South Wales, about 1860

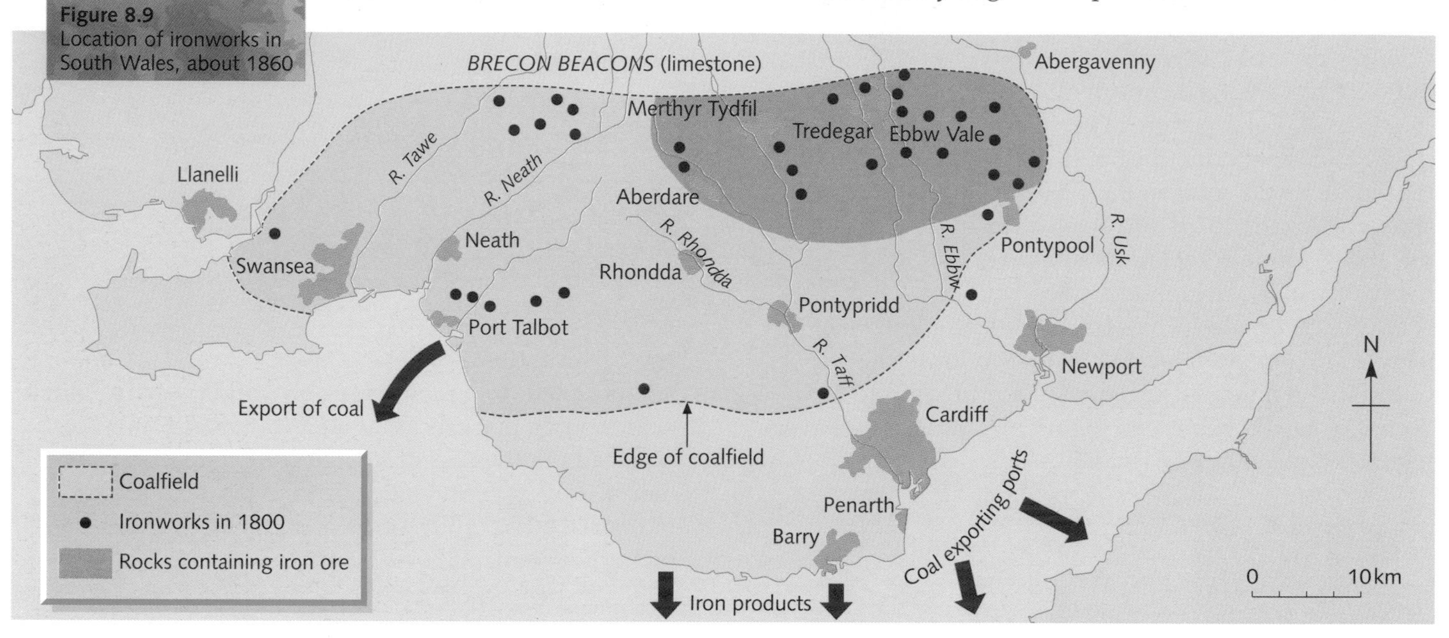

PERIOD OF TIME		(A) LOCATION OF EARLY 19TH-CENTURY FOUNDRIES IN SOUTH WALES (i.e. EBBW VALE)	(B) LOCATION OF INTEGRATED STEELWORKS OF THE LATE 1990S AT PORT TALBOT AND LLANWERN (NEWPORT)
PHYSICAL			
RAW MATERIALS	COAL	MINED LOCALLY IN VALLEYS	ONLY ONE COAL MINE OPEN, MOST OF COAL IMPORTED
	IRON ORE	FOUND WITHIN THE COAL MEASURES	IMPORTED FROM NORTH AFRICA AND NORTH AMERICA
	LIMESTONE	FOUND LOCALLY	FOUND LOCALLY
	WATER	FOR POWER AND EFFLUENT – LOCAL RIVERS	FOR COOLING – COASTAL SITE
ENERGY – FUEL		CHARCOAL FOR EARLY SMELTING, LATER RIVERS TO DRIVE MACHINERY AND THEN COAL	ELECTRICITY FROM NATIONAL GRID (USING COAL, OIL, NATURAL GAS AND NUCLEAR)
NATURAL ROUTES		MATERIALS MAINLY ON HAND. 'EXPORT' ROUTES VIA THE VALLEYS	COASTAL SITES
SITE AND LAND		SMALL VALLEY-FLOOR LOCATIONS	LARGE AREAS OF FLAT, LOW-CAPACITY FARMLAND
HUMAN AND ECONOMIC			
LABOUR		LARGE NUMBER OF UNSKILLED LABOUR	STILL RELATIVELY LARGE NUMBERS BUT WITH A HIGHER LEVEL OF SKILL. FEWER NEEDED BECAUSE HIGH-TECH
CAPITAL		LOCAL ENTREPRENEURS	GOVERNMENT, EU
MARKETS		LOCAL	THE CAR INDUSTRY
TRANSPORT		LITTLE NEEDED, SOME CANALS	M4. PURPOSE-BUILT PORTS
GEOGRAPHICAL INERTIA		NOT APPLICABLE	TRADITION OF HIGH-QUALITY GOODS
ECONOMIES OF SCALE		NOT APPLICABLE	TWO LARGE STEELWORKS MORE ECONOMICAL THAN NUMEROUS SMALL IRON FOUNDRIES
GOVERNMENT POLICY			HAVING THE CAPITAL THEY CAN DETERMINE LOCATIONS AND CLOSURES
TECHNOLOGY		SMALL SCALE – MAINLY MANUAL	HIGH-TECHNOLOGY – COMPUTERS, LASERS, ETC.

Figure 8.10
Reasons for changes in the location of the iron and steel industry in South Wales

By the 1990s there were only two steelworks left in Wales. These were not, however, located in the valleys but on the coast at Port Talbot and Llanwern (Figure 8.11). This was because many of the initial advantages of the area for steelmaking had disappeared (Figure 8.10b). By 1997 only one coal mine remained open in the area, while high-quality iron ore deposits had long since been exhausted. As both of these raw materials had now to be imported, it was logical to build any modern steelworks on the coast at a *break of bulk* location. Break of bulk is when a transported product has to be transferred from one form of transport to another – a process that takes up time and money. It was easier, therefore, to have the steelworks where the raw materials were unloaded, rather than transporting the coal and iron ore to the older, inland works.

Port Talbot (Figure 8.12) has its own harbour and docks for the import of coal and iron ore. These are, with limestone, fed into a blast furnace where the iron ore is smelted and most of the impurities are removed. This produces pig iron to which oxygen is later added. Oxygen reduces the carbon content to give steel. The introduction of the oxygen furnace has reduced the amount of coal needed. The steel is then usually rolled into thin sheets which may be used to make, among other things, car bodies. Port Talbot is one of Britain's four remaining integrated steelworks each using advanced technology. An *integrated works* is where all the stages in the manufacture of steel take place on the same site. The decision to locate at Port Talbot was made by the British government. Whether the works remain open or not is likely to depend on an EU decision. Market demand and government decisions are now more important than nineteenth-century natural physical advantages.

Figure 8.11
Steel production in South Wales, late 1990s

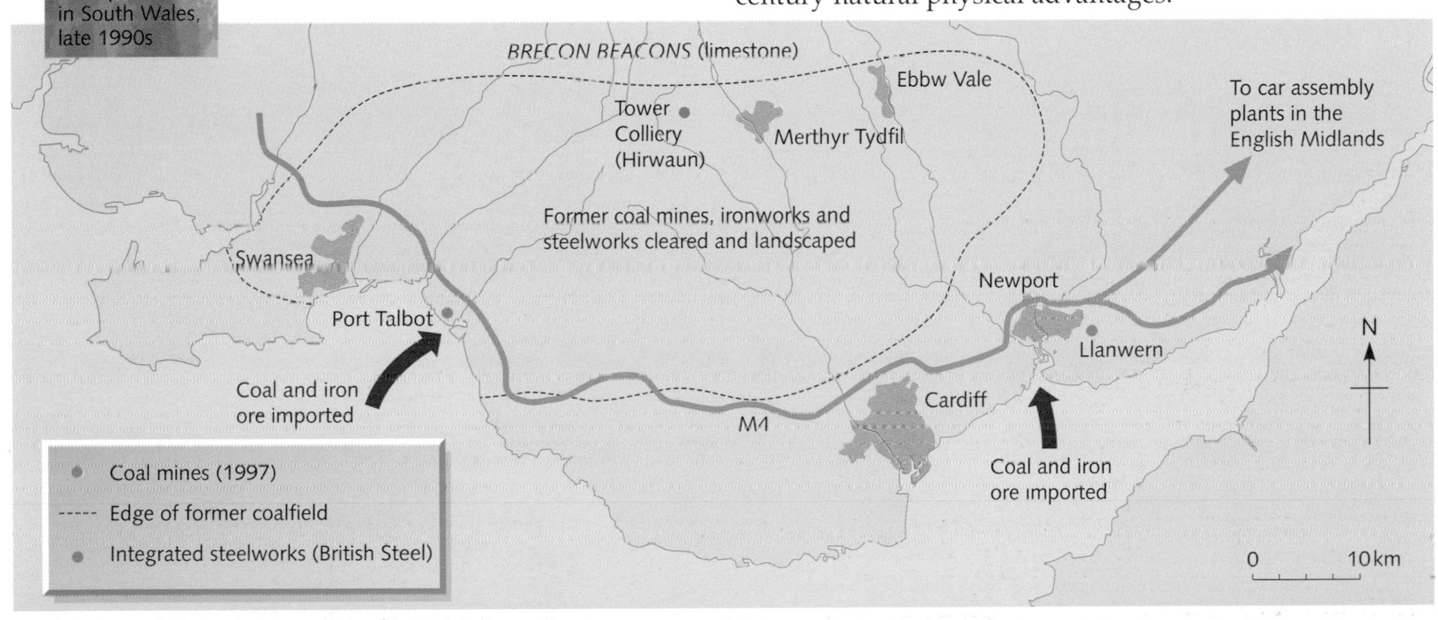

Figure 8.12
Port Talbot steelworks

Growth and changing location

High-technology industries

The term *high-technology industry* (or high-tech) refers, usually, to industries developed within the last 25 years and whose processing techniques often involve micro-electronics. These industries have been the 'growth' industries of recent years though unfortunately they employ few people in comparison with the older, declining heavy industries. Two possible subdivisions of high-tech industries are:

1 The 'sunrise industries' which have a high-technology base.
2 Information technology industries involving computers, telecommunications and micro-electronics.

As a highly skilled, inventive, intelligent workforce is essential, and as access to raw materials is relatively unimportant, these high-tech footloose industries tend to become attracted to areas which the researchers and operators find attractive – from a climatic, scenic, health and social point of view. Such areas include:

- Silicon Glen in central Scotland
- Silicon Valley in California
- Sunrise Strip which follows the route of the M4 from London westwards towards Newbury (locally known as Video Valley), Bristol (Aztec West) and into South Wales (Figure 8.13)
- south of France behind Nice.

Places

M4 Corridor

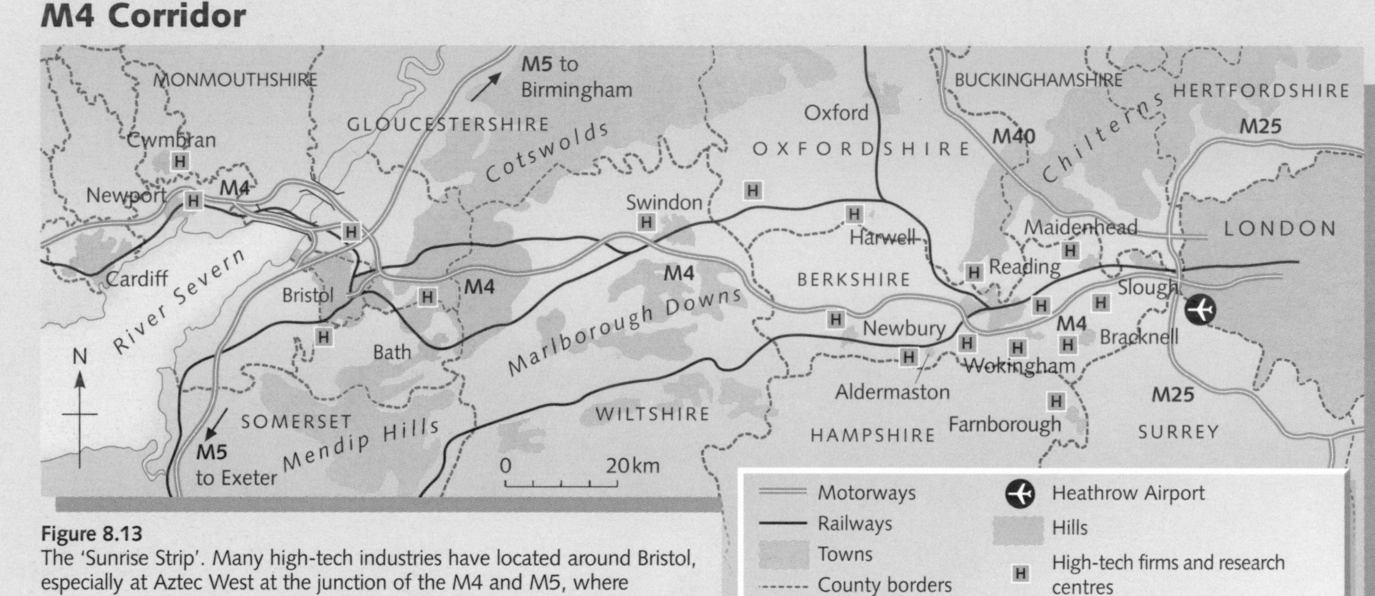

Figure 8.13
The 'Sunrise Strip'. Many high-tech industries have located around Bristol, especially at Aztec West at the junction of the M4 and M5, where expertise had already developed through such firms as Rolls-Royce and British Aerospace. The majority of new industries have tended to locate along the Berkshire section of the M4 where the nearness to Heathrow Airport has been a vital extra advantage.

Figure 8.14
Windmill Hill Business Park, Swindon. In all, three large business parks have been developed, each offering facilities and accommodation in a landscaped, parkland environment.

The advantages of this area for spontaneous, unplanned growth of micro-electronics industries include:

- The proximity of the M4 and mainline railways.
- The presence of Heathrow Airport.
- The previous location and existence of government and other research centres.
- A large labour force, many of whom have moved out of London into new towns and overspill towns.
- The proximity of other associated industries with which ideas and information can be exchanged.
- Nearness to universities with expertise and research facilities available. High-tech firms can work closely with the university campus.
- An attractive environment. Figure 8.13 names the Cotswolds, Mendips, Chilterns and Marlborough Downs. Nearby are the North Downs, three National Parks (Brecon Beacons, Dartmoor and Exmoor) and, through the centre of the area, the Thames Valley.

Business and science parks

Most business parks have grown up on edge-of-city greenfield sites, the remainder as part of inner city redevelopment schemes. The major attractions of greenfield sites are the relatively low cost of land and a pleasant working environment with a low density of buildings. Usually over 70 per cent of the land in business parks is left under grass and trees or converted into ornamental gardens and lakes (Figures 8.14 and 8.16). Business parks form an ideal location for high-tech industries such as electronics, and research institutions. Science parks are similar but with the addition of direct links with universities (Figure 8.15). Some business parks include offices, hypermarkets and leisure complexes.

Why do similar industries locate together?

By locating near to each other, high-tech firms have the advantages of being able to exchange ideas and information with neighbouring companies, sharing maintenance and support services, sharing basic amenities such as connecting roads, and building up a pool of highly skilled, increasingly female labour.

Tsukuba Science City

Tsukuba Science City was specifically created to relieve the pressure on overcrowded, over-expensive Tokyo 40 km to the south-west (Figure 8.17). It has its own university and educational institutions, over 50 national research institutes and laboratories, and more than 150 private firms (Figure 8.15).

Tsukuba's population of 152 000 includes 7000 resident scientists, engineers and researchers, of whom nearly 2000 are non-Japanese. Not only has Tsukuba become a centre of science, it is also in the centre of a scenic environment, part of which has been declared a Quasi National Park (Figure 8.16).

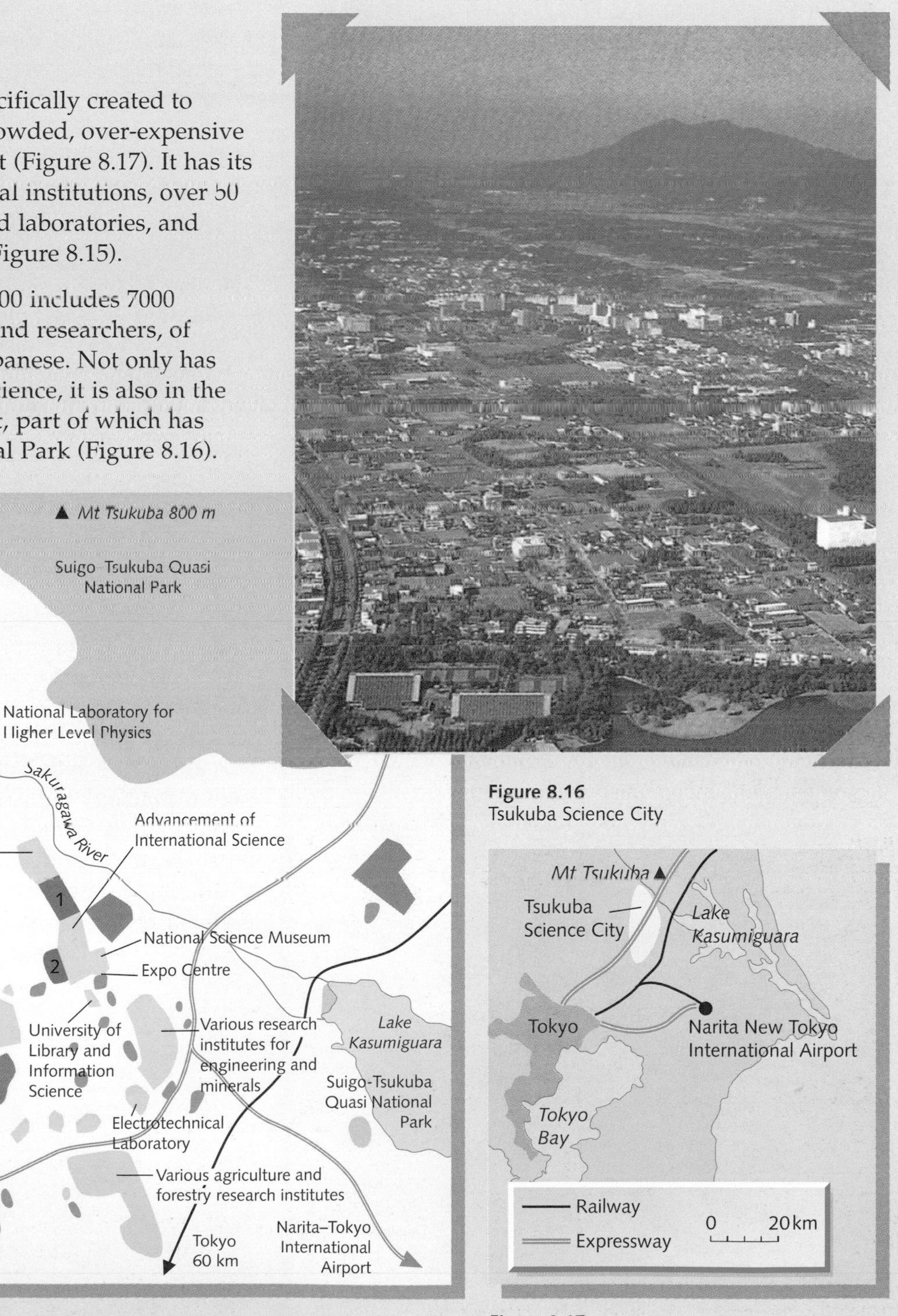

Figure 8.15
Layout of Tsukuba Science City

Figure 8.16
Tsukuba Science City

Figure 8.17
Location of Tsukuba Science City

The Pacific Rim

The Pacific Rim includes all those countries in Australasia, Asia, North America and Latin America that border the Pacific Ocean. The term was originally used to describe a zone of volcanic and earthquake activity which surrounds the Pacific (page 246). More recently it acknowledges the shift in the location of the world's manufacturing industry away from countries bordering the North Atlantic (i.e. in North America and Western Europe) and the emergence of *newly industrialised countries (NICs)*, especially in eastern Asia. This industrialisation began in Japan in the 1950s.

Places

Japan

In 1945 Japan's industry lay in ruins after the Second World War. By 1990 Japan had become, after the USA, the world's second most wealthy and industrialised country. This achievement was even more remarkable because:

- only 17 per cent of the country is flat enough for farming, industry and settlement
- the country has very limited energy resources and has to import virtually all the oil, natural gas and coal it needs
- the country lacks most of the basic raw materials needed by industry. It no longer has workable supplies of iron ore and coking coal (yet it is the world's second largest steel producer), nor has minerals of any significance.

How was Japan's economic miracle achieved?

- Post-war demilitarisation meant money was invested into the economy rather than spent on armaments.
- It had political stability and a government committed to industrialisation.
- Modern machinery and technology was introduced and the profits re-invested into research.
- The workforce were prepared to work long hours, to become better educated and trained, to work as a team, and to give total loyalty to their company.
- The country had, especially along its Pacific coastline, many deep and sheltered harbours which facilitated the import of energy and raw materials and the export of manufactured goods (Figure 8.18).
- The development of transport is based on a first-class rail network, motorway expressways and the use of coastal shipping.
- The domestic demand (market) for high-quality goods increased rapidly as Japan's population and standard of living increased.
- Japan had the wealth and the technology to reclaim land from the sea (Figure 8.19).

These factors have led to the concentration of industry around five main coastal areas (Figure 8.18). The Japanese themselves consider the most important industrial location factors to be the distribution of lowland, the distribution of people, and the availability of sheltered deep-water ports.

Japan began to improve its economy by using its limited supplies of iron ore and coal to produce steel. The steel was used to make ships specifically designed to carry the necessary raw materials (oil tankers and ore carriers). Attention was then turned to making cars and, later, developing electronics and high-technology industries. Although most of the world's largest car and electronic transnationals are Japanese, the vast majority of local firms are still small family units, many of which make component parts for their transnational neighbours.

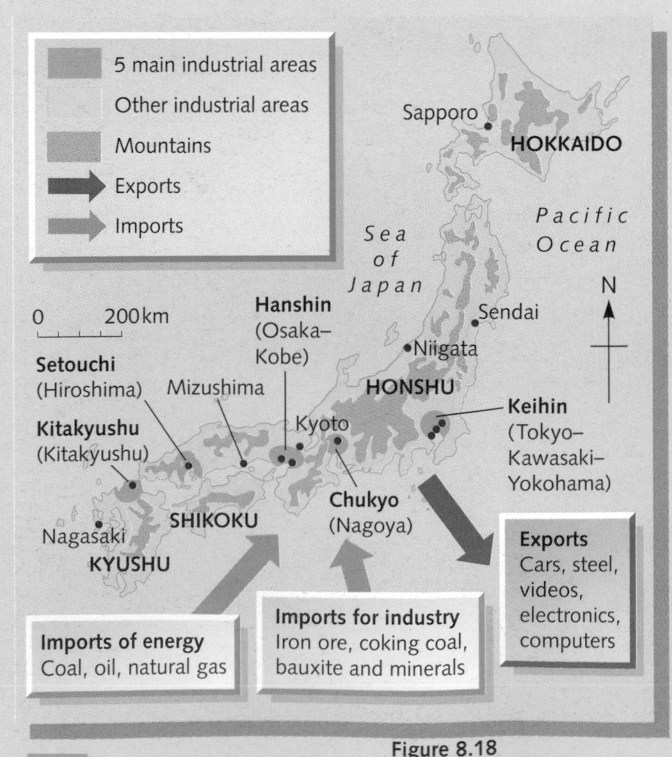

Figure 8.18
Industry in Japan

Figure 8.19
The Mizushima works of the Kawasaki Steel Corporation, built on land reclaimed from the sea

The NICs in East Asia

Encouraged by Japan's success, other governments in eastern Asia set out to improve their standards of living. They did this by investing in manufacturing industry and, early on, developing heavy industries (e.g. steel and shipbuilding). Later, they were to concentrate on high-tech industries (page 126). Manufacturing output rose most rapidly after 1960 in South Korea, Taiwan, Hong Kong and Singapore – four countries which collectively became known as the 'four tigers'. Like Japan, these four countries lacked basic raw materials, had governments that introduced long-term industrial planning and had a dedicated workforce which was reliable and, initially, was prepared to work long hours for relatively little pay. Economic growth in these NICs continued during the 1980s at a time when it was slowing down in the developed economies and world manufacturing was declining.

Since the 1980s, Malaysia (the most successful), Thailand, Indonesia and, to a lesser extent, the Philippines have attempted to join the list of NICs. The next to emerge, and potentially the largest, is likely to be China (Figure 8.20).

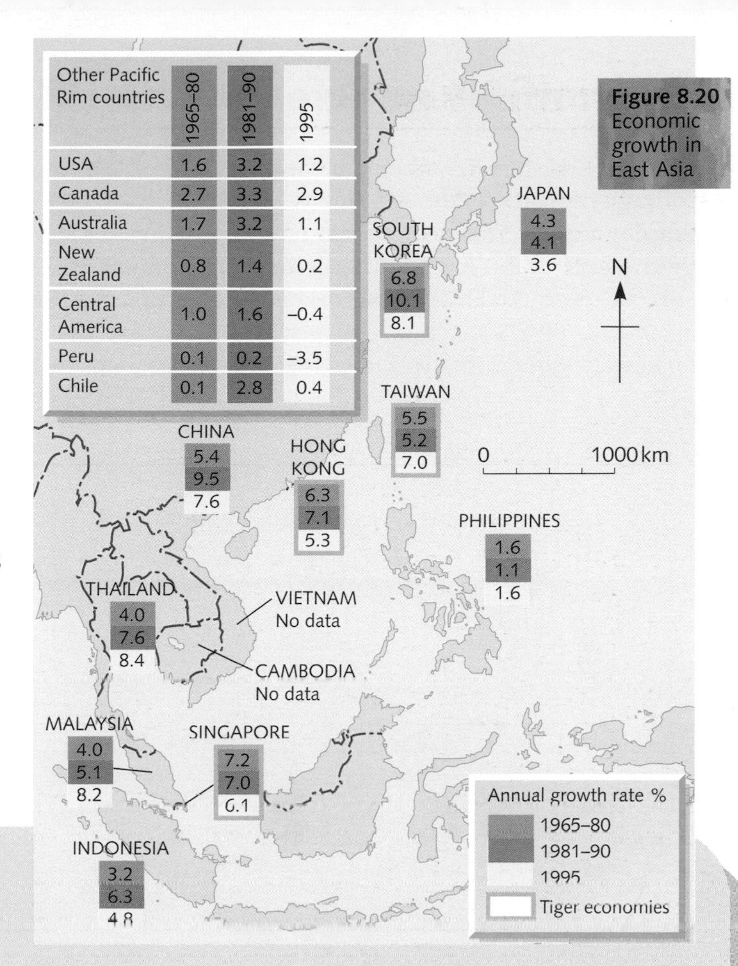

Other Pacific Rim countries	1965–80	1981–90	1995
USA	1.6	3.2	1.2
Canada	2.7	3.3	2.9
Australia	1.7	3.2	1.1
New Zealand	0.8	1.4	0.2
Central America	1.0	1.6	–0.4
Peru	0.1	0.2	–3.5
Chile	0.1	2.8	0.4

Figure 8.20 Economic growth in East Asia

Malaysia

Since 1990, Malaysia's annual economic growth rate has averaged 8 per cent. This has been achieved without high inflation or unmanageable foreign aid. Much of this success is due to the policies of its prime minister, Dr Mahathir, who himself quotes two policies:

1. 'Malaysia Inc', which is the government's aim of turning Malaysia into a fully industrialised country by the year 2020 (i.e. an economically *more* developed country).
2. The decision, made in 1983, to privatise many of the industries and economic sectors which had been set up with strong government support in the early 1960s. This has meant a change in government policy from the original objective of rapid industrialisation under state control to one by which Malaysia's economy will be transformed into one that should be successful and sustainable.

The government is, at present, investing less money in industries that require large workforces and more in ones where the emphasis is on technology. A government 'Technology Action Plan' covers automated manufacturing, micro-electronics, biotechnology and information technology (Figure 8.21).

In 1997, Malaysia had little unemployment – indeed it had to rely on migrant workers from Indonesia and the Philippines (page 78). It has attracted many foreign investors and high-tech firms, and is building a new international airport, a new town and several science parks. One showpiece of success is the Petronas Building in Kuala Lumpur, the tallest in Asia, which will be seen by the world during the 1998 Commonwealth Games (Figure 8.22).

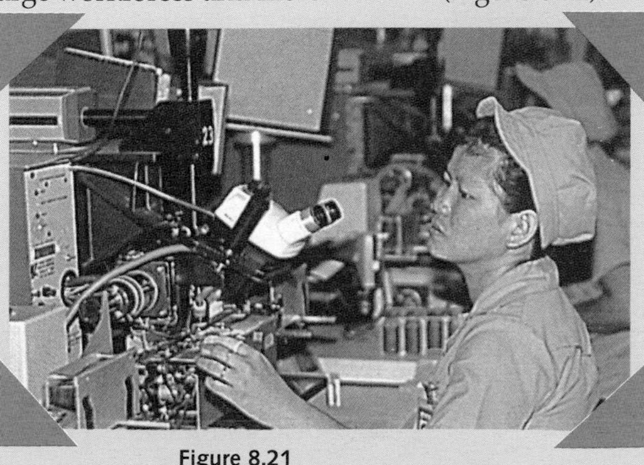

Figure 8.21 High-tech industry in Penang ('Silicon Island')

Figure 8.22 Petronas Towers, Kuala Lumpur

Government policies in the UK

Successive British governments have tried, since 1945, to encourage industry to move to areas of high unemployment. The size and location of these areas of high unemployment have changed over time. Figure 8.23 shows the areas which were regarded as needing the most assistance in 1997. Over the years governments have tried to encourage new industries to reduce unemployment by:

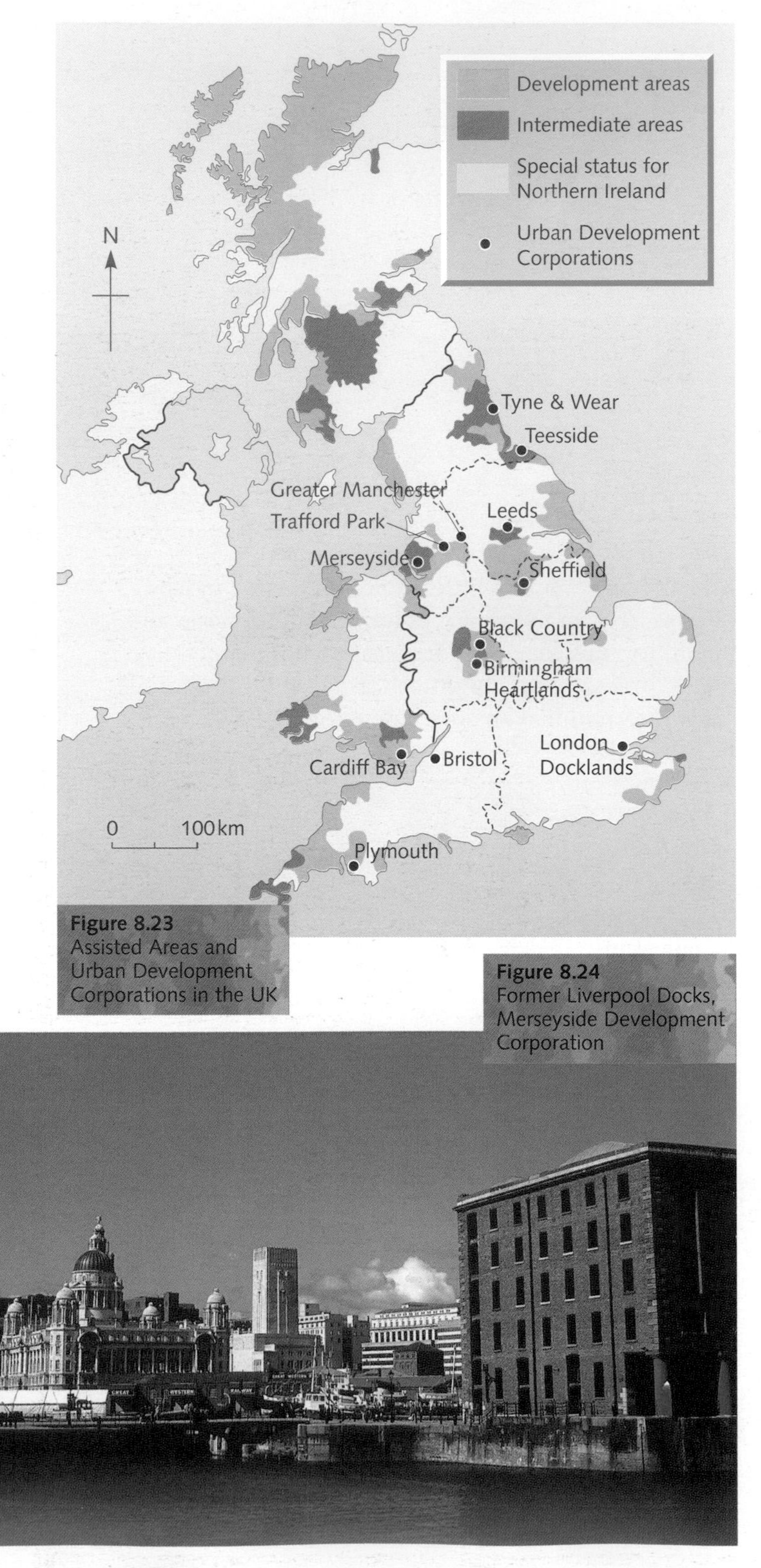

Figure 8.23 Assisted Areas and Urban Development Corporations in the UK

Figure 8.24 Former Liverpool Docks, Merseyside Development Corporation

- industrial development certificates which control where a firm can locate – these were first issued by the British government in 1947
- the creation of new towns in order to take work to the unemployed
- providing 'advanced factories' and industrial estates with services already present (e.g. roads, electricity)
- financial aid in the form of removal grants, rent-free periods, tax relief on new machinery, and reduced interest rates
- decentralising government offices
- improving communications and accessibility
- subsidies to keep firms going which otherwise would close down
- retraining schemes
- job creation, Manpower Creation Schemes (MCS) and Youth Training Schemes (YTS)
- Enterprise Zones
- assistance from the EU.

Enterprise Zones (EZs)

The first Enterprise Zones came into operation in 1981. They were planned for areas in acute physical and economic decay, with the aim of creating conditions for industrial and commercial revival by removing certain tax burdens and administrative controls. They tend to fall into two main groups:

1. Old inner city areas where factories had closed, causing high unemployment, and where old houses had been pulled down and the land left derelict, e.g. Isle of Dogs (pages 34 and 35).
2. Towns that had relied upon one major industry which had been forced to close, e.g. Corby.

Urban Development Corporations (UDCs)

Urban Development Corporations were created by Act of Parliament in 1980. The first two, in London's Docklands (the LDDC – see pages 34 and 35) and the Merseyside Development Corporation (Figure 8.24), were set up in 1981. Although by 1997 there were 13 (Figure 8.23), several are soon to be wound down, beginning with the LDDC by 31 March 1998.

UDCs were an attempt by the government to rejuvenate areas, often in the inner cities, which had undergone economic, social and environmental decay. Four of their main tasks were to:

- reclaim and secure the development of derelict and unused land
- provide land for industry, commerce, housing and leisure
- build roads and improve the quality of the environment
- encourage private investment to protect existing jobs and to create new ones.

These tasks were to be financed jointly by private sector investment and public funds from both the British government and the EU.

Trafford Park Development Corporation

During the early 1980s there was growing concern over the decline of Trafford Park, an area in Manchester which included Britain's first large industrial estate. Early attempts to improve the area included granting it Assisted Area status and, later, parts were designated an Enterprise Zone (Figure 8.23). Although these initiatives produced some results, they failed to tackle the underlying problems of an increasingly outdated infrastructure and the lack of attractive development opportunities to encourage investment. In 1987 Trafford Park, and the former steelworks site at Irlam 4 km to the south-west, became an Urban Development area. The Development Corporation identified four main development areas – Northbank Industrial Park, Village, Wharfside and Hadfield Street (Figure 8.25). Most recent (opening 1998) is the Trafford Centre, a huge two-level 250 unit retail/recreation complex. The Corporation was wound up in March 1998 having exceeded in all eight of its targets for investment, jobs and new companies attracted.

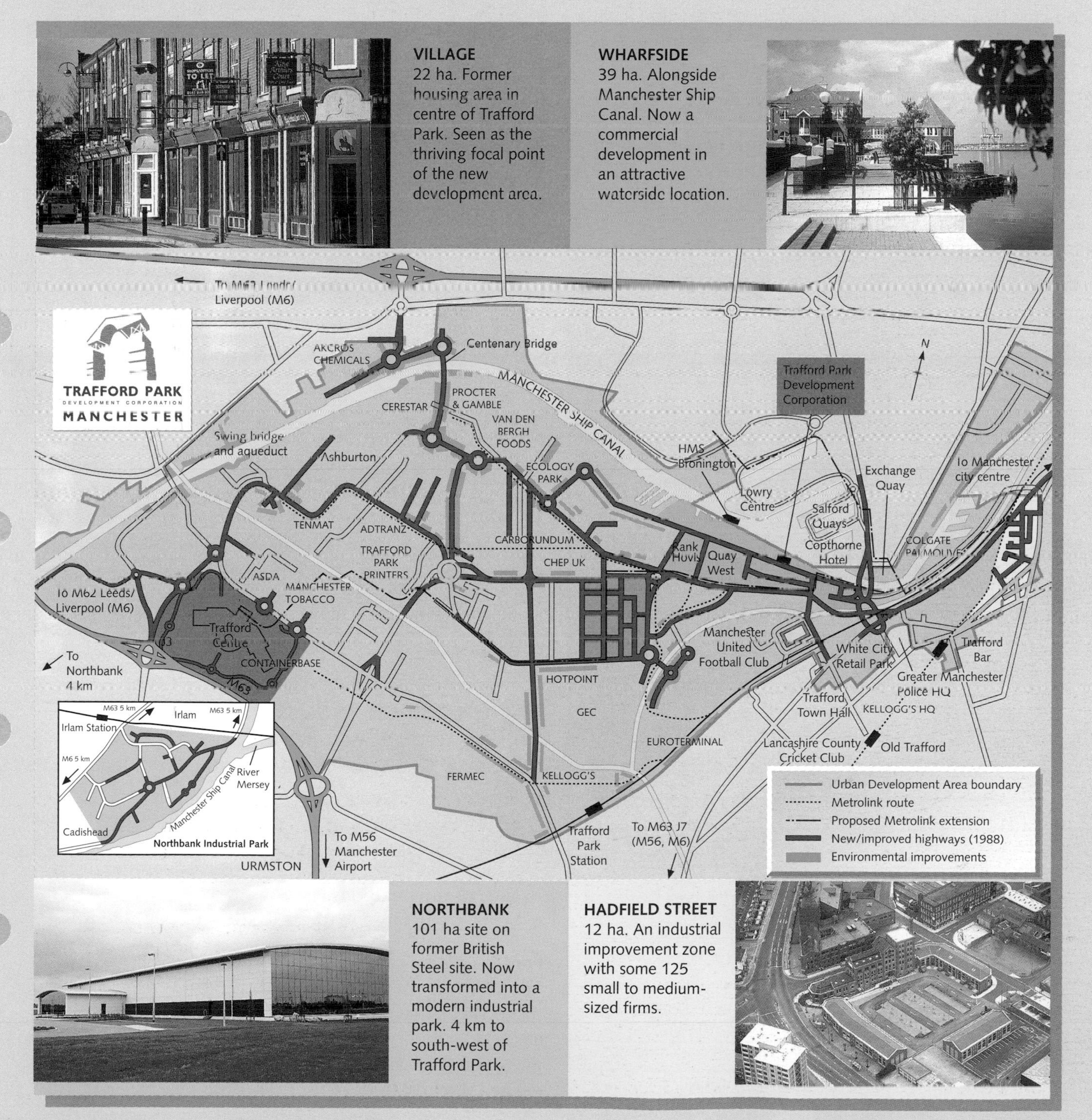

Figure 8.25 The Trafford Park urban development area

Transnational (or multinational) corporations

A transnational corporation, also referred to as a multinational company, is one that operates in many countries regardless of national boundaries. The headquarters and main factory is usually in an economically more developed country with, increasingly, branch factories in economically less developed countries. Transnationals (TNCs) are believed to directly employ some 40 million people around the world, to indirectly influence an even larger number, and to control over 75 per cent of world trade. The largest TNCs have long been car manufacturers and oil corporations but these have, more recently, been joined by electronic and high-tech firms (Figure 8.26a). It has been estimated that TNCs controlled one-fifth of the world's manufacturing in 1966 and over a half by the late 1990s. Several of the largest corporations have a higher turnover than all of Africa's GNP in total (see Figure 8.26b). One predicted change for the early twenty-first century is the expected increase in the number of TNCs based in the Pacific Rim (page 129). Figure 8.26a shows that Samsung (South Korea) has reached 14th place in terms of output.

Figure 8.26 Dominance of transnational corporations

(a)

INDUSTRIAL CORPORATION	COUNTRY	SALES ($ MILLIONS)	EMPLOYEES
1 GENERAL MOTORS	USA	133 622	710 800
2 FORD MOTOR	USA	108 521	322 200
3 EXXON	USA	97 825	91 000
4 ROYAL DUTCH/ SHELL GROUP	UK/ NETHERLANDS	95 134	117 000
5 TOYOTA MOTOR	JAPAN	85 283	109 279
6 HITACHI	JAPAN	68 581	330 637
7 IBM (INTERNATIONAL BUSINESS MACHINES)	USA	62 716	267 196
8 MATSUSHITA ELECTRIC INDUSTRIAL	JAPAN	61 384	254 059
9 GENERAL ELECTRIC	USA	60 823	222 000
10 DAIMLER-BENZ	GERMANY	59 102	366 736
11 MOBIL	USA	56 576	61 900
12 NISSAN MOTOR	JAPAN	53 760	143 310
13 BRITISH PETROLEUM	UK	52 385	72 600
14 SAMSUNG	SOUTH KOREA	51 345	191 303

(b)

COUNTRY (EXAMPLES)	GNP ($ MILLIONS PER COUNTRY, NOT PER PERSON)
RWANDA	1499
AFGHANISTAN	3100
NEPAL	3174
ETHIOPIA	5200
KENYA	6743
GHANA	7036
SRI LANKA	10 688
BANGLADESH	25 882

ADVANTAGES TO THE COUNTRY	DISADVANTAGES TO THE COUNTRY
BRINGS WORK TO THE COUNTRY AND USES LOCAL LABOUR	NUMBERS EMPLOYED SMALL IN COMPARISON WITH AMOUNT OF INVESTMENT
LOCAL WORKFORCE RECEIVES A GUARANTEED INCOME	LOCAL LABOUR FORCE USUALLY POORLY PAID
IMPROVES THE LEVELS OF EDUCATION AND TECHNICAL SKILL OF THE PEOPLE	VERY FEW LOCAL SKILLED WORKERS EMPLOYED
BRINGS WELCOME INVESTMENT AND FOREIGN CURRENCY TO THE COUNTRY	MOST OF THE PROFITS GO OVERSEAS (OUTFLOW OF WEALTH)
COMPANIES PROVIDE EXPENSIVE MACHINERY AND MODERN TECHNOLOGY	MECHANISATION REDUCES THE SIZE OF THE LABOUR FORCE
INCREASED GROSS NATIONAL PRODUCT/PERSONAL INCOME CAN LEAD TO AN INCREASED DEMAND FOR CONSUMER GOODS AND THE GROWTH OF NEW INDUSTRIES	GNP GROWS LESS QUICKLY THAN THAT OF THE PARENT COMPANY'S HEADQUARTERS, WIDENING THE GAP BETWEEN DEVELOPED AND DEVELOPING COUNTRIES
LEADS TO THE DEVELOPMENT OF MINERAL WEALTH AND NEW ENERGY RESOURCES	MINERALS ARE USUALLY EXPORTED RATHER THAN MANUFACTURED AND ENERGY COSTS MAY LEAD TO A NATIONAL DEBT
IMPROVEMENTS IN ROADS, AIRPORTS AND SERVICES	MONEY POSSIBLY BETTER SPENT ON IMPROVING HOUSING, DIET AND SANITATION
PRESTIGE VALUE (e.g. VOLTA PROJECT)	BIG SCHEMES CAN INCREASE NATIONAL DEBT (e.g. BRAZIL)
WIDENS ECONOMIC BASE OF COUNTRY	DECISIONS ARE MADE OUTSIDE THE COUNTRY, AND THE FIRM COULD PULL OUT AT ANY TIME
SOME IMPROVEMENT IN STANDARDS OF PRODUCTION, HEALTH CONTROL, AND RECENTLY IN ENVIRONMENTAL CONTROL	INSUFFICIENT ATTENTION TO SAFETY AND HEALTH FACTORS AND THE PROTECTION OF THE ENVIRONMENT

Figure 8.27 Advantages and disadvantages of transnational corporations

Many organisations and individuals have attacked TNCs as being exploiters of poor people, especially women and children, who live in economically less developed countries. Yet talking to several of these workers in countries as far apart as Brazil, Kenya, Sri Lanka and Malaysia, their attitude was: 'Perhaps, but it is the only way by which we can find full-time work.' Do you think, after studying Figure 8.27, that TNCs are, on balance, a blessing or a curse to an economically less developed country?

The global car industry

Car firms were amongst the first to opt for transnational operations. They found that by locating in different parts of the world they could:

- get around trade barriers which may have been erected to protect home markets
- reduce costs by gaining access to cheaper labour and/or raw materials
- be nearer to large markets (centres of population).

Ford – a global car corporation

The relatively new term 'global factory' is used to describe those TNCs that see the world, rather than the local area, as their supplier of labour, raw materials and component parts and their areas of sales. In other words they have created a world market for their products.

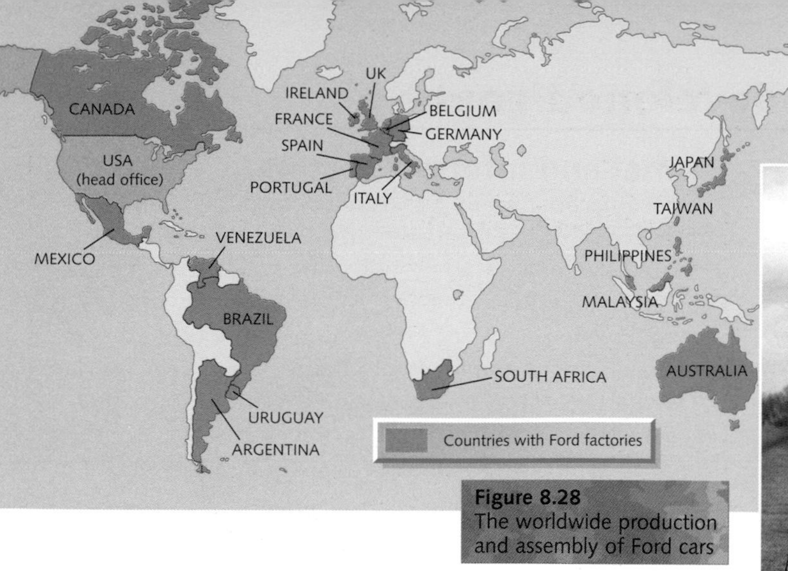

Figure 8.28 The worldwide production and assembly of Ford cars

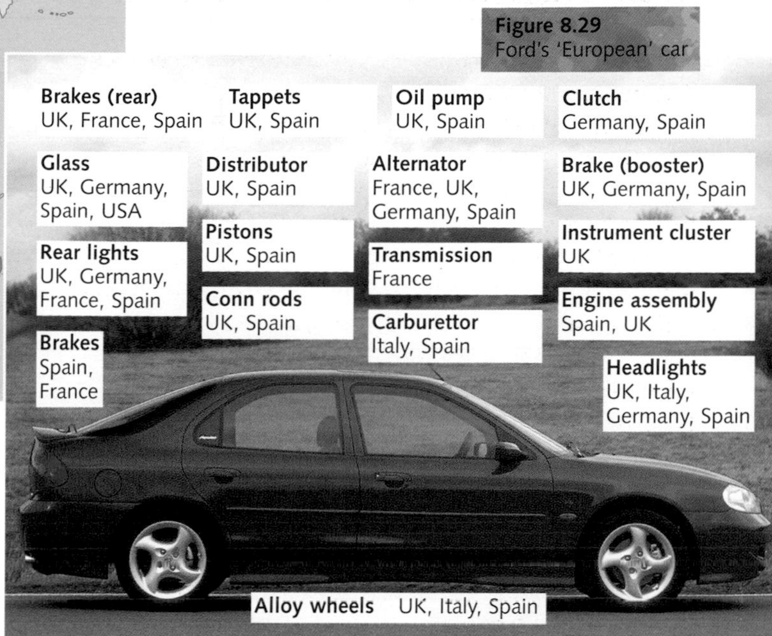

Figure 8.29 Ford's 'European' car

The giant Ford Corporation originally located in Detroit. By the late 1990s, a century later, it was:

- manufacturing and/or assembling its cars worldwide, although the bulk of the parts were still produced in the more industrialised parts of North America, Japan and the EU (Figure 8.28)
- increasingly locating its new factories, which either manufactured cars (e.g. São Paulo) or assembled parts made elsewhere (e.g. Malaysia, the Philippines), in economically less developed countries
- increasingly making parts in several countries (reducing the risk of strikes) so that each particular model is no longer made in one country – Figure 8.29 shows that a Ford car sold in the UK may have been assembled here or elsewhere in the EU, and that its parts may have come from several European countries
- facing increased competition, especially from Japanese manufacturers, at a time of economic recession – any resultant factory closures are more likely in the economically less developed countries where the local market is smaller and where redundancies are of less concern to the parent company
- working in Detroit with its previous rivals, Chrysler and General Motors, to produce a car that will use less fuel, cause less pollution and challenge the dominance of Japanese and, increasingly, Korean cars.

Detroit – motor city

Henry Ford, a local man, saw Detroit as an ideal location for what was to be the world's first mass production line. He built his factory on flat land next to the Detroit River (Figures 8.30 and 8.31) at the heart of the Great Lakes waterway system. Steel was produced on an adjacent site using relatively local iron ore (brought by ship) and coal (brought by train). Ford developed a large local market by paying his workers $5 a day, when the national average was $9 a week, enabling them to buy their own cars. The high wages attracted workers from all over the world, especially from the south-east of the USA. Later, America's two other car giants, Chrysler and General Motors, located their main factories at Detroit. After a serious depression in the 1970s (due to the world oil crisis and competition from Japan), Detroit's car industry is again thriving.

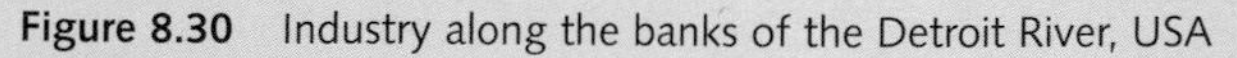

Figure 8.30 Industry along the banks of the Detroit River, USA

Figure 8.31 Ford's car plant, Detroit

Industry in economically less developed countries

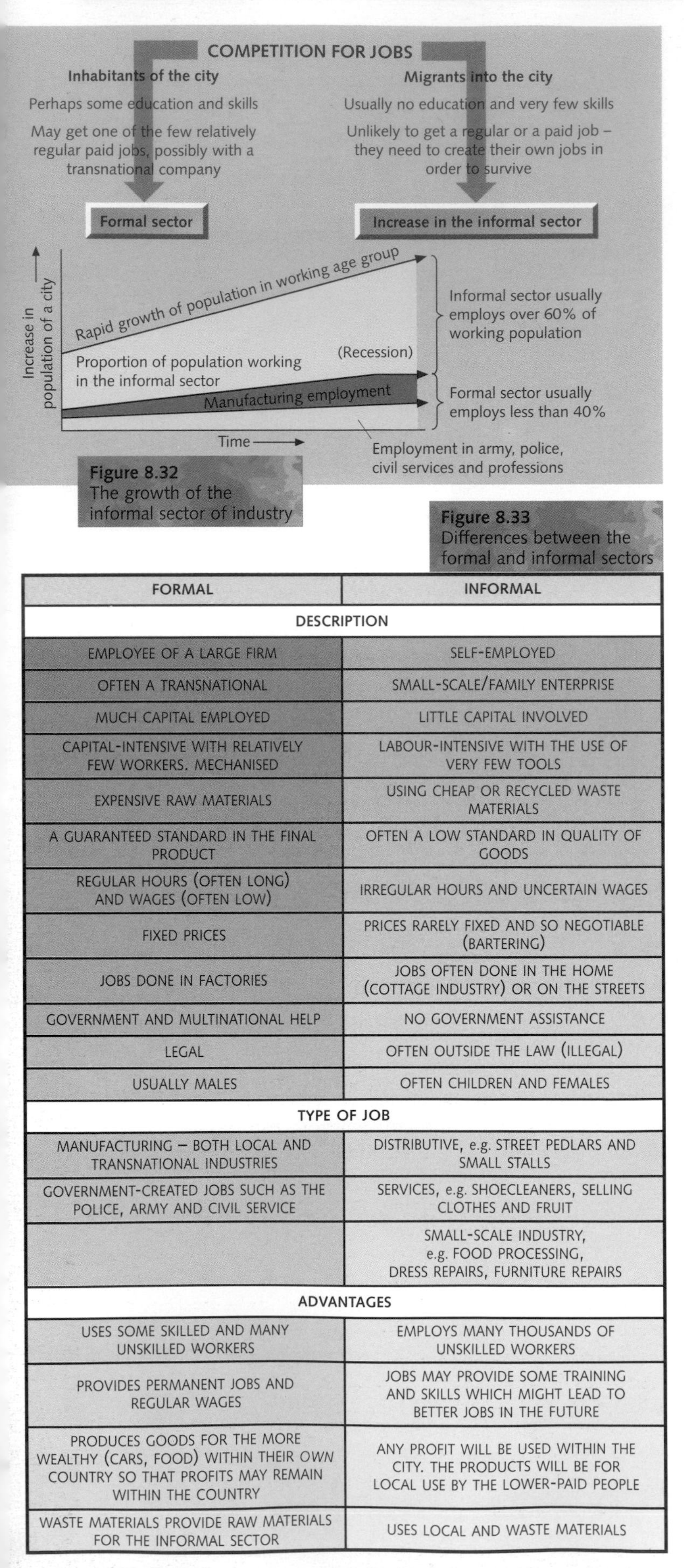

Figure 8.32 The growth of the informal sector of industry

Figure 8.33 Differences between the formal and informal sectors

FORMAL	INFORMAL
DESCRIPTION	
EMPLOYEE OF A LARGE FIRM	SELF-EMPLOYED
OFTEN A TRANSNATIONAL	SMALL-SCALE/FAMILY ENTERPRISE
MUCH CAPITAL EMPLOYED	LITTLE CAPITAL INVOLVED
CAPITAL-INTENSIVE WITH RELATIVELY FEW WORKERS. MECHANISED	LABOUR-INTENSIVE WITH THE USE OF VERY FEW TOOLS
EXPENSIVE RAW MATERIALS	USING CHEAP OR RECYCLED WASTE MATERIALS
A GUARANTEED STANDARD IN THE FINAL PRODUCT	OFTEN A LOW STANDARD IN QUALITY OF GOODS
REGULAR HOURS (OFTEN LONG) AND WAGES (OFTEN LOW)	IRREGULAR HOURS AND UNCERTAIN WAGES
FIXED PRICES	PRICES RARELY FIXED AND SO NEGOTIABLE (BARTERING)
JOBS DONE IN FACTORIES	JOBS OFTEN DONE IN THE HOME (COTTAGE INDUSTRY) OR ON THE STREETS
GOVERNMENT AND MULTINATIONAL HELP	NO GOVERNMENT ASSISTANCE
LEGAL	OFTEN OUTSIDE THE LAW (ILLEGAL)
USUALLY MALES	OFTEN CHILDREN AND FEMALES
TYPE OF JOB	
MANUFACTURING – BOTH LOCAL AND TRANSNATIONAL INDUSTRIES	DISTRIBUTIVE, e.g. STREET PEDLARS AND SMALL STALLS
GOVERNMENT-CREATED JOBS SUCH AS THE POLICE, ARMY AND CIVIL SERVICE	SERVICES, e.g. SHOECLEANERS, SELLING CLOTHES AND FRUIT
	SMALL-SCALE INDUSTRY, e.g. FOOD PROCESSING, DRESS REPAIRS, FURNITURE REPAIRS
ADVANTAGES	
USES SOME SKILLED AND MANY UNSKILLED WORKERS	EMPLOYS MANY THOUSANDS OF UNSKILLED WORKERS
PROVIDES PERMANENT JOBS AND REGULAR WAGES	JOBS MAY PROVIDE SOME TRAINING AND SKILLS WHICH MIGHT LEAD TO BETTER JOBS IN THE FUTURE
PRODUCES GOODS FOR THE MORE WEALTHY (CARS, FOOD) WITHIN THEIR *OWN* COUNTRY SO THAT PROFITS MAY REMAIN WITHIN THE COUNTRY	ANY PROFIT WILL BE USED WITHIN THE CITY. THE PRODUCTS WILL BE FOR LOCAL USE BY THE LOWER-PAID PEOPLE
WASTE MATERIALS PROVIDE RAW MATERIALS FOR THE INFORMAL SECTOR	USES LOCAL AND WASTE MATERIALS

Formal and informal sectors

In cities in economically less developed countries, the number of inhabitants greatly outweighs the number of jobs available. With the rapid growth of these cities the job situation is continually worsening. An increasing number of people have to find work for themselves and thus enter the informal sector of employment (Figure 8.32), as opposed to the formal sector (i.e. the professions, offices, shops and organised modern industry). The differences between the formal and informal sectors are given in Figure 8.33.

In many cities there are now publicly and privately promoted schemes to support these self-help efforts. In Nairobi, for example, there are several *jua kali* (meaning 'under the hot sun') with metal workshops. In one area little more than the size of three football pitches, 1000 workers hammer scrap metal into an assortment of products (Figure 8.34).

Figure 8.34 Workers in the informal sector

Role of children

Children, many of whom are under 10 years old, make up a large proportion of the informal sector workers. Very few of them have schools to go to, and from an early age they go out onto the streets to try to supplement the family income. One such 'worker', who has become well known in British schools, was a shoe-shine boy called Mauru who lived in São Paulo, and was seen in a TV programme *Skyscrapers and Slums*. He would try to earn money during the day, and study to become an airline pilot, at night.

Figure 8.35
The informal sector
Main photo: Copacabana beach and Sugarloaf Mountain
Insets: Girl selling shuttlecocks in Rio de Janeiro; beach vendors on Copacabana beach

❝Our small group of 11 tourists from Britain was staying at the Rio Palace, a five-star hotel overlooking the famous Copacabana beach in Rio de Janeiro. Walking along the beach one afternoon, I noticed numerous beach vendors. These form part of Rio's large informal sector. Several vendors carried large umbrellas from which dangled an assortment of sun hats, suntan lotion or *tangas* (bikinis). One vendor carried pineapples in a basket perched on his head and, in his hand, a large knife with which to cut the fruit. Some vendors carried cool-boxes in which were ice cream, Coca-Cola, coconut water and other drinks, while others carried large metal drums which contained *maté* (a local drink). These people drew attention to themselves by shouting, blowing whistles, beating metal drums or whirling a metal ratchet that clattered loudly. On the pavement next to the beach were small children trying to sell sweets and chocolate, and numerous kiosks with fruit and drinks available.

Returning to the hotel, several of us were tempted to buy a brightly feathered shuttlecock from a girl who had a most enchanting smile (Figure 8.35) but we resisted other sellers of cheap jewellery and Copacabana T-shirts. That evening as our group ate in a restaurant next to the Rio Palace, other pedlars poked their heads through the open door and windows trying to sell not very musical instruments, monkey-puppets and T-shirts. As we left the restaurant at 9.30 pm to visit one of Rio's famous Samba shows, we saw a small boy of four or five years with sad, appealing eyes trying to sell roses individually wrapped in cellophane. How could one fail to buy a rose which, he indicated by his fingers, were only 15 crusadas (about 50p)? However, by the time the money was produced, he had raised his price to 20 crusadas – and shown us that his eyes had been trained and his brain sharpened for business. When we returned at 1.00 am, he was still outside the hotel with the remainder of his roses.❞

David Waugh, 1987

Sustainable development industry in economically less developed countries

In most economically less developed countries, not only are high-tech industries too expensive to develop, they are usually inappropriate to the needs of the local people and to the environment in which they live. An *appropriate technology* is exactly what it says – a technology appropriate or suitable to the place in which it is used.

Figure 8.36
Recycling materials in Nairobi

An appropriate technology can contribute to a more sustainable way of life for people who are rich or poor, living in places which may be developed or developing. If the place is developed and industrialised and its inhabitants are well-off, then the appropriate technology is more likely to be high-tech. If the place is underdeveloped and its inhabitants are poor, then alternative forms of technology should be adopted. These alternative forms may include:

- labour-intensive projects – with so many people likely to be unemployed or underemployed, it is of little value replacing existing workers by machines
- encouraging technology that is sustainable and fully utilises the existing skills and techniques of local people
- using tools designed to take advantage of local knowledge and resources
- developing local crafts and industries by using local natural resources and, where possible, recycling materials (Figure 8.36)
- low-cost schemes using technologies that people can afford and manage (Figure 8.37)
- developing projects that are in harmony with the environment.

Figure 8.37
Appropriate technology in Nepal, one of the poorest countries in the world (see Figure 8.26b)

BEFORE

For women, most tasks are labour-intensive, time-consuming and have to be done by hand

AFTER

Villagers can now hull their rice mechanically with this 3 kW mill driven by a micro-hydro turbine – time is saved and quality and productivity increased

BEFORE

Grinding enough corn to feed a family for just 3 days takes 15 hours when it is done by hand

AFTER

By taking corn to the grinder in the mill-house – usually a popular meeting-place for villagers – 3 days' worth of corn can be ground in just 15 minutes

Intermediate Technology (IT)

Intermediate Technology is a British charitable organisation which works with people in developing countries, especially in rural areas (Figure 8.38). IT helps them to acquire the tools and techniques needed if they are to work themselves out of poverty. IT helps people to meet their needs of food, clothing, housing, farm and industrial equipment, energy and employment. IT uses, and adds to, local knowledge by providing technical advice, training, basic equipment and financial support so that people can become more self-sufficient and independent. Ideally the aim is for people to create a small surplus which can then be invested into their small businesses and communities.

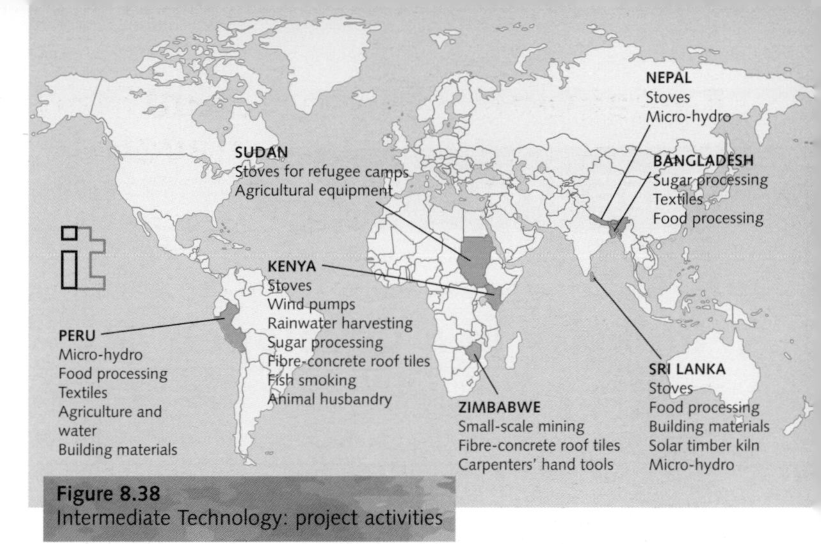

Figure 8.38
Intermediate Technology: project activities

IT's projects in Kenya

Most IT projects in Kenya, as elsewhere in the world, are in response to requests from local groups, often women, who need assistance in one or more stages of their planned work. For example:

1. **The Maasai Housing Project – towards sustainable rural housing improvement** (Figure 8.39) 'The traditional Maasai house, the *enkang*, was made from posts, a soil/cowdung mixture, and grass (Figure 9.31). A narrow opening served as an entrance and source of ventilation. Occupants slept and cooked in the dwelling, usually along with their sheep and goats. *Enkangs* were often characterised by low, leaking roofs; smoky, dark rooms; cramped space; a stench of animal odour; termite-infested posts; no water; and a lack of permanence. They posed health risks, such as eye and respiratory ailments; they were uncomfortable; and they needed constant repair. The community-based Maasai Housing Project is being helped by IT to find sustainable and affordable ways to upgrade and make *enkangs* more permanent. Ferro-cement skin roofs consist of a polythene sheet laid between a twigs/grass base and a thin covering of cement. Water is collected from the now-impermeable roof, saving the women the daily chore of walking several kilometres to the nearest river. Ferro-cement panel walls are made by nailing wire mesh to termite-proof posts, while larger openings and a chimney cowl make the inside of the dwelling lighter, less smoky and more healthy.'
2. **Roofing materials** (Figure 8.40) Improved urban building materials include roofing tiles which are made by adding natural fibres (such as sisal) and lime (or cement) to soil. The resultant blocks are dried, and prove to be far less expensive than commercially produced tiles.
3. **Energy** (Figure 8.41) – training potters in the production of fuel-efficient cooking stoves. The improved stove (*jiko*), drastically reduces the amount of fuelwood needed by rural families and charcoal by urban households. Based on a traditional design, it is made by local metal workers using scrap (recycled) metal. The improved stove reduces the time needed to collect fuelwood (see page 109) and, by reducing smoke, has improved women's health.

Figure 8.39 The Maasai Housing Project

Figure 8.40 Making low-cost tiles

Figure 8.41 A fuel-efficient cooking stove (*jiko*)

Industry in a developed city – Osaka-Kobe 2000

The Osaka-Kobe conurbation is one of Japan's major industrial areas. The many reasons for this industrial growth include the following:

- It is a natural harbour situated within Osaka Bay, an inlet of the Inland Sea. The Inland Sea itself is protected from the worst of the typhoon winds by the island of Shikoku (see page 47).
- Osaka imports many of the raw materials needed by Japan. Many of these raw materials are processed within the port area, e.g. oil and iron ore. The port is also a major outlet for Japanese exports.
- The land around Osaka Bay is one of the relatively few areas of flat land in a country that is 83 per cent too mountainous for development. Even so, Osaka-Kobe has spread outwards as far as the highland allows (Figure 8.42).
- It has a population of over 9 million which provides a highly skilled, dedicated workforce and a large, wealthy domestic market.
- High-tech industries have grown rapidly as a consequence of the inventiveness of the Japanese.

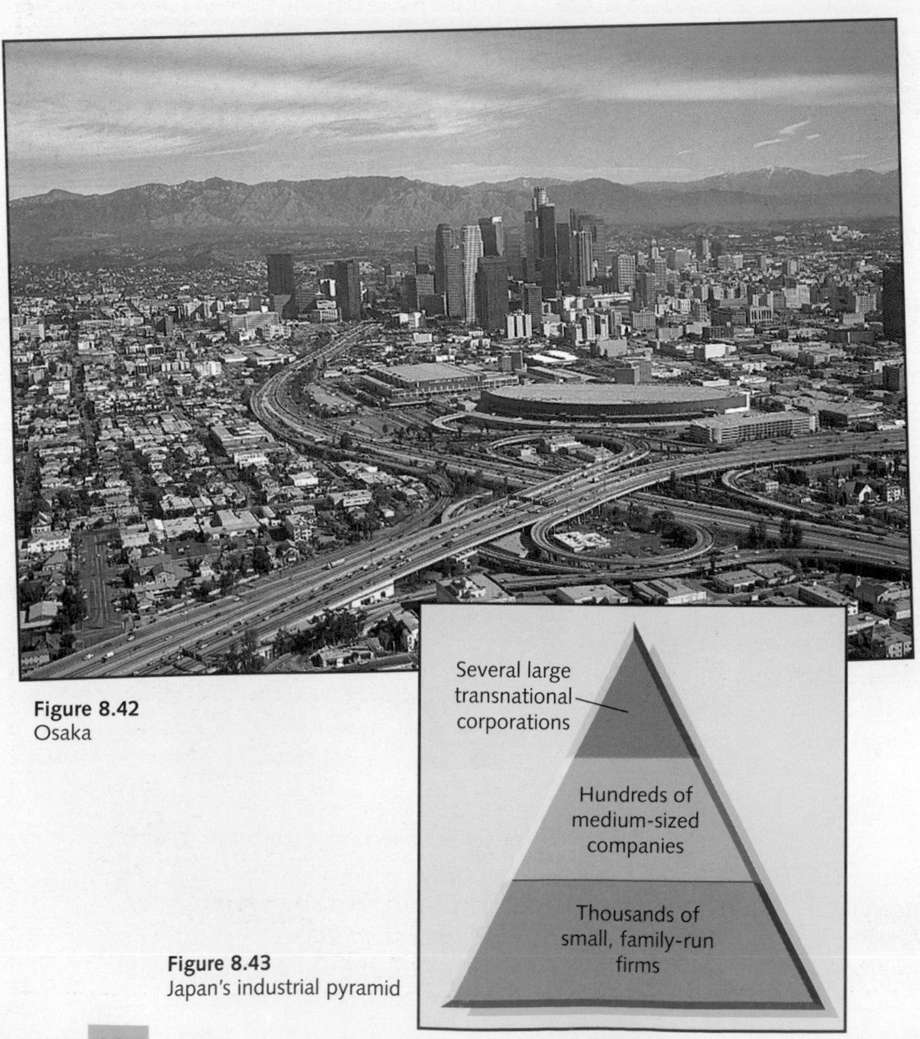

Figure 8.42
Osaka

Figure 8.43
Japan's industrial pyramid

Japanese industry is highly efficient. It is based on hard work, loyalty and trust. The Japanese compare their industrial organisation to a pyramid (Figure 8.43). At the base are thousands of small, often family-run, firms. Small firms, usually employing fewer than six people, account for 90 per cent of Japanese companies. Above them are medium-sized companies which, together with the small firms, produce 60 per cent of Japan's manufactured goods. At the top of the pyramid are the large corporations, many of which, like Toyota, Nissan, Sharp, Panasonic and Mitsubishi, have become major transnationals.

From oil to camcorders

The port petrochemical industry processes oil into many by-products, one of which is plastic. Some of this plastic is taken to the workshop district of Osaka which, for convenience, is located near to the motorway which skirts the east of the city (it can be identified on Figure 3.47). Within this district are numerous small work units (Figure 8.44), including the one run in a converted garage by Mr and Mrs Yamasaki. Mr and Mrs Yamasaki produce plastic parts which they send daily to a nearby medium-sized company run by the Kawasaki brothers. The Kawasakis, who employ 18 people, produce a range of products which include, for example, the tops of car batteries and safety goggles. Recently they have won orders to produce sunglasses for Italy, and 100 000 lenses for Panasonic to use in its camcorders (Figure 8.45). The lenses, when finished, are sent 20 km across the city to the huge Panasonic plant.

Panasonic assembles the component parts made by many small and medium-sized firms. Although the assembly is highly automated with robots doing many very technical jobs, Panasonic still employs several hundred workers (Figure 8.46). Mrs Yamagami, who supervises eight robots, points out that she feels very much part of the company, despite its large size. Along with Japanese companies worldwide, Panasonic encourages close ties between management and staff. Mrs Yamagami, like other employees, is encouraged to suggest ways of improving both production and working conditions. The completed camcorders, along with other Panasonic products which include TVs and videos, are exported through the port at Osaka to 160 countries worldwide (Figure 8.47).

The major problem facing Panasonic at present is competition from the NICs, especially those in South-east Asia, whose labour and other production costs are cheaper (see page 129). Panasonic is trying to meet this challenge by: investing and setting up assembly in other countries; research and development to be ahead in technology; and seeking to develop new products.

Japanese industry does not stand still. Two recent developments include:

- demolishing the 100-year-old Nippon steelworks, located alongside Osaka Bay, and using former employees to create and develop new forms of technology (Japanese industrialists have little sentiment for anything that has outlived its usefulness)
- building up to 11 new science parks in pleasant environments beyond the present urban limits. The science parks, with their research centres paid for by Japan's leading corporations, are being developed to 'make sure that Japan's high-tech stays ahead of the rest of the world'.

Figure 8.44 The working district of Osaka

Figure 8.45 Making camcorder lenses

Figure 8.46 Automation in the Panasonic plant

Figure 8.47 From oil to camcorders

CASE STUDY 8B

Industry in a developing country – São Paulo 2000

Brazil is the most industrialised of the world's economically developing countries. Industrialisation is not, however, evenly spread across the country but is mainly concentrated in and around São Paulo (Figure 1.15). During the nineteenth century São Paulo was only a small town situated in the centre of a major coffee-growing area (Figure 6.49).

Its rapid urbanisation and industrialisation took place during the so-called 'economic miracle' of the 1960s and 1970s. At this time up to half a million people a year migrated to the city looking for work. Today the city has an estimated population of 23 million living in an urban area that extends across 100 km.

The region around São Paulo was well endowed with minerals, including iron ore, and had access to energy resources. This led to the development of the iron and steel and engineering industries (Figure 8.48) and the manufacture of machinery, aircraft and cars. Brazil is the world's ninth largest producer of cars. Four large car transnational corporations – Ford, Volkswagen, General Motors and Mercedes – all have assembly plants in São Paulo (Figure 8.49). They produce cars in sufficiently high numbers to satisfy the domestic market and for export to the rest of South America and even to the USA and the Middle East. Brazilian car workers may earn more than twice the average Brazilian wage, but to international manufacturers they are cheaper to employ than car workers in developed countries.

Industrialisation brought employment and created problems, especially in and around downtown São Paulo. Heavy industry and traffic have caused air pollution; the great number of cars has led to gridlock; and the location of commercial buildings (e.g. banks) and offices (headquarters of large companies) has created a 'sky-scraper jungle', high land prices and a lack of open space. One result has been the movement of industry to new towns. These have been created, mainly to attract industry, by building major roads and locating in a cleaner environment. One such town, Jundaia, is 100 km from São Paulo's city centre.

Figure 8.48 Iron and steel works, São Paulo

Figure 8.49 A car plant in São Paulo

Figure 8.50 Pepsi plant at Jundaia

Jundaia provides employment in the *formal sector* (page 134). Several transnational companies have already located there, including Pepsi (Figure 8.50). Although the plant is heavily automated, it still employs 350 people.

Pepsi's workers not only have a regular job with regular pay, they also get free lunches and free medical care. People living and working in Jundaia are enjoying a rising standard of living.

Nearer the CBD, some of São Paulo's increasing wealth is being used to turn run-down areas into modern business, retail and leisure centres (Figure 8.51). Unfortunately, the development of these new centres with their ultra-modern buildings can only take place by clearing existing favelas such as the one at Edith Gardens (Figure 4.12).

People who live in Edith Gardens favela usually only find work in the *informal sector*, as indeed do over one-third (about 4 million) of São Paulo's working population. The jobs in the informal sector include recycling materials, repairing goods and processing and/or selling food (Figure 8.52). Most of the residents living in Edith Gardens favela are not enjoying a rising standard of living.

Alcione Florencia is one such person. Alcione, who had no schooling, moved to São Paulo with her two small children. For the first three years she was unemployed. She then turned the front of her house into a bar, selling soft drinks to builders working on local construction sites, and took in washing (Figure 8.53). Now the council wish to 'improve' Edith Gardens (that is, to demolish it) and its residents are being offered £1100 to move out. Although Alcione wants to return to her home area, she feels the money is insufficient as it would take more than that to re-start a business elsewhere. To those in Edith Gardens who are unemployed, £1100 (if it is ever paid) is an inducement to dismantle their often flimsy homes and re-erect them in other favelas.

Figure 8.51 New business, retail and commercial centre in São Paulo

Figure 8.52 The informal sector in Edith Gardens favela

Figure 8.53 Alcione Florencia in her 'bar'

QUESTIONS

1

(Page 122)

a Choose and name a manufacturing industry you have studied. Copy and complete the top diagram (right) to show your chosen industry as a system. (6)

b Complete the star diagram to show some of the factors affecting the location of industry. The following list should help you:
- capital
- energy supplies
- labour
- natural routes
- raw materials
- relief
- transport
- markets (8)

c Choose **two** physical and **two** human and economic factors, and explain how they can influence the location of a factory. (2 x 4)

d i) Give three reasons why physical factors were more important than human and economic factors in the location of nineteenth-century industry. (3)

ii) Give three reasons why human and economic factors are usually more important than physical factors in the location of a modern industry. (3)

e Describe how political factors can affect the location of a factory. (3)

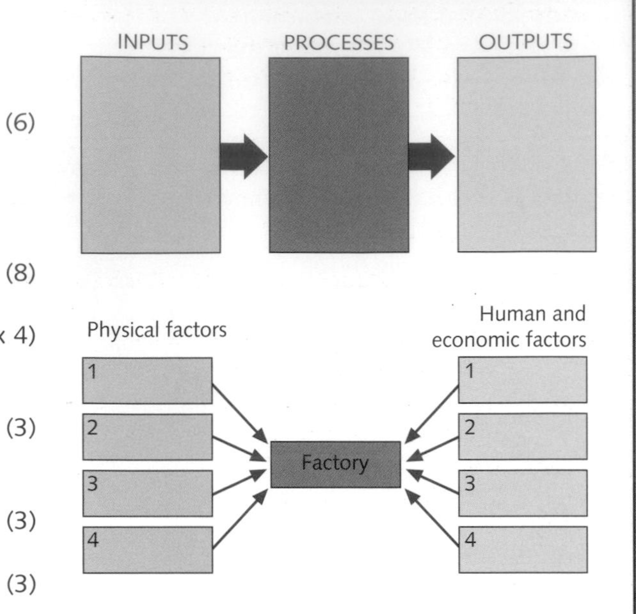

2

(Pages 124 and 125)

a i) Give four reasons why Town X became important for the manufacture of iron in the early nineteenth century. (4)

ii) Why did Town Y become an important exporting port? (2)

b i) Give four reasons why a large modern steelworks has been built at Town A. (4)

ii) Why was the modern steelworks not built at Town B? (2)

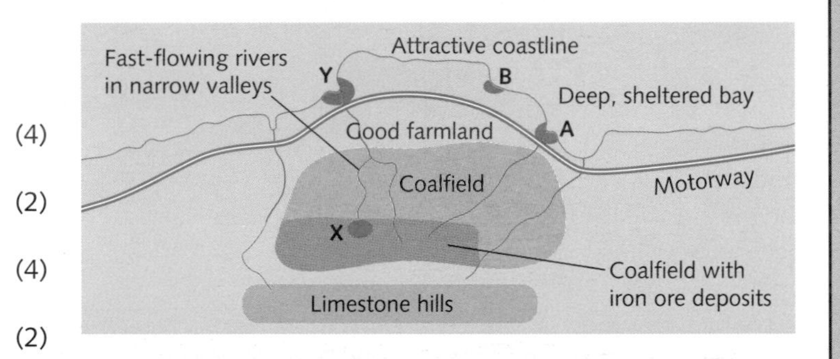

3

(Pages 36, 123 and 126)

Many of Britain's newer industries are said to be 'footloose' and are located on 'greenfield sites' on the edge of urban areas.

a What is meant by the terms:
- footloose
- greenfield site? (2)

b i) Give three examples of footloose industries. (3)

ii) Give three reasons why footloose industries locate on greenfield sites. (3)

c The sketch (right) has been drawn from the photo above which shows new industries located on the edge of a city. Copy and complete the sketch, choosing the correct six labels from the following list:
- nearby motorway
- near to city centre
- cheap land on edge of city
- nearby housing estate for workforce
- attractive greenfield site
- nearby city provides a large market
- space for car parking and future expansion (6)

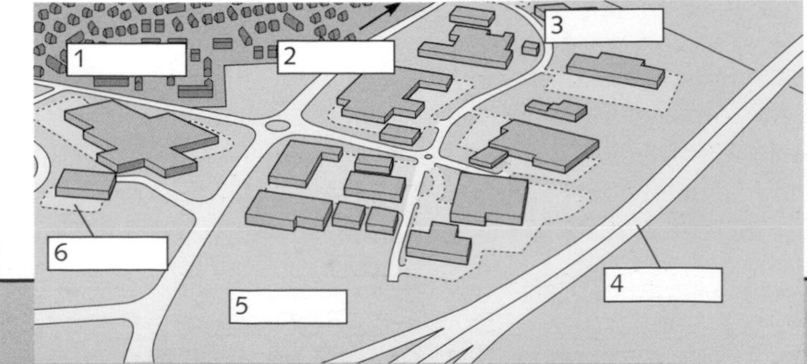

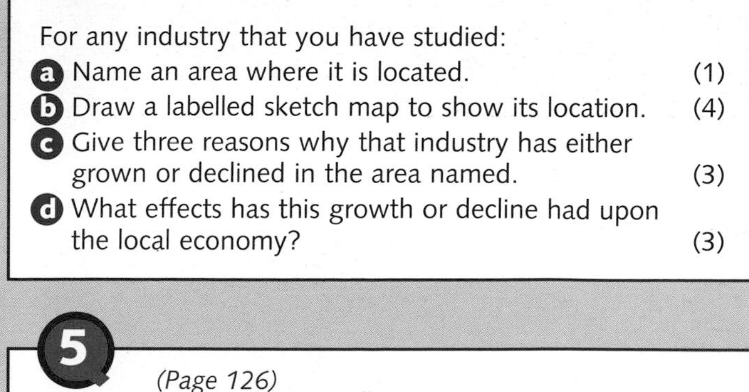

4

(Whole chapter)

For any industry that you have studied:

a Name an area where it is located. (1)

b Draw a labelled sketch map to show its location. (4)

c Give three reasons why that industry has either grown or declined in the area named. (3)

d What effects has this growth or decline had upon the local economy? (3)

5

(Page 126)

If a town wants to attract new industry it has to advertise its advantages.

a i) Give four advantages listed in the advertisement for locating a new factory in Swindon. (4)

ii) Explain why each would be an advantage for a firm seeking a new industrial site. (4)

b Give three other advantages of locating in Swindon that are not listed in the advertisement. (3)

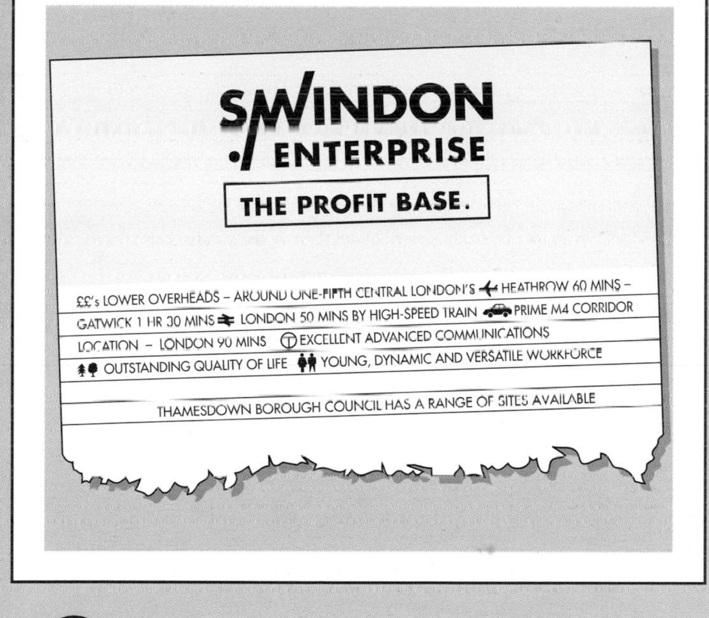

6

(Pages 126 and 127)

In recent years there has been a rapid growth of high-tech industries in the area between London and Bristol.

a i) What is meant by a 'high-tech' industry? (1)

ii) Give two examples of high-tech industries. (2)

b How have the following encouraged the location of high-tech industries between London and Bristol?

i) Accessibility

ii) Labour supply

iii) Universities

iv) Attractive countryside

v) Cultural and social attractions (5)

c What advantages does the photo of Tsukuba Science City (Figure 8.16) show that is likely to attract new industry? (4)

7

(Pages 126 and 127)

Study the information for the Cambridge Science Park below.

a i) What is a science park? (1)

ii) Name three types of company that have located in the Cambridge Science Park. (3)

iii) How many companies have located here? (1)

iv) How many of these companies employ fewer than 10 people? (1)

v) Why do most companies only employ a few people? (1)

vi) What type of people are likely to be employed in a science park? (2)

b i) Why has the science park been built beside the A14 northern by-pass which leads to the M11? (1)

ii) Why have so many car parks been provided? (2)

c Describe the layout of the science park under the following headings:

i) buildings

ii) road pattern

iii) landscaped areas. (6)

CAMBRIDGE SCIENCE PARK

To M11 London
A14 northern by-pass
A10 Milton Road
0 100 metres
N
Railway

Buildings
Car park
Landscaped area
Lake
Trees

Number of employees: 100+, 50–99, 20–49, 10–19, Less than 10
Number of companies: 5, 10, 15, 20, 25

Companies (by types): Electronics, Scientific instruments, Drugs and pharmaceuticals, Others

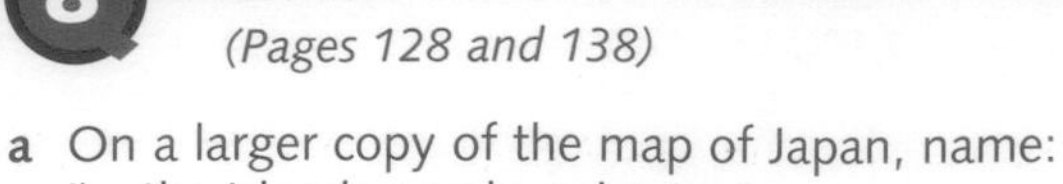

8

(Pages 128 and 138)

a On a larger copy of the map of Japan, name:

i) the islands numbered 1 to 4 (4)
ii) the cities numbered 5 to 8 (4)
iii) two imported types of energy (2)
iv) two imported minerals (2)
v) three major exports. (3)

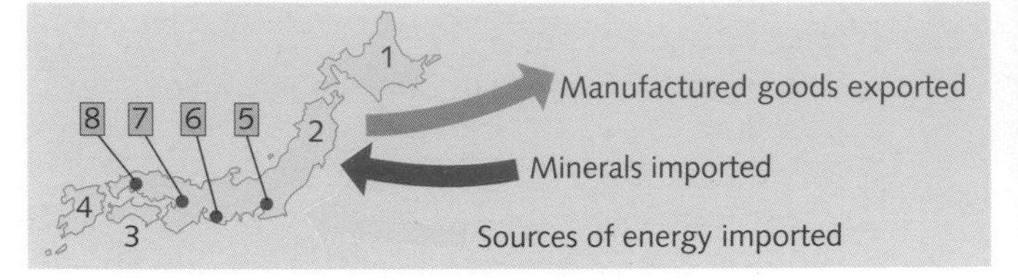

b Japan's industry developed despite three major physical disadvantages. What were these three disadvantages? (3)

c Give five reasons why Japan has become a rich, industrial country. (5)

d The photo shows the Mazda car plant in Hiroshima, Japan.

i) Why has Mazda built its factory on land reclaimed from the sea? (3)
ii) Give five other reasons why Hiroshima was a good site for a car plant. (5)

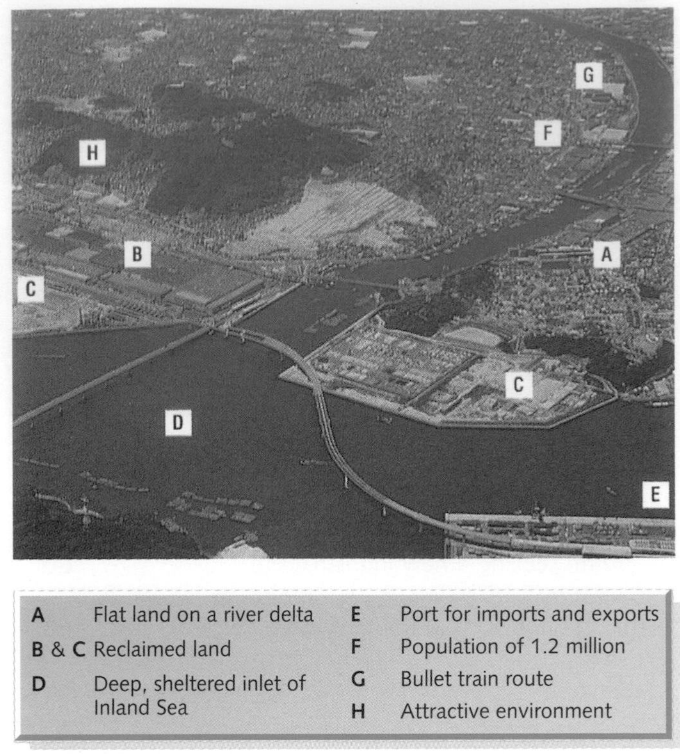

A	Flat land on a river delta	E	Port for imports and exports
B & C	Reclaimed land	F	Population of 1.2 million
D	Deep, sheltered inlet of Inland Sea	G	Bullet train route
		H	Attractive environment

9

(Page 129)

a Taiwan is one of the world's most successful NICs.

i) What is meant by NICs? (1)
ii) Name three other NICs in eastern Asia. (3)

b i) Describe the changes in the graph for Taiwan. (2)
ii) Give two reasons for these changes. (2)

c Explain why there is rapid economic growth in Malaysia. (4)

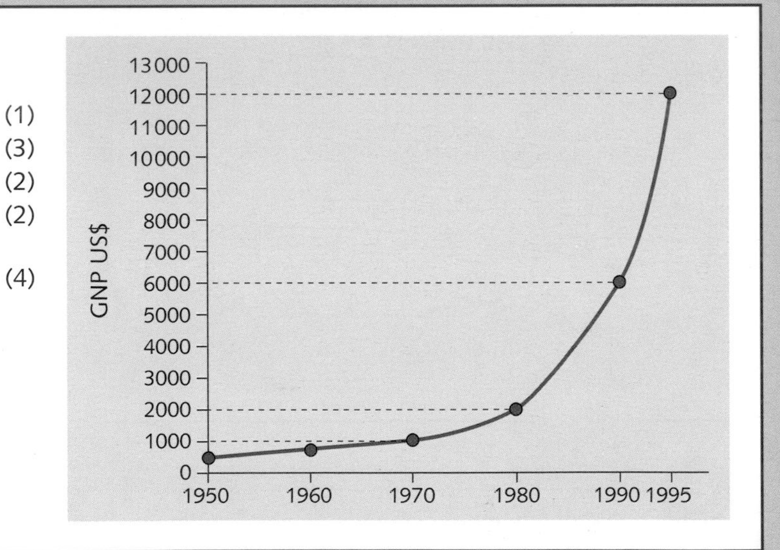

10

(Pages 130 and 131)

a i) In which parts of the UK are most of the Assisted Areas? (2)
ii) Name four Urban Development Corporations in England. (4)
iii) Name one Urban Development Corporation in Wales. (1)

b i) Give three reasons why some parts of the country needed government aid. (3)
ii) Describe three tasks of Urban Development Corporations. (3)

c i) Where is Trafford Park? (1)
ii) What were the main problems in Trafford Park in the 1980s? (3)

d Describe the improvements made so far by the Trafford Park Development Corporation, under the headings:

- transport
- employment
- the environment. (3 x 3)

11

(Pages 132 and 133)

a The map below shows that Japanese firms have invested large amounts of money all over the world.
- i) In which country has most investment been made? (1)
- ii) Which two Asian countries have received most Japanese investment? (2)
- iii) How much money has been invested in the UK? (1)

Answer *either* **b** *or* **c**.

b Why do Japanese firms want to open factories in other countries? Use the headings:
- transport costs
- labour costs. (4)

c Why do Japanese firms invest so much money in:
- i) economically more developed countries like the UK and USA
- ii) economically less developed countries like Malaysia and Hong Kong? (4)

d What is a transnational corporation (company)? (1)

e Name two transnational oil corporations, two transnational car corporations and two electronic/high-tech transnational corporations. (6)

f
- i) Give three possible advantages that a transnational corporation can bring to economically less developed countries. (3)
- ii) Give two possible disadvantages that a transnational corporation can bring to economically less developed countries. (2)

g
- i) Name the city where Ford has its headquarters. (1)
- ii) Why is this city a good site for:
  - Ford's headquarters
  - a major Ford assembly plant? (4)
- iii) Why are parts for Ford cars made in several different countries? (2)

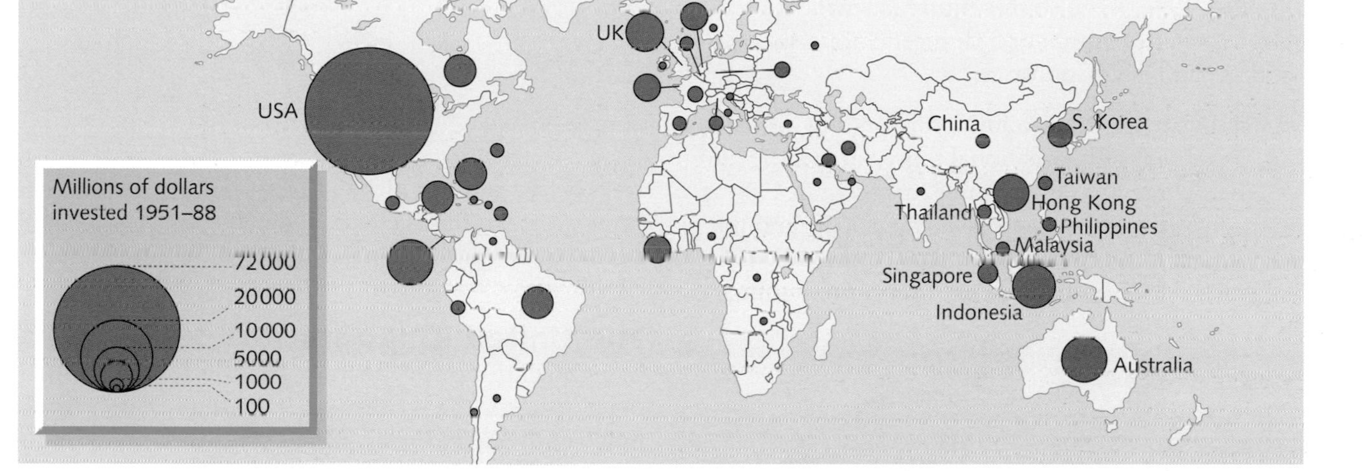

12

(Pages 134 and 135)

a
- i) What is meant by the term 'informal sector'? (1)
- ii) Give four differences between the formal and informal sectors of employment. (4)
- iii) Rearrange the jobs listed below, which were noted during a visit to a country in the economically less developed world, into formal and informal jobs. (8)

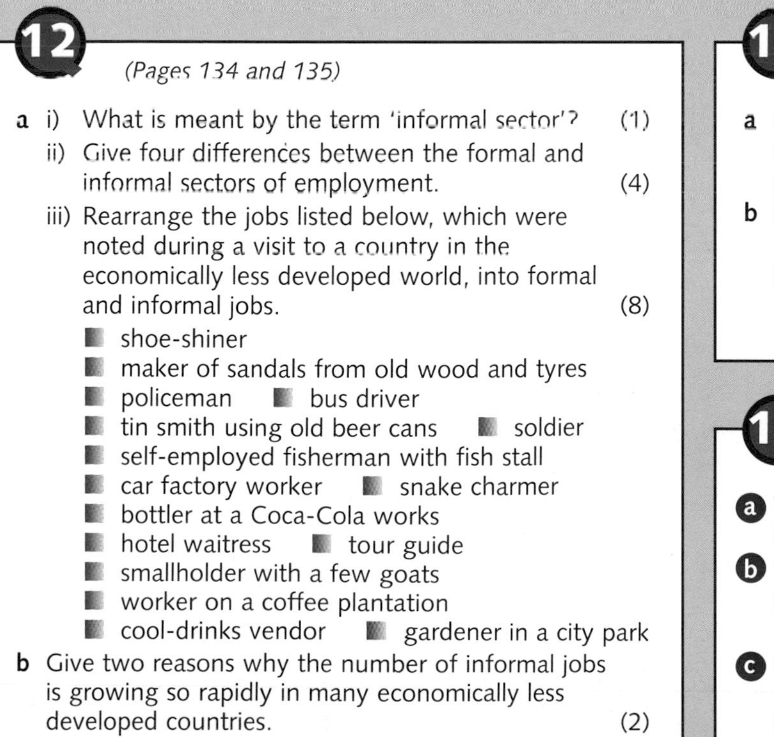

- shoe-shiner
- maker of sandals from old wood and tyres
- policeman
- bus driver
- tin smith using old beer cans
- soldier
- self-employed fisherman with fish stall
- car factory worker
- snake charmer
- bottler at a Coca-Cola works
- hotel waitress
- tour guide
- smallholder with a few goats
- worker on a coffee plantation
- cool-drinks vendor
- gardener in a city park

b Give two reasons why the number of informal jobs is growing so rapidly in many economically less developed countries. (2)

13

(Pages 136 and 137)

a Give five ways by which appropriate technology can provide a sustainable way of living in one of the poorer economically less developed countries. (5)

b
- i) What are the main aims of Intermediate Technology (IT)? (3)
- ii) Describe three of IT's projects in Kenya. How do they help local people? (6)

14

(Pages 138–141)

a Why has industry grown so rapidly in:
- Osaka
- São Paulo? (3 x 2)

b
- i) What is meant by Japan's 'industrial pyramid'? (3)
- ii) How does it work in the case of the Panasonic corporation? (6)

c
- i) In which parts of São Paulo are most:
  - formal
  - informal jobs to be found? (2)
- ii) Describe the importance of formal and informal jobs in São Paulo. (4)

9 TOURISM

Recent trends and changing patterns

Tourism has become the world's fastest-growing industry. It is an important factor in the economy of most developed countries and is seen by many developing countries as the one possible way to obtain income and to create jobs. In 1950, 25 million 'international arrivals' were recorded worldwide – a figure which had risen to over 500 million by 1996. Even so this figure is small in comparison with the number of 'domestic' tourists who travel within their own countries.

Figure 9.1 shows that Europe and North America account for about 75 per cent of international tourist arrivals (1996). However, their share has fallen from 82 per cent in 1980 as destinations in Africa, South-east Asia, the Pacific and the Middle East have become more popular (Figure 9.2). This changing pattern has resulted mainly from price-cutting wars between major airlines, favourable money exchange rates, greater affluence in developed countries, technological advances in transport and changing tourist demands.

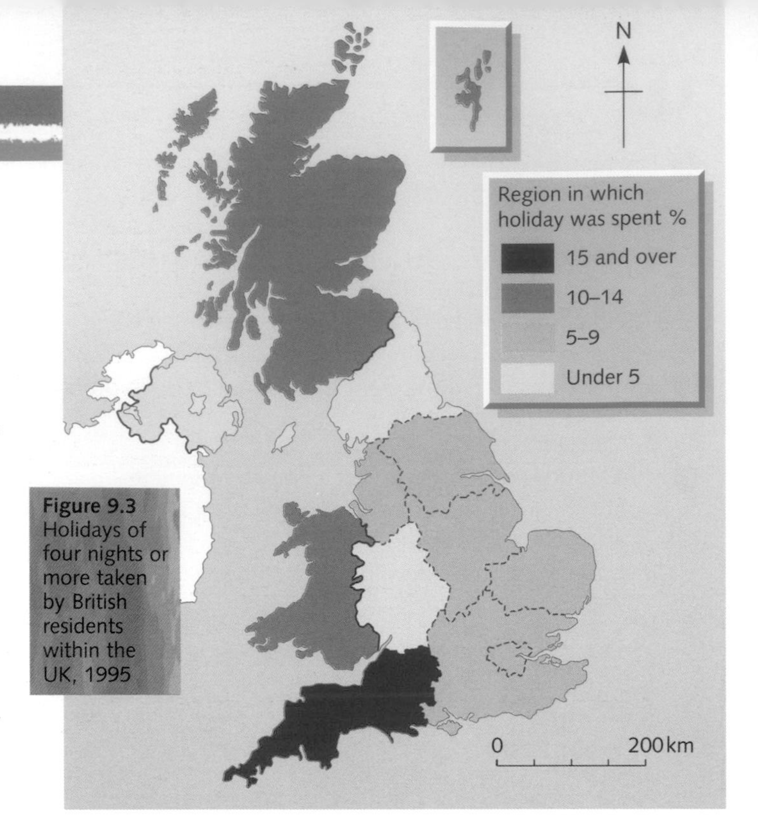

Figure 9.3 Holidays of four nights or more taken by British residents within the UK, 1995

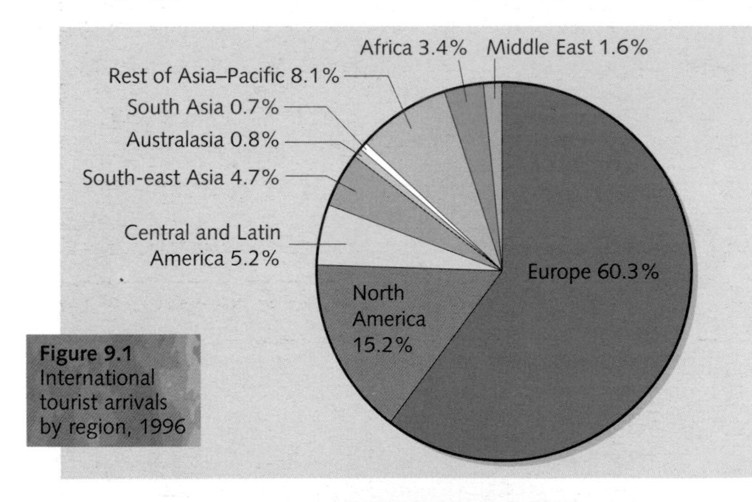

Figure 9.1 International tourist arrivals by region, 1996

The United Kingdom

In eighteenth-century Britain, spa towns (e.g. Bath) developed for the wealthy to 'take the waters'. By the late nineteenth and early twentieth centuries, many industrial workers enjoyed a day, or even a few days, by the seaside (e.g. at Blackpool). Since then, the annual holiday has become part of most British families' way of life, with increasing numbers either travelling abroad (e.g. to Spain) or taking more than one holiday a year. The most popular tourist region in Britain is the South West with its warmer climate and wide range of attractive scenery (Figure 9.3). While Spain remains the most popular overseas destination for British tourists (page 150), its importance has declined as more people are prepared to take 'long-haul' holidays (Figure 9.4).

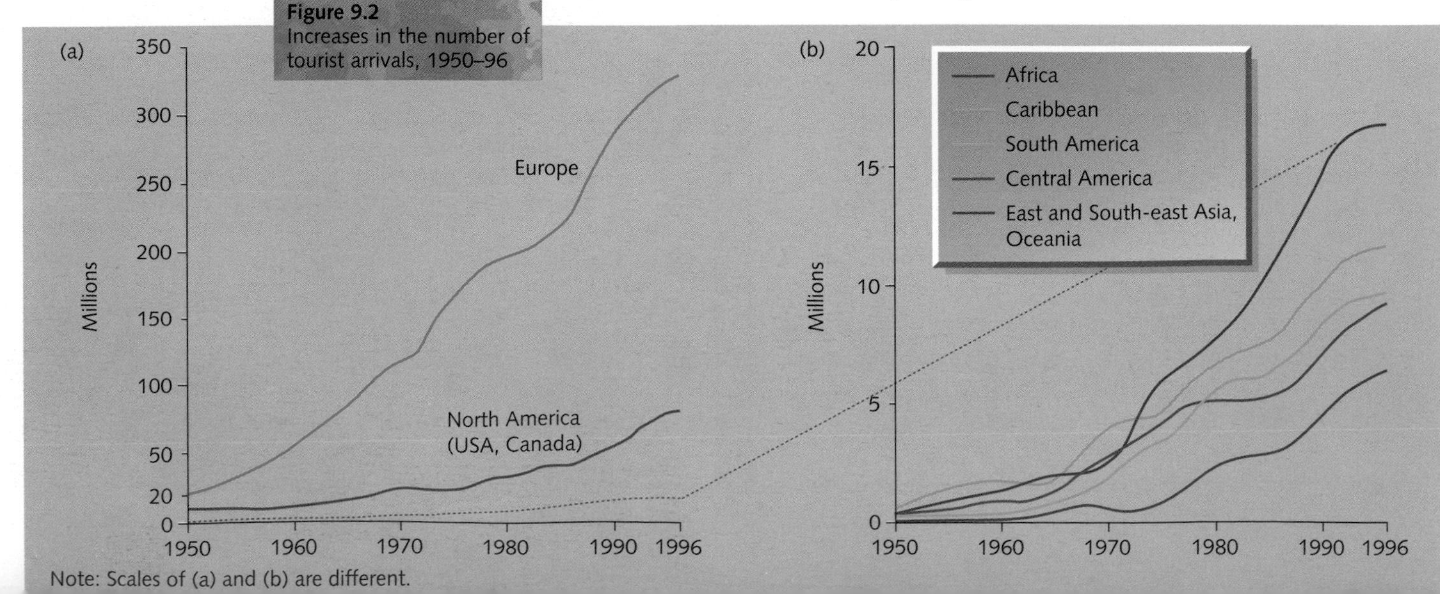

Figure 9.2 Increases in the number of tourist arrivals, 1950–96

Figure 9.4
Foreign holidays taken by UK residents

Recent trends in tourism, and the changing demands made by tourists, have resulted from a wide range of factors. These factors include greater affluence (wealth), increased mobility, improvements in transport, more leisure time, changing lifestyles and fashion, and an increasing awareness of the need for 'green' (sustainable) tourism. These factors are summarised in Figure 9.5.

Figure 9.5
Recent trends in tourism in the UK

GREATER AFFLUENCE	– PEOPLE WHO HAVE EMPLOYMENT WITHIN THE UK USUALLY EARN A HIGH SALARY – CERTAINLY HIGHER THAN SEVERAL DECADES AGO. – PEOPLE IN FULL-TIME EMPLOYMENT ALSO RECEIVE HOLIDAY WITH PAY. THIS MEANS THAT THEY CAN TAKE MORE THAN ONE HOLIDAY A YEAR (FIGURE 9.6) AND CAN AFFORD TO TRAVEL FURTHER.
GREATER MOBILITY	– THE INCREASE IN CAR OWNERSHIP HAS GIVEN PEOPLE GREATER FREEDOM TO CHOOSE WHERE AND WHEN THEY GO FOR THE DAY, OR FOR A LONGER PERIOD. IN 1951, ONLY 1 UK FAMILY IN 20 HAD A CAR. BY 1996, 71 PER CENT HAD AT LEAST ONE CAR. – CHARTERED AIRCRAFT HAVE REDUCED THE COSTS OF OVERSEAS TRAVEL.
IMPROVED ACCESSIBILITY AND TRANSPORT FACILITIES	– IMPROVEMENTS IN ROADS, ESPECIALLY MOTORWAYS AND URBAN BY-PASSES, HAVE REDUCED DRIVING TIMES BETWEEN PLACES AND ENCOURAGE PEOPLE TO TRAVEL MORE FREQUENTLY AND GREATER DISTANCES (FIGURE 9.7). – IMPROVED AND ENLARGED AIRPORTS (ALTHOUGH MANY ARE STILL CONGESTED AT PEAK PERIODS). REDUCED AIR FARES. PACKAGE HOLIDAYS.
MORE LEISURE TIME	– SHORTER WORKING WEEK (ALTHOUGH STILL THE LONGEST IN THE EU) AND LONGER PAID HOLIDAYS (ON AVERAGE 3 WEEKS A YEAR, COMPARED WITH 1 WEEK IN THE USA). – FLEXITIME, MORE PEOPLE WORKING FROM HOME, AND MORE FIRMS (ESPECIALLY RETAILING) EMPLOYING PART-TIME WORKERS. – AN AGEING POPULATION, MANY OF WHOM ARE STILL ACTIVE (FIGURE 1.14).
CHANGING LIFESTYLES	– PEOPLE ARE RETIRING EARLY AND ARE ABLE TO TAKE ADVANTAGE OF THEIR GREATER FITNESS. – PEOPLE AT WORK NEED LONGER/MORE FREQUENT REST PERIODS AS PRESSURE OF WORK SEEMS TO INCREASE. CHANGING FASHIONS, e.g. HEALTH RESORTS, FITNESS HOLIDAYS, WINTER SUN.
CHANGING RECREATIONAL ACTIVITIES	– SLIGHT DECLINE IN THE 'BEACH HOLIDAY' – PARTLY DUE TO THE THREAT OF SKIN CANCER (PAGE 203). – INCREASE IN ACTIVE HOLIDAYS (SKIING, WATER SPORTS) AND IN SELF-CATERING. – MOST RAPID GROWTH IN MID-1990S HAS BEEN IN 'CRUISE HOLIDAYS'. – IMPORTANCE OF THEME PARKS, e.g. ALTON TOWERS, THORP PARK.
ADVERTISING AND TV PROGRAMMES	– HOLIDAY PROGRAMMES, MAGAZINES AND BROCHURES PROMOTE NEW AND DIFFERENT PLACES AND ACTIVITIES.
'GREEN' OR SUSTAINABLE TOURISM	– NEED TO BENEFIT LOCAL ECONOMY, ENVIRONMENT AND PEOPLE WITHOUT SPOILING THE ATTRACTIVENESS AND AMENITIES OF THE PLACES VISITED (PAGE 156).

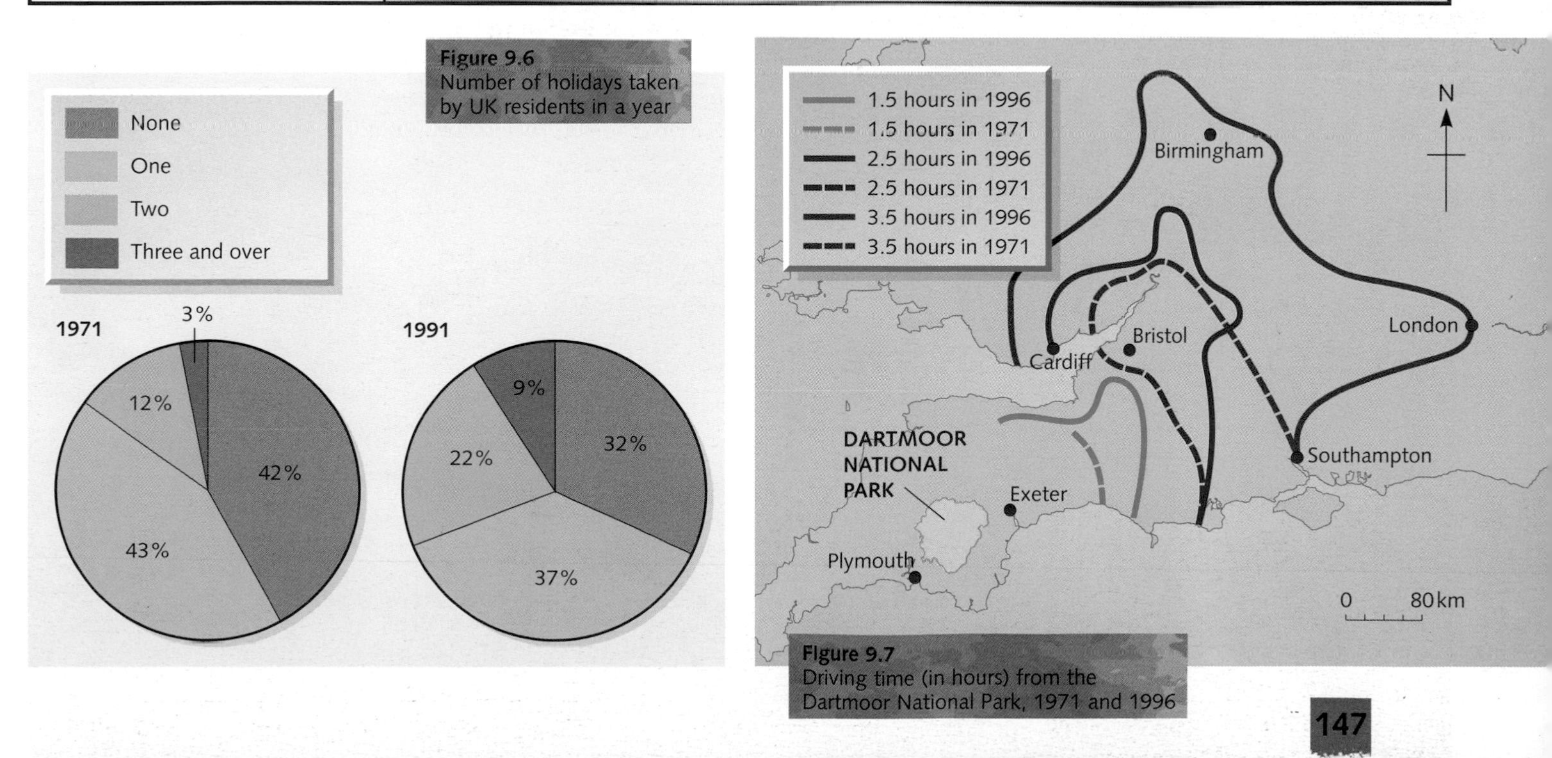

Figure 9.6
Number of holidays taken by UK residents in a year

Figure 9.7
Driving time (in hours) from the Dartmoor National Park, 1971 and 1996

National Parks in the UK

These are defined by Act of Parliament (1949) as *'areas of great natural beauty giving opportunity for open-air recreation, established so that natural beauty can be preserved and enhanced, and so that the enjoyment of the scenery by the public can be promoted'*.

- They contain some of the most diverse and spectacular upland scenery in England and Wales (Pembrokeshire Coast National Park, being coastal, is the exception).
- They are mainly in private ownership, though bodies such as the National Trust, the Forestry Commission and water authorities are important landowners (the parks are not owned by the nation).
- Public access is encouraged, but is restricted to footpaths, bridleways, open fells and mountains (with the exceptions of military training areas and grouse moors).
- They support local populations who are dependent on primary (farming, forestry and mining) and tertiary (tourism) forms of employment.
- The National Parks contain a variety of scenery which in turn provides a wide range of recreational activities (Figure 9.8). All the parks provide basic opportunities for walking, riding and fishing but some provide specialist attractions, e.g. caving and potholing in the limestone areas of the Brecon Beacons and the Peak District. The New Forest, although not a National Park, has National Park status.

Time and distance to National Parks

The National Parks were usually located within easy reach of the major conurbations (Figure 9.8). This enabled the maximum number of people, including those who lived in large urban areas, to escape to a quieter, more pleasant rural environment. Since then the growth of the motorway network has considerably reduced driving times and, in effect, has reduced distances between the conurbations and the National Parks (Figure 9.9).

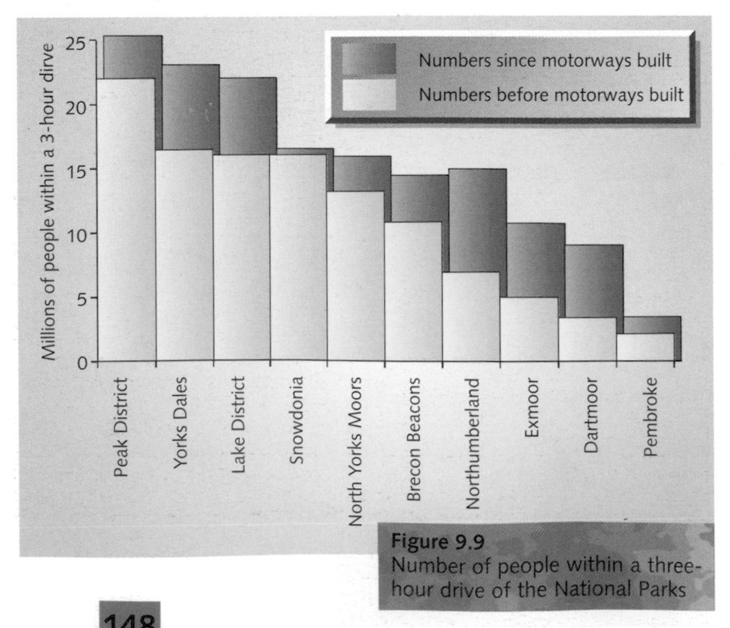

Figure 9.9
Number of people within a three-hour drive of the National Parks

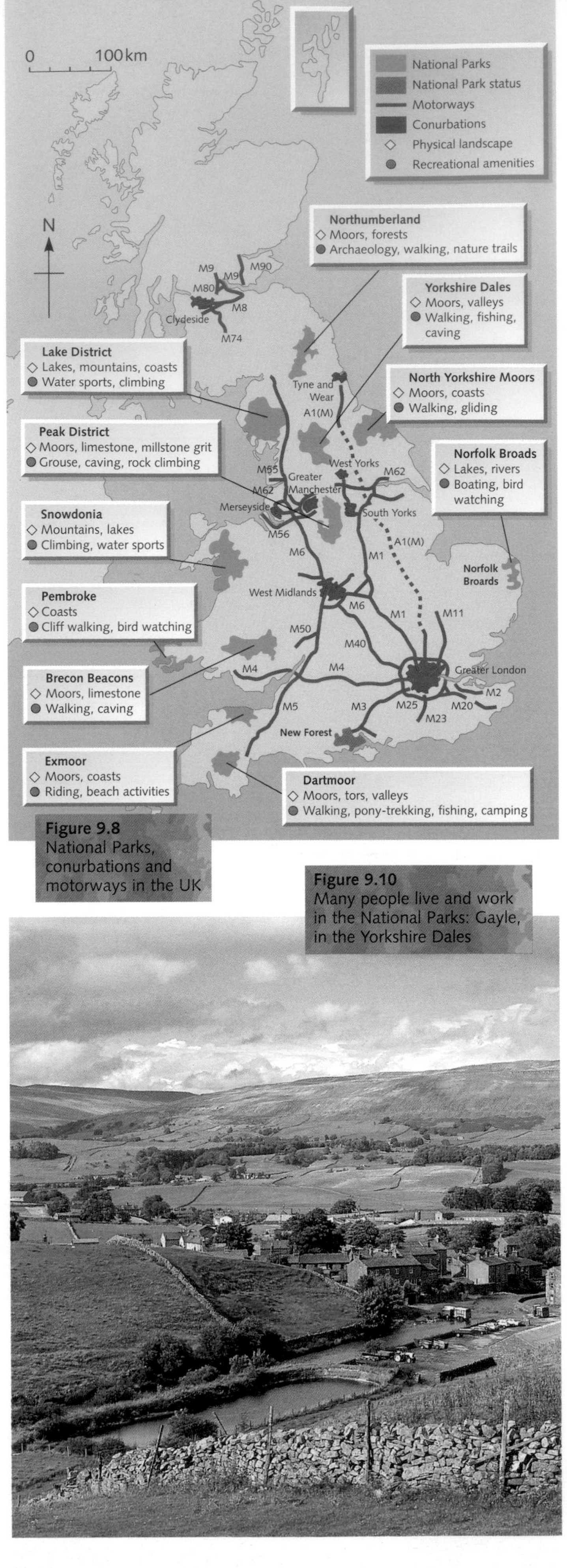

Figure 9.8
National Parks, conurbations and motorways in the UK

Figure 9.10
Many people live and work in the National Parks: Gayle, in the Yorkshire Dales

Figure 9.11
The scars of quarrying in the Peak District

Who owns the National Parks?

In total, 81 per cent of the land is owned privately, mainly by farmers, with 6 per cent belonging to the Forestry Commission, 5 per cent to the National Trust (a charitable organisation earning revenue from membership, admission fees and souvenirs), 3 per cent to water authorities, 3 per cent to the Ministry of Defence, 1 per cent to county councils and 1 per cent to the National Parks themselves. It should be remembered that many people live and work within National Parks (Figure 9.10).

Conflict of users in National Parks

With over two people per hectare, the UK is one of the most densely populated countries in the world, and so there is considerable competition for land. This competition is also seen within the National Parks.

- Town dwellers wish to use the countryside for recreation and relaxation.
- Farmers wish to protect their land and in areas such as Exmoor are ploughing to a higher altitude because they receive government grants.
- The Forestry Commission has planted many hectares of trees in the poorer soils of Northumberland, the North Yorkshire Moors and the Snowdonia Parks.
- The mining and quarrying of slate (Lake District and Snowdonia) and limestone (Peak District) creates local jobs but ruins the environment (Figure 9.11).
- Water authorities have created reservoirs in the Lake District and Peak District Parks.
- The Ministry of Defence owns nearly a quarter of the Northumberland Park.
- Walkers and climbers wish for free access to all parts of the Parks, and campers and caravanners seek more sites for accommodation.
- Despite planning controls, the demand for housing has led to an increased suburbanisation of villages (page 44) and the use of property as 'second homes' for town dwellers.
- Nature lovers wish to create nature reserves and to protect birds, animals and plants from the invading tourists.

Figure 9.12
A badly eroded footpath being repaired

Honeypots

The National Parks include many of the nation's *honeypots* – areas of attractive scenery (Malham Cove in the Yorkshire Dales), or of historic interest (the Roman Wall in the Northumberland Park), to which tourists swarm in large numbers. The problem is how to preserve the honeypots' natural beauty and their unspoilt quality (the essence of their appeal), while providing facilities for the hordes who arrive at peak summer periods. At Malham Cove steps have been cut into the limestone to safeguard paths. It is estimated that £1.5 million is needed to repair the six paths leading to the top of Snowdon where, on a summer's day, 2500 people might reach the summit. Parts of the 400 km Pennine Way have had to be laid with artificial surfaces as the tracks of walkers have penetrated over a metre into the peat in certain places (Figure 9.12). The footpaths on the Roman Wall are being eroded and, as soil is washed away, the foundations of the wall are being exposed.

PROBLEMS	ATTEMPTED SOLUTIONS
FOOTPATHS WORN AWAY (SEE FIGURE 9.12)	NEW ROUTES PLANNED; SIGNPOSTED ROUTES; ARTIFICIAL SURFACES LAID
DESTRUCTION OF VEGETATION, EROSION OF FOOTPATHS	AREAS FENCED OFF; EDUCATION OF VISITORS; LANDSCAPING
LITTER, VANDALISM, TRESPASSING	PROVISION OF PICNIC AREAS WITH LITTER BINS; PARK WARDENS
CARS PARKED ON GRASS VERGES OR IN NARROW LANES	CAR PARKS; ONE-WAY SYSTEMS; PARK AND RIDE SCHEMES
CONGESTION ON NARROW ROADS	ROADS CLOSED TO TRAFFIC IN TOURIST SEASON/AT WEEKENDS; PARK AND RIDE; ENCOURAGEMENT TO USE MINIBUSES, TO CYCLE OR TO WALK
HEAVY LORRIES, LOCAL TRAFFIC AND TOURIST TRAFFIC	SCENIC ROUTES SEPARATING LOCAL AND TOURIST TRAFFIC
'HONEYPOTS' (VIEWS, CAFES) CAUSE CROWDING	DEVELOP ALTERNATIVE HONEYPOTS, DIRECT VISITORS TO OTHER ATTRACTIONS
CONFLICT OF USERS, e.g. A) BETWEEN LOCAL FARMERS AND TOURISTS B) BETWEEN TOURISTS	RESTRICTING TOURIST ACCESS TO FOOTPATHS AND BRIDLEWAYS. SEPARATING ACTIVITIES, e.g. WATER SKIING AND ANGLING
UNSIGHTLY NEW CAFES, CAR PARKS AND CARAVAN PARKS	SCREENED BEHIND TREES. ONLY CERTAIN NATURAL COLOURS ALLOWED IN PAINT SCHEMES

Figure 9.13
How planning in a National Park can help solve problems such as over-use, congestion and conflicts of use

The Costa del Sol

The Costa del Sol (the sun coast) is the most southerly of Spain's many tourist coasts (*costas*). It faces the sun, the Mediterranean Sea and North Africa (Figure 9.14). In the 1950s the area was important only for farming and fishing. Since then, both the landscape and the lives of local people have been transformed by tourism (Figure 9.16). In summer in the main resorts of Torremolinos and Marbella it is more usual to hear English being spoken than Spanish. Why has this become such an important tourist region?

Climate (Figure 9.15) Summers are hot, sunny and dry. Although winters are wet, it rarely rains all day and it is usually mild enough for people to sit out of doors.

Landscape There are long stretches of sandy beach beside the warm, blue Mediterranean Sea. Some beaches consist of shingle; others are artificial. Inland are the spectacular Sierra Nevada Mountains.

Accommodation Torremolinos consists of high-density, low priced, high-rise hotels and apartments (Figure 9.17). Fuengirola's hotels provide the cheapest accommodation along the coast. Marbella has the most modern and luxurious of the hotels, and the hills behind are dotted with many time-share apartments. There are numerous campsites near to the N340 road.

Nightlife and shopping Illuminated shops attract tourists after dark. Numerous restaurants, cafés and bars provide flamenco and disco music, wine and beer, and Spanish, British and other European food. Most resorts have nightclubs. Shops range from the cheaper local bazaars and souvenir shops (mainly Torremolinos area) to chic boutiques with designer clothes (more likely in Marbella). Leather goods, ceramics and perfume can be bought in most places.

Things to do Many activities are linked to the sea, e.g. water sports and Aquapark at Torremolinos, and yachting marinas and harbours at Benalmadena, Puerto Banus and Estepona (Figure 9.18). There are also many golf courses, especially near Marbella. Day visits can be made to the whitewashed village of Mijas (perhaps the only 'real Spain' seen by tourists), to Ronda in the higher mountains, or to the historic centres of Granada and Seville.

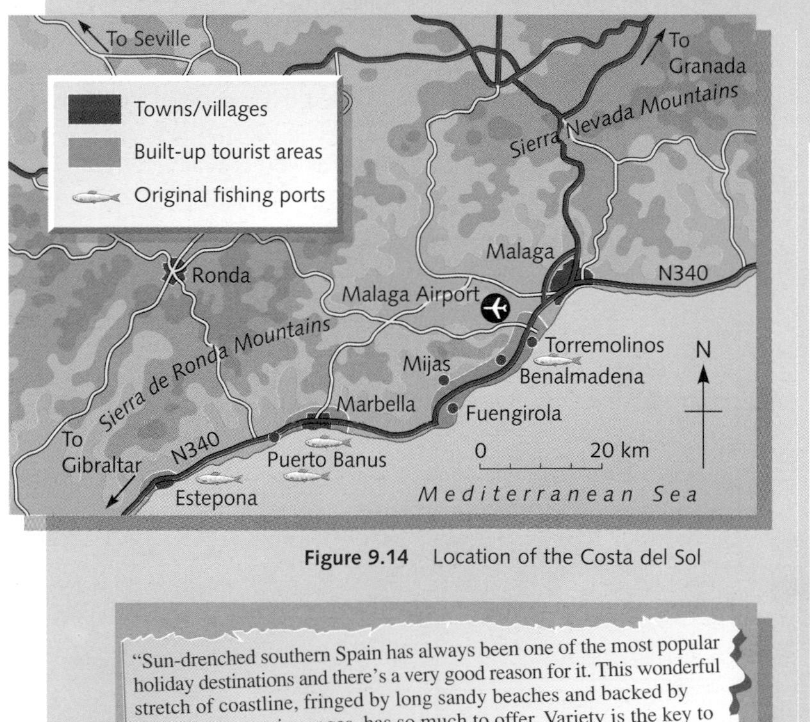

Figure 9.14 Location of the Costa del Sol

"Sun-drenched southern Spain has always been one of the most popular holiday destinations and there's a very good reason for it. This wonderful stretch of coastline, fringed by long sandy beaches and backed by dramatic mountain ranges, has so much to offer. Variety is the key to the coast's success. Dedicate your days to that all-important tan, taking an occasional, refreshing dip in the western Mediterranean. Explore this fascinating part of Spain by visiting romantic Seville, British Gibraltar, stylish Puerto Banus and historic Granada. Once the sun goes down, enjoy a range of nightspots, bars and restaurants from the sophisticated to the informal."

Thomson

Figure 9.16 Extract from a travel brochure

Average hours of sunshine per day

	London	Malaga
J	2	6
F	3	7
M	4	7
A	5	9
M	6	10
J	7	11
J	6	12
A	6	11
S	5	9
O	3	7
N	2	6
D	2	6

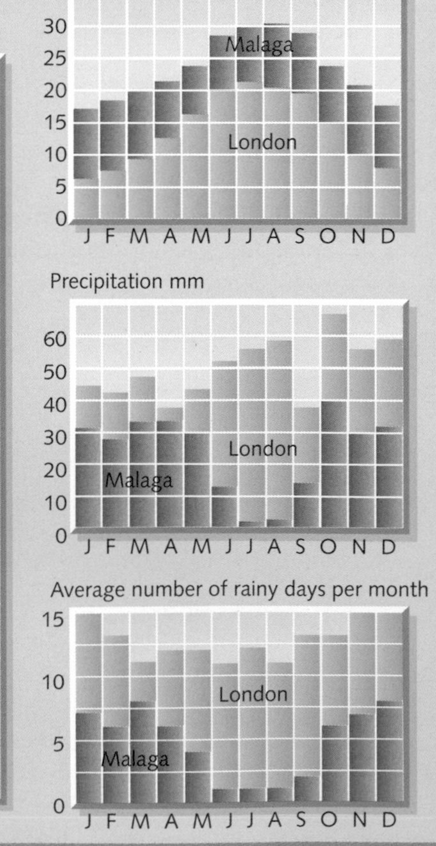

Figure 9.15
Climate graphs for the Costa del Sol and London

Figure 9.17 Torremolinos

Figure 9.18 Yachting marina at Puerto Banus

Growth in tourism	Stage 1 THE TRADITIONAL SOCIETY	Stage 2 TAKE-OFF AND DEVELOPMENT	Stage 3 PEAK PRODUCTION	Stage 4 STAGNATION	REJUVENATION? OR Stage 5 DECLINE?
DATE	1960s	1970s	1980s	1990s	2000s
TOURISTS FROM UK TO SPAIN	1960 = 0.4 MILLION	1971 = 3.0 MILLION	1984 = 6.2 MILLION 1988 = 7.5 MILLION	1990 = 7.0 MILLION	?
STATE OF, AND CHANGES IN, TOURISM	VERY FEW TOURISTS.	RAPID INCREASE IN TOURISM. GOVERNMENT ENCOURAGEMENT.	CARRYING CAPACITY REACHED – TOURISTS OUTSTRIP RESOURCES, e.g. WATER SUPPLY AND SEWERAGE	DECLINE – WORLD RECESSION PRICES TOO HIGH – CHEAPER UPPER-MARKET HOTELS ELSEWHERE. BEACH BOREDOM.	
LOCAL EMPLOYMENT	MAINLY IN FARMING AND FISHING.	CONSTRUCTION WORKERS. JOBS IN HOTELS, CAFES, SHOPS. DECLINE IN FARMING AND FISHING.	MANY IN TOURISM – UP TO 70% IN SOME AREAS.	UNEMPLOYMENT INCREASES AS TOURISM DECLINES (30%). FARMERS USE IRRIGATION.	
HOLIDAY ACCOMMODATION	LIMITED ACCOMMODATION, VERY FEW HOTELS AND APARTMENTS, SOME HOLIDAY COTTAGES.	LARGE HOTELS BUILT (USING BREEZE-BLOCKS AND CONCRETE), MORE APARTMENT BLOCKS AND VILLAS.	MORE LARGE HOTELS BUILT, ALSO APARTMENTS AND TIME-SHARE, LUXURY VILLAS.	OLDER HOTELS LOOKING DIRTY AND RUN DOWN. FALL IN HOUSE PRICES. ONLY HIGH-CLASS HOTELS ALLOWED TO BE BUILT.	
INFRASTRUCTURE (AMENITIES AND ACTIVITIES)	LIMITED ACCESS AND FEW AMENITIES. POOR ROADS. LIMITED STREET LIGHTING AND ELECTRICITY.	SOME ROAD IMPROVEMENTS BUT CONGESTION IN TOWNS. BARS, DISCOS, RESTAURANTS AND SHOPS ADDED.	E340 OPENED – 'THE HIGHWAY OF DEATH'. MORE CONGESTION IN TOWNS. MARINAS AND GOLF COURSES BUILT.	BARS/CAFES CLOSING, MALAGA BY-PASS AND NEW AIR TERMINAL OPENED.	
LANDSCAPE AND ENVIRONMENT	CLEAN, UNSPOILT BEACHES. WARM SEA WITH RELATIVELY LITTLE POLLUTION. PLEASANT VILLAGES. QUIET. LITTLE VISUAL POLLUTION.	FARMLAND BUILT UPON. WILDLIFE FRIGHTENED AWAY. BEACHES AND SEAS LESS CLEAN.	MOUNTAINS HIDDEN BEHIND HOTELS. LITTER ON BEACHES. POLLUTED SEAS (SEWAGE). CRIME (DRUGS, VANDALISM AND MUGGING). NOISE FROM TRAFFIC AND TOURISM.	ATTEMPTS TO CLEAN UP BEACHES AND SEAS (EU BLUE FLAG BEACHES). NEW PUBLIC PARKS AND GARDENS OPENED. NATURE RESERVES.	

Figure 9.19 Changes in tourism

Changes in tourism and to the environment

Places with pleasant climates and spectacular scenery attract tourists. Tourists demand amenities to make their visit more comfortable (e.g. accommodation, car parks) and activities to fill their leisure time (e.g. water sports). As more amenities are added and more leisure activities become available, the environment which first attracted people to these areas either deteriorates or becomes congested. Tourists will therefore seek alternative places to visit. This is happening on the Costa del Sol (Figure 9.19).

Role of the Spanish government

The Spanish government saw tourism as one way to provide jobs and to raise the country's standard of living. In places like the Costa del Sol it encouraged the construction of new hotels and apartment blocks and the provision of leisure amenities such as swimming pools, marinas and entertainment. More recently it has reduced VAT to 6 per cent in luxury hotels to try to maintain cheap holidays. It has also introduced stricter controls to improve the quality of the environment, which include cleaner beaches and reducing sea pollution (Spain now has the most 'Blue Flag' beaches in the EU).

Courmayeur

The winter sports resort of Courmayeur is located in the extreme north-western corner of Italy (Figure 9.20). The town lies at the foot of Mont Blanc. Figure 9.20 is an example of how one British tour company advertises the resort.

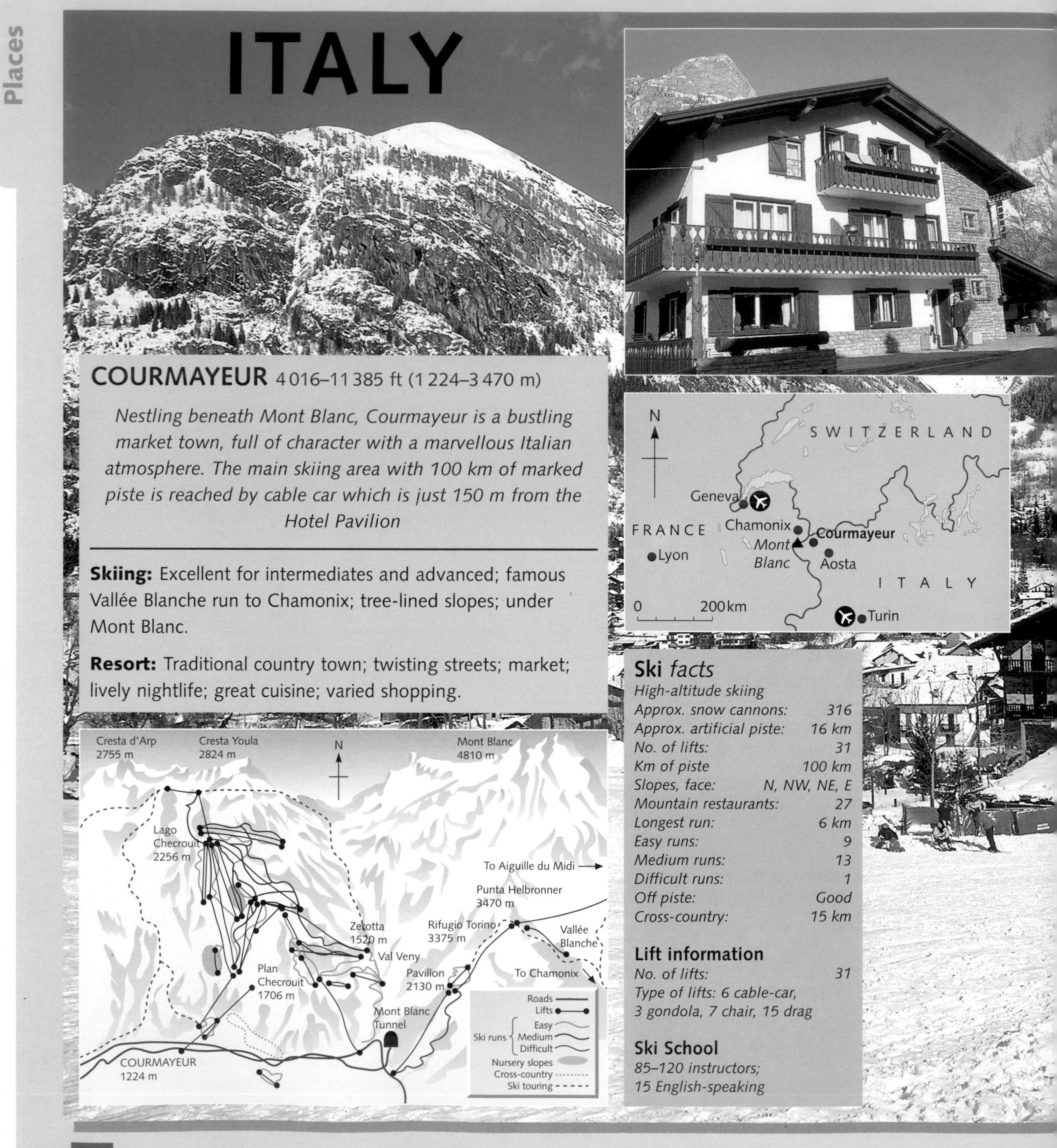

Figure 9.20 Courmayeur

Within Courmayeur there is a weekly market, an ice rink and a swimming pool as well as opportunities for walking and horseriding. There are buses to Chamonix (40 minutes) and Aosta (1 hour).

Ski range: *The high-altitude resort of Courmayeur has a super past snow record. Mont Blanc towers over this impressive region, which takes in 100 km of mainly intermediate pistes. The main ski areas are the massive, sunny plateau of the Checrouit slopes to the north-east, the Val Veny slopes and Mont Blanc, all served by elaborate and extremely well-run lift networks. From Courmayeur, a cable-car (No 1) takes you across the valley to Plan Checrouit, then a gondola and a series of drag and chair lifts ascend further to Colle Checrouit. The Mont Blanc area offers fabulous late-season skiing from the Helbronner peak, including the well-known 20 km Vallée Blanche to Chamonix.*

Included in a special six-day pass is 4 days skiing in the Chamonix area, including Argentière, and one day in Pila, the Aosta ski area. Snow cannons are available to provide extra cover on 16 km of trails.

Apres-ski: *In the evening you can sample some local Aosta specialities such as fontina cheese or beef steak Valdostan. You can also enjoy the many wines from northern Italy, such as Barolo or Prosecco. Check out the best bars – Roma or Steve's Cocktail Bar among them – then go dancing at one of the clubs, like the Clochard or Abat Jour.*

Benefits and problems of winter sports at Courmayeur

Benefits

There has been an increase in employment, especially among younger people, which has reduced the rate of migration away from the area. The standard of living of local people has risen as most jobs in the tourist industry are better paid than the traditional ones in farming and forestry. Roads, water supplies and sewerage have all been improved. The ice rink and swimming pool are available to local residents.

Problems

Alp Action, a conservation group, launched a campaign in 1991 to alert people to the devastating effects of tourism on mountain habitats. The group claimed:

'Every year, 50 million people visit the Alps, two-thirds of them on winter skiing holidays, serviced by 40 000 ski runs. This has resulted in widespread deforestation of mountain slopes to make way for new and enlarged ski resorts and ski runs, while the huge increase in winter sports activities has added to serious erosion of mountain topsoil and a loss of Alpine vegetation. As a result, the danger of floods and avalanches has substantially increased during summer thunderstorms or following snowmelt in spring. Likewise several hundred animal, insect and plant species are threatened with extinction. The several million vehicles which cross the Alps each year are partly blamed for the increase in incidence of acid rain *[page 202]* *which has affected 60 per cent of trees in Alpine Europe.'*

The report could have added the problems of visual pollution caused by the construction of unsightly buildings, as not all have been built to blend in with the natural environment, and ski-lifts. It could also have pointed out that many of the new jobs are seasonal, and have only replaced those in farming and forestry. The change in type of employment, together with the large influx of tourists, has destroyed the community's traditional way of life. The last few winters have been mild, and snowfalls have been light, late in arriving and not lying long at low altitudes. This has increased skiing at higher levels where the environment is most fragile. Artificial snow is being used in some places to prolong the winter season, but this can upset growth and hibernation cycles for plants and animals.

Athens and a classical tour

Culture, in this sense, means learning about a country's civilisation and way of life. Travel programmes on television have made people in economically more developed countries aware of the intellectually satisfying holiday spent in countries and cities where the lifestyle is different to their own (e.g. Thailand). Added to this is the holiday to 'historic' places where tourists can gain an insight into former civilisations (e.g. Egypt). An increasing number of tourists wish to combine a cultural-historic holiday with the more traditional beach holiday. Such visits have become possible due to increases in wealth, longer paid holidays from work, package holidays, and cheaper air flights. Cultural centres tend to attract short-stay visitors and people on touring holidays rather than long-stay tourists. These visits tend to be less seasonal than at coastal or winter resorts.

Although the largest group of tourists to Greece go there to relax on its many islands, an increasing number now visit sites associated with Ancient Greece. The main centre for these visitors is Athens, with the most popular attraction being the Acropolis (Figure 9.21a). From Athens tourists can either make day excursions to the surrounding ancient sites, or join one of several 'classical tours' (Figure 9.22).

The Acropolis

While the increase in tourism is good for local economies, the extra pressure on such sites does create many problems. Many of these can be seen on a visit to Athens' most famous site, the Acropolis:

- Individual buildings like the Parthenon have to be roped off to prevent people climbing over them.
- Several buildings are usually hidden behind scaffolding as attempts are made to repair them from a safety point of view, and to restore them to their original appearance.
- Visitors are not admitted after 3.00 pm partly to give the site 'time to recover'.
- The steps up to the Acropolis and some rocky paths on its summit have become polished and slippery after centuries of use, and dangerous to walk upon.
- Guides, who must be authorised, have whistles to keep people to well-defined paths.
- The many tourist coaches and cars have led to major parking problems and can cause vibrations which affect the foundations of buildings.
- Chemicals released by traffic and local industry are eroding buildings and statues, causing them to crumble away.
- Visual pollution is caused by souvenir and refreshment stalls, and litter left behind by tourists.

Figure 9.21
The sites of Ancient Athens
(a) The Parthenon on the Acropolis
(b) The Caryatids on the Erechtheum (also on the Acropolis)

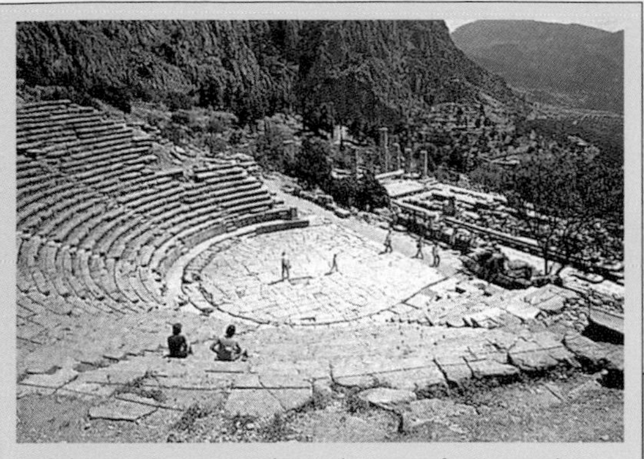

Delphi was famous in classical Greece for its oracle which gave advice and prophecies (although both were ambiguous). The Temple of Apollo and a well-preserved theatre and athletics stadium cling to a mountainside and overlook an olive-forested valley. The museum contains many outstanding works of art.

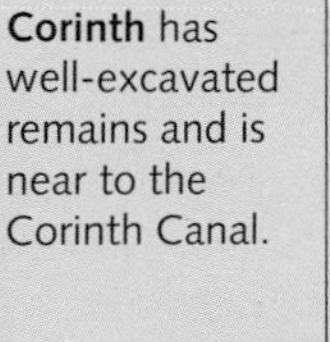

Corinth has well-excavated remains and is near to the Corinth Canal.

Olympia was dominated by the Temple of Zeus, one of the seven wonders of the Ancient World. It later became important (776 BC to AD 393) as the home of the Olympic Games.

Mycenae was the centre of one of Europe's first civilisations. It is noted for the Lion Gate and the discovery of a gold death mask believed to have belonged to Agamemnon, the most famous king of Mycenae.

Sounion has a cliff-top temple and gets spectacular sunsets.

Delphi
Aegean Sea
N
Corinth
ATHENS
Mycenae
Olympia
Sounion
Epidauros
0 50 km
----- Classical tour

Figure 9.22 The classical tour route around Athens

A classical tour

The most popular tour takes place in Delphi and parts of the Peloponnese (Figure 9.22). However, as on the Acropolis, the increasing number of tourists can spoil the environment that is the source of attraction. The impact of tourists is harming both the Greek landscape and the historic buildings. At Delphi, for example, which is built on a steep hillside:

- Footpaths are worn away as tourists walk uphill from the coach park to the temple, theatre and stadium.
- Authorised guides have whistles which are needed to keep people to set paths.
- The guides have to keep a constant watch to ensure that a minority of tourists do not remove parts of buildings to take home as souvenirs.
- The narrow access roads, with their hairpin bends, were not built for large tourist coaches, nor for the increasing volume of cars.
- The small but attractively designed museum is exceptionally congested in summer when each room is likely to be packed with a coachload of visitors. Guides usually have to shout to make themselves heard over 'rival' guides.
- Flash photography is not allowed in the museum as the bright light destroys the natural colours of the exhibits.
- Video cameras are only allowed on payment of an exorbitant price (over £30 in 1996). Other sites have banned the use of video cameras altogether.

Tourism in developing countries

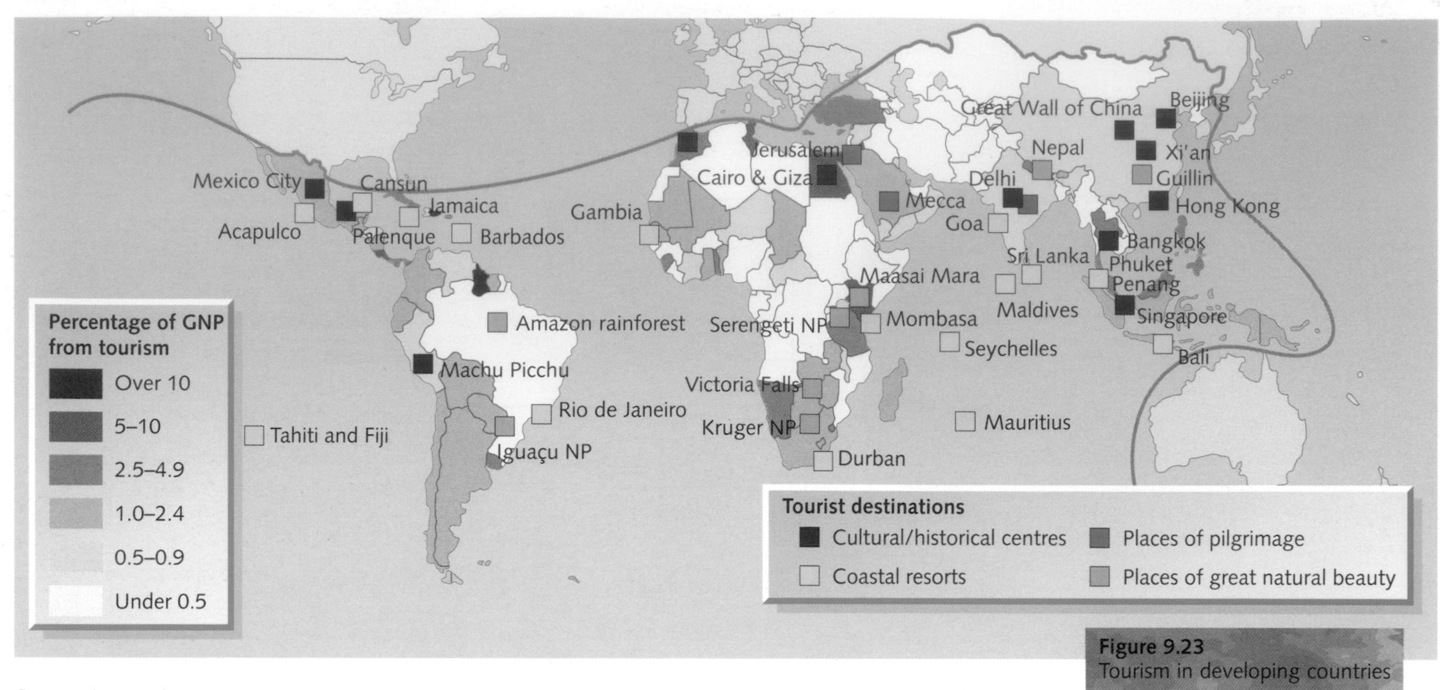

Figure 9.23 Tourism in developing countries

Since the early 1980s, European tourists have become increasingly less satisfied with package tours to Mediterranean coastal resorts. Like tourists from North America and Japan, they have sought holidays in places further afield, where the environment and culture is very different from their own. Long-haul holidays have benefited from travel programmes on television. The chief beneficiaries of this change in holiday fashion have been the tropical, developing countries such as Kenya, Egypt, Sri Lanka, Thailand, Malaysia and Jamaica (Figure 9.23). The attractions of earning money from tourism are considerable to economically less developed countries, many of which see it as the only possible way to increase their standard of living. However, only a limited number of developing countries are able to develop a successful tourist industry, and even then the damage to their culture and environment can, at times, outweigh the benefits (page 157).

Figure 9.24 Tourist sites:
(a) Bali
(b) Machu Picchu

Ecotourism

Ecotourism, sometimes known as 'green tourism', is a sustainable form of tourism which is more appropriate to developing countries than the mass tourism associated with places such as Florida, the Spanish costas (page 150) and the Greek islands in the developed world. Ecotourism includes:

- visiting places in order to appreciate their scenery and wildlife and to understand their culture
- creating economic opportunities (jobs) in an area while at the same time protecting its natural resources (scenery and wildlife) and local way of life (culture).

Ecotourists usually travel in small groups and share specialist interests (e.g. birdwatchers and photographers). They often visit National Parks and game reserves where the scenery (e.g. sandy beaches, waterfalls and forest) and wildlife, which attracted them there in the first place, is carefully protected and managed (Figure 9.24). They are also more likely, by making longer visits, to merge and live with local communities and to appreciate local cultures than larger groups which stop only long enough to 'take a photo and buy a souvenir'. Even so, ecotourists usually pay for most of their holiday in advance so that little of their money is spent in the developing country, visit places where there is less opportunity to spend money, are not all environmentally educated, can cause land prices to rise, congregate at prime sites (honeypots), and may still cause conflict with local people.

West Indies – a beach village

Attractions of the West Indies

- Winters (25°C) are much warmer than those in North America and Europe. Summers are hot (28°C) but not oppressive.
- Most days have over eight hours of sun.
- The scenery is attractive, usually either volcanic mountains covered in forest or coral islands with sandy beaches.
- The warm, clear, blue seas are ideal for water sports (sailing and water skiing).
- There is varied wildlife (plants, birds, animals).
- Customs are different (calypsos, steel bands, food, festivals and carnivals).
- There are many cultural and historic resorts.
- They are situated a relatively short flight time from North America.

The *beach village* is a recent attempt to try to disperse accommodation and amenities so that they merge with the natural environment (Figure 9.25), and to avoid spoiling the advantages which originally attracted tourists to the islands.

Disadvantages of tourism

- Hotels, airports and roads spoil the visual appearance and create noise, air pollution and litter.
- Usually only 10–20% of the income received from tourists stays in the country. Most hotels are foreign-owned and profits go overseas. Tourists spend most of their money in the hotels.
- Much employment is seasonal. Overseas labour may be brought in to fill the better-paid jobs.
- Local craft industries may be destroyed in order to provide mass-produced, cheap souvenirs.
- Farming economy is damaged as land is sold to developers. Much of the food eaten by tourists is imported either because local production is insufficient or to meet the demands for European-type foods (but sold at the developing country's prices).
- Local people cannot afford tourist facilities.
- Borrowed money increases national debt.
- Tourists expect unlimited water – up to 500 litres a day or ten times that used by local people. Many areas may be short of water for domestic and farming use.
- Local cultures and traditions are destroyed. New social problems of prostitution, crime, drugs and drunkenness. Lack of respect for local customs and religious beliefs (e.g. semi-naked tourists into mosques and temples).
- The building of hotels means local people lose their homes, land and traditional means of livelihood (e.g. fishermen, as hotels are built next to beaches) and become dependent on serving wealthy tourists.

..... *Tourism is a form of economic colonialism*

Advantages of tourism

- The natural environment (sun, sand, sea and scenery) is used to attract tourists and their much-needed money.
- Income from tourism is usually greater than the income from the export of a few raw materials.
- Creates domestic employment, e.g. hotels, entertainment and guides. It is labour intensive.
- Encourages the production of souvenirs.
- Creates a market for local farm produce.
- Local people can use tourist facilities.
- Overseas investment in airports, roads and hotels.
- Profits can be used to improve local housing, schools, hospitals, electricity and water supplies.
- Increased cultural links with foreign countries, and the preservation of local customs and heritage.
- Reduces migration.

..... *Tourism raises the standard of living*

Figure 9.25 A beach village merging with the natural environment

Kenya

Few countries in the world can offer the traveller the variety of landscapes that Kenya can. It has mountains, grassy plains, sandy beaches, coral reefs and an abundance of wildlife. Kenya appreciates these natural resources and has set up over 50 National Parks and game reserves to protect and manage its environment. Tourism has become Kenya's major source of overseas income. *Safaris*, meaning 'journeys', are organised so that tourists can be driven around, usually in seven- or nine-seater minibuses with adjustable roofs to allow easier viewing. Unlike on early safaris, today's tourists are only allowed to shoot with cameras. An advertisement for one safari is given in Figure 9.26.

SAFARIWISE

Safari lodges in Kenya provide all modern comforts. While simple in design, your room will have bath or shower (except Shimba Hills and Treetops where shared facilities are provided) and most lodges have a pool. Cuisine, though not *cordon bleu*, is of good standard and sometimes includes game meat.

We use the best available vehicles – 7- or 9-seater safari cruisers with roof hatches and sliding windows for easy game viewing and photography. Journeys, particularly between game reserves, can be long, dusty and tiring but the excitement of seeing wildlife in its natural habitat usually makes it all worthwhile. The occasional change in routeing and/or hotels/lodges may be necessary due to weather conditions or shortage of accommodation.

Tented accommodation is sometimes included at Samburu or Keekorok – but do not be alarmed! The tents have stand-up room, are heavy-duty and erected on a concrete base under an awning, while to the rear (direct access from tent) are simple but private shower and toilet facilities.

Day 1 Nairobi/Samburu (310 km)
After breakfast drive north, cross the Equator and pass Mt Kenya, to Samburu Lodge. After lunch there will be a game drive when you should see elephant, buffalo, lion, reticulated giraffe, zebra, crocodile and many bird species.

Day 2 Samburu
Early morning game drive. Relax at midday around the swimming pool or watch the Samburu perform traditional dances. Late afternoon game drive.

Day 3 Samburu/Treetops (200 km)
Drive south for lunch at the Outspan Hotel. A short journey takes you into the Aberdare Mountains where you will spend the night at Treetops, the world-famous tree hotel. As evening approaches, buffalo, elephant and rhino join other animals at the waterhole.

Day 4 Treetops/Nakuru/Naivasha (240 km)
Transfer to Outspan for breakfast. Drive to the Thomson's Falls, and down into the Rift Valley to Nakuru for lunch. A short drive will let you see vast flocks of flamingos and the endangered Rothschild giraffe. Continue to Lake Naivasha Hotel for the night.

Day 5 Naivasha/Maasai Mara (240 km)
Leisurely morning by the lakeside. After lunch, drive to Keekorok Lodge in the Maasai Mara.

Day 6 Maasai Mara
The huge Mara plain provides some of the best game-viewing in East Africa. During early morning and late afternoon game drives you are likely to see huge herds of wildebeest and zebra, as well as lion, elephant, cheetah, leopard, Maasai giraffe, and hippo. An option is the early morning balloon safari.

Day 7 Maasai Mara/Nairobi (260 km)
Early morning departure arriving at Nairobi for lunch. Afternoon flight to Mombasa to continue your holiday at a beach hotel.

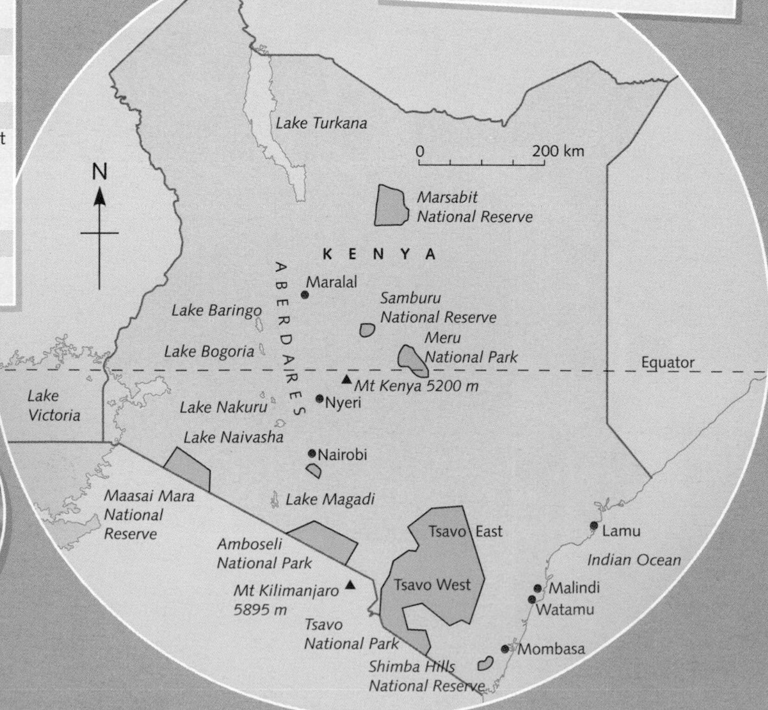

Figure 9.26 Safari in Kenya

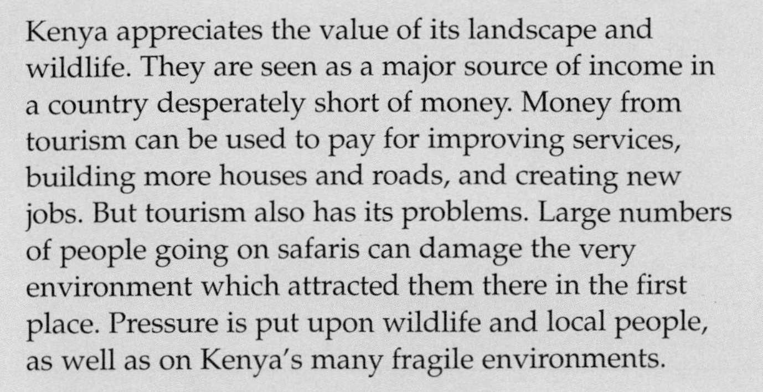

Kenya appreciates the value of its landscape and wildlife. They are seen as a major source of income in a country desperately short of money. Money from tourism can be used to pay for improving services, building more houses and roads, and creating new jobs. But tourism also has its problems. Large numbers of people going on safaris can damage the very environment which attracted them there in the first place. Pressure is put upon wildlife and local people, as well as on Kenya's many fragile environments.

The environment Safari minibuses are meant to keep to well-defined tracks in National Parks and game reserves. However, drivers often form new routes, either to enable their passengers to get as close as possible to wildlife, or to avoid wet season marshy areas. Minibuses can get stuck in the mud, ruining vegetation (Figure 9.27), or widening existing tracks (Figure 9.28). In Amboseli, as in other parks, the wind, minibuses and herds of animals all cause mini dust storms which increase the rate of soil erosion (Figure 9.29).

Wildlife Minibuses are not meant to go within 25 metres of animals, but their drivers often ignore this as they are unlikely to get good tips from their passengers if the best close-up views of wildlife are not obtained. Animals may be prevented from mating, making a kill, or forced to move to less favourable areas. Balloon safaris (Figure 9.30) cause controversy as conservationists claim that the intermittent release of hot air and the shadow of passing balloons disturb wildlife.

People Today, apart from employees at safari lodges, nobody is allowed to live in National Parks. Even game reserves only permit a limited number of herders and their cattle. The setting up of National Parks meant that nomadic tribes, such as the Maasai, had to be moved away from their traditional grazing grounds. Many now have to live a more permanent life, earning money by selling small artefacts to, or performing traditional dances for, the tourists (Figure 9.31). Recently the government has begun to work with the Maasai, allocating them a share of the wealth obtained from tourism to help improve their education, housing and water supply (page 137).

Figure 9.27 Erosion by minibuses

Figure 9.30 Balloon safari

Figure 9.28
Dust track in Amboseli

Figure 9.29
A dust storm causing soil erosion

Figure 9.31 Maasai people selling artefacts to tourists

CASE STUDY 9

The Lake District National Park

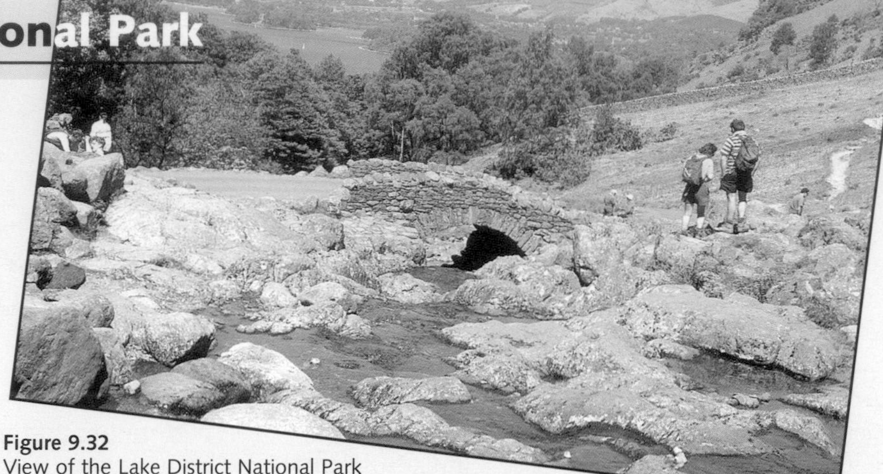

Figure 9.32
View of the Lake District National Park

Each year, 12 million people visit the Lake District. They visit the National Park because of its mountain, valley and lake scenery and its centuries of human history (Figure 9.32). Some of the early tourists came for spiritual refreshment, while most of the present-day visitors seek a range of recreational activities. What attracts the tourist also attracts second-home owners and retired people. These people, together with the long-established local population, add up to a population of over 40 000 living, and often working, within the National Park.

The latest Lake District National Park Plan (1997) identifies several special qualities (Figure 9.33).

- The mixture of natural and farmed landscapes provide a release from the pressures of urban living.
- There is a diversity of landscapes – sandy coasts and clear rivers, rugged mountains and deep lakes, woodlands and moorland, wildlife and historic remains.
- Its most spectacular scenery results from glaciation (page 233). The work of frost and ice has created landforms which include tarns (cirques), narrow ridges (arêtes), and U-shaped valleys (glacial troughs) with waterfalls (hanging valleys), cliff-like sides (truncated spurs) and lakes (ribbon lakes).
- The wide range of ecosystems (Chapter 12) include freshwater habitats, ancient broad-leaved woodland, limestone grassland, heath, arctic–alpine communities, heathland, coastal marsh and estuary. The National Park includes 101 Sites of Special

Figure 9.33 Lake District National Park

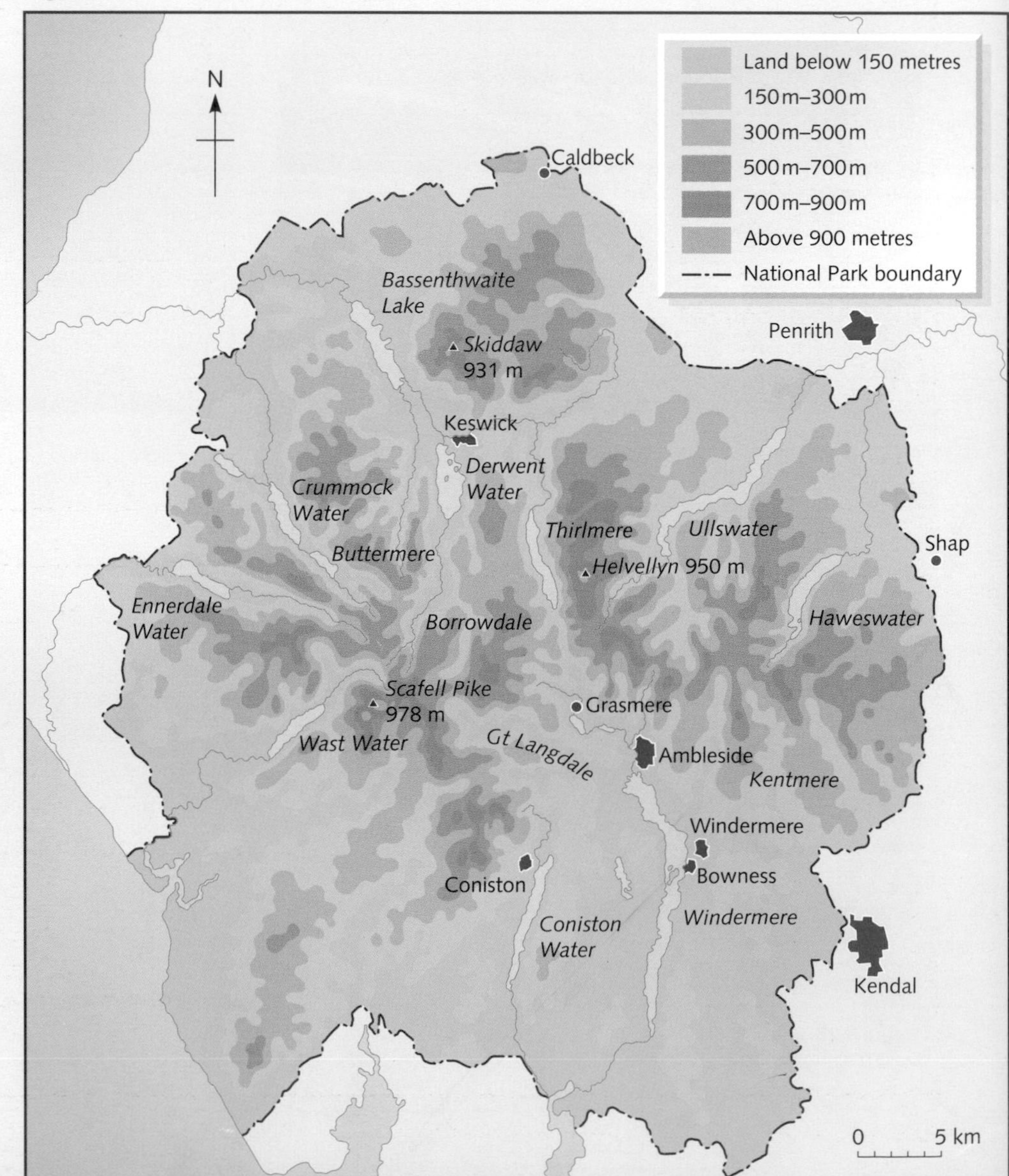

Scientific Interest (SSSIs), six national nature reserves and eight areas of protected limestone pavement (page 233).

- The 3200 km of footpaths, bridleways and green lanes give access to large areas and create a sense of freedom and discovery.
- There are numerous local settlements, each with its own architecture and character. The area, with its human history extending back for 12 000 years, includes Neolithic sites, Roman remains and more recent mining and industrial activities. Stone-walled fields have developed since the twelfth century, and parklands from medieval times.
- The rural character of Lake District life with its dialect, sports and links with literature and artistic movements, has created a strong local culture.

Figure 9.34 gives information about the Lake District National Park collected during the 1991 census. It also gives some of the results of a national survey of National Parks in England and Wales carried out between March and November 1994. During that time, almost 4000 visitors to the Lake District National Park (business trips were excluded) were interviewed at four entry/exit points to the National Park and 12 key sites within the Park. The results suggested that there were 22 million visitor-days that year. This consisted of approximately 2 million holiday visitors (each staying an average of 6 days giving a total of 12 million visitor-days) and 10 million day visitors.

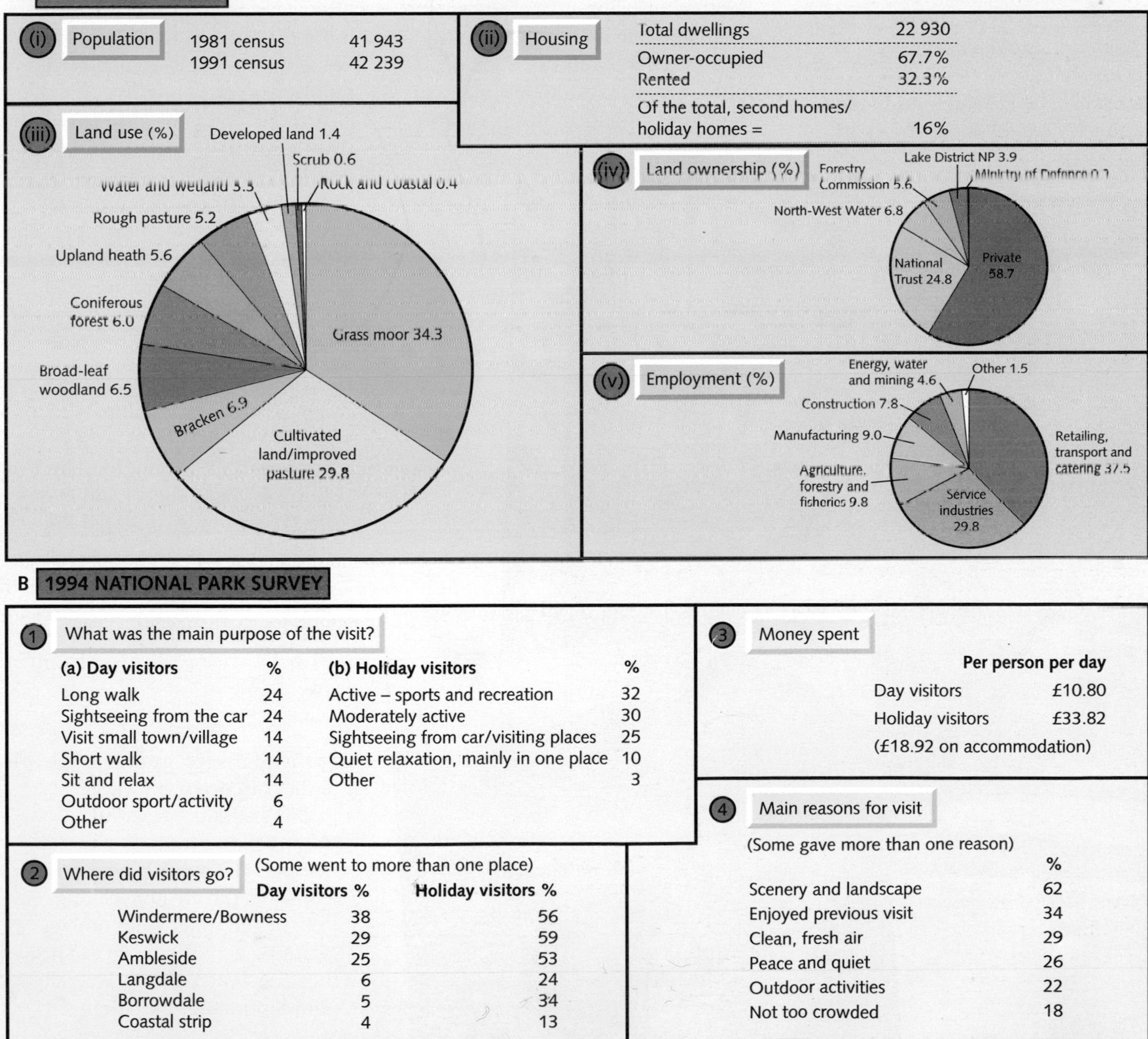

1 What was the main purpose of the visit?

(a) Day visitors	%
Long walk	24
Sightseeing from the car	24
Visit small town/village	14
Short walk	14
Sit and relax	14
Outdoor sport/activity	6
Other	4

(b) Holiday visitors	%
Active – sports and recreation	32
Moderately active	30
Sightseeing from car/visiting places	25
Quiet relaxation, mainly in one place	10
Other	3

2 Where did visitors go? (Some went to more than one place)

	Day visitors %	Holiday visitors %
Windermere/Bowness	38	56
Keswick	29	59
Ambleside	25	53
Langdale	6	24
Borrowdale	5	34
Coastal strip	4	13

3 Money spent

	Per person per day
Day visitors	£10.80
Holiday visitors	£33.82
(£18.92 on accommodation)	

4 Main reasons for visit (Some gave more than one reason)

	%
Scenery and landscape	62
Enjoyed previous visit	34
Clean, fresh air	29
Peace and quiet	26
Outdoor activities	22
Not too crowded	18

Figure 9.34 Statistics for the Lake District National Park

CASE STUDY 9 continued

Problems created by visitors to the Lake District National Park

The 12 million annual visitors to the Lake District National Park are bound to cause problems both to the environment and for the 42 000 local residents. These problems include the following.

Traffic Nearly 90 per cent of visitors arrive by car, and traffic levels are projected to go on rising. Congestion is often heavy on approach roads to the Park, especially at weekends and in the summer. Limited public transport within the Park means that visitors and local people alike have to rely upon the car. This creates periodic congestion in bottlenecks such as Bowness and Ambleside (Figure 9.35), parking problems, and noise and air pollution. Many roads outside the built-up areas are narrow, steep and winding, making them unsuitable for large volumes of traffic and especially for tourist buses.

Footpath erosion The 1994 survey estimated that 4 million visitors a year now take walks of 6 km or more. This has resulted in many of the more popular routes becoming severely eroded and ugly scars being created on hillsides (Figure 9.36). A constant stream of new guidebooks is attracting more people to previously remote areas.

Second homes and holiday homes One in six properties in the Lake District (16 per cent) are either second homes or holiday cottages, and in places in the north-east and south-east (nearer to M6 interchanges) the proportion is more than one in four. A second home is one that is owned by a family who normally live elsewhere, and is used primarily by that family and their friends. A holiday cottage is rented out on short-term lets on a commercial basis to many different people. The increase in this type of accommodation, and in retirement homes, has forced house prices to rise beyond the reach of most local people.

Honeypots (page 149) People congregate in large numbers at:

- scenic points such as Aira Force (waterfall), the Langdales and Borrowdale (valleys) and even on the summits of Scafell Pike and Helvellyn
- important and historic buildings such as Beatrix Potter's house and Wordsworth's cottage
- car parks, especially in places like Ambleside
- cafés.

Conflict between groups of people Conflict can occur between:

- local residents (especially farmers) and tourists
- people wishing to protect and conserve the area or seeking 'peace and quiet', and others wanting activities such as speed boating, water-skiing, car rallying and mountain bike riding.

Figure 9.35 Traffic congestion in Bowness

Figure 9.36 Footpath erosion on Catbells

Possible solutions to the problems

The latest management plan put forward by the Lake District National Park Authority (1997) sees, as one of its main aims, the need to balance the needs of local residents with those of visitors. It must ensure that '*the Lake District remains a beautiful place, where a range of landscapes retain their character and give shelter to a diversity of flora and fauna. It must continue to be an area for recreation and a source of relaxation and inspiration. The future must see the landscape maintained, visitors catered for and the local community flourishing*.' Two important aspects of the plan are landscaping and traffic management.

Landscaping

This includes:

- repairing eroded footpaths and damaged stone walls (Figure 9.37)
- planting trees to screen, wherever possible, car parks and quarries.

The Lake District Traffic Management Initiative

This initiative was a joint project between Cumbria County Council, the National Park Authority, the Countryside Commission and both the English and Cumbria Tourist Boards. Its initial main objectives were to:

- reduce periodic traffic and parking congestion and to offer alternative modes of transport to the car, mainly by improving public transport
- reduce the impact of increasing levels of traffic on the countryside and ensure that the National Park remains accessible for quiet enjoyment, irrespective of income or disability
- tailor traffic to the availability of the existing roads and allow the local community to proceed with its normal business.

a Road hierarchy It was felt that the current roads classification did not reflect the ability to accommodate traffic and gave the wrong impression to motorists. It has been proposed that all the roads within the Park, including feeder roads, be reclassified according to their type and strategic importance. A six-layer hierarchy has been proposed (Figure 9.38) with design and traffic control standards (e.g. speed and weight limits) based upon the volume and type of traffic and the destinations served. A 40 mph (65 km/hr) restriction is proposed for all but the trunk roads, and many minor roads would be downgraded to cater for the needs and safety of walkers, cyclists and horse-riders using roads.

b Integration of rail, bus and lake steamer transport

c Local initiatives

A number of individual local schemes have been implemented based on the need, local support, cost and achievability. Some of these are listed and described in Figure 9.39.

Figure 9.37
Repairing a wall

- **Trunk roads** – intended principally to carry long-distance/through traffic.
- **County strategic roads** – linking the main population and activity centres to the motorway and trunk road network.
- **Local distributor roads** – carrying traffic between the main centres and linking all other centres of activity. Local distributors would be subdivided into grade 1, grade 2 or grade 3, depending on their function, and with regard to their characteristics (particularly width and visibility standards) and the amount of traffic they would be expected to carry.
- **Local access roads** – remainder of the network, which would be relatively traffic-free routes for walkers, cyclists and horse riders.

Town and village streets, other than the main through routes, are omitted from these designations.

Figure 9.38
Proposed road hierarchy

- Restricting on-road car parking.
- Making roads 'access only'.
- Village traffic-calming schemes.
- Bus priority lane to the Windermere ferry.
- Cattle grids and speed limits on open common land.
- Improved bus services in some valleys.
- Village schemes for parking, traffic management and environmental improvement.

In Grasmere, for example, traffic restrictions, improved coach parking and a link between the bus operators and the main attractions, have been introduced.

Consensus is not always easy. In Elterwater, the Tourist Board is concerned that traffic problems may have been exaggerated, that commercial interests have not been reflected, and that off-road parking should be provided to compensate for loss of on-road spaces. In some places, concerns about loss of trade have been met by allowing signs towards commercial premises.

Figure 9.39
Local initiatives

QUESTIONS

1

(Page 147)

How has tourism been affected by changes in:
- affluence (wealth)
- mobility
- retirement age
- transport facilities
- leisure time
- green tourism? (6)

2

(Pages 147–149)

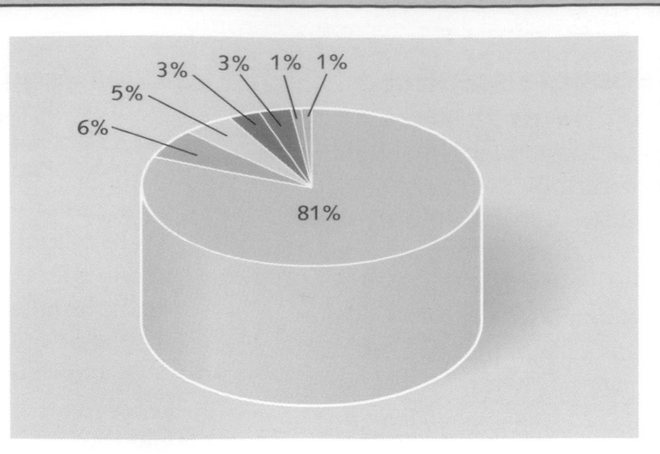

a What is a National Park? (2)

b The pie-graph (right) shows the different land owners in the National Parks. Match up each segment of the graph with its correct land owner. (6)

c 'National Parks contain a variety of scenery which in turn provides a wide range of recreational activities'. Copy and complete the two tables below. (10)

d i) Which conurbation is not near to a National Park? (1)

ii) Which National Park is surrounded by the most motorways and conurbations? (1)

iii) Which two parks have been least affected by the construction of motorways? (2)

e Many more people now visit National Parks than in 1950. Give four reasons (other than the building of motorways) for the increase in numbers. (4)

A SCENERY	NATIONAL PARK
COASTS	PEMBROKE
FORESTS	
LAKES	
LIMESTONE	
MOORS	
MOUNTAINS	

B ACTIVITY	NATIONAL PARK
ARCHAEOLOGY	NORTHUMBERLAND
BIRD-WATCHING	
CAVING	
CLIMBING	
GLIDING	
RIDING	

3

(Pages 148 and 149)

a Name two towns within a short distance of the Peak District National Park. (2)

b The map (right) shows the Peak District National Park. Using *only* this map:

i) List three different physical attractions. (3)

ii) Name three different types of summer activity. (3)

iii) Give two pieces of evidence to show that the Park attracts visitors on cultural (historic) trips. (2)

iv) Why might some visitors be attracted to the Park in winter? (2)

c Choose the correct word from each pair of words given in brackets:

'Most visitors to the Peak District National Park will go in (summer/winter). They will tend to visit mainly (in the week/at the weekends) and stay there for (a week/a day). Most will travel for less than (1 hour/3 hours).' (4)

d Castleton is said to be a 'honeypot'.

i) What is meant by the term 'honeypot'? (1)

ii) Give three problems that may result from the increased number of visitors to a honeypot. (3)

iii) Describe and explain three ways in which these problems may be solved by the National Park planners. (3)

iv) Give three ways in which the pressure on a honeypot may be reduced. (3)

v) Name two famous buildings in Britain and two places of natural beauty which have become honeypots. (4)

e Visitors to a National Park may come into conflict with the local inhabitants. Suggest three conflicts which might occur between the visitors and different groups of residents in the Park. (3)

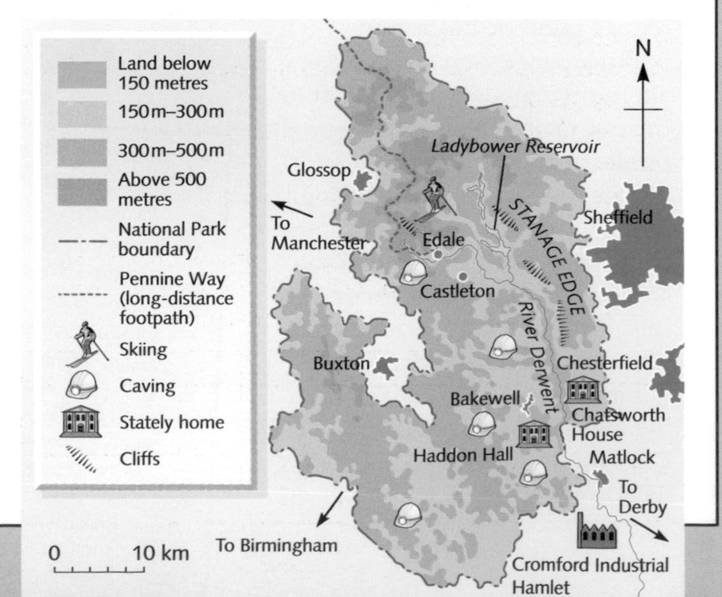

4

(Pages 150 and 151)

Study the map showing the number of tourists visiting Spain.

a i) Name the area marked A. (1)

ii) Name the sea area marked B. (1)

iii) From which country do most tourists come? (1)

iv) From which of the named countries do fewest visitors come? (1)

b i) Why do many people from the UK, France and Germany visit Spain? (3)

ii) Why do relatively few people from Portugal and Italy visit Spain? (3)

Look back at Figure 9.15.

c i) How many hours of sunshine do Malaga and London get on an average day in:
- July
- January? (4)

ii) Apart from hours of sunshine, give two other differences in the weather between Malaga and London in summer. (2)

Study the landsketch which shows part of the southern coast of Spain.

d i) Apart from the weather, what other physical attractions does this coast possess? (3)

ii) Name five amenities added for the benefit of tourists. (5)

iii) It is proposed to build a new hotel and leisure complex at location A on the sketch. Why might this development be:
- opposed by conservationists and some local residents?
- welcomed by many local residents? (4)

e i) How did the growth of tourism between 1960 and 1990 affect:
- the local environment
- the local community? (6)

ii) What problems are facing the tourist industry in this part of Spain today? (3)

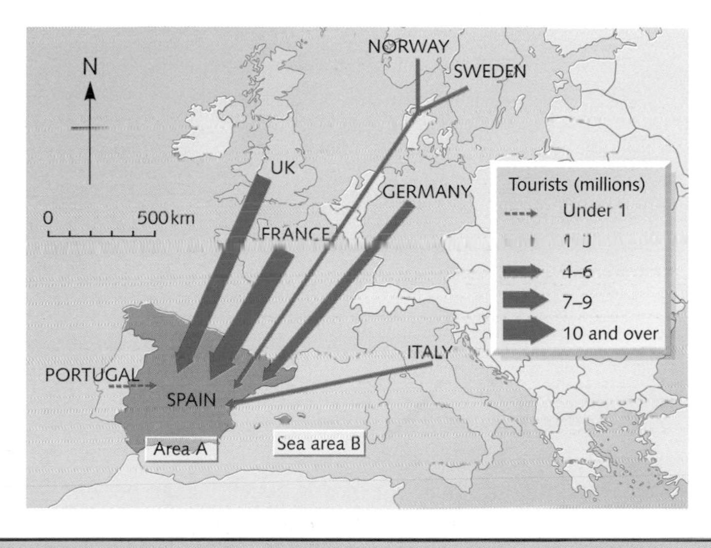

5

(Pages 152 and 153)

a There is a proposal to build a new ski-resort in the Alps.

i) Where, on the sketch, would you build the resort? Give two reasons for your answer. (2)

ii) Explain why some of the villagers might be in favour of the proposal. (3)

iii) Explain why some of the villagers might be alarmed by the proposal. (3)

b Courmayeur, in the Italian Alps, is becoming increasingly dependent upon tourism.

i) Describe its
- natural environment
- traditional way of life. (4)

ii) Complete the following table, listing ways in which tourism has harmed or benefited the local economy, the local community and the natural environment. (6)

	HARMED	BENEFITED
LOCAL ECONOMY		
LOCAL COMMUNITY		
NATURAL ENVIRONMENT		

Slopes to suit the season

Metres

3500 SNOW GUARANTEED
Glacier often bleak and cold until February

3000

2500 SNOW SPLENDID
December to late April — High-level ski resorts

2000 SNOW SURE
Christmas to early April — Mid-height resorts

1500 SNOW SAFE
Mid-January to late March

1000 SNOW UNRELIABLE
To mid-January and after mid-March — Low resorts

500 Village on lower slopes

6

(Page 156)

The map shows the number of holiday destinations in a 1995 holiday brochure.

a In which country are most holiday centres offered? (1)

b Describe the location of the holiday centres offered. (3)

c Name a tourist area that you have studied in a less economically developed country (LEDC). What attractions does the area have for European tourists? (5)

d i) What advantages can tourism bring to an LEDC? (4)

ii) What disadvantages can tourism bring to an LEDC? (4)

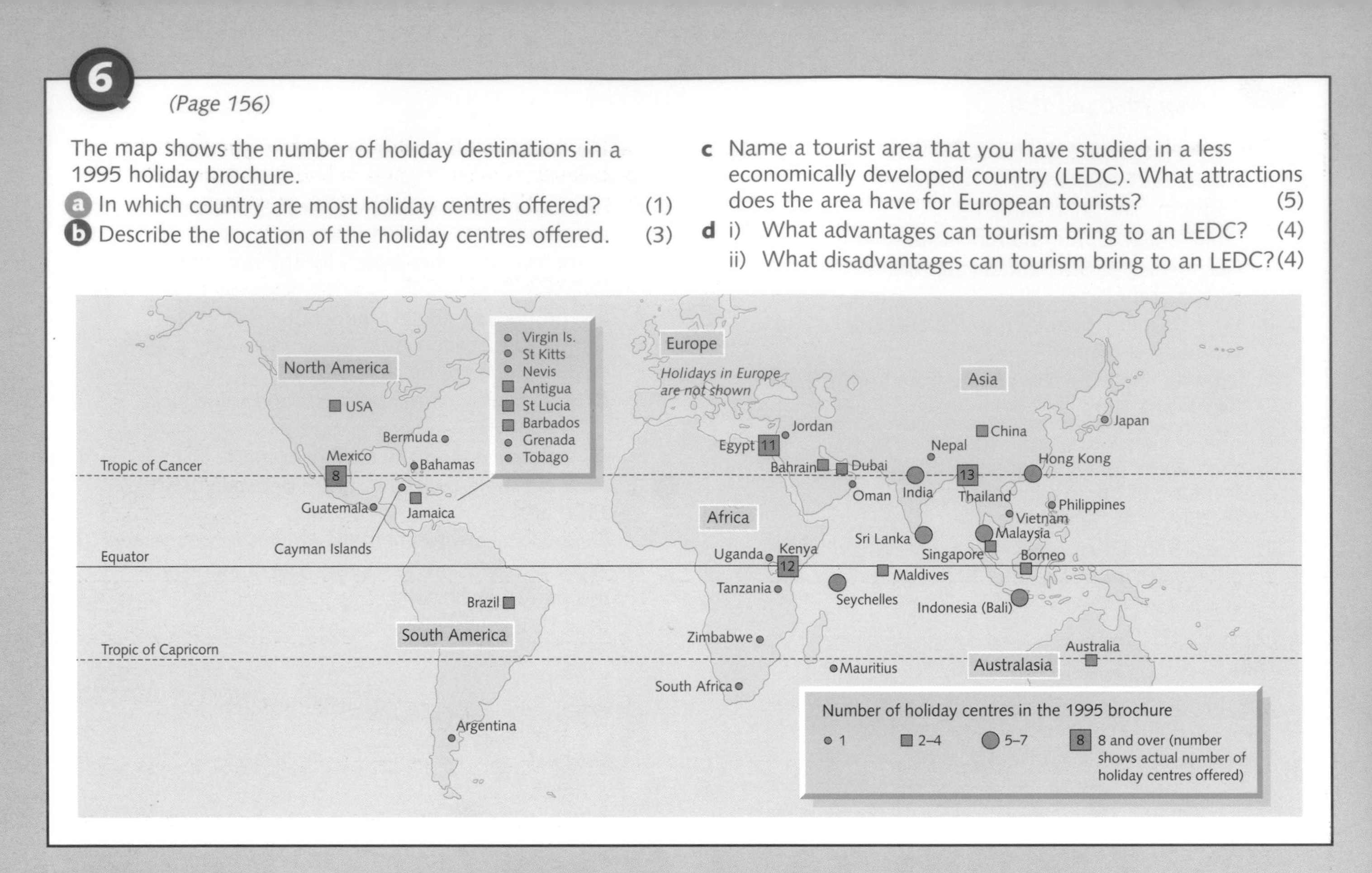

7

(Page 157)

a Using the climate graph of Montego Bay, Jamaica (West Indies):

i) What is the maximum temperature? (1)

ii) What is the annual range of temperature? (1)

iii) Which month has both the lowest rainfall and most hours of sunshine? (1)

iv) Which month has both the highest rainfall and the fewest hours of sunshine? (1)

b Draw a star diagram to show the main features of climate and scenery in the West Indies. (6)

c i) At which two airports in Jamaica do tourists arrive? (2)

ii) Do most visitors to Jamaica stay in hotels or resort cottages? (1)

iii) Name three physical attractions of the area around Montego Bay. (3)

d It is proposed to build a beach village betwen Montego Bay and Negril.

i) What is a beach village? (1)

ii) How may the building of this beach village benefit and harm the natural environment, the local economy, and the local community? (6)

iii) Who else might benefit from the building of the beach village? Give a reason for your answer. (2)

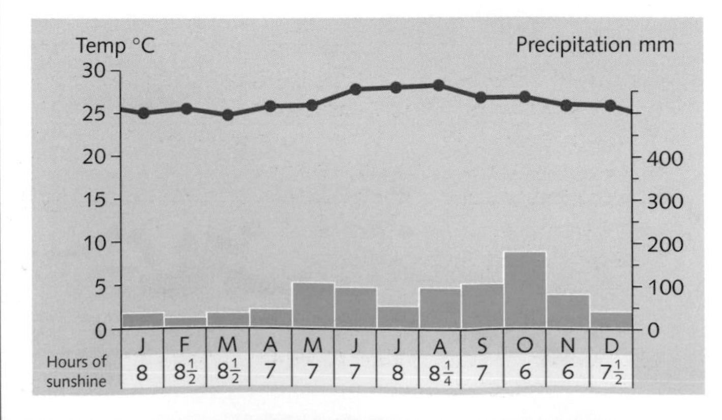

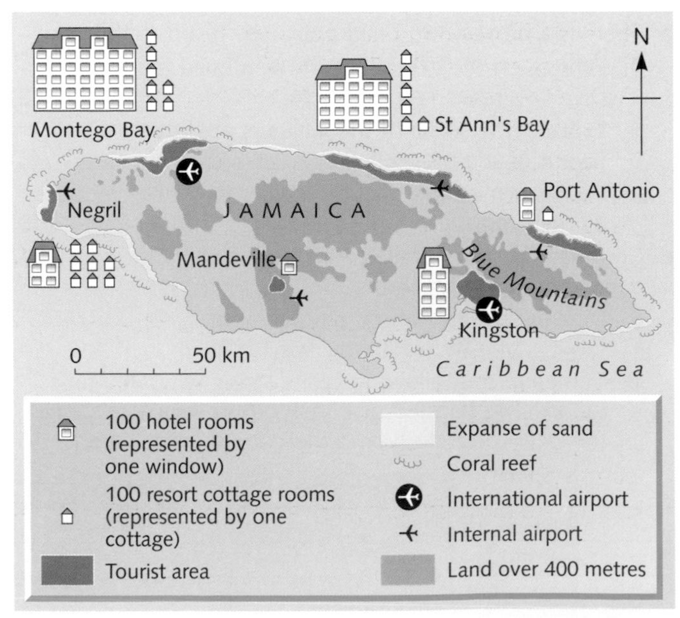

8

(Page 157)

a The two photos were taken in Rio de Janeiro in Brazil. One shows luxury hotels along the coast, the other a small favela (shanty) settlement hidden away from the main tourist area. Look at these photos, read the extract beside them, and then decide whether you think that, on balance, tourism is good or bad for people living in places like Rio. Give reasons for your answer. (8)

Tourism – benefits and costs

A conference held in 1985 concluded that 'International tourism is causing severe damage to the culture and economies of many developing countries and contributes little to their development.' Yet to many developing countries tourism appears as their only hope of escape from the vicious circle of poverty. Figures A and B are intended to help you to draw your own conclusions.

9

(Pages 158 and 159)

a Why is tourism important to economically less developed countries like Kenya? (4)

b Give three natural advantages that attract tourists to Kenya. (3)

c i) Many people go to Kenya for a safari holiday. What is a safari? (1)

ii) How has the growth of safari holidays in Kenya affected the:

- landscape
- wildlife
- local people? (6)

10

(Pages 160–163)

a With reference to the map of the Lake District National Park:

i) Name three of the lakes numbered 1 to 5

ii) Name three of the towns numbered 6 to 10

iii) Name the three mountains lettered A to C. (9)

b The latest Lake District Plan identifies several special qualities of the National Park. Describe these under the following headings:

- landscape (scenery)
- vegetation and wildlife (ecosystems)
- human settlement and culture. (9)

c What, according to the 1991 census, was the:

i) resident population

ii) percentage of dwellings used as second homes

iii) major type of land ownership

iv) two most important types of employment in the Lake District? (5)

d According to the 1994 visitor survey:

i) What is the difference between a day visitor and a holiday visitor?

ii) Why did 48 per cent of day visitors go to the Lake District?

iii) Why did 32 per cent of holiday visitors go to the Lake District?

iv) Did most tourists visit small towns, or valleys? (4)

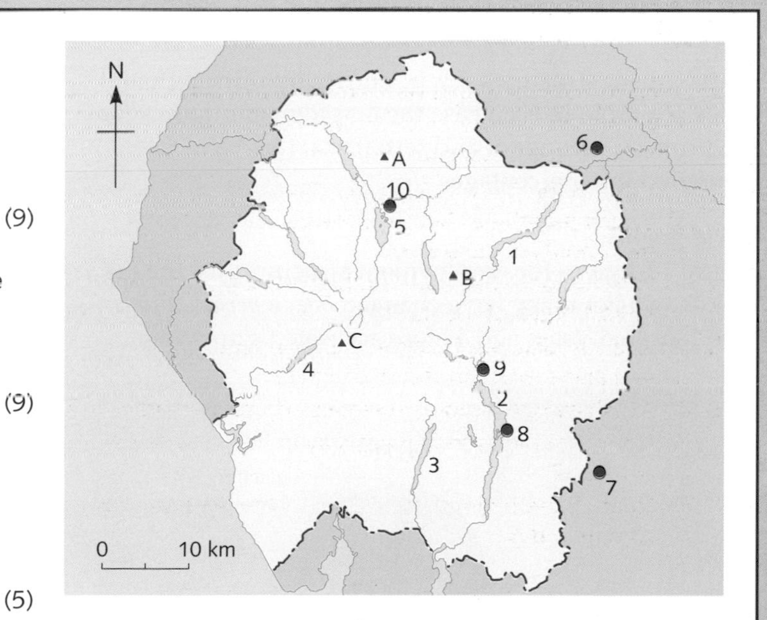

e For *three* of the following, describe how visitors to the Lake District have added to problems of:

- traffic congestion
- footpath erosion
- second homes
- honeypots
- conflict between different groups of people. (9)

f i) What are the main objectives of the Lake District Traffic Management Initiative? (3)

ii) Describe why a new road hierarchy has been proposed and explain how it would work. (5)

iii) Describe one local initiative aimed at reducing traffic congestion. (3)

Employment structures

Classification of industries

Traditionally, industry has been broken down into the three groups of *primary*, *secondary* and *tertiary* although, since the 1980s, *quaternary* has been added to make a fourth group (Figure 10.1).

- **Primary** industries extract raw materials directly from the earth or sea. Examples are farming, fishing, forestry and mining.
- **Secondary** industries process and manufacture the primary products (e.g. steelmaking and furniture making). They include the construction industry and the assembly of components made by other secondary industries (e.g. car assembly).
- **Tertiary** industries provide a service. These include education, health, office work, retailing, transport and entertainment.
- **Quaternary** industries provide information and expertise. The relatively new micro-electronics industries fall into this category.

Employment structures can be shown by three different types of graph. In each case the employment figures for the primary, secondary and tertiary sectors have to be converted into percentages.

1 Pie graphs

Figure 10.2 gives the employment structures for the UK and shows how they have changed *over a period of time*. Two hundred years ago, before accurate figures were

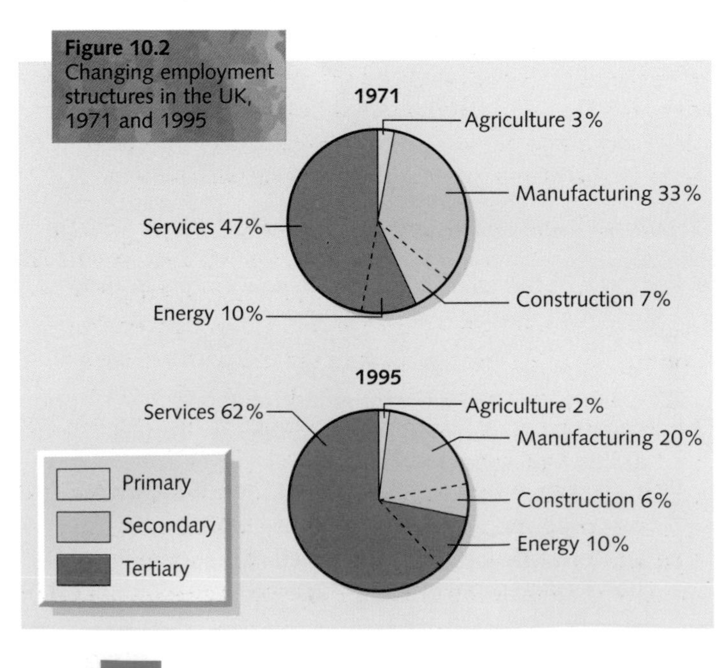

Figure 10.2 Changing employment structures in the UK, 1971 and 1995

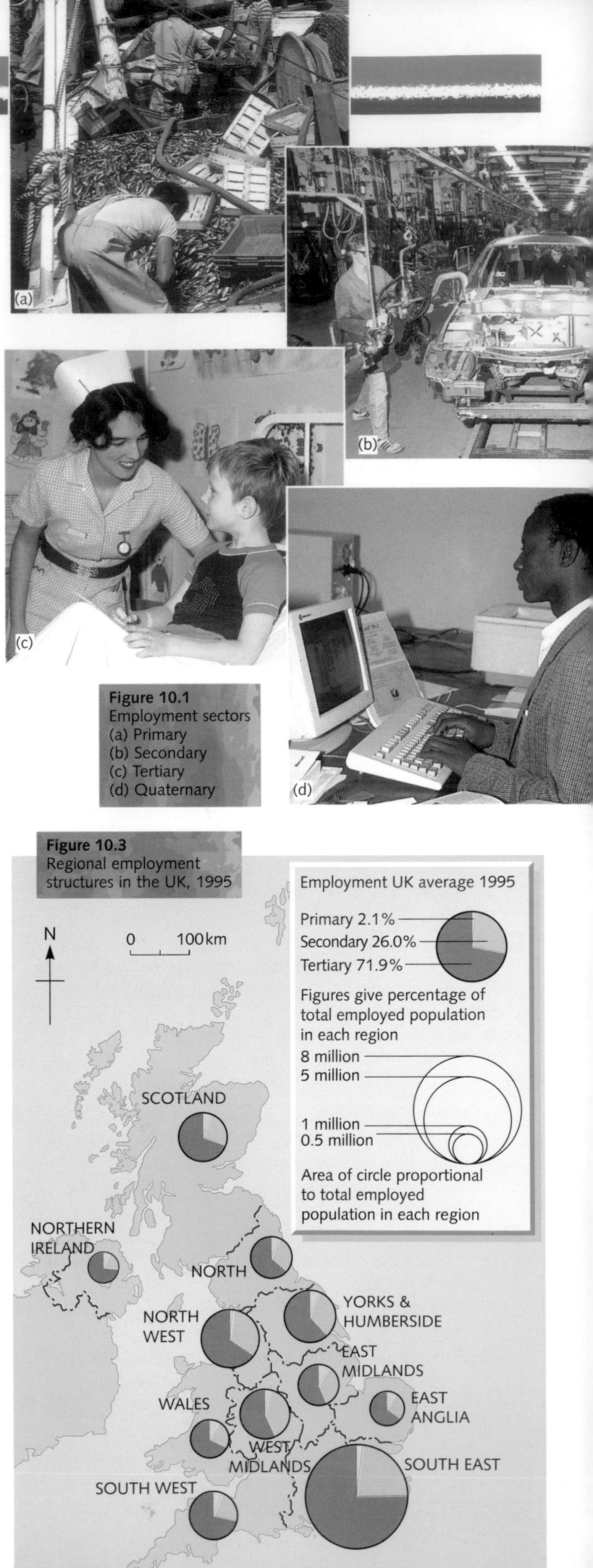

Figure 10.1 Employment sectors (a) Primary (b) Secondary (c) Tertiary (d) Quaternary

Figure 10.3 Regional employment structures in the UK, 1995

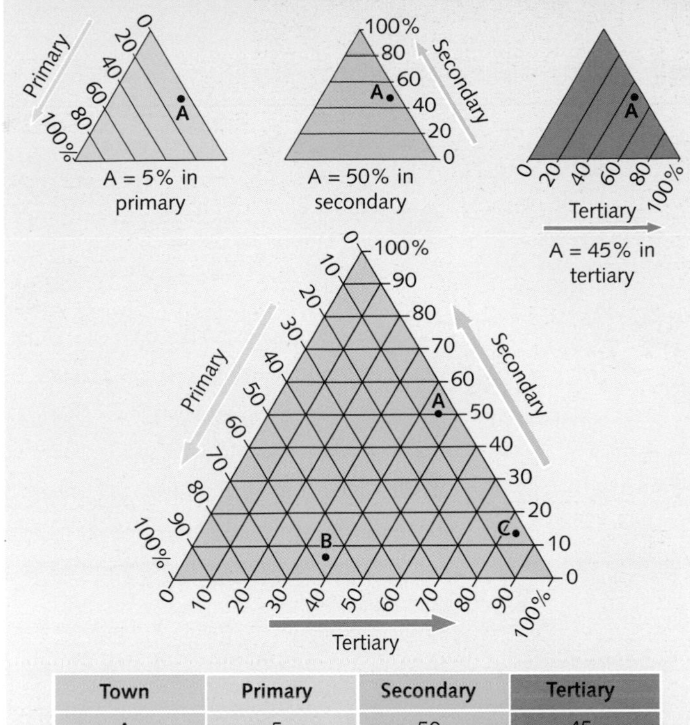

Town	Primary	Secondary	Tertiary
A	5	50	45
B			
C			

*Note that the numbers in each row add up to 100.

Figure 10.4
Employment structures: the triangular graph

% employed

0 10 20 30 40 50 60 70 80 90 100

Nepal, Ethiopia, Kenya, China, Sierra Leone, India, Bangladesh, Pakistan, Egypt, Peru, Malaysia, Brazil, Mexico, Italy, Japan, France, Australia, Germany, USA, UK

Primary · Secondary · Tertiary

Figure 10.5
Employment structures for countries at different levels of economic development.

available, most working people in Britain were engaged in the primary sector. At the beginning of the twentieth century most people found employment in the secondary sector, while today most people work within the tertiary (service) sector.

Employment structures can also be used to show changes *between places*. Figure 10.3 shows regional employment in the UK. Notice that it is the North of England and the Midlands that have the highest proportion in the secondary sector, and the South East, which includes London, in the tertiary sector.

Usually there is a link between employment structures and levels of economic development (page 170). Generally it is places, either regions or countries, with a higher percentage in secondary and tertiary activities that are considered to be the more economically developed, and those with a higher proportion in primary activities that are regarded as economically less developed.

2 Triangular graphs

A triangular graph is an equilateral triangle with each of its three 'bases' divided into percentage scales. Each base represents one of the three variables of primary, secondary and tertiary activities. It is convenient, though not essential, to make the sides of the triangle 10 cm long (so that 1 cm = 10%).

Figure 10.4 shows how the three variables are plotted to show the employment structure of Town A. The figure for the primary sector is found by using the left-hand scale (see the yellow graph), for secondary activities by using the right-hand scale (blue graph) and for tertiary activities the base (red graph). The answer for Town A is given underneath the graph. Try to complete the table for Towns B and C. These three towns represent (but not necessarily in this order) a small market town, a holiday resort and an industrial town. Match the figures with the appropriate letter (and be able to justify your answer).

3 Percentage bar graphs

A percentage bar graph is a horizontal bar, again ideally drawn 10 cm long. The bar is divided into three, to show the percentage of the working population employed in each of the primary, secondary and tertiary sectors.

Figure 10.5 shows the employment structure for selected countries at different levels of economic development. The countries have been ranked according to the proportion of people engaged in the primary sector. Remember that countries with a high proportion in the primary sector and a small proportion in the secondary and tertiary sectors are more likely to be economically less developed and to have a lower standard of living than those countries with a low proportion in the primary sector and a higher proportion in the secondary and tertiary sectors. You will be able to see how accurate this statement is when you look at the GNP of those countries on page 170.

Differences in world development

Figure 10.6
World GNP, 1994

Development means growth. Geographers are interested in differences in levels of development and rates of growth between places across the world and within a continent or country (regional). However, they find it difficult to find an acceptable and accurate method of measuring the levels of development.

1 Economic wealth

This book has, so far, referred to just one measure of development – that of economic development. This is because the traditional, and easiest, method to measure development is to compare wealth. Based on wealth, the world can be divided into the:

- **economically more developed countries (EMDCs)** which include the richer, more industrialised nations of the so-called developed 'North'
- **economically less developed countries (ELDCs)** which include the poorer, less industrialised nations of the so-called developing 'South' (Figure 10.6).

The wealth of a country is measured by its *gross national product per capita*, i.e. its *GNP per person*. The GNP per capita is the total value of goods produced and services provided by a country in a year, divided by the total number of people living in that country. To make comparisons between countries easier, GNP is given in US dollars (US$).

Figure 10.7 gives the GNP per capita for selected countries at different levels of economic development. It shows, for example, that in 1993 every person in the UK would have received US$ 17 970 if the wealth had been shared out evenly. It also shows that countries with the highest GNP are those said to be economically more developed. The disadvantage of using GNP is that it does not show differences in wealth between people and places *within* a country.

Figure 10.7
Measures of world development, 1993

	COUNTRY	GNP US$ PER CAPITA	BIRTH RATE PER 1000	DEATH RATE PER 1000	INFANT MORTALITY PER 1000	LIFE EXPECTANCY	DIET, CALORIES PER PERSON PER DAY	ENERGY CONSUMPTION*	% IN AGRICULTURE	NO. PEOPLE PER DOCTOR	ADULT LITERACY %
DEVELOPED COUNTRIES	JAPAN	31 450	12	8	5	80	2903	4.74	7	600	99
	USA	24 750	14	9	8	77	3732	10.74	3	420	99
	ITALY	19 620	11	11	9	77	3561	4.02	9	211	97
	UK	17 970	14	12	8	76	3317	5.40	2	300	99
OPEC	KUWAIT	23 350	26	2	15	75	2523	4.04	0	600	77
MIDDLE-INCOME DEVELOPING COUNTRIES	MEXICO	3750	17	6	36	71	3146	1.89	23	621	87
	MALAYSIA	3160	28	5	20	71	2888	1.80	26	2564	82
	BRAZIL	3020	26	8	57	67	2824	0.81	25	1010	81
LOW-INCOME DEVELOPING COUNTRIES	EGYPT	660	31	9	57	62	3335	0.70	42	1300	49
	INDIA	290	31	10	88	61	2395	0.35	62	2439	50
	KENYA	270	47	10	64	61	2075	0.11	81	10 000	75
	BANGLADESH	220	41	14	108	53	2019	0.08	59	12 500	36
	ETHIOPIA	100	48	18	122	47	1610	0.03	88	33 000	33

* TONNES OF COAL EQUIVALENT PER PERSON PER YEAR

2 Social factors

Although economic development to people living in a Western society tends to mean a growth in wealth, other criteria have also been suggested. Figure 10.7 suggests links, or correlations, between wealth and a range of social factors. In general, countries that are economically more developed have lower birth and infant mortality rates, a longer life expectancy, a slower natural population increase, a smaller proportion aged under 15 and a higher proportion aged over 65 than countries that are less economically developed (Figure 10.8 and pages 6 to 11).

Notice, however, that many of these criteria are themselves dependent upon the wealth of a country. For example, the more wealthy, and therefore economically developed, a country is, the smaller the proportion of its population in agriculture (with more in services); the higher its energy consumption; the greater its calorie intake (better diet); the lower its infant mortality rate, the longer its life expectancy, the fewer of its population per doctor (better health care); and the higher its adult literacy rate (better education).

3 Human Development Index (HDI)

In 1990, the United Nations replaced GNP as the measure of development with the *Human Development Index* (*HDI*) (Figure 10.9). The HDI is a social welfare index measuring the adult literacy rate (education), life expectancy (health) and the real GNP per person – that is, what an income will actually buy in a country (economic). The HDI is an attempt to compare the *quality of life* between people and places and, unlike the GNP, it can measure differences within a country.

	ECONOMICALLY MORE DEVELOPED COUNTRIES	ECONOMICALLY LESS DEVELOPED COUNTRIES
GROSS NATIONAL PRODUCT	MAJORITY OVER US$5000 PER PERSON PER YEAR; 80% OF WORLD'S TOTAL INCOME	MAJORITY UNDER US$2000 PER PERSON PER YEAR; 20% OF WORLD'S TOTAL INCOME
POPULATION GROWTH	RELATIVELY SLOW PARTLY DUE TO FAMILY PLANNING; 25% OF WORLD'S POPULATION; POPULATION DOUBLES IN 80 YEARS	EXTREMELY FAST, LITTLE OR NO FAMILY PLANNING; 75% OF WORLD'S POPULATION; POPULATION DOUBLES IN 30 YEARS
HOUSING	HIGH STANDARD OF PERMANENT HOUSING; INDOOR AMENITIES, e.g. ELECTRICITY, WATER SUPPLY AND SEWERAGE	LOW STANDARD, MAINLY TEMPORARY HOUSING; VERY RARELY ANY AMENITIES
TYPES OF JOBS	MANUFACTURING AND SERVICE INDUSTRIES (75% OF WORLD'S MANUFACTURING INDUSTRY)	MAINLY IN PRIMARY INDUSTRIES (25% OF WORLD'S MANUFACTURING INDUSTRY)
LEVELS OF MECHANISATION	HIGHLY MECHANISED WITH NEW TECHNIQUES; 96% OF WORLD SPENDING ON DEVELOPMENT PROJECTS AND RESEARCH	MAINLY HAND LABOUR OR THE USE OF ANIMALS
EXPORTS	MANUFACTURED GOODS	UNPROCESSED RAW MATERIALS
ENERGY	HIGH LEVEL OF CONSUMPTION; MAIN SOURCES ARE COAL, OIL, HEP AND NUCLEAR POWER. USE 80% WORLD'S ENERGY	LOW LEVEL OF CONSUMPTION; WOOD STILL A MAJOR SOURCE. USE 20% WORLD'S ENERGY
COMMUNICATIONS	MOTORWAYS, RAILWAYS AND AIRPORTS	ROAD, RAIL AND AIRPORTS ONLY NEAR MAIN CITIES, RURAL AREAS HAVE LITTLE DEVELOPMENT
DIET	BALANCED DIET; SEVERAL MEALS PER DAY; HIGH PROTEIN INTAKE	UNBALANCED DIET; 20% OF POPULATION SUFFERS FROM MALNUTRITION; LOW PROTEIN INTAKE
LIFE EXPECTANCY	OVER 75 YEARS	OVER 60 YEARS
HEALTH	VERY GOOD, LARGE NUMBERS OF DOCTORS AND GOOD HOSPITAL FACILITIES	VERY POOR, FEW DOCTORS AND INADEQUATE HOSPITAL FACILITIES
EDUCATION	MAJORITY HAVE FULL-TIME SECONDARY EDUCATION (16+)	VERY FEW HAVE ANY FORMAL EDUCATION; FEMALES DISADVANTAGED

Figure 10.8
Differences between economically more developed and economically less developed countries

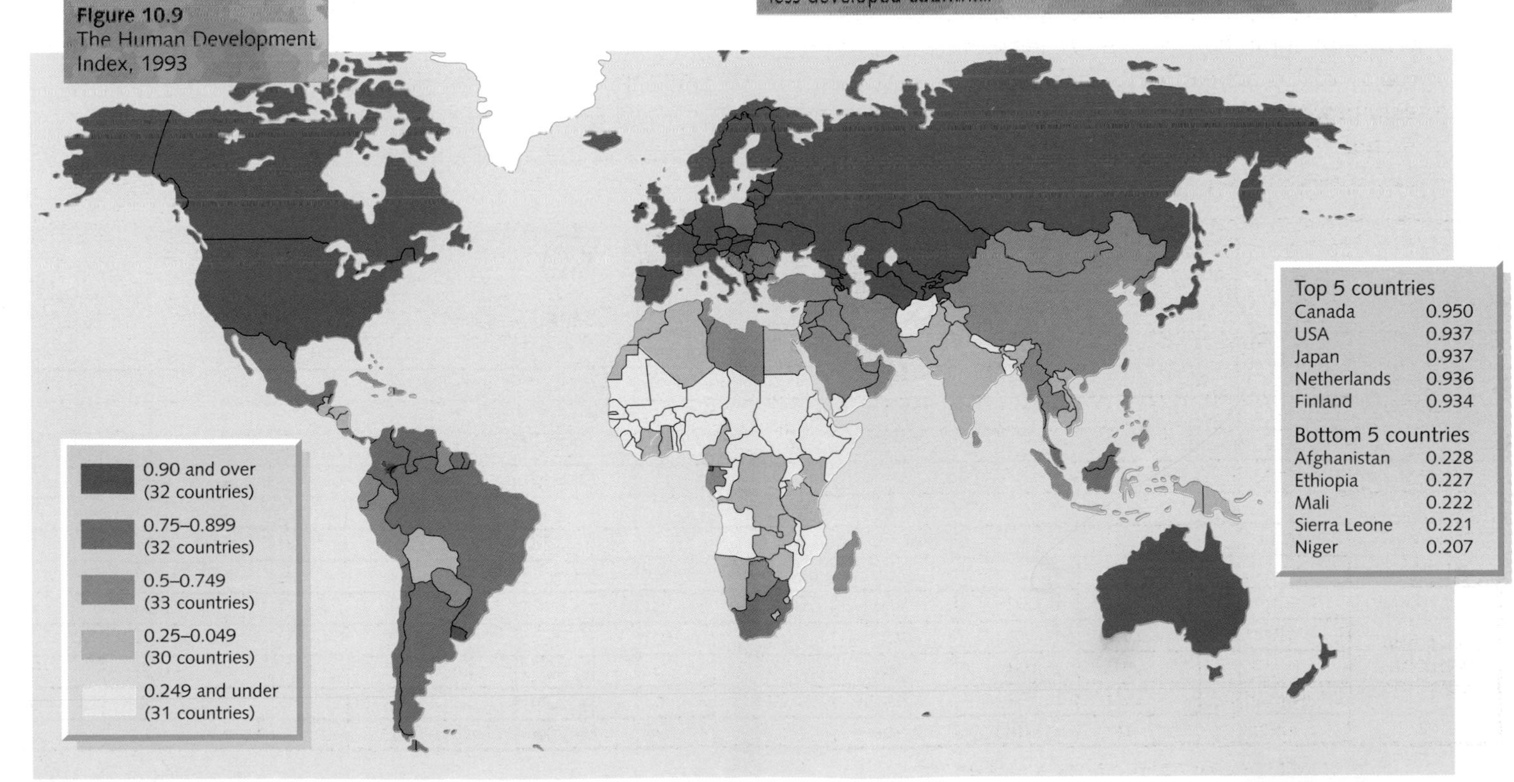

Figure 10.9
The Human Development Index, 1993

Differences in regional development

Economic development is rarely evenly distributed, either at a local, regional, national or international level. Growth becomes concentrated in a few favoured locations, leaving other places relatively poor and underdeveloped in comparison.

LEVEL	MORE DEVELOPED/ BETTER OFF	LESS DEVELOPED/ WORSE OFF
LOCAL/BRITISH CITY	SUBURBIA	OLD INNER CITY
REGIONAL/NATIONAL	CAPITAL CITY SOUTH-EAST ENGLAND NORTH OF ITALY	ISOLATED, RURAL VILLAGE NORTH AND WEST BRITAIN SOUTH OF ITALY
INTERNATIONAL	EU – GERMANY, NETHERLANDS	EU – PORTUGAL AND IRELAND

Core–periphery

The most prosperous part of a country can be referred to as the *core*. This region is likely to contain the capital city and the country's major industrial areas. These provide a large local market, and attract other industries and services such as banking, insurance and government offices. As levels of capital, technology and skilled labour increase, the region becomes more wealthy. It can afford to provide schools, hospitals, shopping centres, modern transport networks and better-quality housing. These pull factors encourage the in-migration of people from surrounding areas.

In many countries the level of prosperity decreases with distance from the core region. The poorest places are therefore found towards the *periphery* of a country. Here jobs are fewer in number, poorly paid and, probably, mainly in the primary sector. There is often a lack of opportunity, poor service provision and insufficient government investment. These push factors force many people to migrate to the core region.

Figure 10.11
Levels of wealth in the EU, 1990

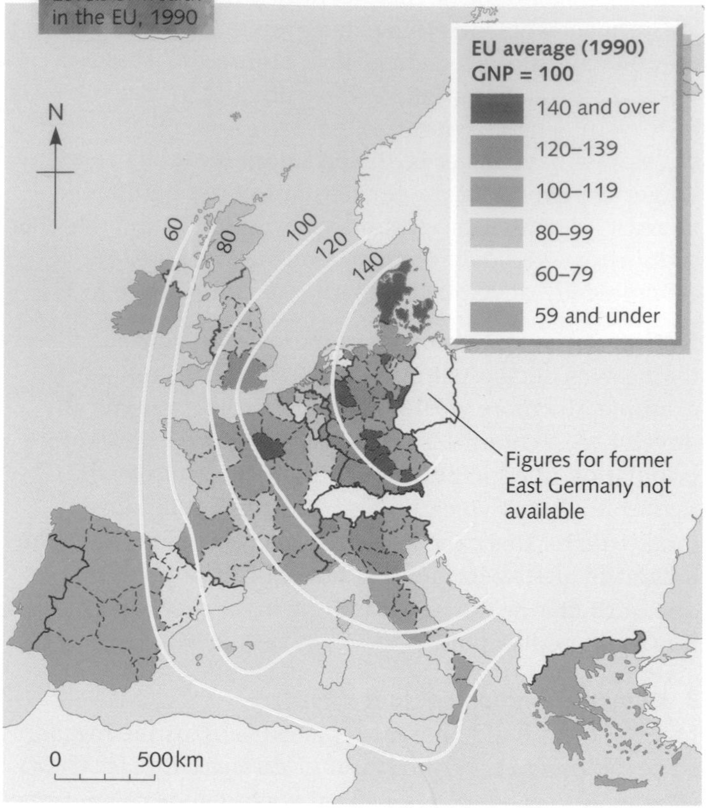

As a country develops economically, industry and wealth often begin to spread out. Initially a second core region develops, followed, in time, by others. In the UK, the South-east has become the core region, with the West Midlands, parts of Lancashire and Yorkshire, and Clydeside the secondary core regions. Places furthest from London, especially Northern Ireland, form the periphery (Figure 10.10). Similarly in the EU, there is a more wealthy core which includes Denmark and most of Germany. Generally, though there are exceptions, prosperity decreases towards the fringe regions such as Portugal, Greece and Ireland (Figure 10.11).

Figure 10.10
Uneven regional development in the UK

UK REGION	GNP EU AVERAGE = 100	AVERAGE INCOME FOR FAMILY £	AVERAGE HOUSE PRICE £	UNEMPLOYMENT %	PROPORTION OF 16–18 YEAR OLDS IN HIGHER EDUCATION	5 OR MORE GCSE GRADES A–G	FREE SCHOOL DINNERS %	ONE OR MORE CAR PER FAMILY %	HOMES WITH COMPUTER %	HOMES WITH DISHWASHER %
SOUTH EAST	116	22 640	80 000	7.9	70	86.5	17	71	27	22
SOUTH WEST	94	19 700	64 000	7.0	68	89.4	13	75	24	22
EAST ANGLIA	101	17 970	59 000	6.2	64	88.8	12	76	23	18
EAST MIDLANDS	93	19 020	53 300	7.7	66	86.6	15	69	22	16
WEST MIDLANDS	91	16 730	60 500	8.4	72	84.7	19	69	24	14
YORKS & HUMBERSIDE	91	17 850	52 600	8.8	75	82.7	19	67	23	13
NORTH WEST	90	17 730	55 500	8.8	68	83.4	23	64	22	14
NORTH	89	15 800	46 100	10.6	68	83.1	22	59	25	11
WALES	84	14 700	52 500	8.5	65	79.0	22	67	23	16
SCOTLAND	97	18 870	52 700	8.2	71	85.4	10	62	21	15
NORTHERN IRELAND	79	16 970	42 400	11.4	79	82.5	25	65	16	15
UK	99	19 200	63 200	8.2	70	85.2	18	68	24	17

Source: *Regional Trends 1996* (for 1994/95)

Core–periphery in Japan

The core – Tokyo

Land values in Tokyo are the highest in the world. The Kanto region, which surrounds Tokyo Bay and where over 30 per cent of Japanese live, is the centre of the country's industry, commerce and services (Figure 10.12). Tokyo Bay provides a deep, sheltered harbour for large ships to bring in raw materials and energy supplies (page 128) and to take out manufactured goods. Land is constantly being reclaimed from the sea for the building of new factories, offices, houses and port facilities. Tokyo is at the centre of land communications, with the Shinkansen, or bullet train (page 49), and other main rail and road routes meeting here. Despite the city's already high population density (Figure 10.13) and high cost of living, the promise of highly-paid jobs and the perception of Tokyo's 'bright lights' continue to act as a magnet to many younger people living in the surrounding rural areas.

The periphery – Hokkaido

Hokkaido is much poorer than Tokyo, although it is still as well off as the UK and many other economically more developed countries. Nevertheless, by Japanese standards, and to Japanese perception, it is a peripheral region. The island used to be important in providing primary products such as food, fish, timber and coal. This led to the development of a steel and shipbuilding industry. There has been a recent decline in both fishing and forestry. Due to the closing of all but one coal mine, and the distance from Japan's major ports and internal markets, steel and shipbuilding have also declined. Hokkaido tends not to attract new industries as Japanese industrialists perceive the island to be too cold, prone to earthquakes and isolated, despite the building of a rail tunnel linking the island with the mainland. Although Hokkaido already has the lowest population density in Japan (Figure 10.12), and has wages that are high by world standards, many younger people choose to move away to live and work in Japan's core regions.

Figure 10.12
Location of major industries in Japan

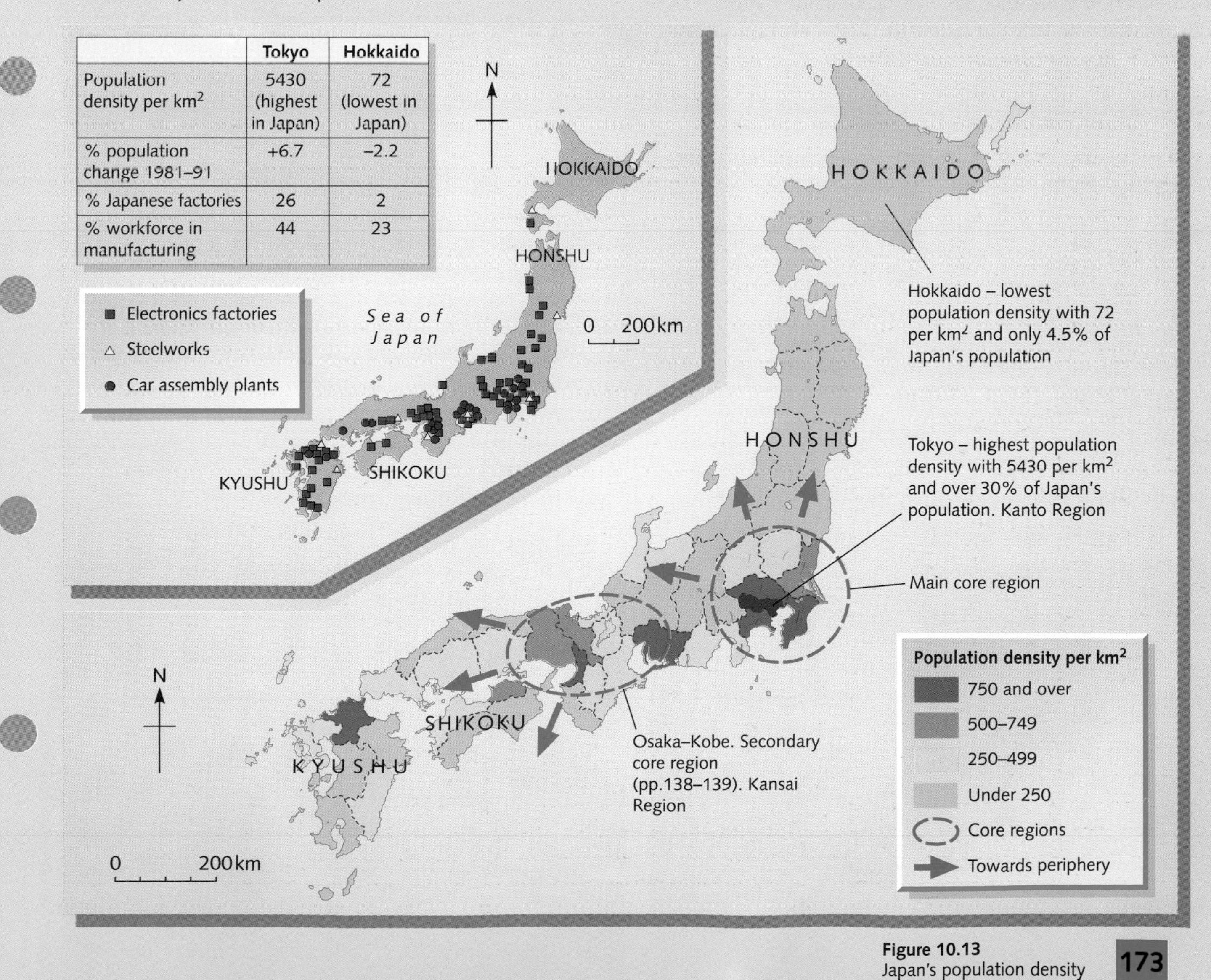

	Tokyo	Hokkaido
Population density per km²	5430 (highest in Japan)	72 (lowest in Japan)
% population change 1981–91	+6.7	−2.2
% Japanese factories	26	2
% workforce in manufacturing	44	23

Figure 10.13
Japan's population density

Stages in economic development

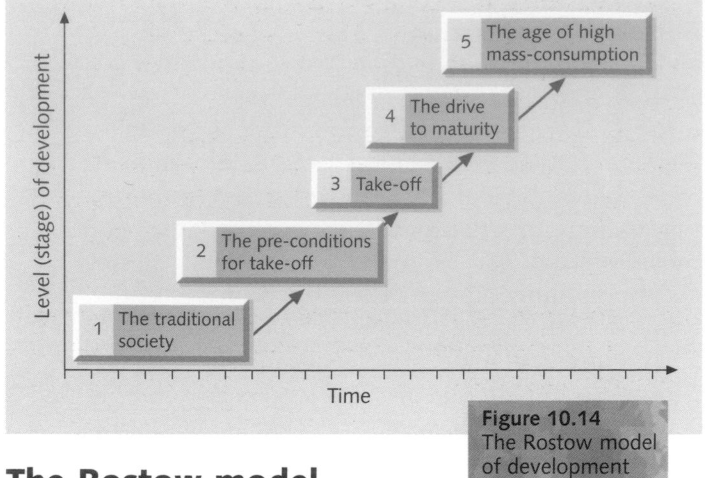

Figure 10.14 The Rostow model of development

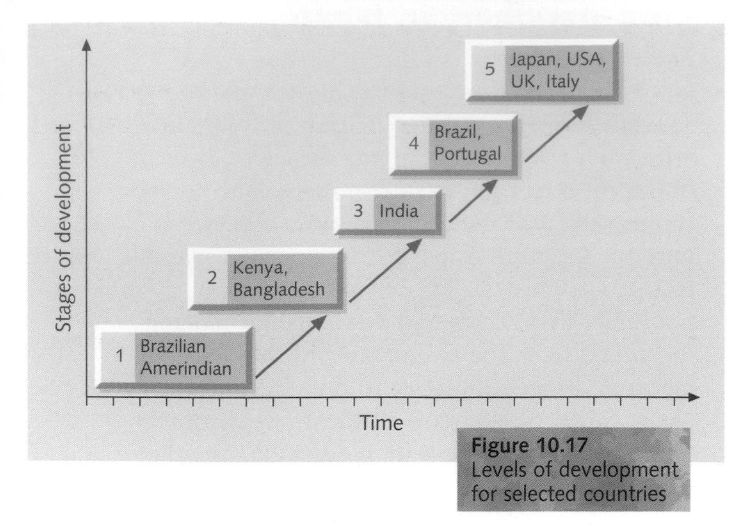

Figure 10.17 Levels of development for selected countries

The Rostow model

Walt Rostow was an economist. After studying several economically more developed countries, he suggested a model for economic growth. Remember that a *model* (page 28) is a theoretical framework which may not actually exist, but which helps to explain reality in a more simplified way. Rostow suggested that all countries had the potential to pass through a series of stages of growth until they became fully industrialised and economically developed (Figure 10.14).

Stage 1 This is usually a subsistence economy based mainly upon farming. There is insufficient technology and capital to process raw materials or to develop industries and services (Figure 10.15).

Stage 2 A country usually needs external help to move into this stage, e.g. a colony. Primary activities are developed although most products are exported. There are some technological improvements, and the development of a transport network and one or two industries. There is also a slow improvement in the standard of living (GNP).

Stage 3 The manufacturing industry grows rapidly as there is improved technology and capital to process raw materials. There is increasing investment in agriculture, transport and services. However, economic development is likely to be confined to one or two 'core' regions, e.g. around the capital city or chief port. A rapid improvement in the standard of living is seen.

Stage 4 Economic growth spreads to most parts of the country. A more complex transport network develops, as does a wider range of industry, including several using higher levels of technology and mechanisation. This is often a time of rapid urbanisation and declining primary activity. Standards of living continue to improve.

Stage 5 There is a rapid expansion of service and high-tech industries, but a decline in manufacturing (Figure 10.16).

Rostow's model, like all models, can be criticised. He suggested that capital was needed from a more economically developed country before 'take-off' could begin. It is now accepted that many countries, even with aid, are unlikely to become industrialised. This may simply be due to a lack of raw materials or the necessary capital. It has also been suggested that many developed countries were once colonial powers and developed at the expense of their 'satellites'. Figure 10.17 suggests different levels which several selected countries have reached based upon the Rostow model.

Figure 10.15 Subsistence farming – a stage 1 or 2 activity

Figure 10.16 Medical research – a stage 5 activity

Sustainable development

Sustainable development should lead to an improvement in people's:

- **quality of life** – allowing them to become more content with their way of life and the environment in which they live
- **standard of living** – enabling them, and future generations, to become better off economically.

This may be achieved in a variety of ways:

- By encouraging economic development at a pace a country can afford so as to avoid that country falling into debt.
- By developing technology that is appropriate to the skills, wealth and needs of local people, and developing local skills so that they may be handed down to future generations.
- By using natural resources without spoiling the environment, developing materials that will use fewer resources, and using materials that will last for longer. When using resources, remember the 3 Rs – renew, recycle, replace.

Sustainable development needs careful planning and, as it involves a commitment to conservation, the co-operation of different countries.

Ladakh (northern India)

Global Concerns is a British charitable organisation, based in Edinburgh, which supports sustainable development projects in Ladakh. Ladakh, in the foothills of the Himalayas, is very mountainous. It also has an extreme climate with, on average, temperatures for over half the year below or near to 0°C, a growing season of less than 5 months, an annual precipitation of under 20 cm (making it a desert) most of which falls as snow, 325 days of sunshine, and strong winds sweeping down its valleys. People in Ladakh do not want development based on the Western model with, as they see it, its resultant socio-cultural, urbanisation and environmental problems. Instead they are trying to *'promote a development policy based on ecological principles, with an emphasis on local conditions, resources and traditions. This should help protect our environment and preserve the foundations of our traditional culture'* (Figure 10.18).

Leh, the capital, is 3500 m above sea-level. Like other settlements it is next to a river.

Water from summer snowmelt can be used for:
- irrigation
- micro-hydro power.

Strong winds and the sun are important renewable sources of energy.

Houses
- Solar power heats water and houses.
- Longest wall faces south to get most sunshine.
- Solar cookers and new locally-made metal braziers (compare Figure 8.41) avoid use of precious dung and fuelwood, and improve health as less smoke.
- Cavity walls filled with mud and stones to conserve heat.

Solar greenhouses allow vegetables to grow throughout the year.

Small, locally-made hydraulic rams can pump water to a higher level, extending the area of cultivation and allowing Ladakh to remain self-sufficient in food.

The ecological development centre trains people in traditional handicrafts.

Small micro-hydro schemes generate electricity for lighting and for grinding corn (see Figure 8.37).

Figure 10.18 Features of Ladakh

Places

Trade and interdependence

No country is self-sufficient in the full range of raw materials (food, minerals and energy) and manufactured goods which are needed by its inhabitants. To try to achieve sufficiency, countries must trade with each other. *Trade* is the flow of commodities from producers to consumers, and is important in the development of a country. One way for a country to improve its standard of living and to grow more wealthy is to sell more goods than it buys. Unfortunately, for every country that exports more than it imports, at least one other country will have to import more than it exports. The result is that some countries have a *trade surplus* allowing them to become richer, while others have a *trade deficit* making them poorer.

Most of the world's population live in economically less developed countries. They produce many of the primary products (agricultural and mineral) which are needed by the economically more developed countries. The developed countries then process many of these primary products into manufactured goods which are needed by themselves and the developing countries.

COUNTRY AND STAGE OF ECONOMIC DEVELOPMENT	RANK ORDER			
	WEALTH (GDP)	IMPORT OF RAW MATERIAL	EXPORT OF RAW MATERIAL	EXPORT OF MANUFACTURED GOODS
USA – WEALTHY INDUSTRIALISED COUNTRY	1	1	4	1
UK – ECONOMICALLY DEVELOPED COUNTRY	2	2	3	2
BRAZIL – ECONOMICALLY DEVELOPING COUNTRY	3	3	2	3
SIERRA LEONE – ECONOMICALLY LESS DEVELOPED COUNTRY	4	4	1	4

Figure 10.19
Stages of economic development

TRADE OF DEVELOPING COUNTRIES	TRADE OF DEVELOPED COUNTRIES
A LEGACY OF FORMER COLONIAL ECONOMIES WHERE A MINERAL ONCE MINED, OR A CROP ONCE GROWN IS EXPORTED IN ITS 'RAW STATE'. MOST EXPORTS ARE PRIMARY PRODUCTS.	MAINLY MANUFACTURED GOODS ARE TRADED, AS THESE COUNTRIES HAVE BECOME INDUSTRIALISED. CEREALS ALSO EXPORTED.
OFTEN ONLY TWO OR THREE ITEMS ARE EXPORTED.	A WIDE RANGE OF ITEMS ARE EXPORTED.
PIECES OF, AND DEMAND FOR, THESE PRODUCTS FLUCTUATE ANNUALLY. PRICES RISE LESS QUICKLY THAN FOR MANUFACTURED GOODS.	PRICES OF, AND DEMAND FOR, THESE PRODUCTS TEND TO BE STEADY. PRICES HAVE RISEN CONSIDERABLY IN COMPARISON WITH RAW MATERIALS.
THE TOTAL TRADE OF THESE COUNTRIES IS SMALL.	THE TOTAL TRADE OF THESE COUNTRIES IS LARGE.
MOST EXPORTS COME FROM TRANS-NATIONAL COMPANIES WHICH TEND TO SEND PROFITS BACK TO THE PARENT COMPANY.	PROFITS ARE RETAINED BY THE EXPORTING COUNTRY.
TRADE IS HINDERED BY POOR INTERNAL TRANSPORT NETWORKS.	TRADE IS HELPED BY GOOD INTERNAL TRANSPORT NETWORKS.
TRADE IS SEVERELY HIT AT TIMES OF WORLD ECEONOMIC RECESSION.	TRADE IS BADLY AFFECTED AT TIMES OF WORLD ECONOMIC RECESSION.

	% WORLD'S POPULATION	% WORLD'S EXPORTS OF PRIMARY PRODUCTS	% WORLD'S EXPORTS OF MANUFACTURED GOODS
DEVELOPED (EMDC)	25	24	61
DEVELOPING (ELDC)	75	76	39

Figure 10.21
Differences between the trade of developing and developed countries

However, although prices of raw materials have increased over the years, the prices of manufactured goods have increased much more. This means that a developed country exporting manufactured goods earns increasingly more than a developing country selling raw materials. The result is a widening trade gap between the developed and developing countries. The differences in the type of trade between four countries at different stages of economic development are shown in Figures 10.19 and 10.20 (the graphs do not show the great contrast in the volume of trade), and the effect of these differences in Figure 10.21.

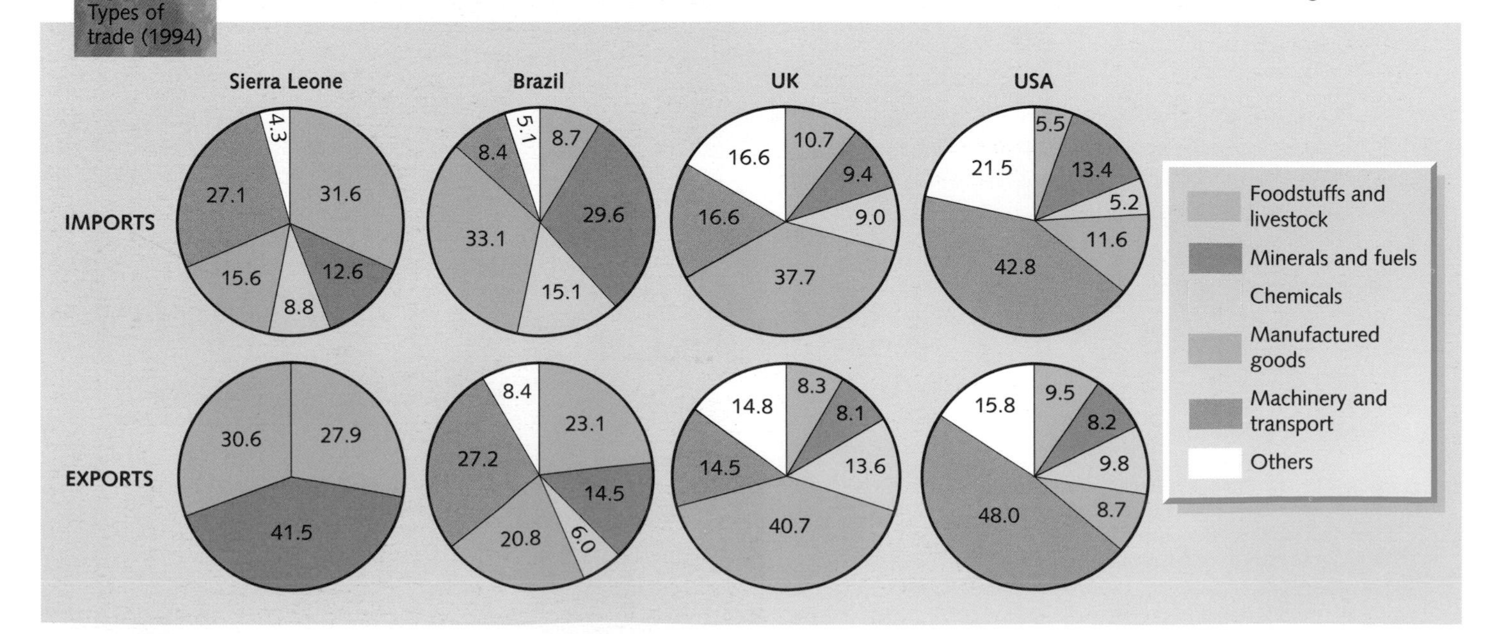

Figure 10.20
Types of trade (1994)

Dependence upon:

(a) a single commodity	(b) two or three commodities
Nigeria 90% oil	Kenya 52% coffee and tea
Ghana 68% coal	Egypt 83% oil, cotton and textiles
Zambia 88% copper	Mauritania 84% iron ore and fresh fish
Uganda 86% coffee	Algeria 98% oil, oil products and natural gas

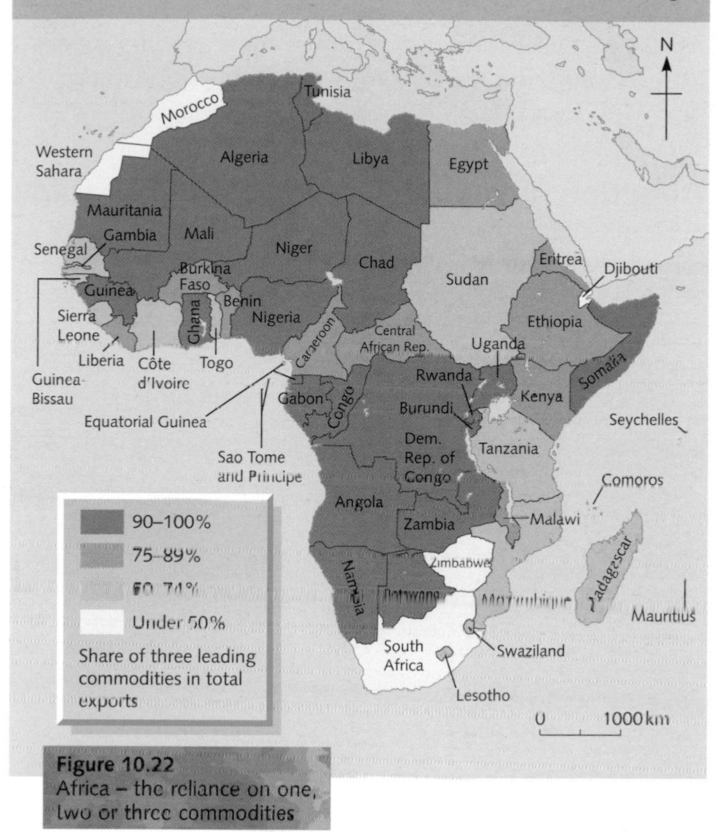

Figure 10.22
Africa – the reliance on one, two or three commodities

Many developing countries rely heavily on the export of only one or two major commodities (Figure 10.22). The price paid for these commodities is fixed by the developed countries. If there is a world recession, an overproduction of a crop or mineral, a change or a fall in demand for a product, or a crop failure, then the economy of the producing country can be seriously affected. The developed countries try to depress the prices of raw materials which they have to buy from developing countries. They also try to protect their own manufacturing industries by suppressing industrial development and limiting the imports of processed goods from developing countries. In many cases, it is large transnational corporations (page 132) which finance and organise either the growing of crops, the mining of minerals or the processing of the raw materials. This means that, as profits are sent overseas, those countries that supply the raw materials do not get their fair share of the wealth.

Changes in Britain's trade

Figure 10.23 shows how Britain's trading partners have changed since 1973.

- The biggest increase has been with countries in the EU. Whether we like being in the EU or not, the graph shows just how dependent we have become upon it for trade.
- In 1994, 80 per cent of Britain's trade was with economically more developed countries (the EU, North America and Japan) compared with 74 per cent in 1973 and 68 per cent in 1951.
- Despite our need for raw materials, there had been, until the 1990s, a decline in trade between ourselves and the economically less developed countries. Most of these links were originally formed during colonial times, e.g. cocoa from the Gold Coast (Ghana), tea from Ceylon (Sri Lanka) and rubber from Malaya (Malaysia). Recently, trade has increased with the 'tiger' economies (page 129) of eastern Asia, especially South Korea, Singapore and Taiwan, which have themselves set up factories within the UK.
- Britain's links with oil-producing countries (OPEC) have declined since the exploitation of North Sea oil.
- There has always been minimal trade between Britain and the former centrally planned economies of the former USSR and Eastern Europe.

Despite Britain's position as an economically more developed country, we have developed a trade deficit that is proving increasingly hard to reduce.

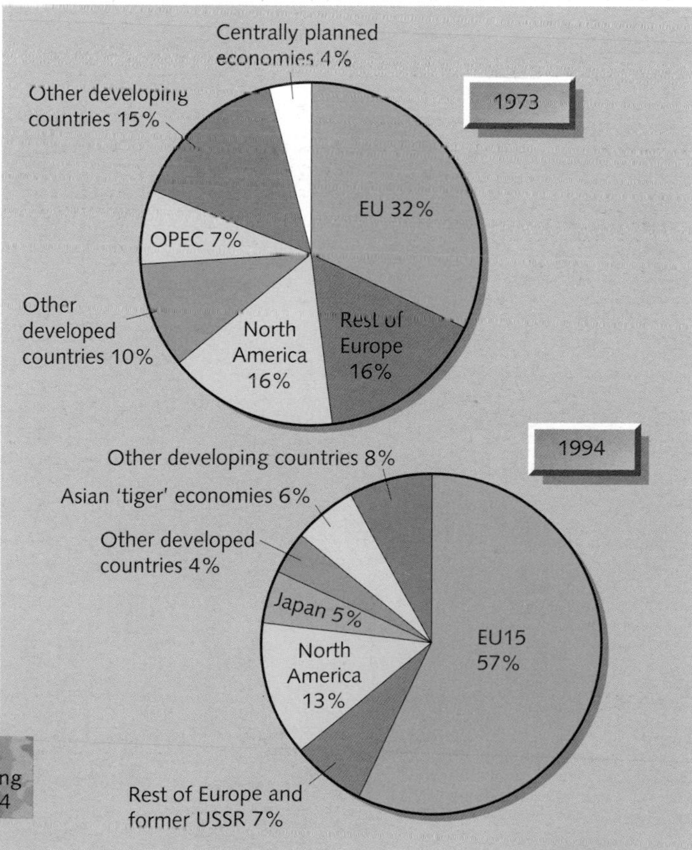

Figure 10.23
Changes in Britain's trading partners – 1973 and 1994

World trade

Regional trade groups

The EU is one example where several countries have grouped together for the purpose of trying to increase trade (Figure 10.24). The EU was, until the more recent formation of NAFTA, the world's largest single market with 370 million people (NAFTA has 390 million). By eliminating import duties (*tariffs*) between members, the EU reduced the cost of products sold between member countries and increased its number of potential customers. However, while this made the EU more competitive against its major rivals, Japan and the USA, it also created restrictions (*trade barriers*) which protected goods made in the EU against cheaper imports from economically less developed countries. This meant that developing countries found it increasingly difficult to sell their products on the world market.

The future seems to indicate the enlargement of regional trading blocks. For example, the EU is considering accepting membership of former East European states, and by the year 2005, NAFTA may embrace all of the Americas. The question is, 'Will the super-trading blocks increase or hinder the move towards global free trade? (See 'GATT and the WTO', opposite.)

Direction, value and volume of world trade

- Despite the growing number of trading groups, most of the world's trade is dominated by the relatively few 'market economies' of the more industrialised, developed countries (Figures 10.25 and 10.26). In 1995, the EU's share of world trade was 38 per cent, the USA's 14 per cent and Japan's 8 per cent (compared with 44, 13 and 9 per cent respectively in 1990). Japan had a significant, though slightly declining, trade surplus, the USA an increasing trade deficit and the EU a slight trade surplus (the UK had a deficit).
- In recent years, the older industrialised countries of the USA and Western Europe have faced increasing competition, initially from Japan but more recently from the newly industrialised countries (NICs) of Asia's Pacific Rim (page 129). These emerging trading nations now have over 10 per cent of world trade and, in the case of Singapore and Hong Kong, the highest trade per capita (Figure 10.27). The amount of trans-Pacific trade (USA and eastern Asia) now exceeds that which crosses the Atlantic Ocean (USA and Europe).

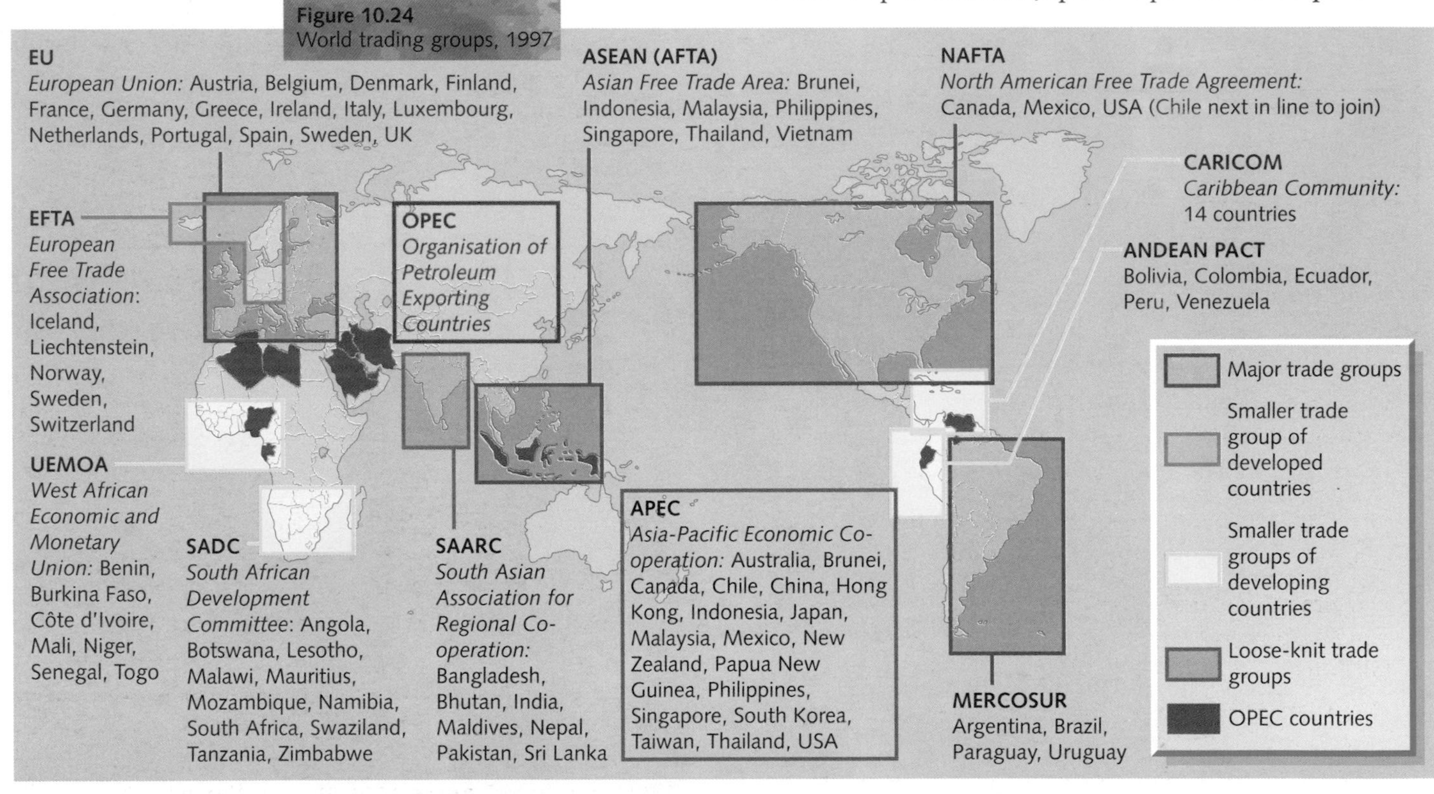

Figure 10.24 World trading groups, 1997

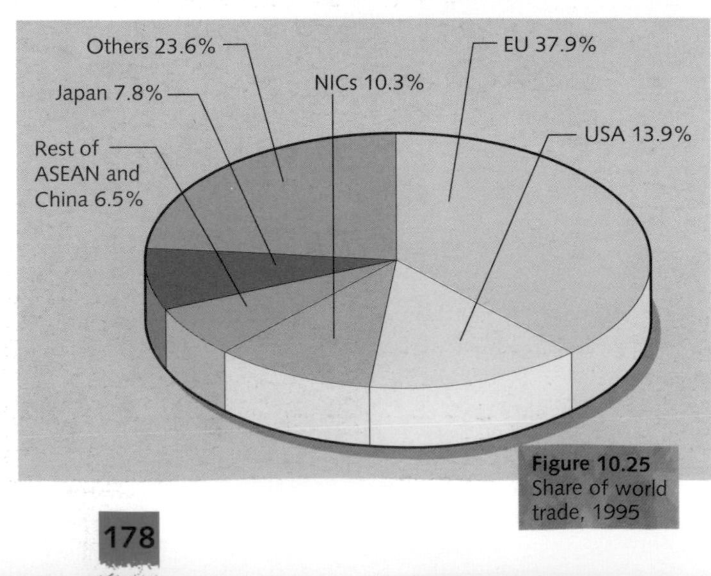

Figure 10.25 Share of world trade, 1995

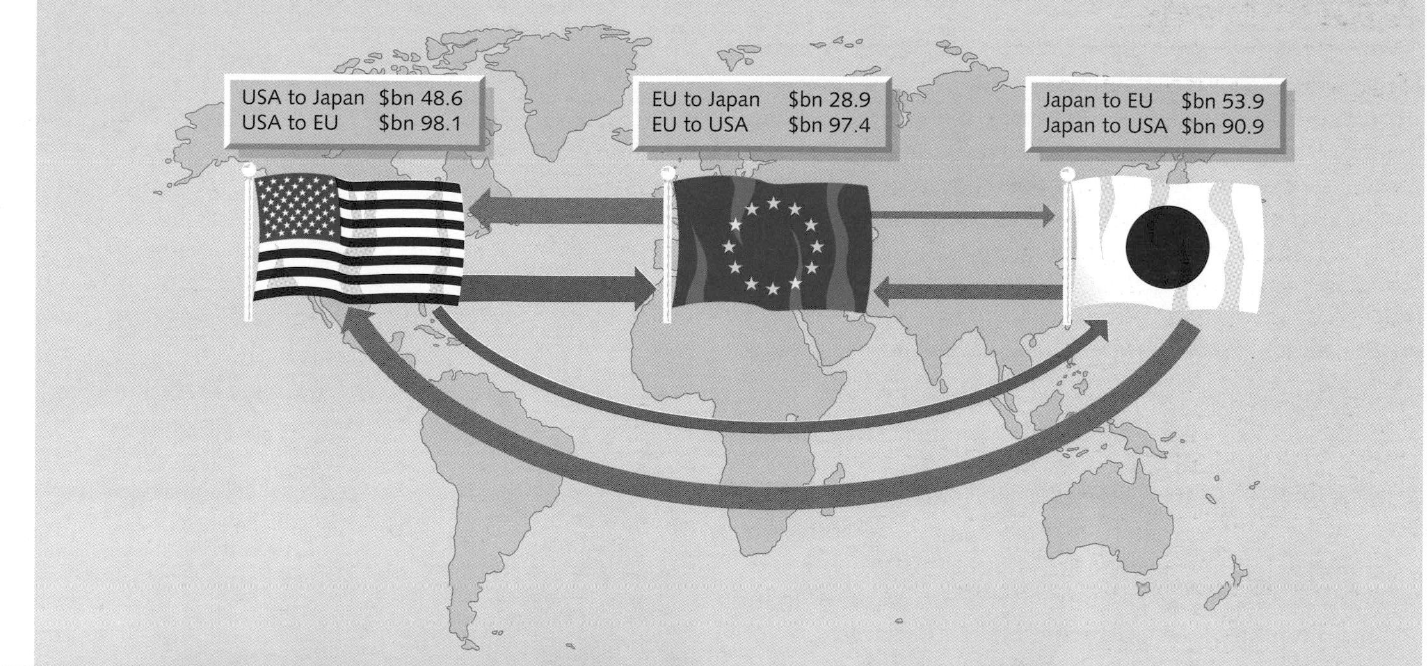

	JAPAN	EU	USA
JAPAN	–	LARGE SURPLUS	LARGE SURPLUS
EU	LARGE DEFICIT	–	VERY SMALL DEFICIT
USA	LARGE DEFICIT	VERY SMALL SURPLUS	–

Figure 10.26
The main players in international trade

- Although trade between economically less developed countries has increased steadily since the 1970s, in 1995 over 60 per cent of this trade was shared by only eight nations (e.g. Brazil and Mexico). The trade gap for most developing countries remains wide as they continue to export unprocessed raw materials and to import manufactured goods. During any world recession, as in the early 1990s, raw materials become even more vulnerable to changes in market price and demand. As the trade gap of developing countries widens, they fall further into debt and, as they cannot buy as many overseas goods, the volume of world trade decreases.
- The oil producing countries (OPEC) have seen their share of world trade decline since 1990 due to the Gulf War and the world recession.
- The world's trade is becoming dominated by the increasing number of large, powerful, transnational corporations.

GATT and the WTO

GATT (General Agreement on Trade and Tariffs) was established in 1947 to encourage free and multiple trading between countries. It was hoped that GATT could stimulate world trade in goods and services by finding ways of reducing tariffs (import duties), quotas and other trade barriers set up by countries to protect their domestic producers. In Britain, for example, we complained about the loss of jobs caused by cheap imports (e.g. textiles), yet without being able to sell their goods, how else can developing countries earn sufficient money to buy our goods?

It took until April 1994 before some agreement was signed. Tariffs were immediately reduced on many industrial products. This was to the benefit of the NICs and, as increased competition should lower prices, consumers. Farm subsidies and tariffs on agricultural goods were also to be cut, but only over six years. This delay, resulting from the tactics of strong farming lobbies in the EU and the USA, was to the detriment of many developing countries which relied on the export of agricultural products.

In January 1995, GATT was replaced by the World Trade Organisation (WTO). The WTO, with 108 members, was set up to supervise the implementation of trade agreements and to settle trade disputes.

Figure 10.27
Trade per capita (US$)

DEVELOPED COUNTRIES		NICS		OPEC		MIDDLE-INCOME DEVELOPING COUNTRIES		LOWER-INCOME DEVELOPING COUNTRIES	
GERMANY	9815	SINGAPORE	55 483	UAE	16 000	MEXICO	1507	KENYA	110
CANADA	9355	HONG KONG	45 761	KUWAIT	12 009	CHILE	1221	SIERRA LEONE	62
UK	6697	TAIWAN	7713	BRUNEI	11 108	ARGENTINA	877	NEPAL	61
AUSTRALIA	4993	MALAYSIA	4870	VENEZUELA	1224	PERU	369	BANGLADESH	54
JAPAN	4849	SOUTH KOREA	3767	INDONESIA	293	BRAZIL	343	INDIA	50
USA	4611			NIGERIA	197			ETHIOPIA	16

Aid

Many economically developing countries have come to rely upon aid. *Aid* is the giving of resources by one country, or an organisation, to another country. The resource may be in the form of money, goods, food, technology or people. The basic aim in giving aid is to help poorer countries to develop their economy and services in order to improve their standard of living and quality of life. In reality, the giving of aid is far more complicated and controversial, because it does not always benefit the country to which it is given.

Why do many developing countries need aid?

- Many countries have large and often increasing trade deficits (page 176). They need to borrow money in order to buy goods from the richer industrialised countries. Unfortunately, by borrowing money, these countries fall further into debt (Figure 10.28).
- Aid is needed to try to improve their standard of living. This often involves borrowing money for large prestigious schemes, e.g. improving an international airport or constructing a large hydro-electricity scheme which, in reality, benefits relatively few people.
- Many are prone to either natural disasters (e.g. drought, flooding, earthquakes), or suffer as a result of human induced disasters (e.g. desertification (page 242), civil war). This aid is often only needed for a short period of time.

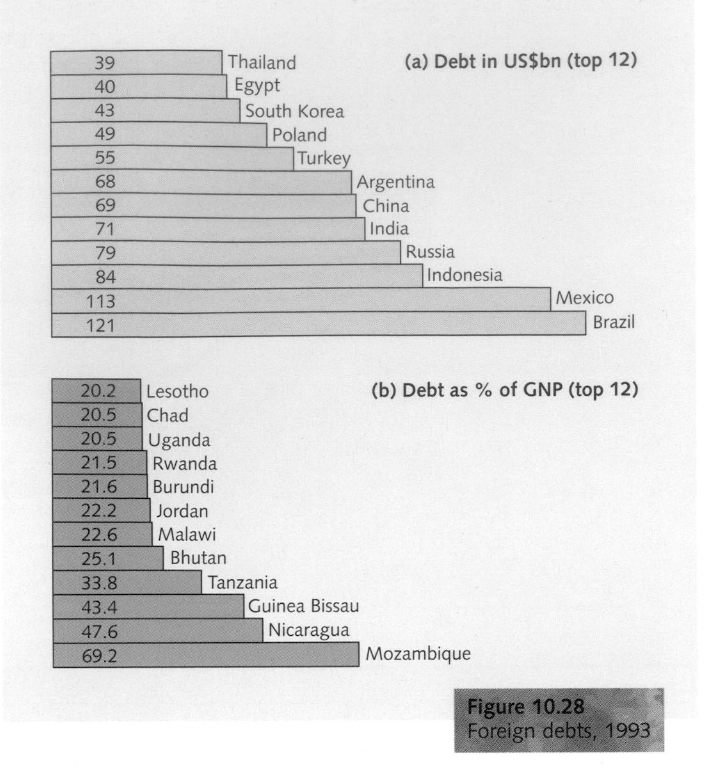

Figure 10.28 Foreign debts, 1993

Types of aid

1 **Bilateral aid** This type of aid is between two countries. Resources are 'given' directly by a rich 'donor' country to a poor 'recipient' country. In reality the aid is not 'given' but is often 'tied', i.e. there are 'strings attached'. This means that the donor imposes conditions upon the recipient country, e.g. building contracts are given to the donor country, or goods and services can only be bought from the donor country. The donor country benefits by increasing its trade and extending its economic influence. The recipient country falls further into debt as it has, eventually, to repay the loan, at a relatively high rate of interest. Developing countries regard this form of aid as 'economic colonialism'.

2 **Multilateral aid** This is when the richer countries give money to international organisations such as the World Bank, the IMF (International Monetary Fund), or the EU Development Fund. These organisations then redistribute the money to poorer countries. Although, theoretically, there should be no political ties, recently these organisations have taken upon themselves to withhold aid if they disagree with the economic and political system within a recipient country. Aid is unlikely to be given to countries that do not have a market economy or a democratically elected government. The UN recommends that rich countries spend 0.7 per cent of their GNP on aid – a figure that is rarely reached (Figure 10.29).

3 **Voluntary aid** Voluntary organisations raise money from the general public in rich countries and send it for use on specific projects in poorer countries. There are no political ties. Projects are on a small scale and use a more appropriate type of technology (page 137). These organisations are usually the first to provide food, clothing and shelter following a major disaster within a country.

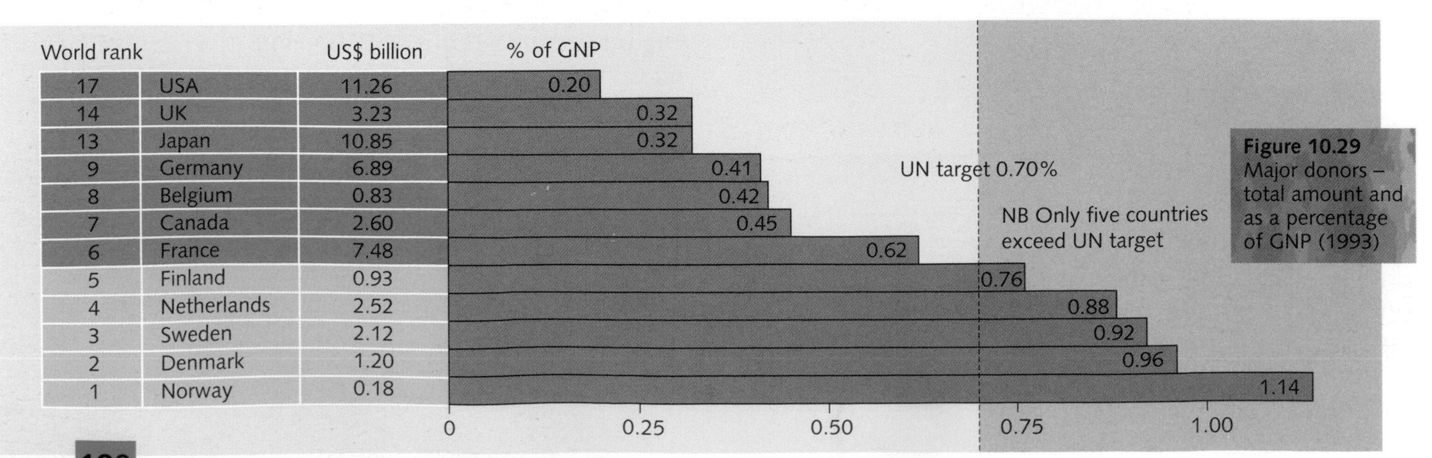

Figure 10.29 Major donors – total amount and as a percentage of GNP (1993)

The disadvantages to a recipient country

Aid rarely reaches the poorest people who live in the rural areas. Inefficient and corrupt officials direct it to themselves and the urban areas where many of them live. The gap between the wealthier town dweller and the poorer rural dweller increases. Aid also forces countries to produce raw materials for richer countries rather than growing food or developing industries for themselves. In time these countries come to rely upon aid. As aid comes in the form of loans on which interest has to be paid, the poorer countries get into permanent and ever increasing debt (Figures 10.30a and 10.31).

There might have been some justification for the steady increases from year to year in external aid if there was evidence that the battle against poverty was being won. All the evidence, however, clearly demonstrates that the poor are getting poorer.

The reason why there is little effect on poverty is that most aid is spent on heavily capitalised infrastructure projects such as railways, bridges and roads. These may have some indirect effect on the lives of poor people, but what people really want is better health and education services, improvements in village roads rather than highways and access to proper credit facilities to help them improve agriculture and raise their incomes. Although there are some enlightened donors, most big Western government and UN agencies find these types of project too small-scale and difficult to measure and administer; they would rather give money for power stations and fertiliser plants since such things can be seen and their performance measured. They also look impressive in glossy magazines telling the tax-payers back home how the government is spending their money.

A CHARITY STATEMENT

'Aid is primarily a mechanism for transferring money from poor people in rich countries into the pockets of rich people in poor countries.'

THE CUMULATIVE effect of the way in which the developing world is portrayed by charitable organisations and the media is grossly misleading. And it results in deeply held public misconceptions that are ultimately damaging to the understanding they seek to promote.

Ninety per cent of people's knowledge and impressions of the developing world come from two sources: the news media and fund-raising agencies. Both are responsible for distorting the public's perception of the developing world – the one because it is in the business of reporting the exceptional; the other because it is in the business of raising money.

One of the worst examples of a distorted message came during the massive campaign mounted in response to the African emergency in the mid-eighties. The public in the industrialised world donated roughly half a billion dollars in one 12-month period – one of the greatest fund-raising responses of all time. At the same time, the industrialised world's governments gave \$2bn in extra emergency aid to Africa. However, almost three times that sum was paid to the industrialised world by Africa in debt and interest repayments. The net flow of finance in that year was therefore *out* of Africa. This year the outflow is at least \$10bn in interest repayments alone – more than Africa spends on its health and education services. **UNESCO**

Figure 10.31 The aid/debt relationship

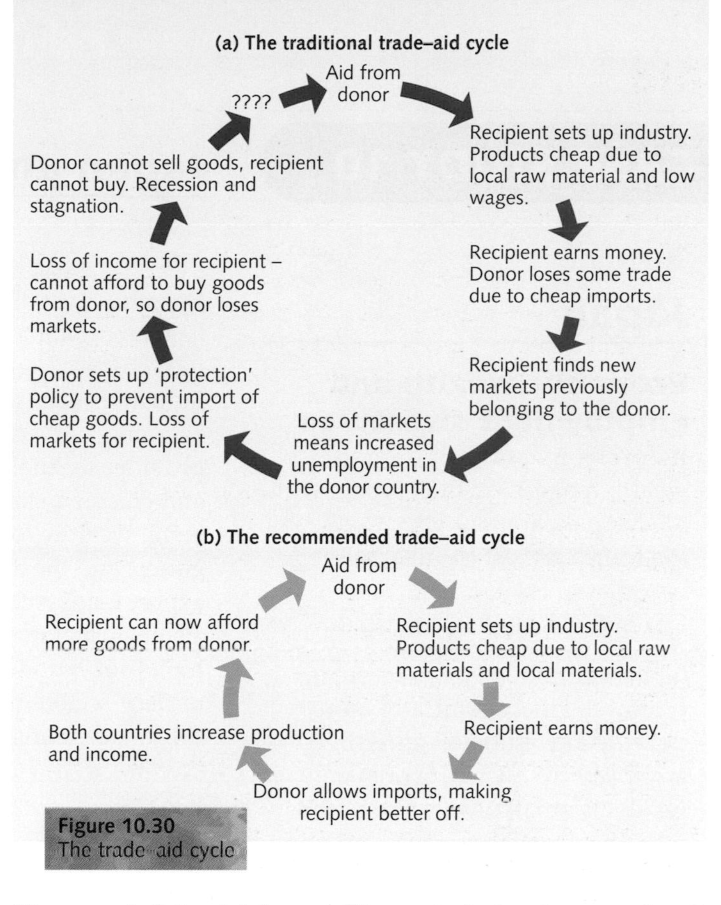

Figure 10.30 The trade–aid cycle

How might aid benefit a recipient country?

Short-term aid can include food, clothing, shelter and medical care after a natural disaster or a civil war. Long-term aid should encourage poorer countries to become increasingly self-sufficient and independent. This can be achieved by helping to improve education standards and to develop skills, to grow higher-yielding crops for their own consumption rather than for export, to develop small-scale, sustainable industries using appropriate technology, and to improve water supplies as one method of improving health standards. The rich countries can also help by buying products rather than setting up tariffs to protect their own industries (Figures 10.30b and 10.32).

Figure 10.32 Aid for self-sufficiency and independence

Japan and Kenya

Japan

Economic wealth and employment structures

If development is based upon wealth (page 170), then Japan, with a GNP per capita of US$ 31 450, is the fourth richest country in the world after Switzerland, Luxembourg and Lichtenstein.

Japan's employment structure is typical of an economically more developed country (Figure 10.33 and page 169). There is only a relatively small proportion of the working population in the primary sector. This is because most young people prefer to live and work in urban areas rather than on farms (Figure 6.21); farming itself is highly mechanised (pages 92 and 93); Japan has few minerals and so there is very little mining; and there is little forestry, as most domestic forests are protected. Japan, like other developed countries, has a high proportion of its workforce engaged in the secondary sector. Despite a lack of raw materials, it has the capital to set up large, modern and highly mechanised industries (e.g. high-tech and car assembly); an education system that provides technological knowledge and creates a skilled workforce; a large, wealthy local market to buy goods; and the ability and ports to export manufactured goods. Japan also has, mainly due to its wealth, a high proportion employed in the tertiary sector, in health, education, commerce, transport and recreation.

Social measures and the HDI

Figure 10.34 shows how population data for Japan corresponds closely with that expected in an economically more developed country (page 170). This is particularly true in the case of life expectancy, as the Japanese live longer than people in any other country. Figure 10.34 also lists other measures often used to show a country's level of development. Japan has, after Canada and along with the USA, the world's highest HDI – 0.937 (page 171).

Figure 10.33
Japan's GNP and employment structure

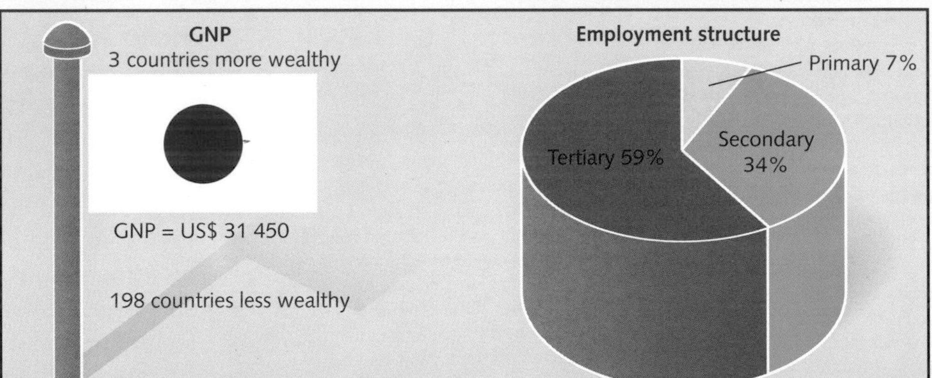

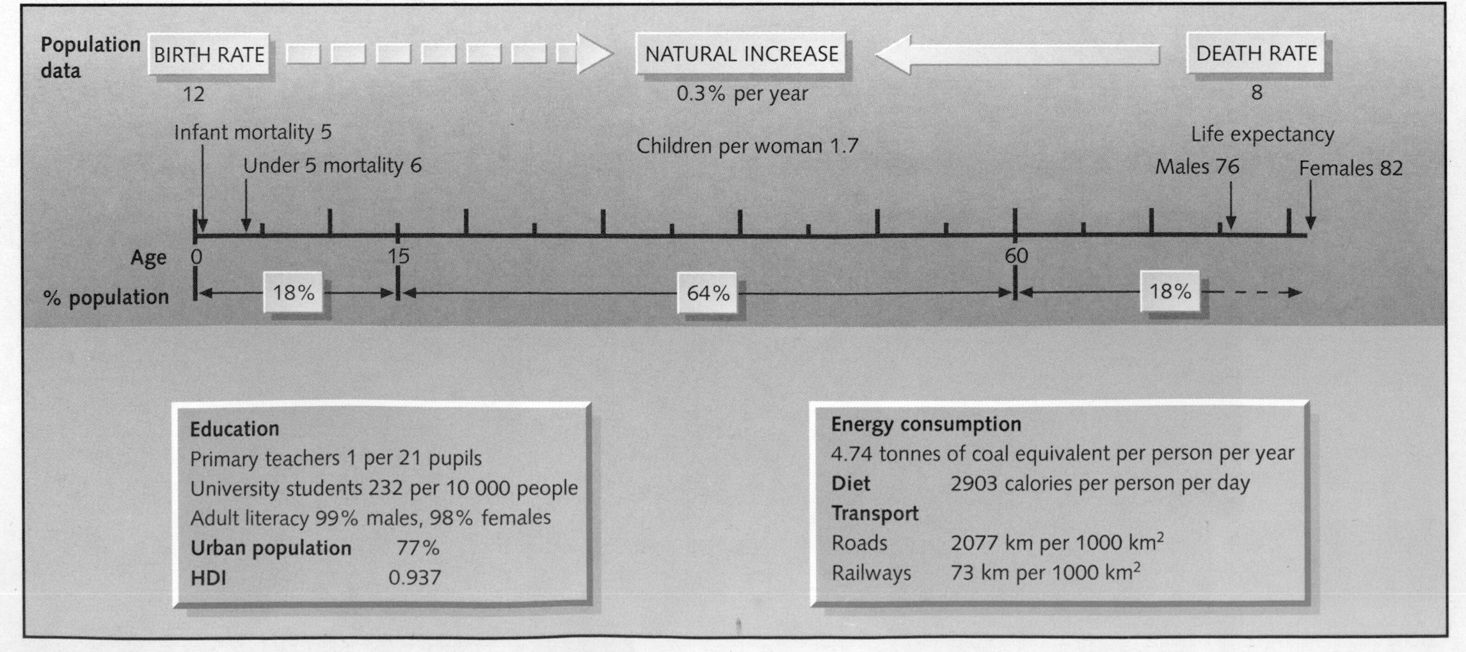

Figure 10.34 Japan – facts and figures

Kenya

Economic wealth and employment structures

If development is based purely upon wealth, then Kenya, with a GNP per capita of US$ 270 (giving it ranking of equal 184), is one of the world's poorest and least economically developed countries.

Kenya's employment structure is typical of an economically less developed country (Figure 10.35 and page 169). There is a high proportion of the working population in the primary sector. Most of these people are engaged in farming, which is labour-intensive and often at a subsistence level (page 94). Smaller numbers are employed in mining, forestry and fishing. Kenya, like other developing countries, has a low proportion engaged in the secondary sector. This is mainly due to a lack of capital, energy supplies and technical knowledge to establish industry; a limited education system leaving a less skilled workforce; the export of raw materials and agricultural produce; and a relatively small local market unable to afford to buy manufactured goods. Mainly due to its lack of wealth, Kenya has a relatively small proportion employed in the tertiary sector with limitations in its health, education, commerce and transport services. The exception is tourism (page 158), which is the country's main money earner. It should also be remembered that large numbers are employed in the informal sector (page 134).

Figure 10.35 Kenya's GNP and employment structure

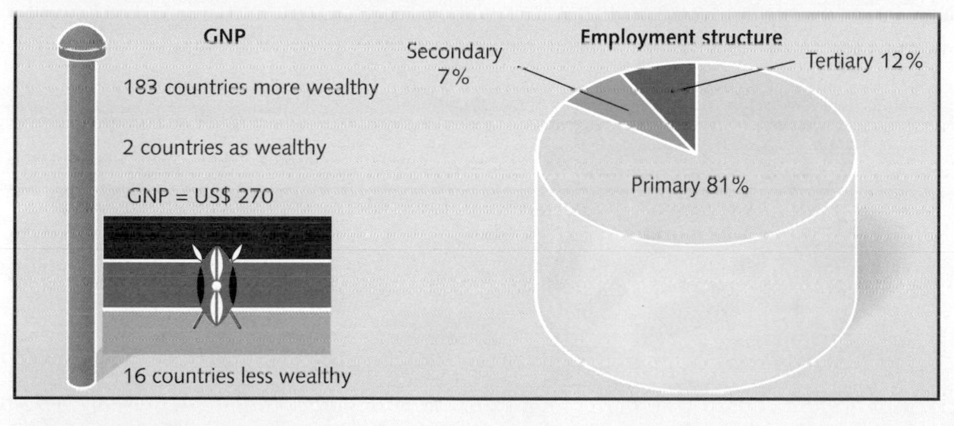

Social measures and the HDI

Figure 10.36 shows how population data for Kenya corresponds closely with that expected in an economically less developed country This is particularly true in the case of natural increase: especially during the early 1990s, when Kenya had the highest natural increase in the world. Figure 10.36 also lists other measures often used to show a country's level of development. Kenya has an HDI of 0.481. Although this still gives the country a low world ranking of 151, it is, nevertheless, appreciably higher than that based on GNP.

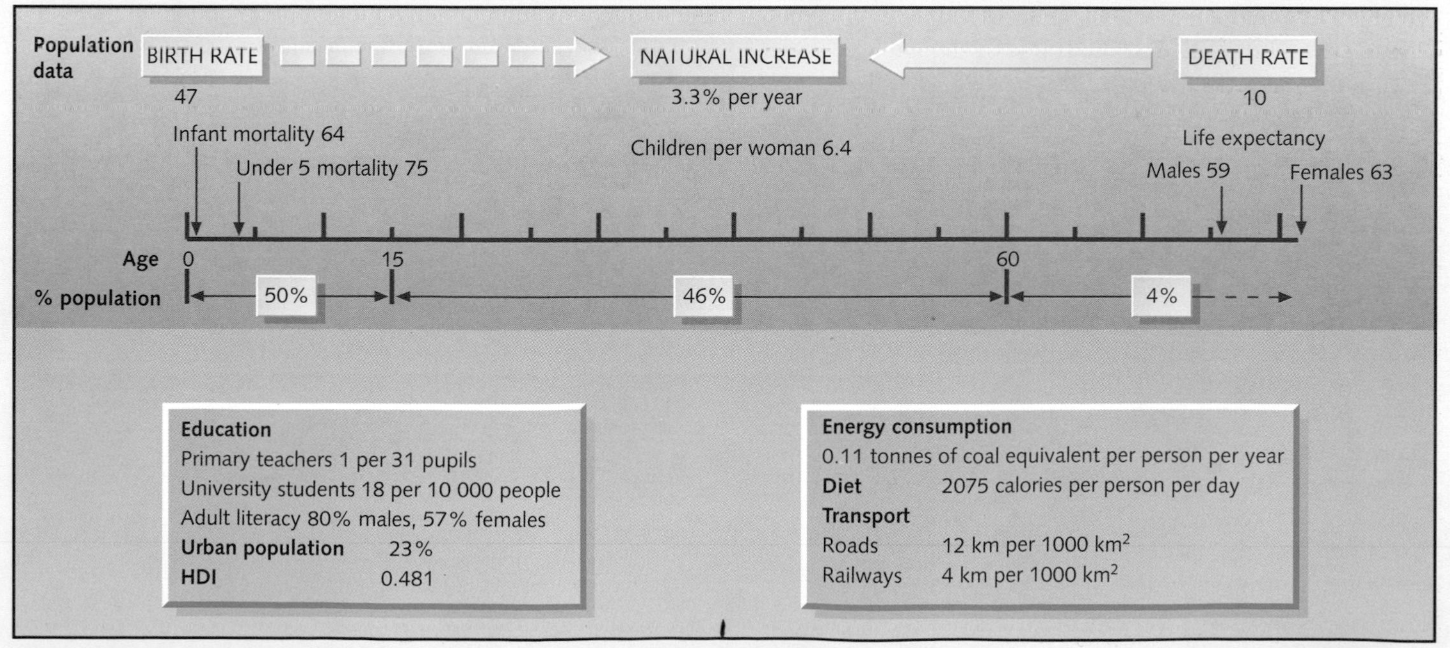

Figure 10.36 Kenya – facts and figures

CASE STUDY 10 Japan and Kenya

Japan

Japan has a large population, limited amounts of flat land and few natural resources of its own (page 128). It has, therefore, to import:

- virtually all of its energy supplies (e.g. oil and coal), which are expensive, as well as various raw materials (e.g. timber) and minerals (e.g. iron ore) needed by its industries (Figure 10.37)
- considerable amounts of foodstuffs because although Japanese farming is intensive and highly mechanised, there is insufficient space to grow enough for the country to be self-sufficient (page 92).

On the other hand, by working long hours, introducing modern machinery and developing high levels of technology, the Japanese are able to produce and export across the world a range of goods noted for their high quality and reliability (e.g. electrical goods, cars and high-tech products).

This has resulted in Japan becoming the world's third largest trader, after the USA and the EU. Indeed 41.8 per cent of Japan's trade in 1993 was with the USA and the EU (Figure 10.38), although during the 1990s there has been a noticeable increase in trade with the NICs of Asia's Pacific Rim (page 129). Almost every year since the mid-1960s Japan has exported more, in value, than it has imported, and since 1983 it has had the world's largest trade surplus (Figure 10.26). This healthy trade surplus is due to Japan:

- reducing its previously high energy bill by changing from expensive imports of oil to the more controversial use of nuclear power
- importing relatively cheap raw materials (often from poorer, developing countries) and exporting expensive processed goods (usually to richer, developed countries)
- protecting – until the 1994 GATT agreement (page 179) – its domestic industries by imposing tariffs on imported goods, and gaining foreign markets by overseas investment and building new factories abroad (e.g. Nissan, Toyota and Honda in the UK).
- financing scientific research projects and building projects in developing countries in return for their often non-renewable resources (e.g. Japanese logging companies are responsible for much deforestation in South-east Asia in order to protect Japan's own forests).

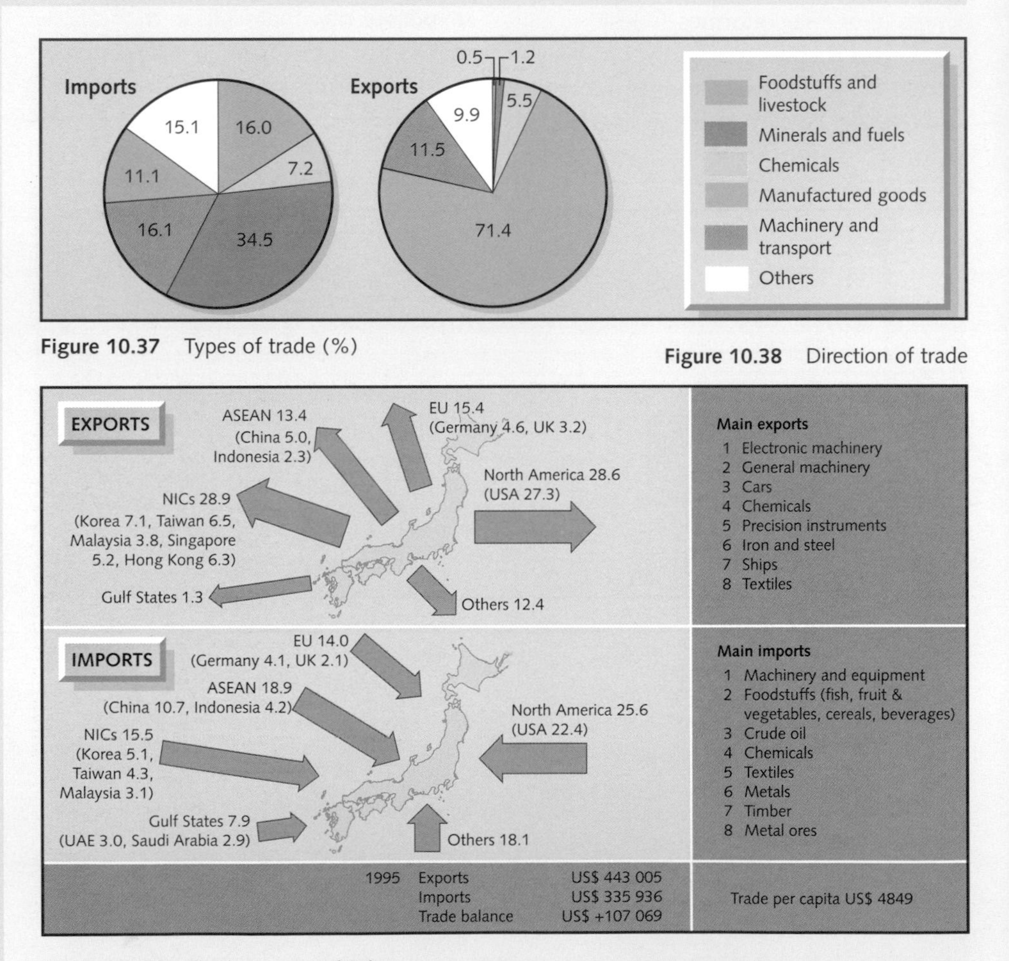

Figure 10.37 Types of trade (%)

Figure 10.38 Direction of trade

Figure 10.39 The busy port of Tokyo

B Trade and interdependence

Kenya

Although parts of northern Kenya are desert, many areas to the south are well suited to agriculture. Where the climate and soil is more favourable, subsistence farmers can, in a normal year, grow sufficient crops to be self-supporting. Where rainfall is abundant and the volcanic soils are fertile, crops such as tea, coffee and fruit can be grown commercially for export. Unfortunately, foodstuffs, and raw materials such as soda ash, are low in value and do not earn the country much money. In contrast, Kenya has little formal industry and so it has to import most of the manufactured goods, machinery and cars it needs – all of them expensive to buy. The result is that Kenya has a large trade deficit (Figure 10.41).

Over 40 per cent of Kenya's trade is with the EU. The UK remains, as in colonial times, both the most important single market for Kenyan exports and the country's main supplier, although the UAE briefly assumed the latter position in the late 1980s when it secured the country's oil supply contract.

Three significant changes have occurred during the 1990s.

- Japan has become Kenya's largest overseas investor, which has meant that, in return, Kenya has had to buy more Japanese goods (see 'Bilateral aid' page 180).
- Kenya has developed a trade surplus within Africa, exporting mainly cement and refined oil to countries such as Uganda, Tanzania and Zambia (Figure 10.41). However, although Kenya earns four times more through exports to African countries than it pays out in imports, the volume of trade is too small to reduce significantly its balance of trade deficit (compare the amount of activity in the ports of Mombasa and Tokyo, Figures 10.39 and 10.42).
- Air freight is used to export perishable goods to Europe. Places near to Nairobi airport can cut flowers (e.g. carnations and roses) and pick vegetables (e.g. peas and beans) one day and have them sold in Europe (e.g. in London and Amsterdam) the next morning.

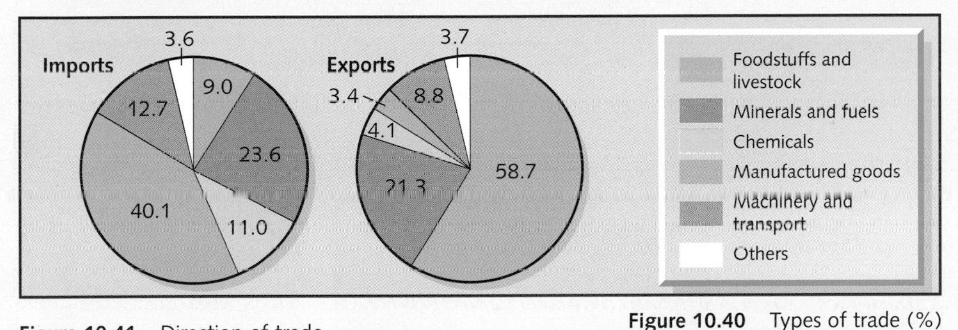

Figure 10.40 Types of trade (%)

Figure 10.41 Direction of trade

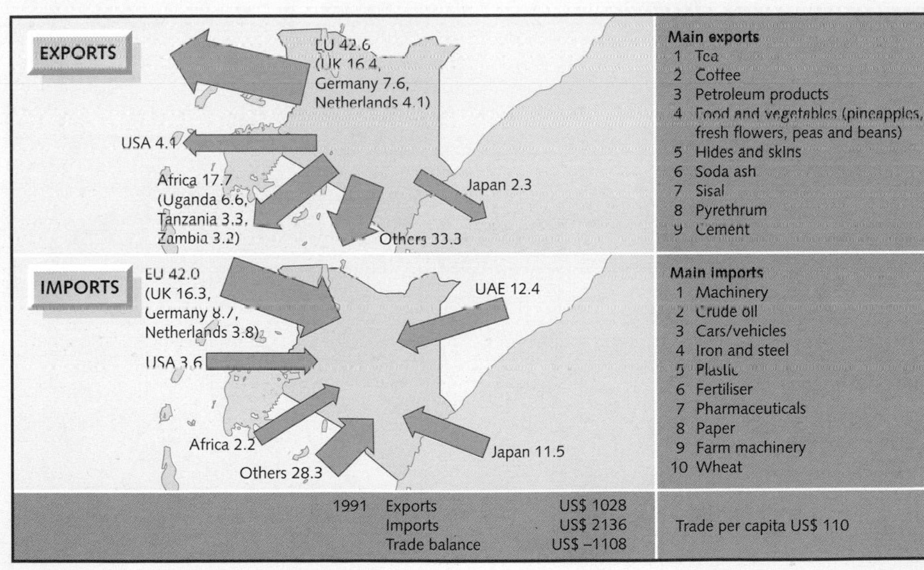

Figure 10.42 The quieter port of Mombasa

QUESTIONS

1

(Pages 168 and 169)

a What is meant by the terms:
- primary
- secondary
- tertiary sectors of employment? (3)

b The triangular graph shows the percentage of the working population in each of the three main sectors of activity for selected countries. Using the graph:

i) Complete the table above the graph. (5)

ii) Which country has 42 per cent in primary, 21 per cent in secondary and 37 per cent in tertiary? (1)

iii) Name the four economically most developed countries. (4)

iv) Name the four economically least developed countries. (4)

v) Give three differences between the figures for the two sets of countries. (3)

vi) Give three reasons for the differences between the two sets of figures. (3)

	JAPAN	BRAZIL	KENYA
% PRIMARY SECTOR	3		81
% SECONDARY SECTOR		43	
% TERTIARY SECTOR	56		

Primary
Secondary
Tertiary
0 10 20 30 40 50 60 70 80 90 100

1 Bangladesh
2 Brazil
3 Egypt
4 Nepal
5 India
6 Italy
7 Japan
8 Kenya
9 UK
10 USA

2

(Pages 170 and 171)

a Name three continents in the 'richer North' and three in the 'poorer South'. (6)

b i) State how the five basic indicators on the map show that Africa is economically the least well-off. (5)

ii) Suggest three other indicators which may be used to measure differences in living standards. (3)

c Make a copy of the table headings below and add, in the correct columns, the information given below them. (7)

d i) What is the difference between GNP and the HDI? (2)

ii) Why is GNP not always a good measure of development? (1)

Economically more developed	Economically less developed

◆ High GNP ◆ Low GNP ◆ Low birth rate ◆ Long life expectancy ◆ Short life expectancy ◆ High percentage in agriculture ◆ Low percentage in agriculture ◆ High percentage in services ◆ Low percentage in services ◆ High percentage of urban dwellers ◆ Low percentage of urban dwellers ◆ High literacy levels ◆ Low literacy levels

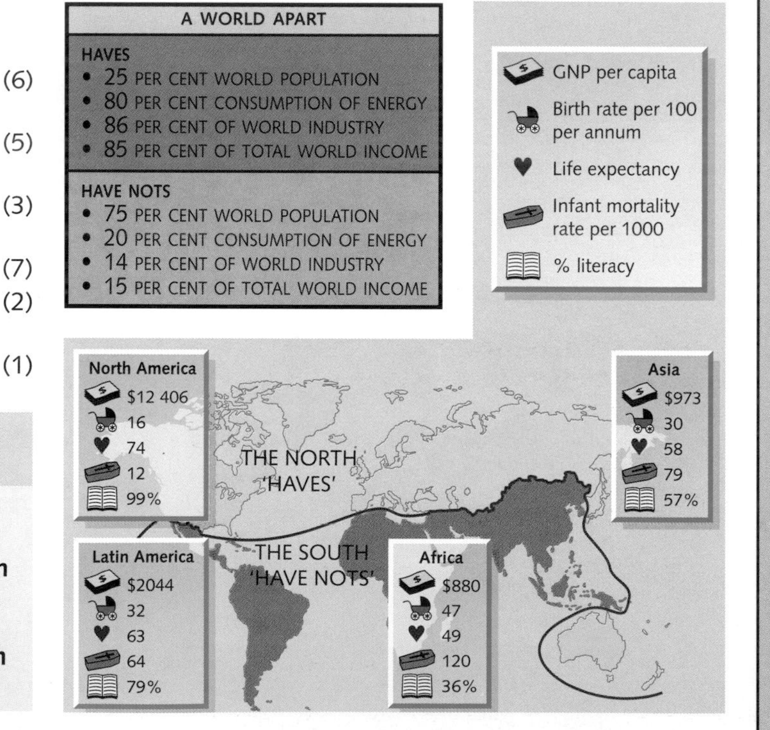

3

(Pages 172 and 173)

a What is meant by the terms *core* and *periphery*? (2)

b For any country that you have studied (e.g. UK, Japan, Italy, Brazil, etc.):

i) Draw a sketch map to show its core and its periphery regions. (3)

ii) Give reasons for the development of the core region. (3)

iii) Give reasons why the periphery region developed less rapidly. (3)

4

(Pages 176 and 177)

The two pie charts show the pattern of trade for a typical economically less developed country.

a i) Name, in rank order, the three main exports. (3)
ii) Are these exports raw materials or manufactured goods? (1)
iii) Why is this typical of a developing country? (2)

b i) Name, in rank order, the three main imports. (3)
ii) Are these imports raw materials or manufactured goods? (1)
iii) Why is this typical of a developing country? (2)

c i) By how much do imports exceed exports (in US dollars)? (1)
ii) What problems will this create for the developing country? (2)

d i) What is meant by the term 'balance of trade'? (1)
ii) What might happen to a country with a deficit in its balance of trade? (1)
iii) How might industrialisation help the balance of trade? (2)

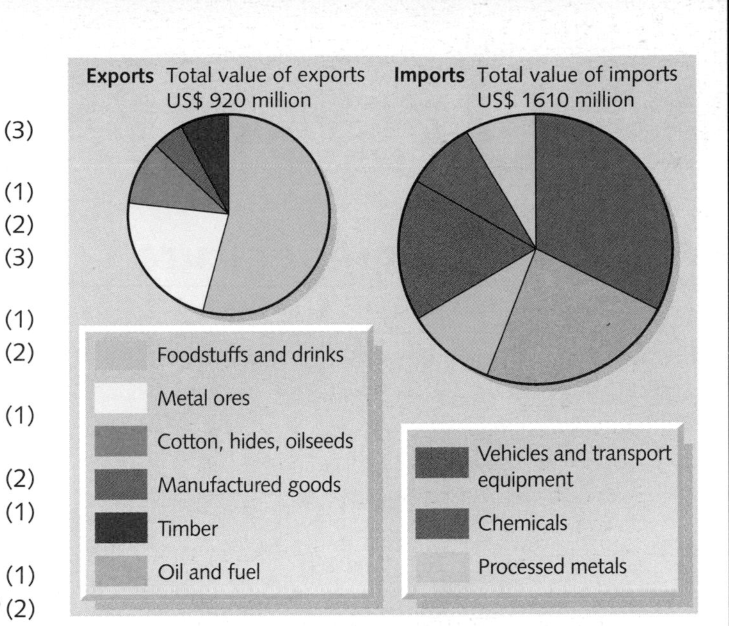

5

(Pages 178 and 179)

a Why do countries form trade groups? (1)

b i) Name one trade group in each of:
- North America
- Europe
- Asia. (3)

ii) Name three countries in each of the trade groups that you have just named. (9)
iii) Which trade group has become increasingly more important during the 1990s? (1)

c i) Which two countries have the highest 'trade per capita'? (2)
ii) What do you notice about their location? (1)

d i) What does GATT stand for? (1)
ii) Give one advantage and one disadvantage of tariffs. (2)
iii) Why were developing countries keen to see the end of tariffs? (1)

6

(Pages 180 and 181)

a i) What is the meaning of the term 'short-term aid'? (1)
ii) Give two conditions when short-term aid may be needed. (2)
iii) Why might long-term aid be more useful to a developing country? (2)

b i) What term is used for aid that has conditions linked with it? (1)
ii) What are the advantages and disadvantages of this type of aid? (3)

c Using the data for 1993:
i) Which four countries received most aid? (2)

SHORT-TERM AID	LONG-TERM AID
FOOD AID, BLANKETS, TENTS	EDUCATION AND TRAINING
LORRIES, MEDICINES	CAPITAL INVESTMENT
	TRANSPORT IMPROVEMENT SCHEMES

ii) Which four countries gave most aid? (2)
iii) How much aid did the UK give in 1993? (1)

d Some people criticise Britain for giving relatively little aid. Others claim that if we stopped giving aid then every person in the UK would be £50 a year better off. What do you think Britain should do? (3)

7

(Pages 182–185)

a Describe, and give reasons for, the differences in the level of development between Japan and Kenya, under the following headings:
- GNP
- Employment structures
- Population data
- Education
- Energy consumption
- HDI (12)

b Describe, and give reasons for, the differences in trade between Japan and Kenya under the following headings:
- Main types of imports
- Main types of exports
- Main trading partners
- Value of imports and exports
- Balance of trade (10)

WEATHER AND CLIMATE

Weather and temperature

Weather is the hour-to-hour, day-to-day state of the atmosphere. It includes temperature, sunshine, precipitation and wind. It is short-term and is often localised in area.

Climate is the average weather conditions for a place taken over a period of time, usually 30 years. It is the expected, rather than the actual, conditions for a place. It is long-term and is often applied to sizeable parts of the globe (e.g. the equatorial or Mediterranean climate).

Britain's climate

Britain has:

- a variable climate, which means that the weather changes from day to day, and this makes it difficult to forecast
- an equable climate, which means that extremes of heat or cold, or of drought or prolonged rainfall, are rarely experienced.

If we wish to generalise about Britain's climate, we can say that it has cool summers, mild winters, and a steady, reliable rainfall which is spread fairly evenly throughout the year. However, there are, even across an area as small on a global scale as the British Isles, significant differences:

- Seasonally, between summer and winter.
- Between places in the extreme north and south, and places on the east and west coasts (Figure 11.1).

Why is it:

- That places in the south are warmer and sunnier than places to the north in summer (page 189)?
- That places in the west are milder and cloudier than places to the east in winter (page 189)?
- That places in the west are wetter, with a winter maximum of rainfall, than places in the east which are drier, and have a summer maximum of rainfall (pages 190 and 191)?

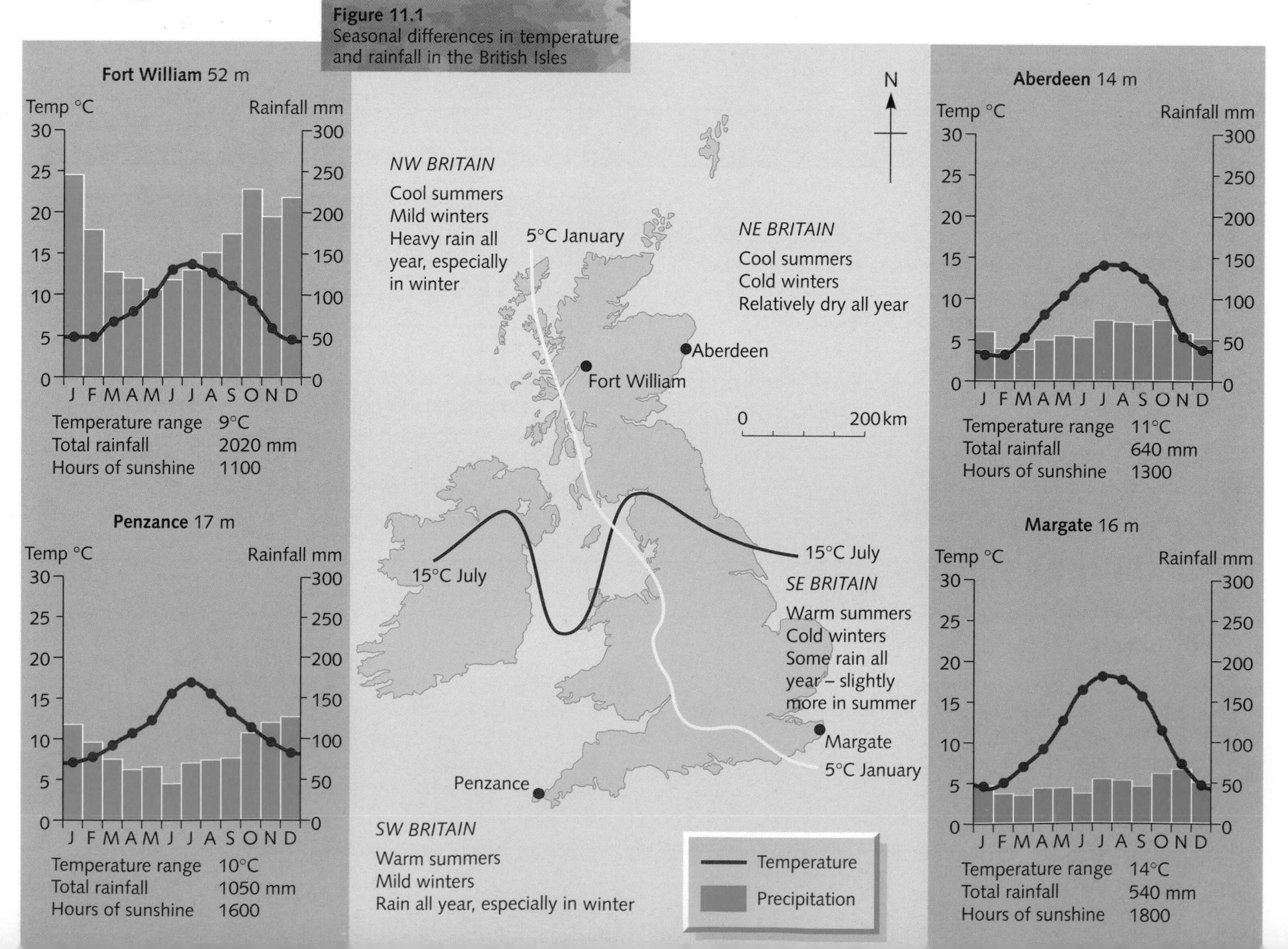

Figure 11.1 Seasonal differences in temperature and rainfall in the British Isles

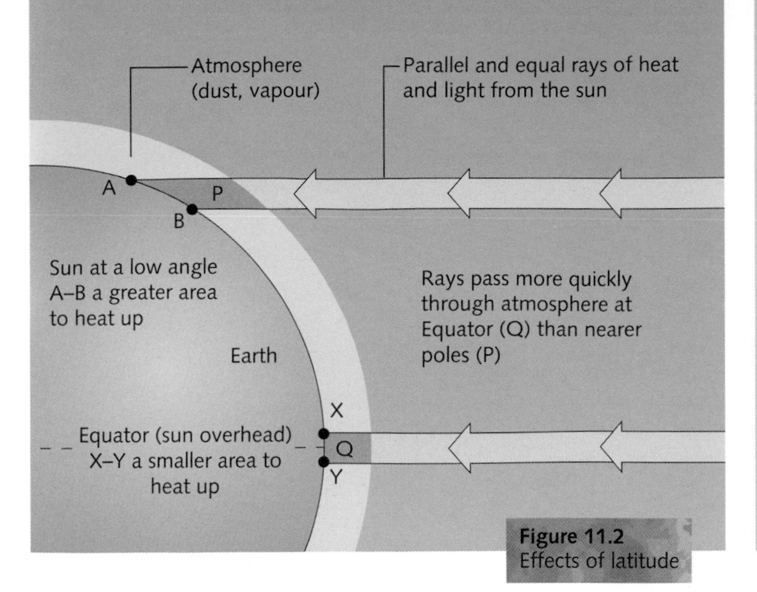

Figure 11.2
Effects of latitude

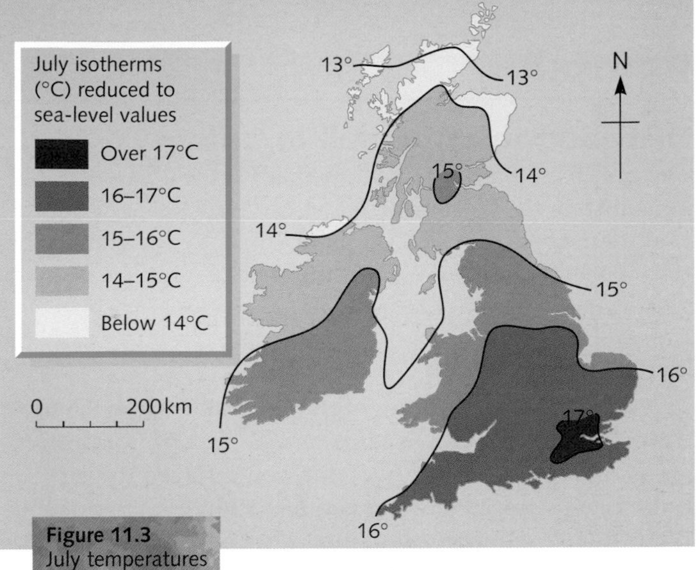

Figure 11.3
July temperatures

Factors affecting temperature

Latitude Places nearer to the Equator are much warmer than places nearer to the poles. This is due to the curvature of the Earth and the angle of the sun (Figure 11.2). At the Equator the sun is always high in the sky. When it is overhead it shines directly downwards, concentrating its heat into a small area which will become very hot. In contrast, the sun is always low in the sky towards the poles. This means that its heat is spread over a wide area, and so temperatures remain lower. Notice also that the lower the angle of the sun, the greater the amount of atmosphere through which the rays have to pass. This means that more heat will be lost to gases, dust and cloud in the atmosphere. This is why places in the south of Britain can expect to be warmer, especially in summer, than places further north (Figure 11.3).

Distance from the sea The sea (a liquid) is less dense than the land (a solid) and can be heated to a greater depth. This means that the sea takes much longer to heat up in summer than does the land. Once warmed, however, the sea retains its heat for much longer, and cools down more slowly than the land in winter. This is why places that are inland are warmer in summer but colder in winter than places on the coast. As Britain is surrounded by the sea, it tends to get cool summers and mild winters. The largest reservoir of heat in winter is the Atlantic Ocean, even though it is still cold enough to die from hypothermia within a few minutes should you fall into it. This explains why western parts of Britain are warmer than places to the east in winter (Figure 11.4).

Prevailing winds Prevailing winds will bring warm weather if they pass over warm surfaces (the land in summer, the sea in winter) and cold weather if they blow across cold surfaces (the land in winter, the sea in summer). As Britain's prevailing winds are from the south-west, they are cool in summer but warm (mild) in winter.

Ocean currents Many coastal areas are affected by ocean currents. The North Atlantic Drift is a warm current of water which originates in the Gulf of Mexico. It keeps the west coast of Britain much warmer in winter than other places in similar latitudes.

Altitude Temperatures decrease, on average, by 1°C for every 100 metres in height. As many parts of the Scottish Highlands are over 1000 metres, they will be at least 10°C cooler than coastal places. In fact, the windchill factor will make them even colder, and enables snow to lie for long periods during winter (Figure 11.5).

Figure 11.4
January temperatures

Figure 11.5
Influence of altitude on length of snow cover in winter

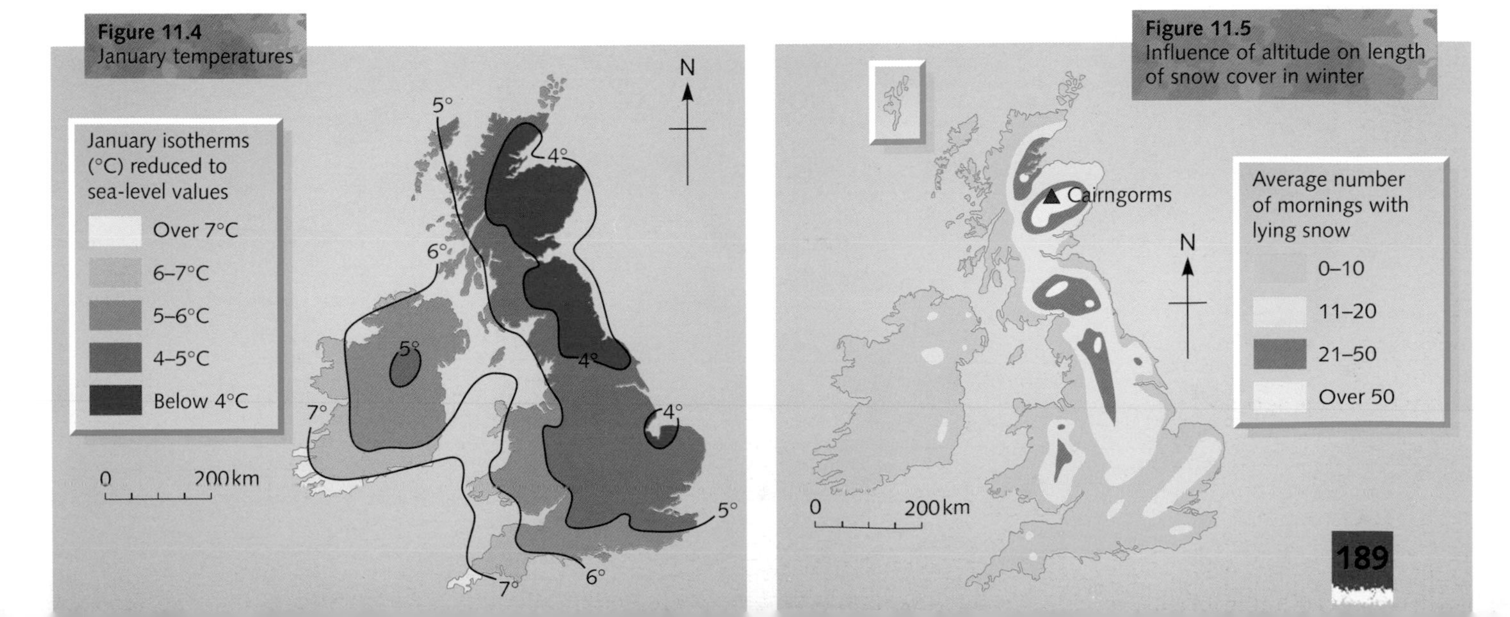

Rainfall

Distribution of rainfall in Britain

The graphs in Figure 11.1 show that Fort William and Penzance in the west of Britain receive appreciably more rain than Aberdeen and Margate which are located on the east coast. This uneven distribution between the east and west is confirmed in Figure 11.6.

Types of rainfall

There are three main types of rainfall: relief, frontal and convectional. In all three cases rainfall results from warm air, which contains water vapour, being forced to rise until it cools sufficiently for condensation to take place (Figure 11.7). Condensation can only occur when two conditions are met:

i) Cold air cannot hold as much moisture as warm air. As the warm air and water vapour rises, it cools until a critical temperature is reached, at which point the air becomes saturated. This critical temperature is called the *dew point*. If air continues to rise and cool, some of the water vapour in it condenses back into minute droplets of water.
ii) Condensation requires the presence of large numbers of microscopic particles known as *hygroscopic nuclei*. This is because condensation can only take place on solid surfaces such as volcanic dust, salt or smoke (or on windows and walls in a bathroom or kitchen).

The difference between the three types of rainfall is the condition that forces the warm air to rise in the first place.

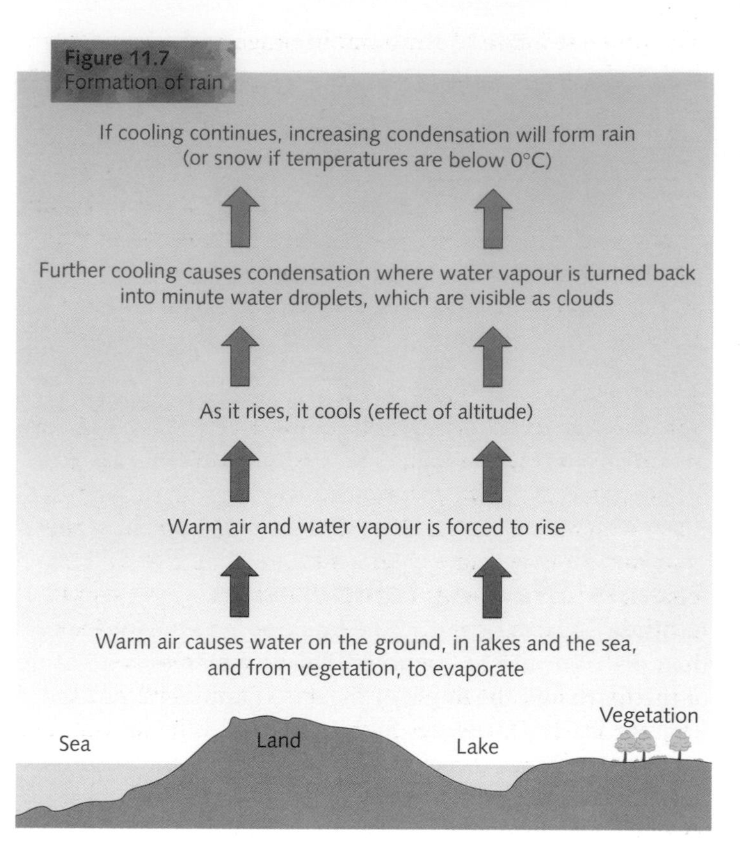

Figure 11.7
Formation of rain

1 Relief rainfall (Figure 11.8)

Relief rain occurs when warm, almost saturated air from the sea is blown inland by the wind. Where there is a coastal mountain barrier, the air will be forced to rise over it. The rising air will cool and, if dew point is reached, condensation will take place. Once over the mountains the air will descend, warm and, therefore, the rain is likely to stop.

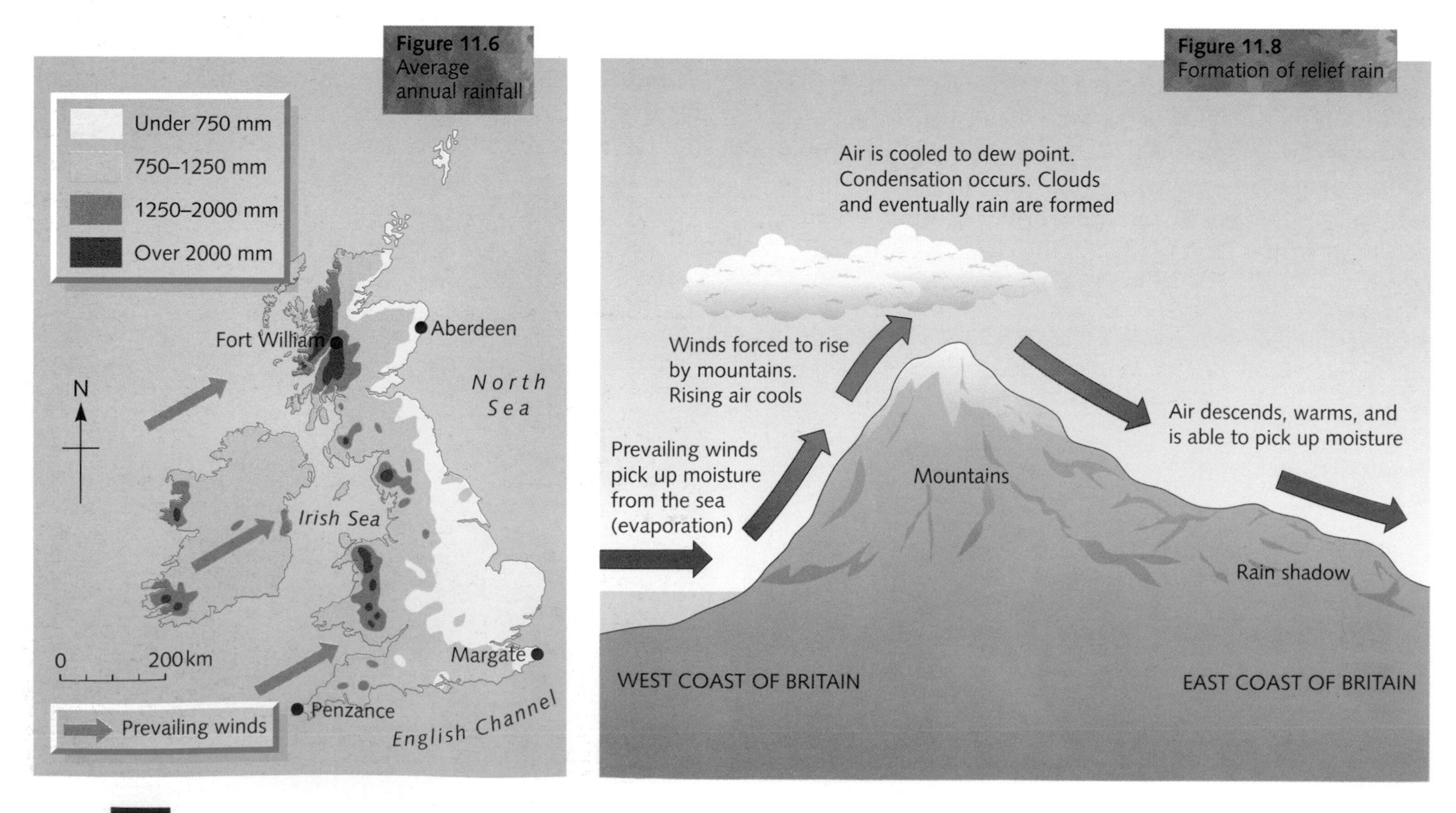

Figure 11.6
Average annual rainfall

Figure 11.8
Formation of relief rain

The protected side of a mountain range is the *rain shadow*. In Britain the prevailing winds come from the south-west, collecting moisture as they cross the Atlantic Ocean. They bring heavy rainfall to western parts as they cross the mountains of Scotland, Wales and northern England. Eastern areas receive much less rain as they are in the rain shadow area. Places like Fort William and Penzance get heavy rainfall in late autumn when the sea is at its warmest, and winds blowing over it can pick up most moisture.

2 Frontal rainfall (Figure 11.9)

Frontal rain is associated with depressions (page 192) and results from warm, moist air from the tropics meeting colder, drier air from polar areas. As the two air masses have different densities, they cannot merge. Instead the warmer, moister and lighter air is forced to rise over the colder, denser air, setting the condensation process into motion. The boundary between the warm and cold air is called a *front*. Most depressions have two fronts, a warm and a cold front, giving two periods of rainfall (Figure 11.11). Britain receives many depressions and their associated fronts each year. Depressions usually come from the Atlantic Ocean, increasing rainfall on the west coasts. Depressions are more common in winter, as illustrated by the winter rainfall maximum.

3 Convectional rainfall (Figure 11.10)

Convectional rain occurs where the ground surface is heated by the sun. As the air adjacent to the ground is heated, it expands and begins to rise. If the ground surface is wet and heavily vegetated, as in equatorial areas (page 196), there will be rapid evaporation. As the air rises, it cools and water vapour condenses to form towering cumulonimbus clouds and, later, heavy thunderstorms. Equatorial areas, where the sun is constantly at a high angle in the sky, experience convectional storms most afternoons. Convectional rain is less frequent in cooler Britain, and is most likely in south-east England in summer when temperatures are at their highest (Figure 11.1). This also accounts for the summer rainfall maximum in this region.

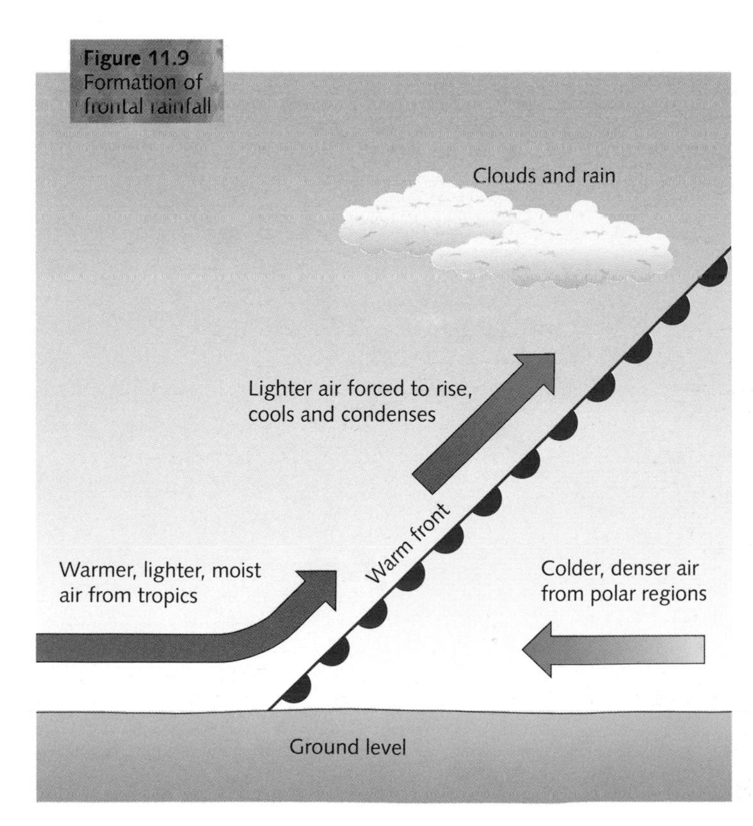

Figure 11.9 Formation of frontal rainfall

Figure 11.10 Formation of convectional rainfall

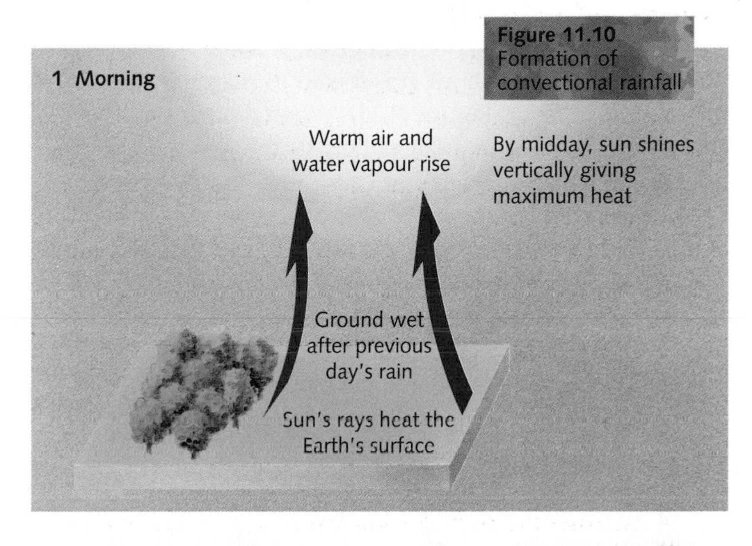

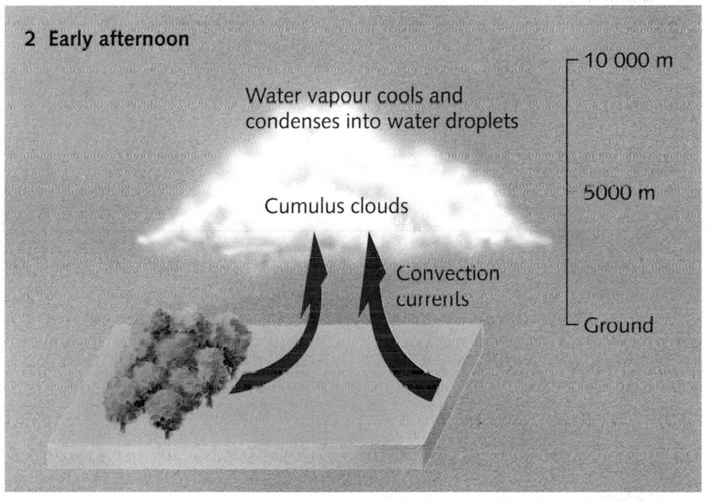

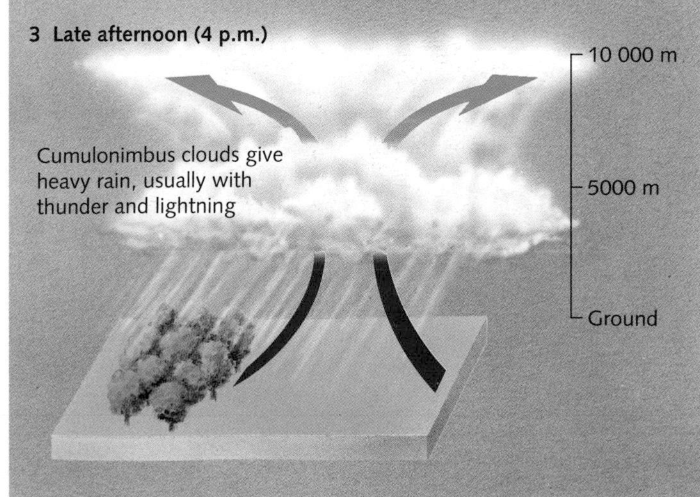

Depressions

Britain's weather changes from day to day. For much of the year our climate is dominated by the passing of depressions. *Depressions* are areas of low pressure which bring rain, cloud and wind. They form over the Atlantic Ocean when a mass of warm, moist tropical air from the south meets a mass of colder, drier, heavier polar air from the north. The two masses of air do not easily mix due to differences in temperature and density. The boundary between two air masses is called a *front*. When lighter, warmer air moves towards denser, colder air, it is forced to rise over the cold air at a warm front (Figure 11.11). When denser, colder air moves towards warm air, it undercuts the warm air forcing it to rise at a cold front. In both cases the rising warm air is cooled and some of its water vapour content condenses, producing cloud and frontal rain (page 191). The cold front travels faster than the warm front, catching it up to form an *occluded front*. Although each depression is unique, the weather they bring to Britain as they travel eastwards tends to have an easily recognisable pattern.

As a warm front approaches clouds begin to form. They get lower, and thicken (Figure 11.11). Winds blow from the south-east, in an anticlockwise direction, and slowly increase in strength. As the air rises, atmospheric pressure drops. The passing of the warm front is usually characterised by a lengthy period of steady rainfall, low cloud and strong winds. As the warm front passes there is a sudden rise in temperature and the wind turns to a south-westerly direction. The warm sector of a depression is usually a time of low and sometimes broken cloud, decreasing winds, and drizzle or even dry weather. As a cold front passes the weather deteriorates rapidly. Winds often reach gale force and swing round to the north-west. Rainfall is very heavy, though of relatively short duration, and temperatures fall rapidly. After the cold front passes, the weather slowly improves as pressure increases. The heavy rain gives way to heavy showers and eventually to sunny intervals. Winds are cold and slowly moderate, but still come from the north-west. Most depressions take between one and three days to pass over the British Isles.

Depressions can be seen on satellite images as masses of swirling cloud (Figure 11.12). Satellite images are photos taken from space and sent back to Earth. They are invaluable when trying to produce a weather forecast or predicting short-term changes in the weather. The state of the weather at any one given time is shown on a *synoptic chart* (a weather map). The three synoptic charts in Figure 11.13 were meant to match the satellite images in Figure 11.12. The daily weather map as shown on television or in a newspaper aims to give a clear, but very simplified, forecast. Synoptic maps produced by the Meteorological Office use official symbols to show conditions at specific weather stations (Figure 11.14). The weather stations on Figure 11.13 show five elements: temperature, wind speed, wind direction, amount of cloud cover and type of precipitation, while a sixth, atmospheric pressure, can be obtained by interpreting the isobars.

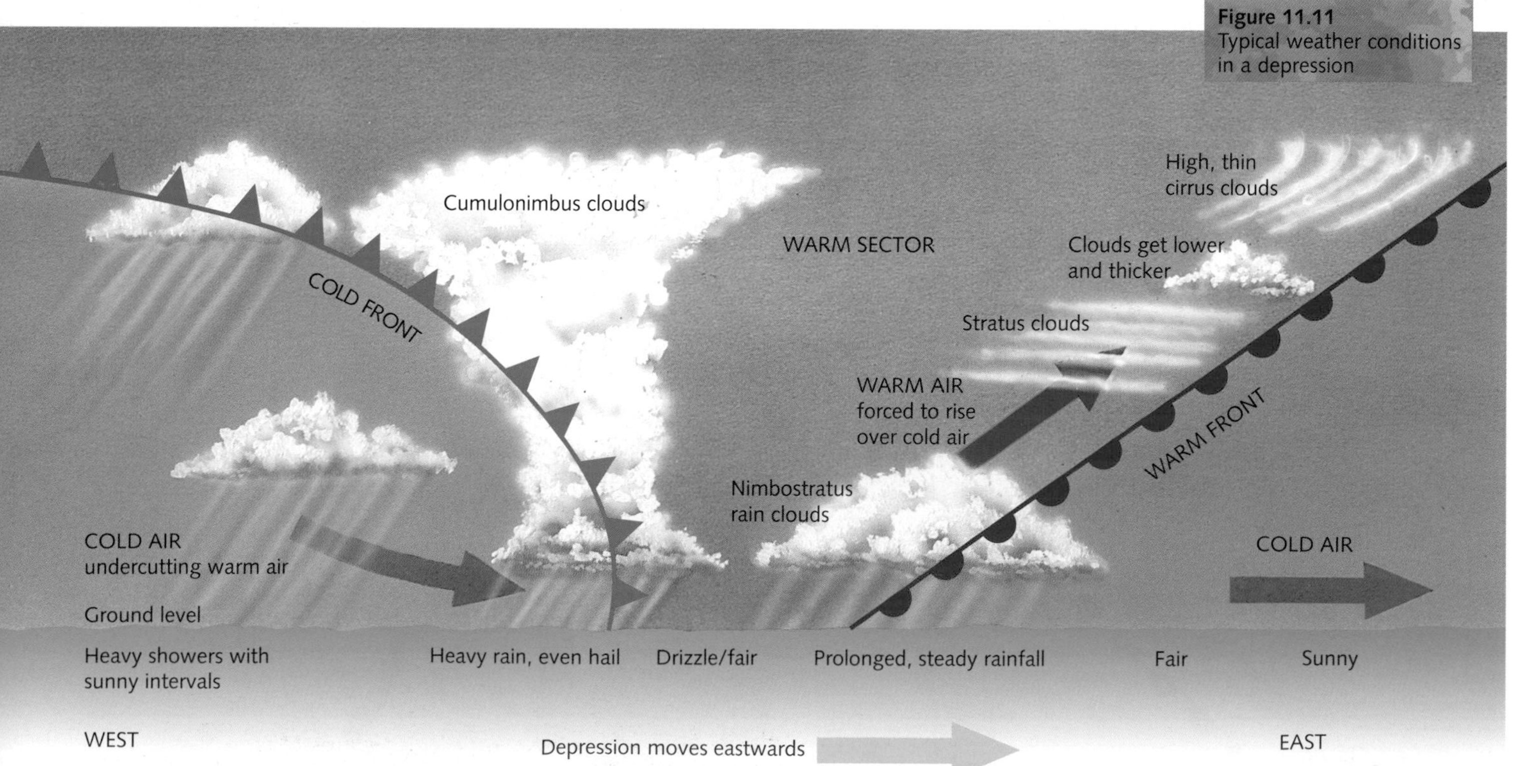

Figure 11.11 Typical weather conditions in a depression

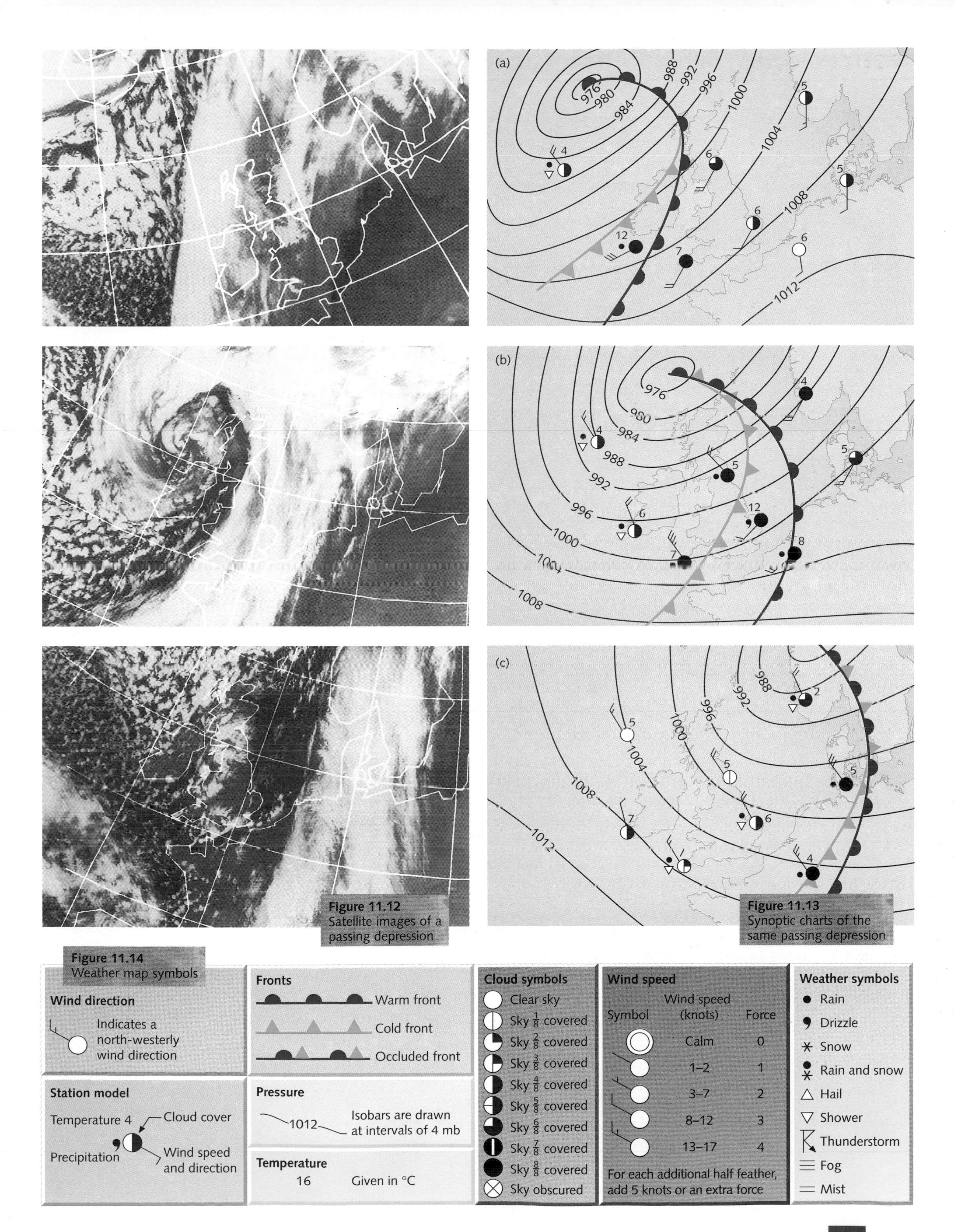

Figure 11.12 Satellite images of a passing depression

Figure 11.13 Synoptic charts of the same passing depression

Figure 11.14 Weather map symbols

Anticyclones, world climates and atmospheric circulation

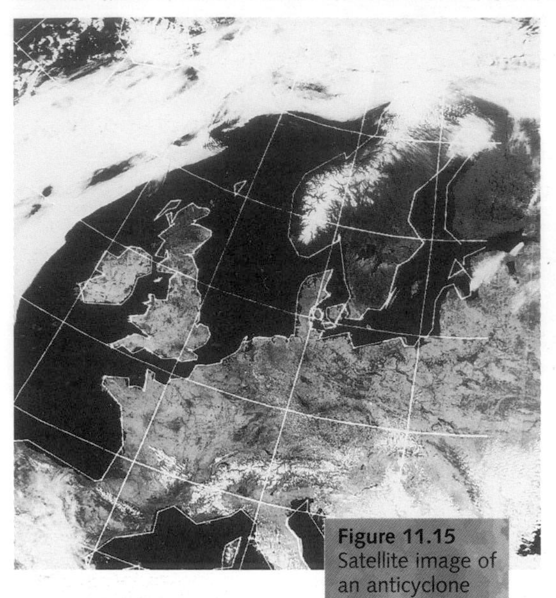

Figure 11.15
Satellite image of an anticyclone

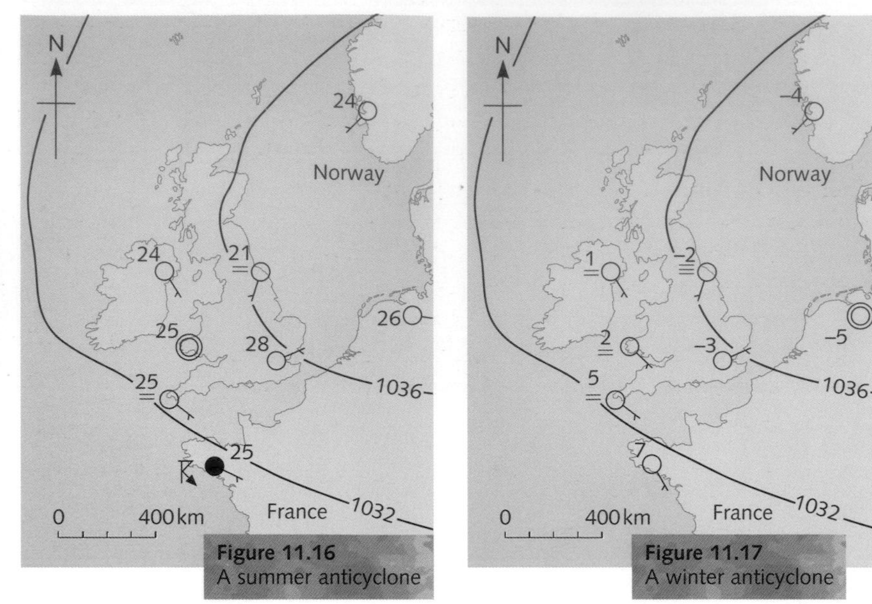

Figure 11.16
A summer anticyclone

Figure 11.17
A winter anticyclone

Anticyclones

In Britain anticyclones are experienced far less frequently than depressions but, once established, they can remain stationary for several days and, under extreme conditions, even weeks. Their main characteristics are opposite to those of depressions (Figure 11.15). In an anticyclone, air descends and pressure increases. Winds are very light (notice the wide spacing of isobars) and blow in a clockwise direction. At times they may even be non-existent and give periods of calm. As the air descends it warms and is able to pick up more moisture through evaporation. This usually results in settled conditions with clear skies and a lack of rain. However, there are differences between summer and winter anticyclones.

Summer (Figure 11.16) The absence of cloud gives very warm, sunny conditions during the day although at night, when clear skies allow some of this heat to escape, temperatures can fall rapidly. As air next to the ground cools, condensation can occur and dew and mist may form. Thunderstorms are also a risk (convectional rainfall) under 'heat-wave' conditions (page 191).

Winter (Figure 11.17) Although temperatures remain low during the day, due to the sun's low angle in the sky (Figure 11.2), the weather is likely to be dry and bright. The rapid loss of heat under the clear evening skies means that nights can be very cold. Condensation near to the ground can produce frost and fog which may, due to the sun's lack of heat, persist all day.

World climates

Figure 11.18 shows the location of the six types of climate described in this chapter. Maps that show the location and extent of the world's main climatic types are very generalised (simplified) as, due to their scale, they cannot show local variations. They also suggest that the boundary between any two climates is a thin line, whereas in reality any change is often gradual and extends across a wide transition zone.

Figure 11.18
World climates

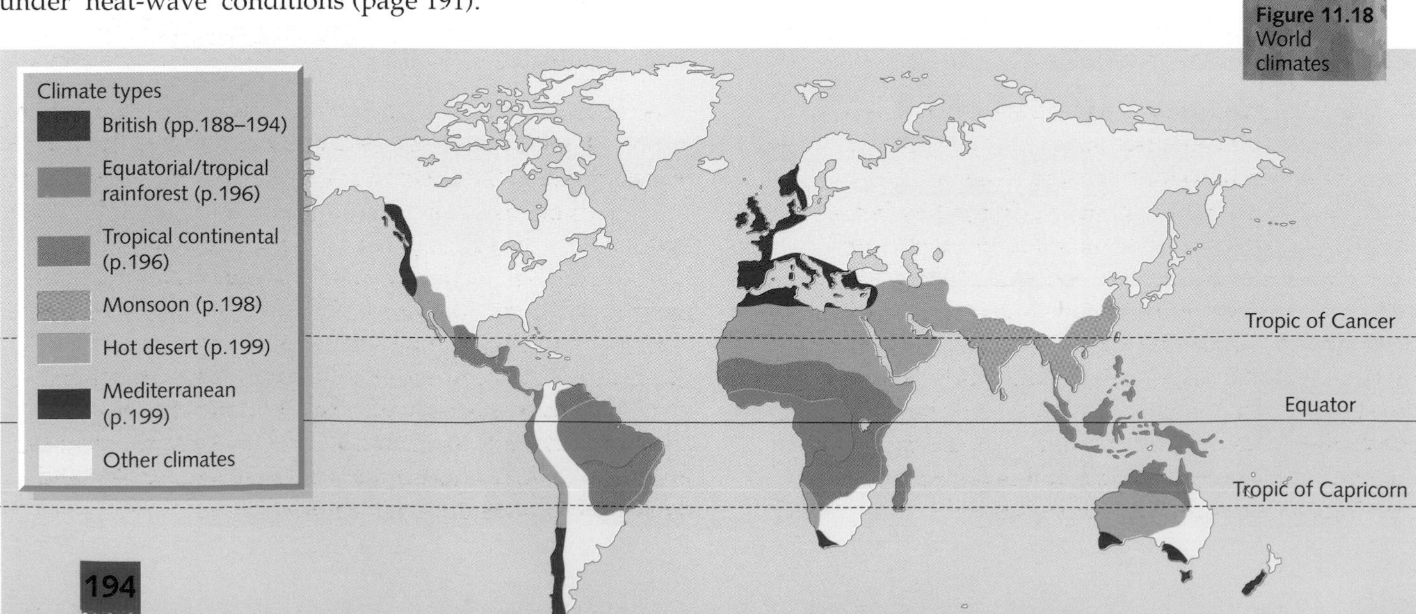

Atmospheric circulation

Although it may not be obvious at first glance, Figure 11.18 does show distinct patterns as to the location of the main climatic types. Take the British type as an example. Notice that it is located on the west coast of continents and between latitudes 40° and 60° north and south of the Equator. Can you, at this stage, identify patterns for the other five named climatic types? Any pattern that you could identify is mainly due to the circulation of air in the atmosphere (Figure 11.19). In order to understand this circulation, you need to be aware of three processes:

1. If air next to the ground is heated, it expands, gets lighter and rises. This causes a decrease in the amount of air at ground level and the formation of an area of *low pressure*.
2. If air in the atmosphere is cooled, it becomes denser and sinks. This results in an increase in the amount of air at ground level and the creation of an area of *high pressure*.
3. Wind, which is air in motion, blows from areas of high pressure to areas of low pressure (in reality other factors, including the Earth's rotation, prevent air from moving directly from areas of high pressure to areas of low pressure – otherwise the prevailing wind in Britain would be from the south, not the south-west).

Figure 11.19 shows the major areas of high and low pressure, and the general circulation of the atmosphere.

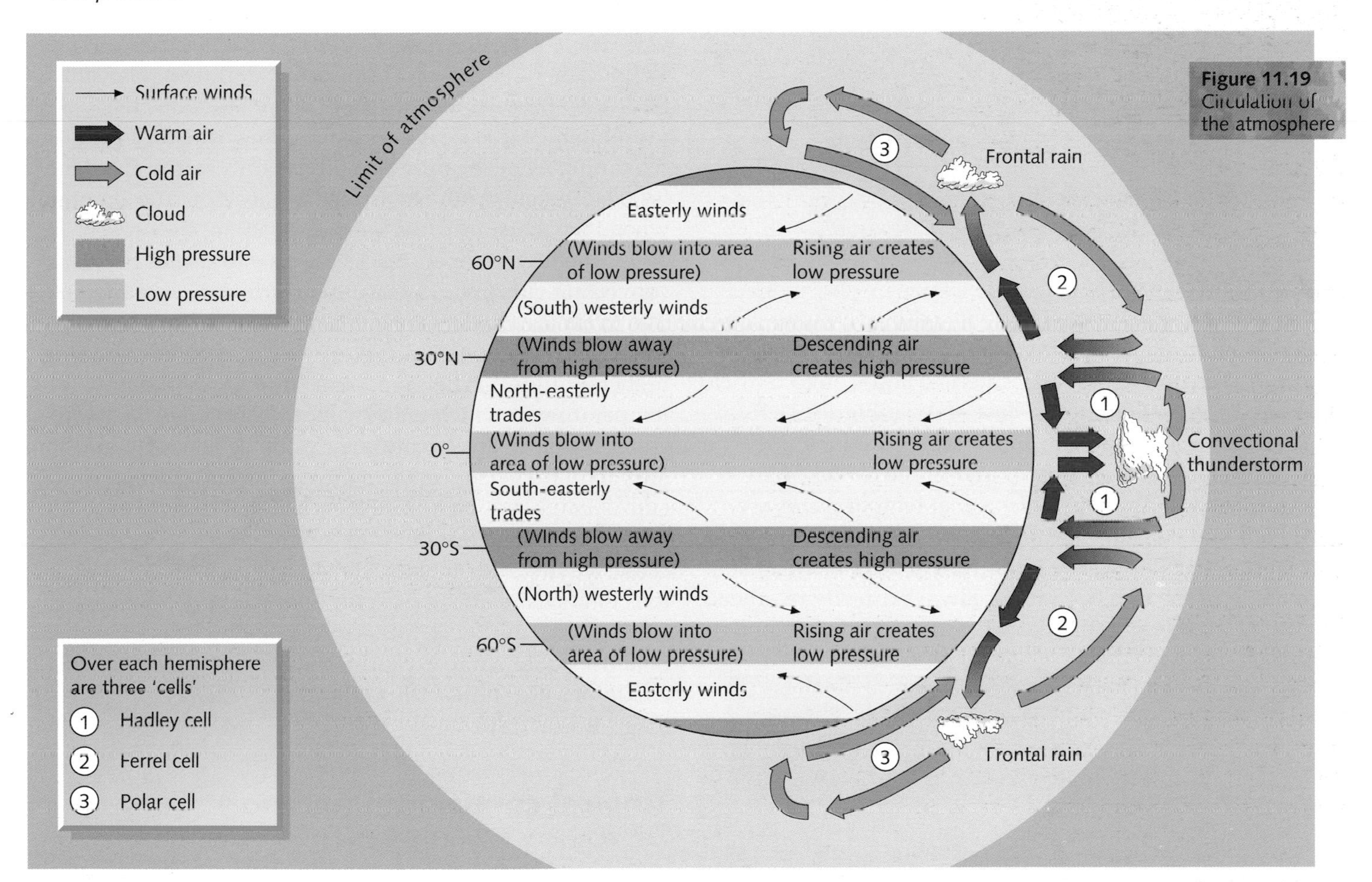

Figure 11.19 Circulation of the atmosphere

The circulation is controlled by the build-up of heat at the Equator, which causes air to rise (convection currents) and low pressure to form. As the rising air cools, it condenses to give thunderstorms (page 196) before spreading outwards from the Equator. As the air becomes denser, it descends to form a high pressure belt and, as it gets warmer and drier, an area of desert. Some of the descending air, on reaching the ground, returns to the Equator as the trade winds. The remainder moves away from the tropics (the westerlies) towards places like Britain where, as warm, moist tropical air, it meets with colder polar air and is forced to rise, creating low pressure and depressions.

This explanation is an oversimplified one (if it was not, then weather forecasting would be quite easy!). Due to the Earth's tilt and rotation, the position of the overhead sun appears to change, giving us our seasons. The northern hemisphere is hottest when the sun appears to be above the Tropic of Cancer (21 June). At this time, the pressure and wind belts move northwards. In contrast, the southern hemisphere is hottest when the sun appears to be over the Tropic of Capricorn (21 December), by which time the pressure and wind belts have moved south. The resultant changes in pressure and wind are responsible for the seasonal contrasts in climates, such as the tropical continental (page 196) and the Mediterranean (page 199) climate types.

Equatorial and tropical continental climates

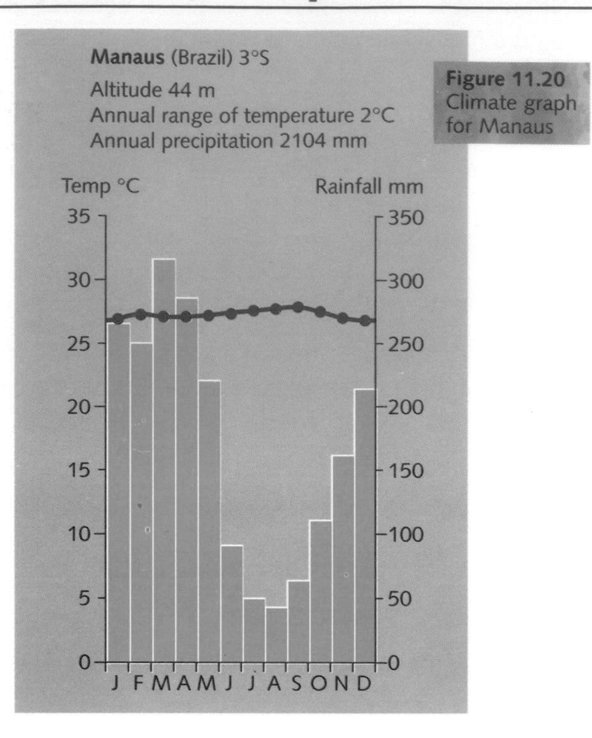

Figure 11.20 Climate graph for Manaus

Equatorial

Places with an equatorial climate lie within 5° either side of the Equator. The main areas are the large drainage basins of the Amazon (South America) and Congo (Africa) rivers and the extreme south-east of Asia (Figure 11.18). Figure 11.20 is a climate graph for Manaus which is located 3° south of the Equator and in the centre of the Amazon basin in Brazil. It shows that temperatures are both high and constant throughout the year. The small annual range of 2°C is due to the sun always being at a high angle in the sky, even if not always directly overhead (Figure 11.2). Equatorial areas have annual rainfall totals in excess of 2000 mm, mainly due to convectional thunderstorms which occur during most afternoons throughout the year (Figure 11.10). Some places, like Manaus, may have two or three drier, but not dry, months when the sun is overhead at the opposite tropic (which is Cancer in the case of Manaus), and most rain when the sun is closer to being overhead. Winds are generally light (the doldrums are areas of calm over equatorial oceans) and variable (there are no prevailing winds). The equatorial climate is characterised by its high humidity, a lack of seasonal change, and a daily weather pattern that remains remarkably uniform throughout the year.

Figure 11.21 The rainforest is a store for vast amounts of water. This water is recycled daily due to evapotranspiration and afternoon convectional storms

The daily pattern

One day is very much like another, with places receiving 12 hours of daylight and 12 hours of darkness. The sun rises at 0600 hours and its heat soon evaporates the morning mist, the heavy overnight dew, and any moisture remaining from the previous afternoon's storm. Even by 0800 hours temperatures are as high as 25°C and by noon, when the sun is near to a vertical position, they reach 33°C. The high temperatures cause air to rise in powerful convection currents. The rising air, which is very moist due to rapid evapotranspiration from swamps, rivers and the rainforest vegetation (Figure 11.21), cools on reaching higher altitudes. When it cools to its *dew point*, the temperature at which water vapour condenses back into water droplets, large cumulus clouds develop. By mid-afternoon these clouds have grown into black, towering cumulonimbus which produce torrential downpours, accompanied by thunder and lightning. Such storms soon cease, leaving the air calm. By sunset, at about 1800 hours, the clouds have already begun to break up. Nights are warm (23°C) and very humid. As one nineteenth-century botanist wrote while exploring the Amazon area, 'Such conditions are similar throughout the year, seemingly one long tropical day'.

Tropical continental (interior)

This climate is found in central parts of continents, away from coasts, and mainly between latitudes 5° and 15° north and south of the Equator (Figure 11.18). This includes the Brazilian Highlands and parts of Venezuela (South America), northern Australia, and a large semi-circle surrounding the Congo basin (Africa). This last area includes Kenya which, although straddling the Equator, lies at an altitude too high to share the main features of the equatorial climate. The main characteristic of the tropical continental climate, as illustrated by the climate graph for Kano in northern Nigeria (Figure 11.22), is alternate wet and dry seasons. This seasonal variation in rainfall is caused by the apparent movement of the overhead sun (page 195 and Figure 11.23).

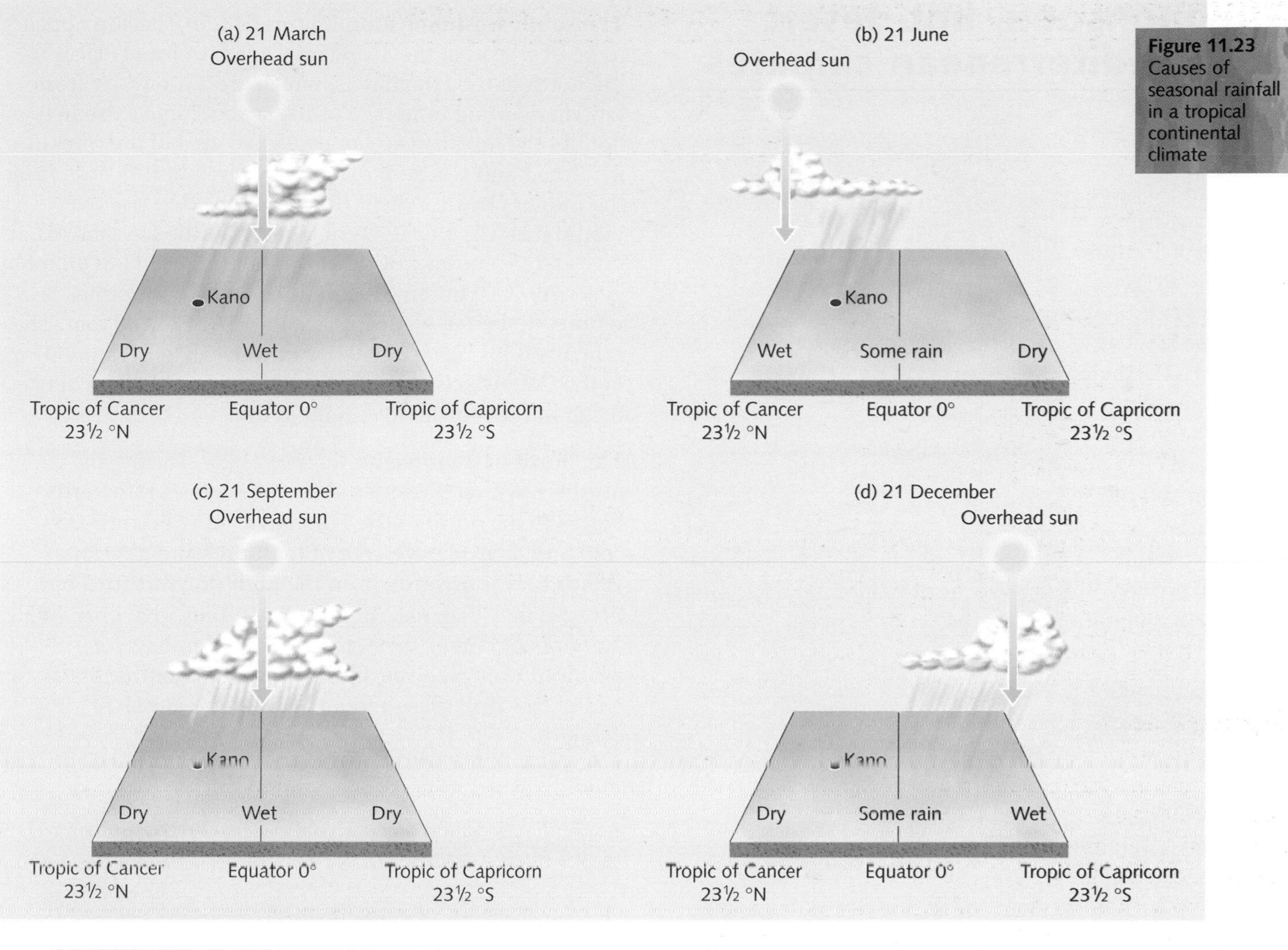

Figure 11.23 Causes of seasonal rainfall in a tropical continental climate

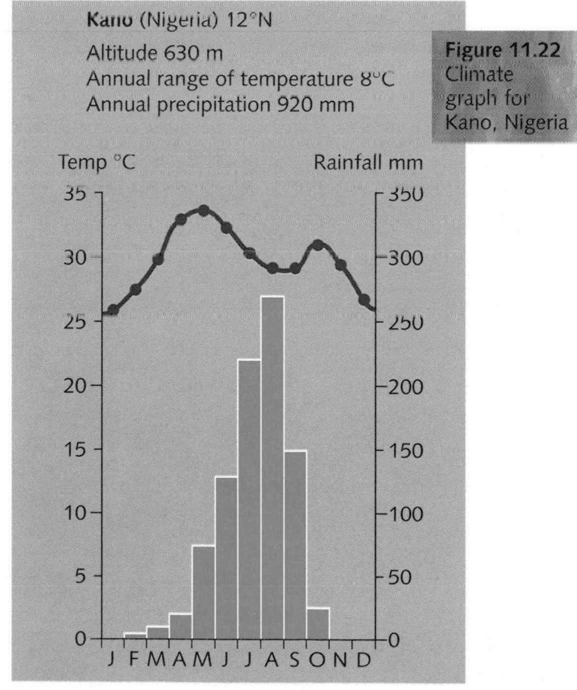

Figure 11.22 Climate graph for Kano, Nigeria

The hot, wet season occurs at Kano when the sun is overhead in the northern hemisphere (21 June – see Figure 11.23b). Temperatures rise as the sun takes a more vertical position and because places are too far inland to be affected by any moderating influence of the sea. However, temperatures fall slightly (July and August) as cloud cover and rainfall increase. The frequent afternoon convectional thunderstorms mean that the climate resembles, at this time of year, that of equatorial areas. Unfortunately both the length of the rainy season and the total amounts of rain are unreliable and many areas, especially those with increasing distance from the Equator in Africa, have experienced severe drought in recent years (Case Study 11B).

The dry, slightly cooler season at Kano occurs when the sun is overhead in the opposite (i.e. southern) hemisphere (21 December on Figure 11.23). The apparent movement of the sun southwards is accompanied by the movement of the equatorial low pressure and the global wind belts (Figure 11.19 and page 195) so that the area is now influenced by the prevailing trade winds. The trade winds, blowing from the east, will have shed any moisture long before they reach these inland areas. Indeed, as the sun is still at a sufficiently high angle to give quite high temperatures, the climate here at this time of year is more similar to that of the hot deserts (page 199). (Remember that for places south of the Equator, the seasonal pattern described here is reversed.)

The monsoon, hot desert and Mediterranean climates

The monsoon

The word *monsoon* means 'a season'. In much of South-east Asia, as seen in the climate graph for Bombay (Figure 11.24), there are two seasons. These, the so-called south-west monsoon and the north-east monsoon, result from the reversal in the direction of the prevailing wind.

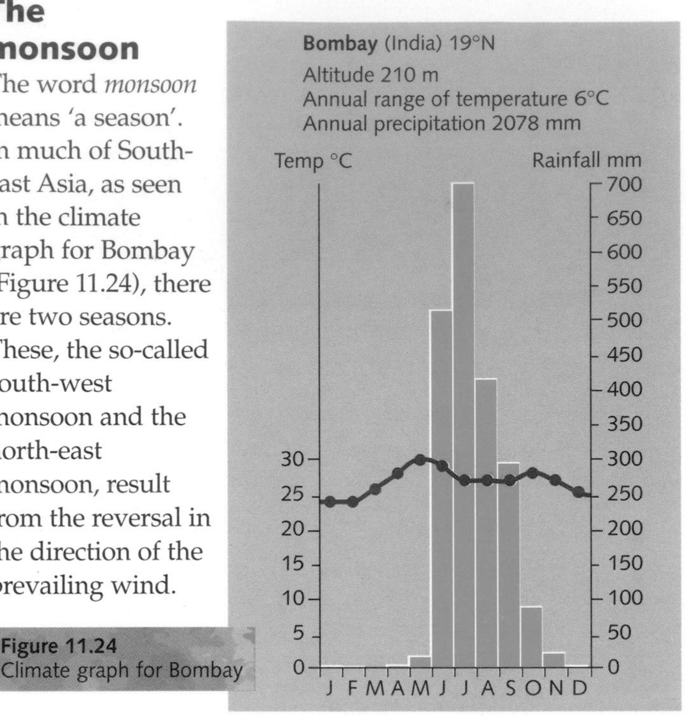

Figure 11.24
Climate graph for Bombay

The south-west monsoon (Figure 11.25) The sun appears to be overhead at the Tropic of Cancer in June. Places to the north of the Himalayas, which are a long way from any moderating influence of the sea, become extremely hot. As the hot air rises, an extensive area of low pressure is formed and warm, moist air is drawn northwards from the Indian Ocean. Where the air is forced to rise over mountains (e.g. the Western Ghats and the Himalayas), it gives large amounts of relief rainfall (Figure 11.8). Bombay gets over 2000 mm in five months and Cherrapunji, reputed to be the wettest place on Earth, 14 000 mm. The rain, ideal for rice (page 94), can cause extensive flooding in the Ganges basin, while places north of the Himalayas, being in the rainshadow, remain dry.

The north-east monsoon (Figure 11.26) During the northern winter, the overhead sun moves southwards. Places to the north of the Himalayas now become very cold and an extensive area of high pressure develops. Winds blow outwards from the high pressure area but, because they originate in a dry area, they give only small amounts of rain as they cross India – Bombay, for example, only receives 45 mm in seven months. By the end of the 'dry' season, many places are at risk of drought.

Figure 11.25
South-west monsoon

Figure 11.26
North-east monsoon

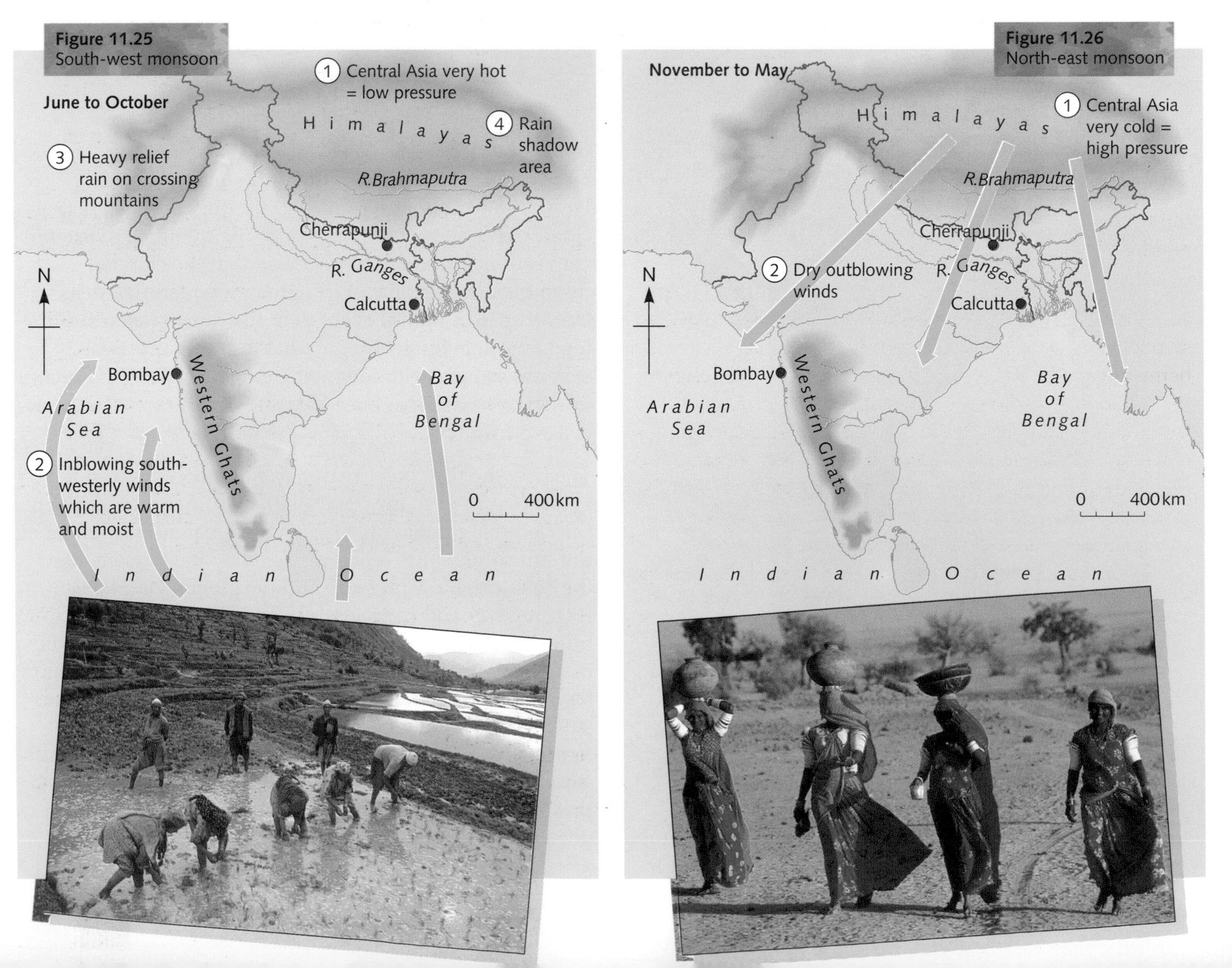

Hot deserts

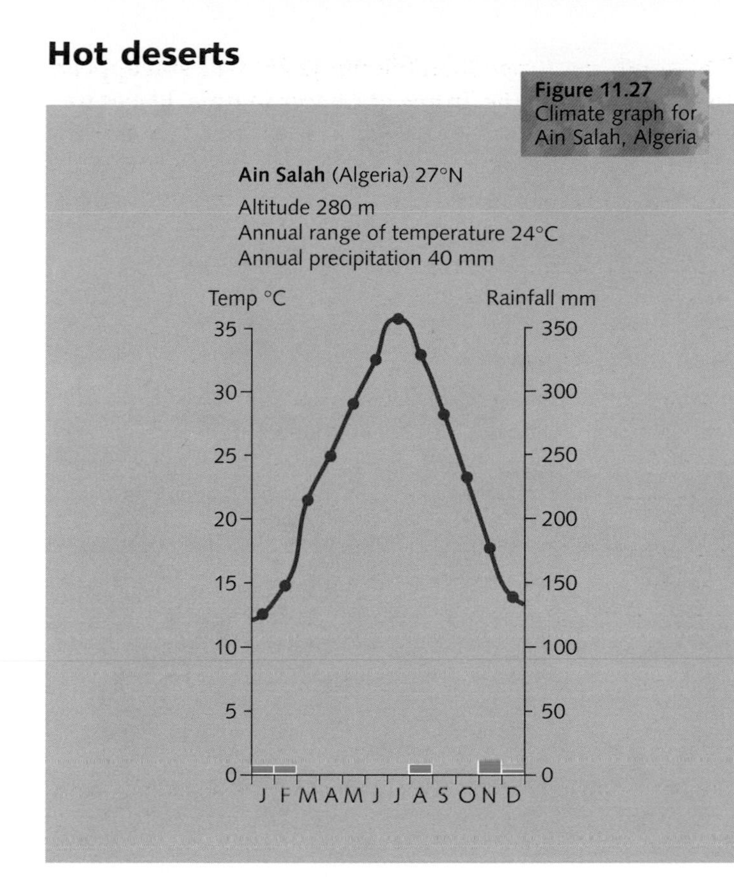

Figure 11.27 Climate graph for Ain Salah, Algeria

Hot deserts are places with high annual temperatures, less than 250 mm of rain a year and high evapotranspiration rates. Apart from the Sahara, which extends across Africa, most deserts are located on the west coast of continents, in the subtropical high pressure belt (Figure 11.19), and between 5° and 30° north and south of the Equator (Figure 11.18). They include the Atacama (South America), the Kalahari–Namib (southern Africa) and the Australian and Mexican deserts.

As the climate graph for Ain Salah in the Sahara Desert shows (Figure 11.27), temperatures are highest when the sun is overhead, but lower when it is in the opposite hemisphere. Coastal areas are much cooler, partly due to the moderating influence of the sea and partly because of cold offshore currents. Inland, and away from the influence of the sea, cloudless skies allow daytime temperatures to rise to 50°C and night-time temperatures to fall to around freezing. Although deserts are very dry, all have some precipitation even if, as in the case of the Atacama and Namib, it comes mainly in the form of coastal fog. The lack of rain is due to a combination of factors:

- Prevailing winds blow from the dry land and so cannot pick up moisture.
- Prevailing winds have to cross mountain barriers which create rain shadows.
- Air that rose as convection currents at the Equator descends in these latitudes. As it descends it warms, can hold more moisture, and gives clear skies.
- When winds do blow from the sea, they are cool and unable to pick up much moisture.

Mediterranean

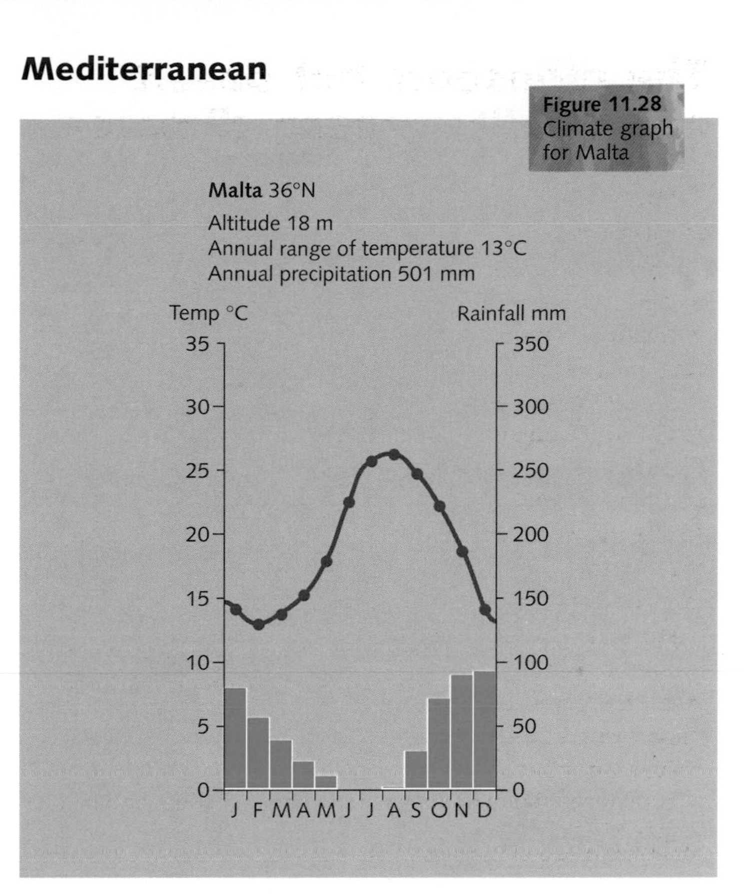

Figure 11.28 Climate graph for Malta

Places with a Mediterranean climate are found on the west coast of continents between latitudes 30° and 40° north and south of the Equator (Figure 11.18). Apart from the area surrounding the Mediterranean Sea, where the climate does extend inland, this climate also occurs in California, central Chile, around Cape Town in South Africa and in parts of southern Australia. The climate is characterised by hot, dry summers and warm, wet winters, as shown by the climate graph for Malta (Figure 11.28).

Summers are hot. This is partly because the sun is at a high angle in the sky (though never directly overhead) and the prevailing winds blow from the warm land. Places next to oceans (but not those around the Mediterranean Sea) are cooler due to the moderating influence of the sea (San Francisco is no warmer than southern England). As the wind also blows across a dry land surface, it is unable to pick up much moisture. Apart from an occasional thunderstorm, most places have several months of dry and sunny weather.

Winters are warm, for although the sun is lower in the sky, it is still higher than places further from the Equator (like Britain). However, the prevailing winds have reversed direction and now blow from the sea. This helps to keep coastal areas warm, and also brings warm, moist air which, as it is forced to rise over the many coastal mountains, gives large amounts of relief rainfall. Even so, wet days are usually separated by several that are warm and sunny.

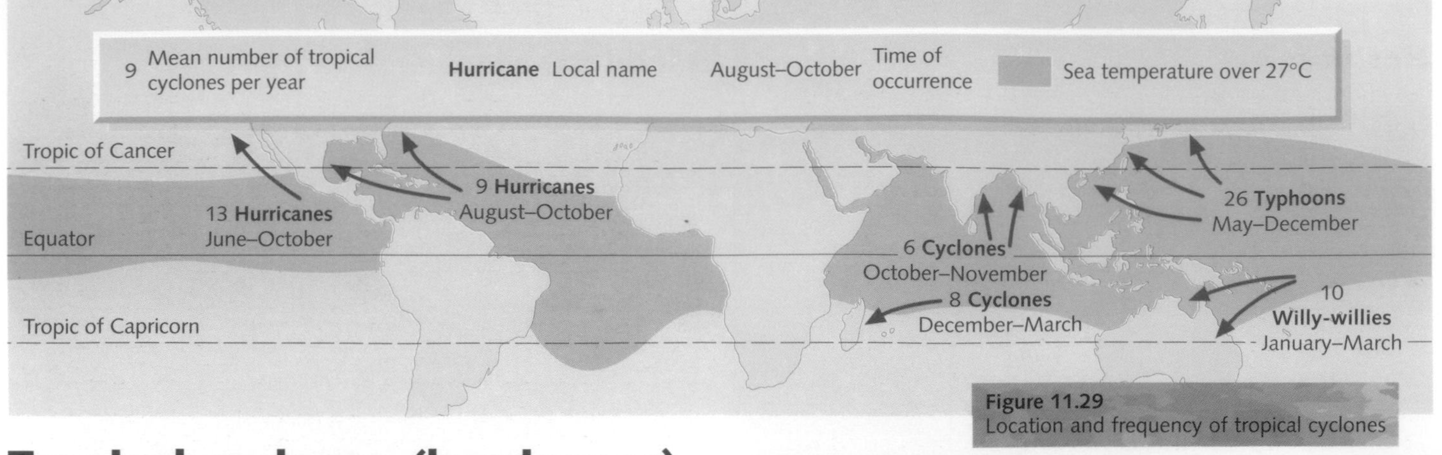

Figure 11.29
Location and frequency of tropical cyclones

Tropical cyclones (hurricanes)

Formation

Tropical cyclones are areas of intensive low pressure known locally as *hurricanes*, typhoons, cyclones or willy-willies (Figure 11.29).

Hurricanes tend to develop:

- over warm tropical oceans, where sea temperatures exceed 27°C over a vast area, and where there is a considerable depth of warm water
- in late summer and early autumn, when sea temperatures are at their highest
- in the trade wind belt between latitudes 5° and 20° north and south of the Equator.

Figure 11.30
A hurricane with its eye to the west of Florida

Although their formation is not yet fully understood, they appear to originate when a strong vertical movement of air draws with it water vapour from the ocean below. As the air rises, in a spiral movement, it cools and condenses – a process that releases enormous amounts of heat energy. (It has been estimated that the heat energy released in a single day in a hurricane is equivalent to 500 000 atomic bombs the size of those dropped on Japan during the Second World War.) It is this heat energy that powers the storm and which must be maintained if the hurricane is to move westwards on a course that is usually erratic and difficult to predict. In time an area of colder air sinks downwards through the centre of the hurricane to form a central *eye* (Figure 11.30). Once the hurricane reaches land, and its source of heat energy and moisture is removed, it rapidly decreases in strength. Its average lifespan is 7 to 14 days.

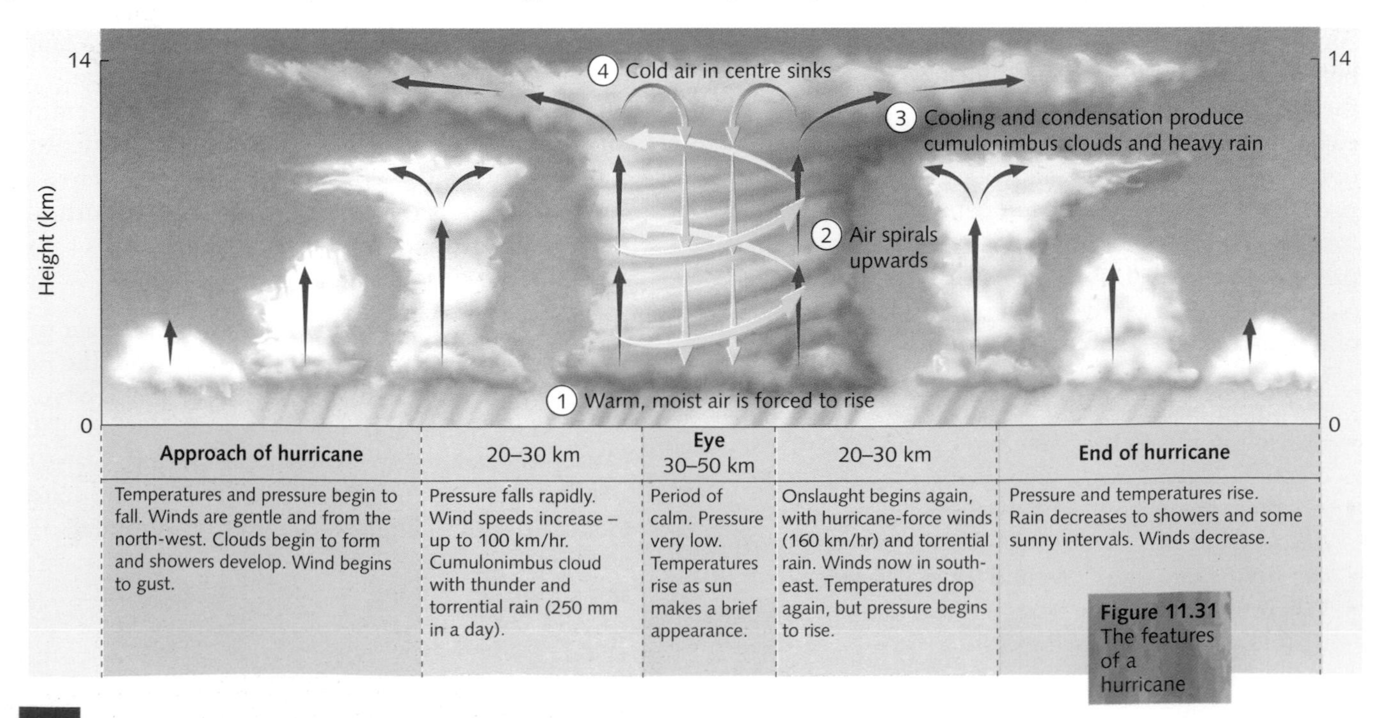

Approach of hurricane	20–30 km	Eye 30–50 km	20–30 km	End of hurricane
Temperatures and pressure begin to fall. Winds are gentle and from the north-west. Clouds begin to form and showers develop. Wind begins to gust.	Pressure falls rapidly. Wind speeds increase – up to 100 km/hr. Cumulonimbus cloud with thunder and torrential rain (250 mm in a day).	Period of calm. Pressure very low. Temperatures rise as sun makes a brief appearance.	Onslaught begins again, with hurricane-force winds (160 km/hr) and torrential rain. Winds now in south-east. Temperatures drop again, but pressure begins to rise.	Pressure and temperatures rise. Rain decreases to showers and some sunny intervals. Winds decrease.

Figure 11.31
The features of a hurricane

Weather

The characteristic weather associated with the passing of a typical hurricane is described in Figure 11.31.

Effects

Hurricanes, and other tropical cyclones, are a major hazard which can cause considerable loss of life and damage to property and to the economy of a country.

- **High winds** often exceed 160 km/hr and can, in extreme cases, reach 250 km/hr. In economically less developed countries, of which there are many in the tropical cyclone belt (e.g. in Central America and the West Indies), whole villages may be destroyed. Even in the southern states of the USA (e.g. Florida), where people have the money to erect reinforced buildings, houses and coastal developments can be severely damaged. High winds uproot trees and disrupt telephone and electricity power lines. Countries whose economies rely upon the production of just one or two crops (e.g. bananas in Nicaragua and the Dominican Republic) may suffer serious economic problems.
- **Storm (tidal) surges**, sometimes up to 5 m and heightened by storm waves, flood low-lying coastal areas. Where these areas are densely populated, there may be a major loss of life, as at Galveston (Texas), where 6000 people drowned in 1900, and in Bangladesh, where 40 000 lives were lost in 1985 (page 294). Flooding also blocks coastal escape and relief roads.
- **Flooding** can also result from rivers swollen by the torrential rain. In 1974, flash floods in Honduras caused 8000 deaths as people's flimsy homes were washed away. Flooding may pollute water supplies, increasing the risk of cholera.
- **Landslides** may occur where heavy rainfall washes away buildings erected on steep, unstable slopes.

Figure 11.32 describes some of the effects of the 1995 hurricane season in the south-east USA, Central America and the West Indies.

Predictions and precautions

At present it is very difficult predict the speed and path of a hurricane. Figure 11.33 shows the advice given to people on what to do before, during and after a storm. Warnings are given, using the five-point Saffir–Simpson scale, which is based on the potential damage of a predicted hurricane (Figure 11.32).

In the USA, a new aeroplane has been specially designed to fly through hurricanes to try to obtain more information on how they operate, money has been spent on improving early warning systems and in trying to maintain communications during a hurricane, emergency services are better trained in relief work and more buildings have been reinforced.

Saffir–Simpson scale

	SCALE	PREDICTED DAMAGE	PREDICTED PRESSURE (MB)	WIND SPEED (KM/HR)
TROPICAL STORM	–	–	–	55–119
HURRICANE	1	MINIMAL	>980	120–149
HURRICANE	2	MODERATE	965–979	150–179
HURRICANE	3	EXTENSIVE	945–964	180–209
HURRICANE	4	EXTREME	920–944	210–249
HURRICANE	5	CATASTROPHIC	<920	>250

The 1995 Atlantic hurricane season became the second stormiest season since the late 1880s, with 11 hurricanes and 8 tropical storms. They killed at least 121 people (36 in the USA) and caused damage amounting to US$ 8 billion (US$ 5 billion in the USA).
Hurricanes are named alphabetically each year, using male and female names alternately. The 1995 hurricanes included the following:

ALLISON Early July. Caused US$ 800 000 damage in Florida.

ERIN 31 July. Devastated parts of The Bahamas and Florida. 11 deaths. Power lines brought down, affecting thousands of people. Damage US$ 350000.

IRIS Affected cruise liners.

LUIS Hit the northernmost Caribbean islands from Antigua to St Maarten. Many deaths. US$ 300000 damage on Antigua alone. Two weeks later came . . .

MARILYN 9 deaths and 80% houses destroyed or damaged on St Thomas. 200 yachts sunk on St Maarten. US$ 1 billion damage in Virgin Islands. Dominica lost 90% of its main crop, bananas.

OPAL Left 50 dead from flash floods and landslides in Mexico's Yucatan peninsula. Doubled back to devastate parts of Alabama, Georgia and Florida, with winds up to 240 km/hr. Storm surge of 5 m. 21 more lives lost. Damage US$4 billion – the third costliest storm ever to hit the USA.

Figure 11.32
The Saffir–Simpson scale and the 1995 Atlantic hurricane season

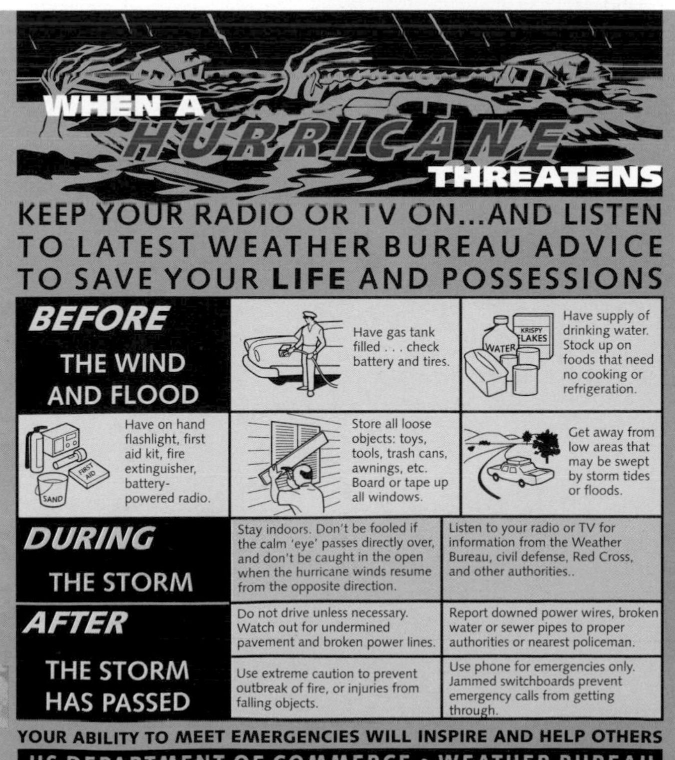

Figure 11.33
Hurricane warning

Acid rain

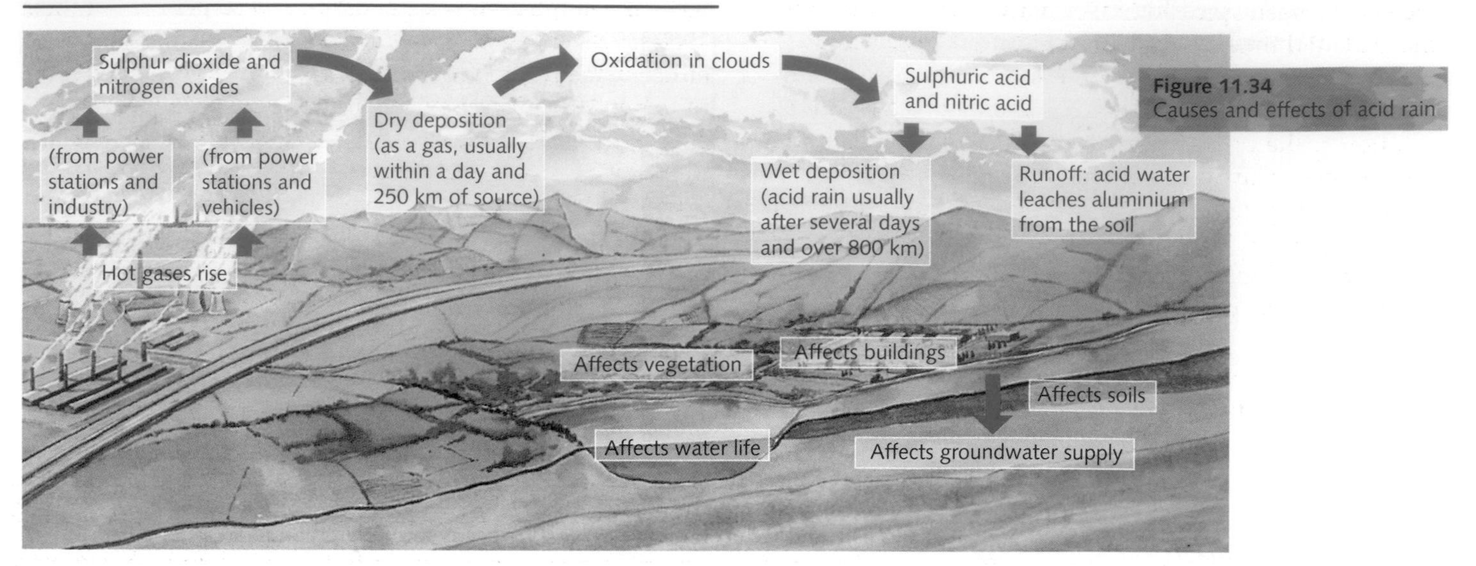

Figure 11.34
Causes and effects of acid rain

Acid rain was first noticed in Scandinavia in the 1950s when large numbers of freshwater fish died. Research showed that the water in which these fish had lived contained more than average amounts of acid. Later it was discovered that this extra acid had been carried by rain, hence the term *acid rain*. The acid is formed in the air from sulphur dioxide and nitrogen oxide which are emitted by thermal power stations, industry and motor vehicles (Figure 11.34). These gases are either carried by prevailing winds across seas and national frontiers to be deposited directly on to the Earth's surface (dry deposition), or are converted into acids (sulphuric and nitric acid) which then fall to the ground in the rain (wet deposition). Clean rainwater has a pH value of between 5.5 and 6 (pH7 is neutral). Today the pH readings are between 4 and 4.5 through much of north-west Europe, with the lowest ever recorded being 2.4 (Figure 11.35). A falling pH is the sign of increasing acidity. (Remember: when pH falls by one unit it means that the level of acid has increased ten times.) The lowest pH readings in European lakes were recorded in 1985, since when acidity has slowly decreased.

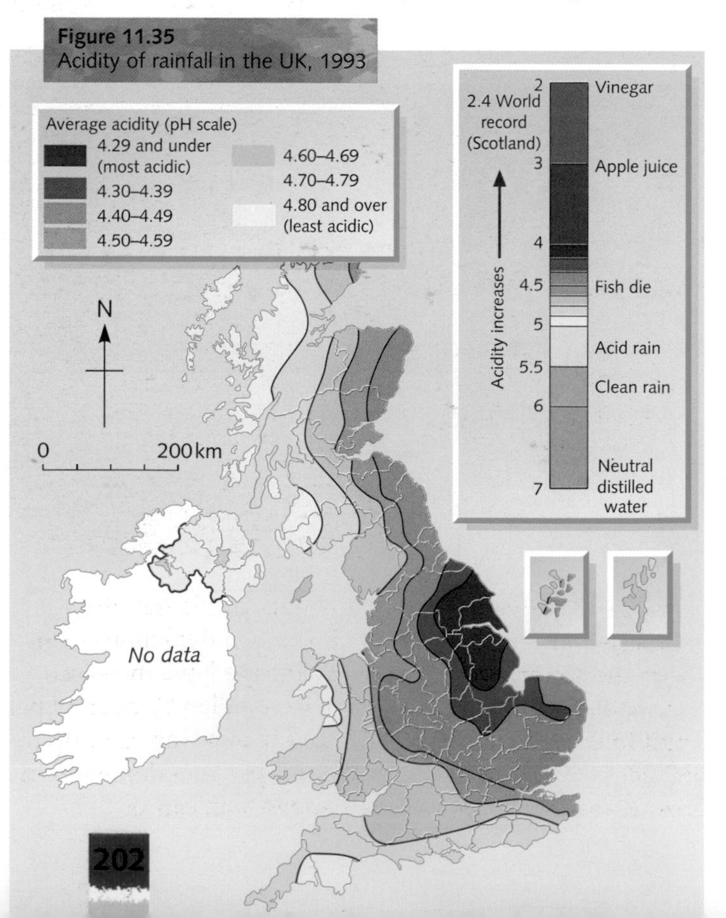

Figure 11.35
Acidity of rainfall in the UK, 1993

Europe's pollution budget

Most European countries add acids to the air, with Britain being one of the major culprits. However, only about one-third of Britain's contribution actually lands back on British soil. Some falls into the North Sea but most is carried by the prevailing south-westerly winds towards Scandinavia. Despite being one of the least of the offenders, Scandinavia is one of the main sufferers from acid rain. It is because acid rain crosses national frontiers that it is an international problem which can only be solved through global management and co-operation.

The effects of acid rain

- The acidity of lakes has increased. Large concentrations kill fish and plant life.
- An increase in the acidity of soils reduces the number of crops that can be grown.
- Forests are being destroyed as important nutrients (calcium and potassium) are washed away (leached). These are replaced by manganese and aluminium which are harmful to root growth. In time, the trees become less resistant to drought, frost and disease, and shed their needles (Figure 7.2b).
- Water supplies are more acidic and this could become a future health hazard. For example, the release of extra aluminium has been linked to Alzheimer's disease.
- Buildings are being eroded by chemical action caused by acid rain. The Acropolis in Athens and the Taj Mahal in India have both deteriorated rapidly in recent years.

Prevention or cure?

Trees have been sprayed in Germany to try to wash off the acid, and lime has been added to soils, lakes and rivers in Scandinavia to try to reduce acidity. However, these procedures are expensive and are not a sustainable solution as the processes have to be continually repeated. Prevention, as always, is preferable to the cure, and can be achieved by reducing the emissions of sulphur dioxide and nitrogen oxides. In Britain, over 60 per cent of these gases are produced by thermal (fossil-fuel burning) power stations. Several options as to how existing power stations can reduce emission are given in Figure 11.36. Although the British government has increased the money available to combat these emissions, many people consider that far more needs to be done. The government is committed to reducing sulphur dioxide emissions to 60 per cent of their 1980 levels by the year 2000. This could be achieved by either using improved technology to reduce the release of sulphur dioxide from existing power stations, turning coal-fired stations into gas-fired stations, or becoming more reliant upon nuclear power. Each method meets with considerable opposition, from consumers (increase in electricity prices), coal miners (more pit closures) and conservationists (anti-nuclear). In reality, Britain could meet its commitments and still burn domestic coal, provided that power stations are fitted with equipment to remove sulphur dioxide. At present, the generators prefer to switch to gas, which is cheaper (page 108).

OPTIONS FOR REDUCING SULPHUR DIOXIDE (SO_2) EMISSIONS	PROBLEMS
1 BURN NON-FOSSIL FUELS (NUCLEAR/RENEWABLE).	WOULD MEAN CLOSURE OF EVEN MORE OF THE FEW REMAINING BRITISH COAL MINES.
2 BURN COAL WHICH CONTAINS LESS SULPHUR.	THIS TYPE OF COAL WOULD HAVE TO BE IMPORTED, FORCING BRITISH PITS TO CLOSE.
3 REMOVE SULPHUR FROM COAL BEFORE IT IS USED. DESULPHURISATION CAN BE DONE EITHER BY (A) WASHING FINELY GROUND COAL OR (B) TREATING COAL WITH CHEMICALS.	EXTRA PROCESSES MEANS EXTRA TIME AND COSTS.
4 INSTALL NEW BOILERS IN POWER STATIONS WHICH ALLOW SO_2 TO REMAIN IN ASH. 'FLUIDISED BED TECHNOLOGY' BURNS COAL AND LIMESTONE TOGETHER SO THAT THE SULPHUR 'STICKS' TO THE LIMESTONE.	EXTRA PROCESSES ADD TO THE EXPENSE. PROBLEM OF DISPOSING OF WASTE (ASH). EXPENSIVE.
5 REMOVE SO_2 FROM WASTE GASES AFTER USE. *FLUE GAS DESULPHURISATION*: GAS IS SPRAYED WITH WATER, TURNING IT INTO SULPHURIC ACID, WHICH IS THEN NEUTRALISED AFTER BEING TREATED WITH LIME.	EXTRA PROCESS. EXTREMELY EXPENSIVE.

Figure 11.36
How emissions can be reduced

The thinning ozone layer

Ozone is a gas. It is concentrated in a layer 25–30 km above the Earth's surface and acts as a shield, protecting the Earth from the damaging effects of ultraviolet radiation from the sun. Ultraviolet radiation is responsible for sunburn and skin cancer, snow blindness, cataracts and eye damage, ageing and wrinkling of skin, and reduced immunity to disease (in October 1997, doctors suggested a link with blood cancer).

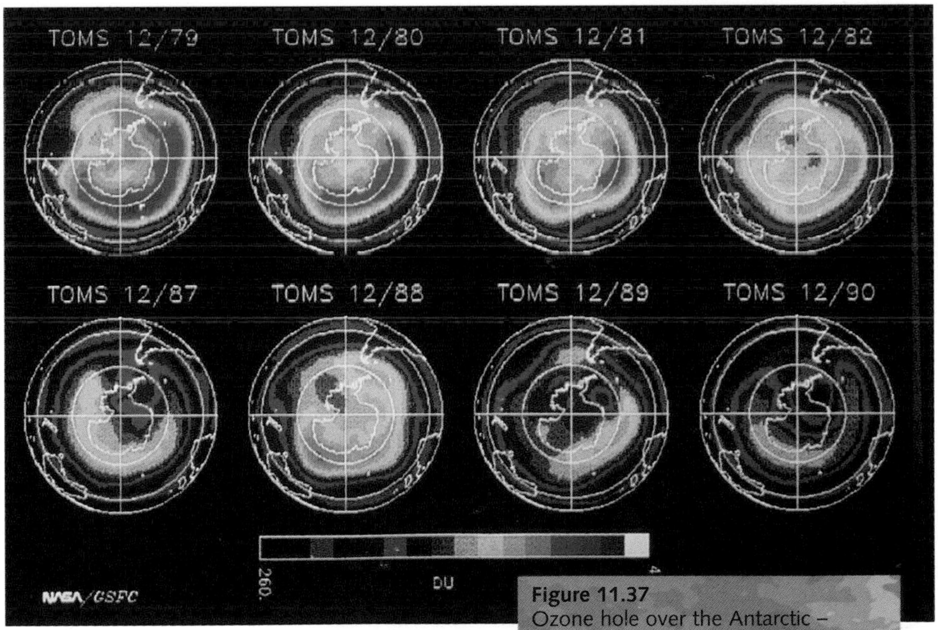

There is serious concern as parts of the shield are breaking down. A depletion of ozone above the Antarctic was first observed by the British Antarctic Survey in 1977 and the first 'hole' noted in 1985 (a 'hole' is where ozone depletion is over 50 per cent). This hole, which appears around September–October, develops when very low temperatures allow the ozone to be destroyed in a chemical reaction with chlorine (Figure 11.37). The main sources of chlorine are:

- the release of chlorofluorocarbons, especially from aerosols, by humans into the atmosphere (a long-term effect)
- from major volcanic eruptions (a short-term effect).

Figure 11.37
Ozone hole over the Antarctic – notice how it increases in size annually

The first observed hole over the Arctic followed the coldest-ever January in 1989. Since then depletions over both the Antarctic and northern Europe have increased annually. Levels over the Arctic have fallen by over 10 per cent in the last decade, with losses in some areas of up to 60 per cent. Scientists claim that a 1 per cent depletion in ozone causes a 5 per cent increase in skin cancer.

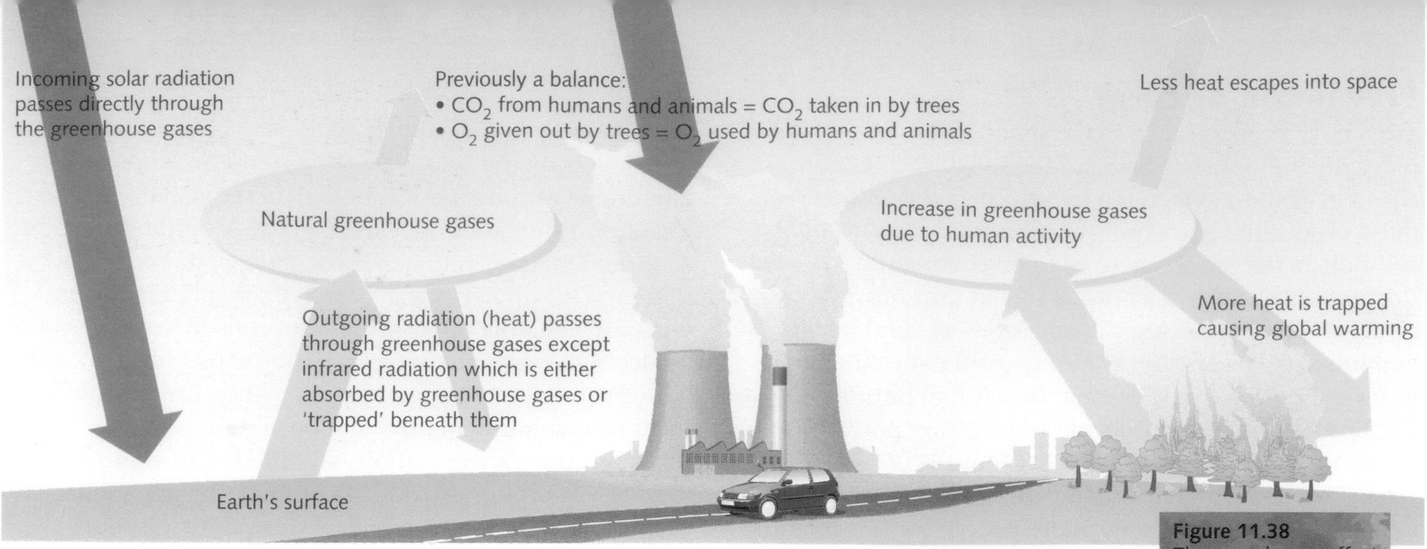

Figure 11.38
The greenhouse effect

Global warming

The greenhouse effect

The Earth is warmed during the day by incoming radiation from the sun. The Earth loses heat at night through outgoing infrared radiation. Over a lengthy period of time, because there is a balance between incoming and outgoing radiation, the Earth's temperatures remain constant.

On cloudy nights, temperatures do not drop as low as on clear nights. This is because the clouds act as a blanket and trap some of the heat. Greenhouse gases in the atmosphere also act as a blanket, as they prevent the escape of infrared radiation (Figure 11.38). Without these greenhouse gases, which include carbon dioxide, the Earth's average temperature would be 33°C colder than it is today (during the Ice Age, temperatures were only 4°C lower than at present). Recent human activity has led to a significant increase in the amount, and type, of greenhouse gases in the atmosphere. This is preventing heat from escaping into space, and is believed to be responsible for a rise in world temperatures. World temperatures have risen by 0.5°C this century (Figure 11.39). Seven of this century's warmest ten years were in the 1980s. Estimates suggest that a further rise of between 1.5°C and 4.5°C could take place by the end of the next century. The process by which world temperatures are rising is known as *global warming*.

Figure 11.39
The rise in global temperatures since 1860

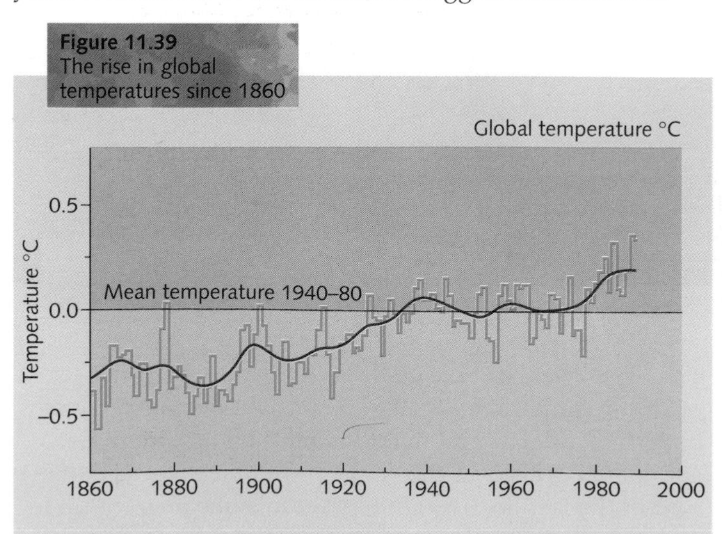

Figure 11.40
Greenhouse gases

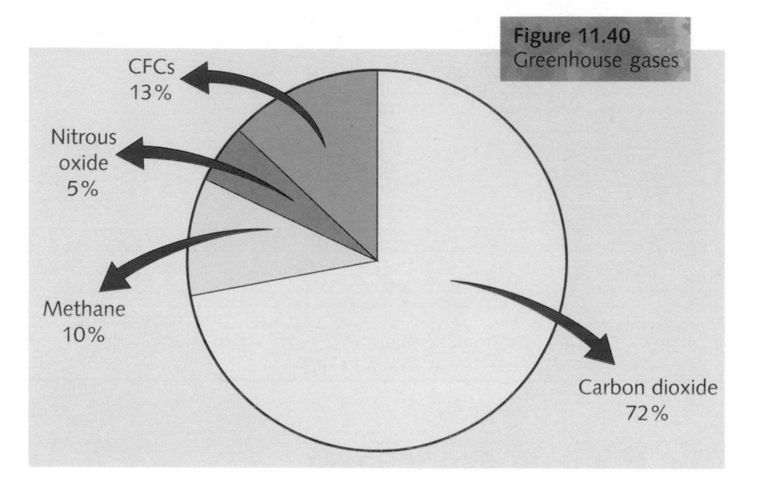

Causes of global warming

The major contributors to global warming are carbon dioxide and other pollutants released into the atmosphere (Figure 11.40).

- Carbon dioxide is the most important single factor in global warming. It is produced by road vehicles and by burning fossil fuels in power stations, in factories and in the home (Figure 11.41a). Since the economically more developed countries consume three-quarters of the world's energy, they are largely responsible for global warming. A secondary source of carbon dioxide is deforestation and the burning of the tropical rainforests (page 223).
- CFCs (chlorofluorocarbons) from aerosols, air conditioners, foam packaging and refrigerators are the most damaging of the greenhouse gases.
- Methane is released from decaying organic matter such as peat bogs, swamps, waste dumps, animal dung and farms (e.g. ricefields in South-east Asia).
- Nitrous oxide is emitted from car exhausts, power stations and agricultural fertiliser.

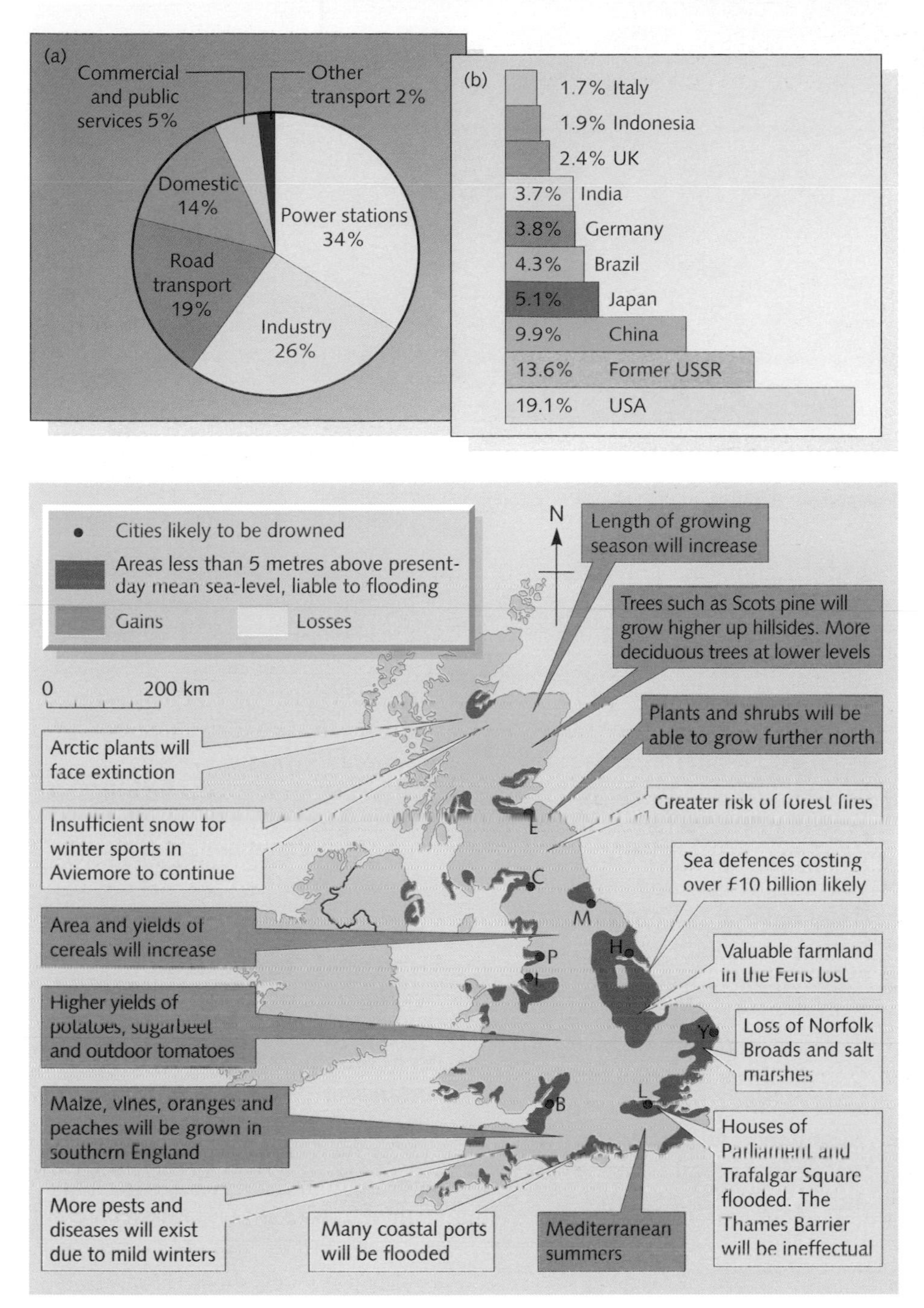

Figure 11.41 Gas emissions
(a) UK carbon dioxide emissions 1990
(b) Main greenhouse gas emitters

Effects of global warming

The major consequences of global warming are the predicted world changes in climate and sea-levels. Scientists are suggesting that as air temperatures increase:

- Sea temperatures will also rise. As the sea gets warmer it will expand causing its level to rise by between 0.25 and 1.5 metres.
- Ice caps and glaciers, especially in polar areas, will melt. The release of water at present held in storage as ice and snow in the hydrological cycle (Figure 15.3) could raise the world's sea-level by another 5 metres. Even a rise of one metre could flood 25 per cent of Bangladesh (page 295), 30 per cent of Egypt's arable land and totally submerge several low-lying islands in the Atlantic and Pacific Oceans (e.g. Maldives).
- The distribution of precipitation will alter, with some places becoming wetter and stormier, others becoming drier and with a less reliable rainfall.

Figure 11.42 shows some predicted effects of global warming on the British Isles, and Figure 11.43 on the wider world.

Figure 11.42 Predicted UK gains and losses resulting from the greenhouse effect

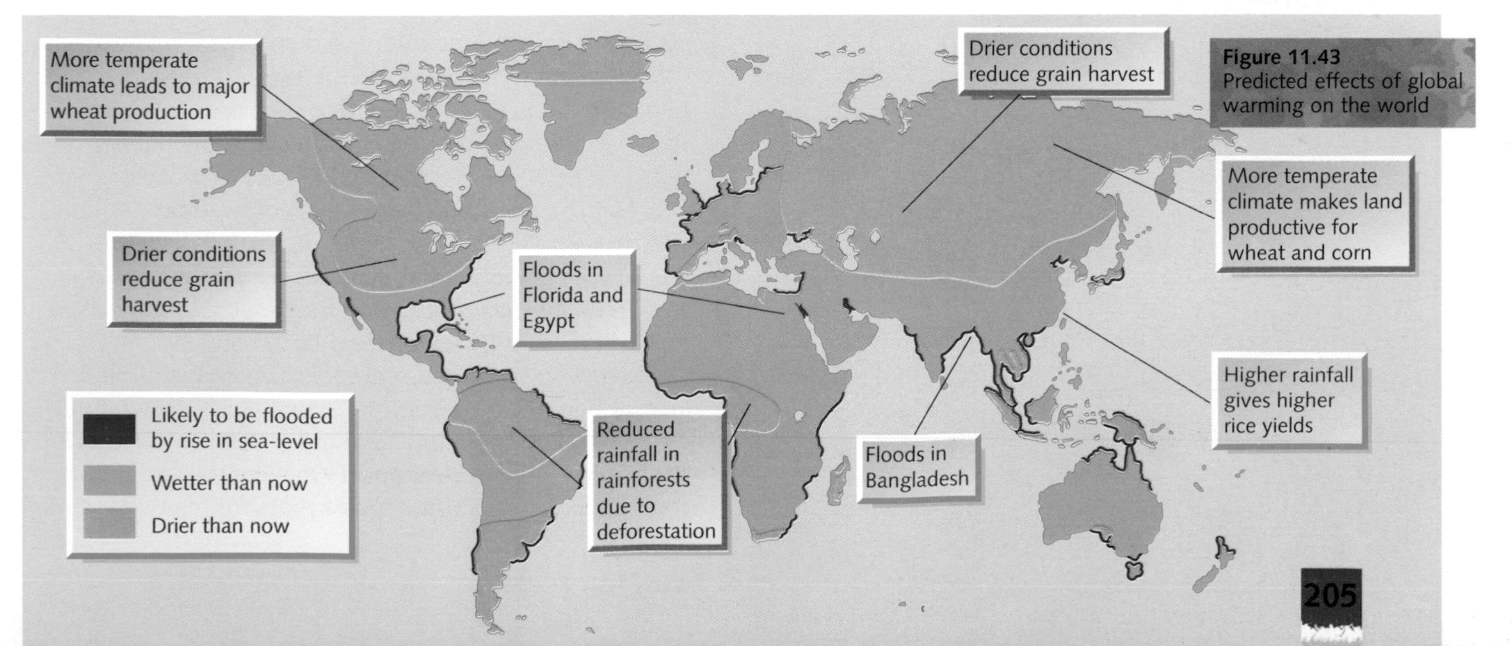

Figure 11.43 Predicted effects of global warming on the world

CASE STUDY 11A

Drought and water supply in the UK

Drought 1975–76

During the summer of 1975, depressions, which usually bring cloud and rain from the Atlantic, were diverted to the north of Britain. The result was a hot, dry summer and a depletion of water supplies. The winter of 1975–76, when normally it would be expected that water supplies would be replenished, remained mild and dry. Apart from a wet May in the north, the dry weather continued over England and Wales until the end of August 1976. Many places in the south received less than 50 per cent of their normal rainfall and over 50 per cent more than their expected sunshine.

Figure 11.44
Haweswater Reservoir in the Lake District during the 1976 drought

Consequences

- The high temperatures and lack of rain caused reservoirs to dry up and underground water supplies to run low. Two reservoirs, Haweswater in Cumbria (Figure 11.44) and Derwent in Derbyshire, fell to their lowest levels since they had been created, exposing former villages drowned when the reservoirs had begun to fill.
- A garden hosepipe ban was imposed in some areas as early as June 1976 and by August several places were affected by rationing (Figure 11.45).
- Clay soils in southern England dried out and, as they shrank, buildings were damaged as their foundations moved.
- Farmers were affected because there was insufficient grass for their cattle as it turned brown and stopped growing, crops wilted and yields fell, and there was a shortage of winter fodder due to a poor hay harvest.
- Heathland in southern England became tinder-dry and large areas were destroyed by fires. People further north were asked to avoid visiting coniferous forest due to the fire risk.
- Recreation and sport were affected, for example bowling greens could not be watered and reservoirs fell too low for water-based activities.

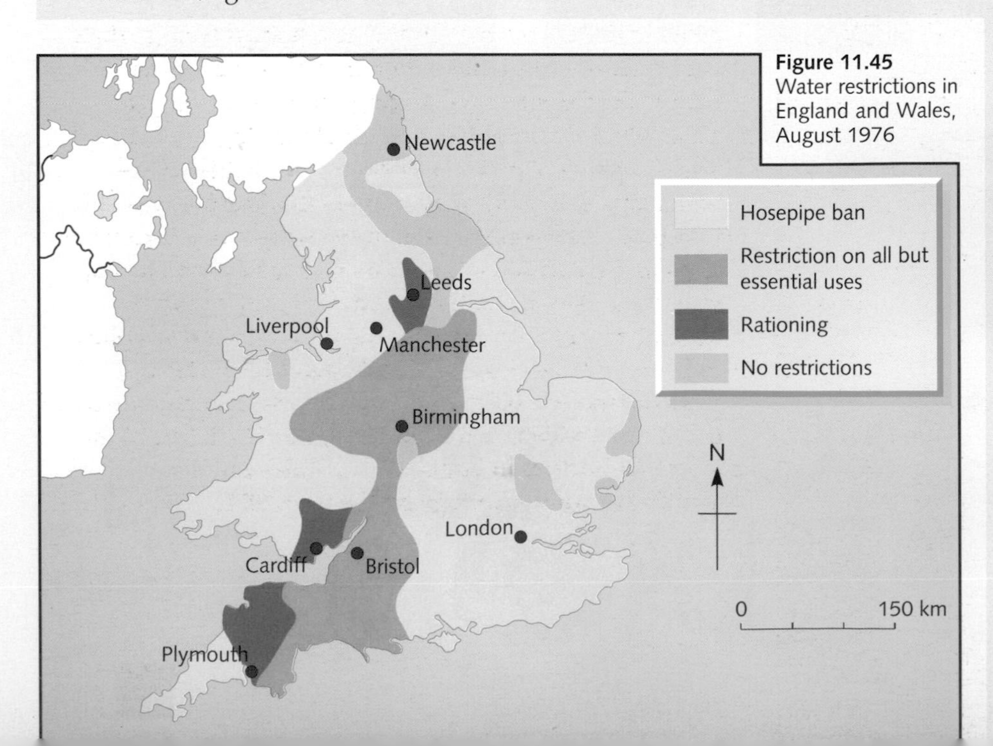

Figure 11.45
Water restrictions in England and Wales, August 1976

Drought 1995–96

Despite the problems of 1975–76, water authorities seemed to have been unprepared for the hot, dry summer of 1995. By late August, after six months of below average rainfall, many rivers were flowing at less than half their average for that time of year (e.g. Lune 6%, Wye 12%, Ribble 13% and the Exe 21%). Low water levels meant less water could be extracted by water companies, there was less water to dilute pollutants and less oxygen for fish. The dry spell continued throughout the autumn and by November reservoirs in West Yorkshire were only 10 per cent full. Yorkshire Water had to buy, and transport, surplus water from Northumbria Water. Although 1996 was wetter, the irregular showery outbreaks were insufficient to fill reservoirs or replenish the aquifers (underground water-bearing rocks). 1995 and 1996 were the two driest consecutive years for over 200 years.

Water supplies and transfers

The UK receives more than enough rainfall in an average year to meet the present demand, and predicted future demand. Unfortunately this rainfall does not always fall where and when it is needed (Figure 11.46). Most rain falls in the higher north and west (where fewer people live and work) and in the winter (when demand is lowest). This means that surplus water has to be *stored* (on the surface or underground) until it is needed and then *transferred* (naturally or artificially) to areas which have a deficit (Figure 11.47).

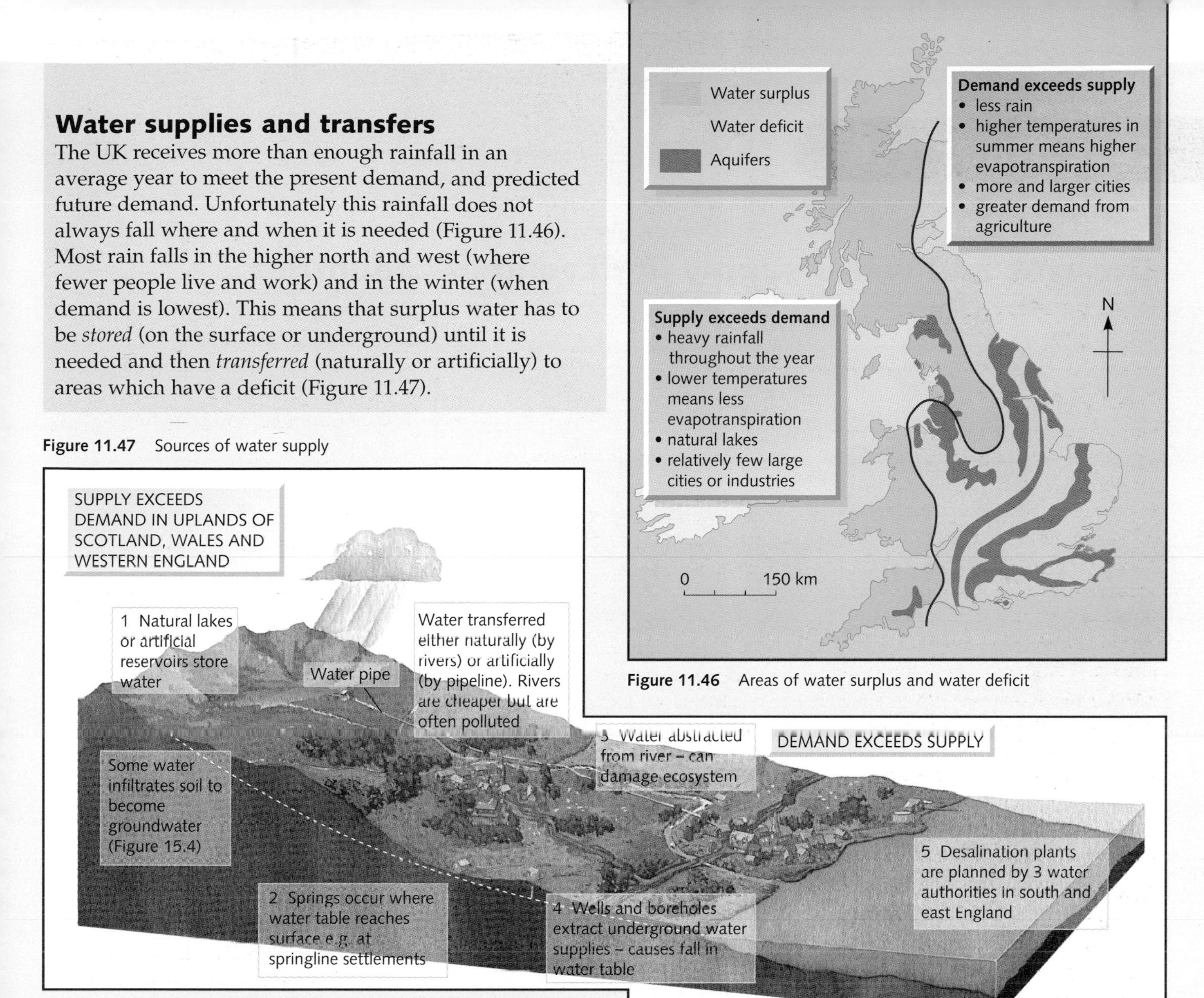

Figure 11.46 Areas of water surplus and water deficit

Figure 11.47 Sources of water supply

Figure 11.48 Estimated changes in demand for water, 1990–2021

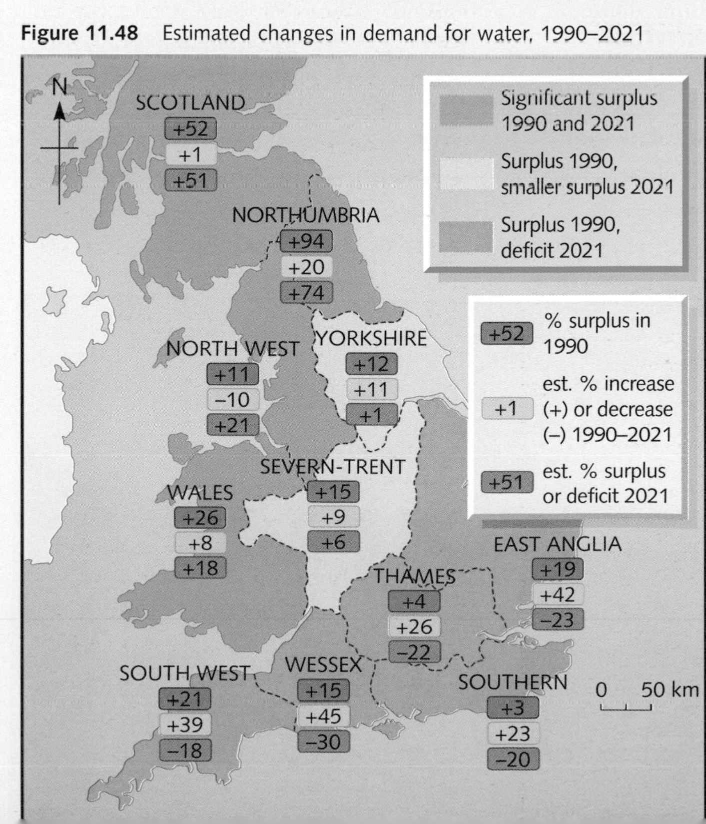

Most of the UK's water supply is obtained from surface waters in rivers and lakes but an increasing amount is being taken from water-bearing rocks called *aquifers*, especially in the south-east of England (Figure 11.47). The dry years of 1995 and 1996 meant that far more water was extracted from aquifers than nature was able to replace. The resultant fall in the water table is causing trees to die and some rivers, such as the Darent, to stop flowing altogether. Although rainfall may return to normal, it will take years for the underground supplies to be renewed – especially as demand for water is growing more rapidly in the south-east than elsewhere in Britain. Indeed, as shown on Figure 11.48, five of the water authorities in England and Wales predict a significant water shortage by the year 2021.

CASE STUDY 11B

Drought and water supply in developing countries

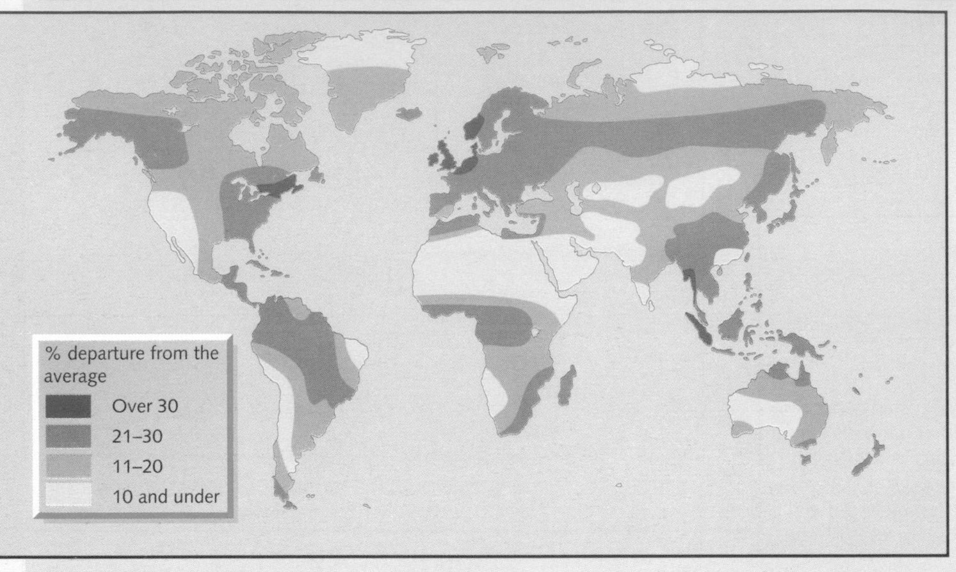

Figure 11.49 Rainfall reliability

Although there is more water than land on the Earth, 97 per cent of the total is found in seas and oceans and 2 per cent is stored as ice and snow. The remaining 1 per cent is constantly recycled in the hydrological (or water) cycle (Figure 15.3). As water is neither added to nor lost from the hydrological cycle, it is said to be a *closed system*, with the amount of fresh water available for life on Earth remaining constant.

Over the last 300 years, when the world's population has increased sevenfold, the demand for water for domestic, agricultural and industrial use has increased by 35 times. This increased use of water has been accompanied by an increase in its pollution – so much so that the UN estimate that each day 25 000 people die from using contaminated water. This, coupled with the uneven distribution of fresh water, means that some 80 countries (40 per cent of the total) and 1.5 billion people are already experiencing 'severe water stress' either within certain regions or at certain times of the year. By 2025, according to UN estimates, two-thirds of the world's population, mainly those living in the low-income, developing countries in Africa, Asia and Latin America, will be short of reliable, clean water.

Reliability of rainfall

In Britain it is taken for granted that it is likely to rain every few days, and that rain will fall fairly evenly throughout the year. The situation in many developing countries is very different. Many countries experience a pronounced wet and dry season. If the rains fail during one year, the result for crops, animals and sometimes people, can be disastrous. The most vulnerable areas are desert margins and tropical interiors where average annual amounts are low (page 197). Here, where a small variation of 10 per cent below the average total can be critical, many countries experience a variation of over 30 per cent (Figure 11.49). The rain, when it does fall, often comes in torrential downpours which give insufficient time for it to infiltrate into the ground (Figure 15.4). Instead, surface runoff causes flash floods and water is lost to the local community. Few developing countries have the technology to build dams to store water, and where they have been built, it has been with foreign financial 'aid' which has put the recipient country further into debt (page 180). Relatively few places, especially in rural areas, have piped water, and the daily walk to the local well or river can be a distance of several kilometres, taking several hours (Figure 11.50).

Figure 11.50 Carrying water, Burkina Faso

Clean water

According to the UN, an estimated 1200 million people lack a satisfactory or safe water supply (Figure 11.51). Rarely do people, even in developed countries, take sufficient care of this the most essential of resources. In developing countries, water supplies may be contaminated:

- in rural areas where rivers and streams are used for drinking, washing and the disposal of sewage (e.g. Nile, Figure 15.43)
- in urban areas where, especially in shanty settlements, the absence of drains means that sewage may pollute water supplies (e.g. São Paulo page 58, Calcutta page 61)
- due to a lack of government legislation and implementation.

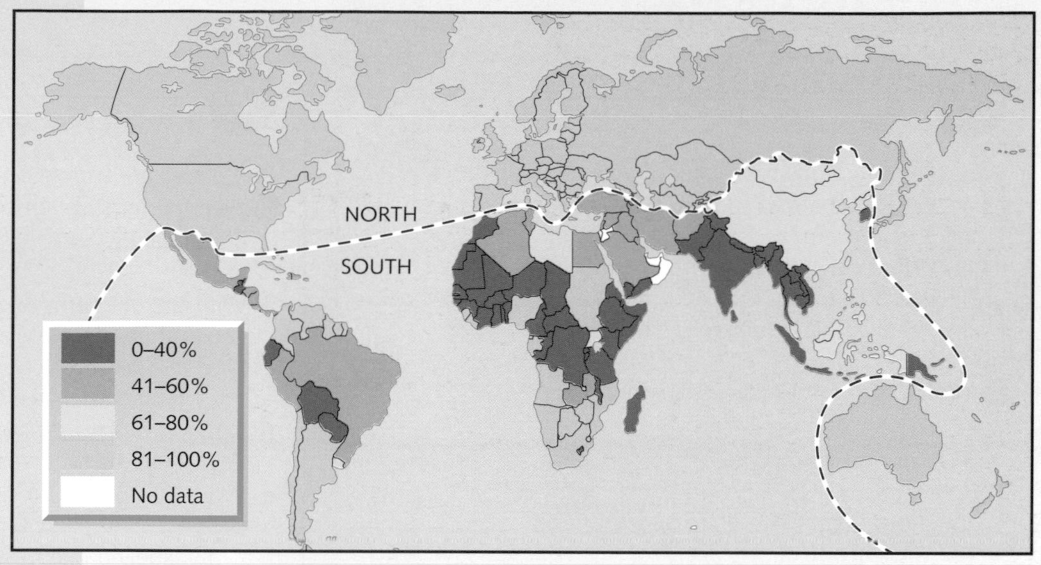

Figure 11.51 Percentage of the population with access to safe water

The result is often outbreaks of diseases such as cholera, typhoid, dysentery and diarrhoea.

The UN 'International Drinking Water Supply and Sanitation Decade' was launched in 1980 to provide 'water and sanitation for all by 1990'. This ambitious target was never realised. Although there was some progress (Figure 11.52), especially with water supply 50 per cent of people in rural areas now have an adequate supply compared with only 30 per cent in 1980 – the rapid growth in population and urbanisation meant that 300 million more people lacked sanitation in 1990 than in 1980. The improvements in rural water supplies were mainly due to international charity organisations like Oxfam, ActionAid and Intermediate Technology (page 137) which have helped local communities to introduce appropriate technology. These included self-help schemes such as:

- digging wells to reach permanent underground supplies
- lining the sides of the well with concrete (to prevent seepage) and adding a cover (to reduce evaporation)
- using modern pumps (Figure 11.53).

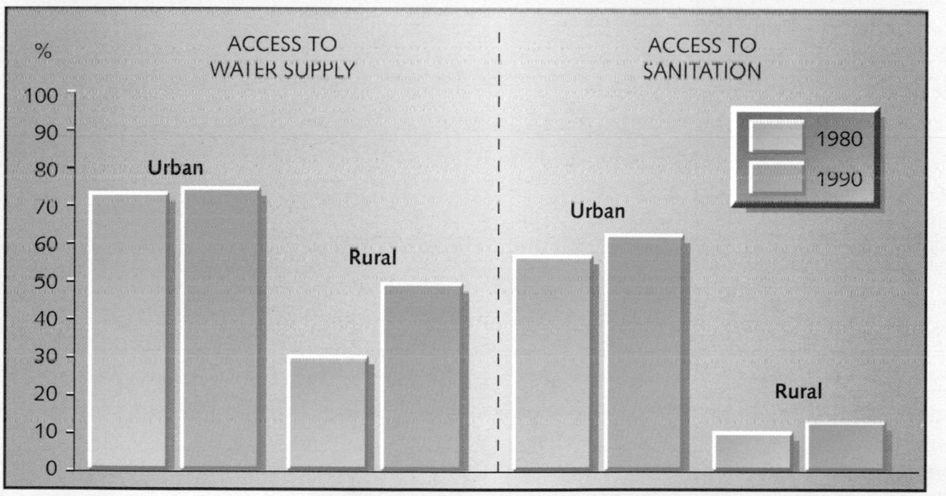

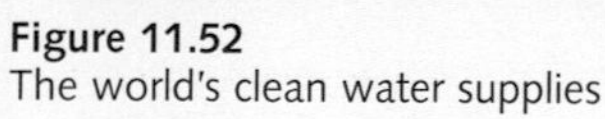
Figure 11.52
The world's clean water supplies

Figure 11.53
A water pump in Burkina Faso

QUESTIONS

1

(Page 188)

Four climate graphs have been drawn in Figure 11.1.

a i) What is the maximum temperature for Fort William? (1)
ii) What is the minimum temperature for Fort William? (1)
iii) What is the annual range of temperature for Fort William? (1)

b Which of the four places has the:
i) coolest summer
ii) warmest summer
iii) mildest winter
iv) coldest winter
v) most sunshine in a year? (5)

c i) What is the annual average rainfall for Penzance? (1)
ii) Which is the wettest season in Penzance? (1)
iii) Which is the wettest month in Penzance? (1)

d Compare the climate of Fort William and Margate. (4)

2

(Pages 188–191)

Five places are shown on the map below.

a Give the missing temperature for:
i) January
ii) July. (2)

b With the help of a diagram explain why place B is cooler than place A in summer. (3)

c Give two reasons why place C is warmer during the winter than place D. (4)

d Why does place E have snow lying for over 50 days in an average year? (2)

e Why does place C receive more rainfall throughout the year than place D? (3)

f Draw labelled diagrams to show the formation of:
i) relief rainfall
ii) frontal rainfall
iii) convectional rainfall. (9)

g Describe and account for the distribution of rainfall over the UK. (4)

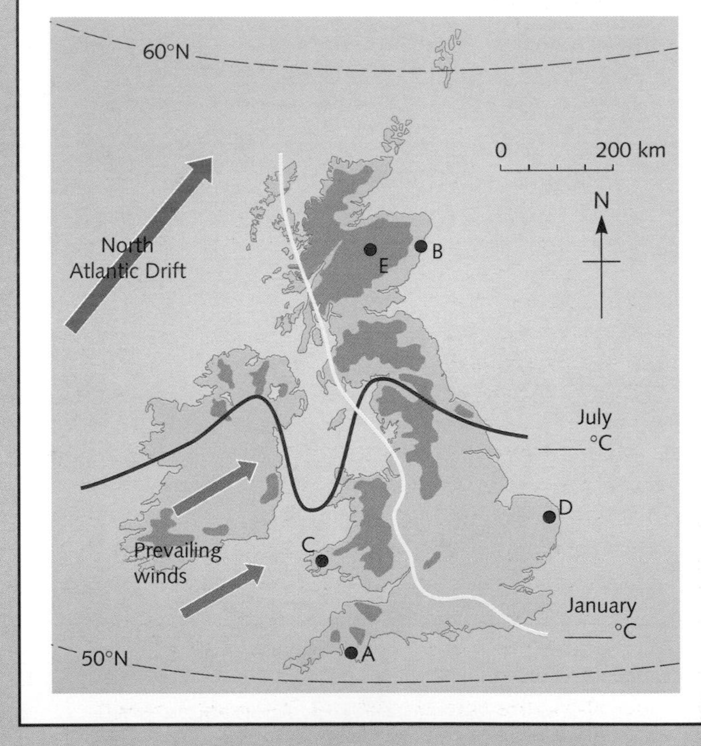

3

(Pages 192–194)

a i) Why do depressions form over the Atlantic Ocean? (2)
ii) What happens to warm and cold air at:
a) a warm front
b) a cold front? (2)

b There is usually a recognisable weather pattern as a depression passes over the British Isles. Describe, on an enlarged copy of table A below, the likely weather conditions:
i) as a warm front passes
ii) in a warm sector
iii) just after a cold front passes. (18)

c Using table B as a guide, list six differences between the expected weather conditions in **summer** between a depression and an anticyclone. (6)

d What are three likely differences in the weather of an anticyclone in winter and an anticyclone in summer? (3)

e How does the weather in i) a depression
ii) an anticyclone affect people's lives in the UK? (6)

WEATHER CONDITIONS	A AS A WARM FRONT PASSES	IN A WARM SECTOR	JUST AFTER A COLD FRONT PASSES	B DEPRESSION IN SUMMER	ANTICYCLONE IN SUMMER
TEMPERATURE					
CLOUD COVER					
PRECIPITATION					
WIND SPEED					
WIND DIRECTION					
PRESSURE					

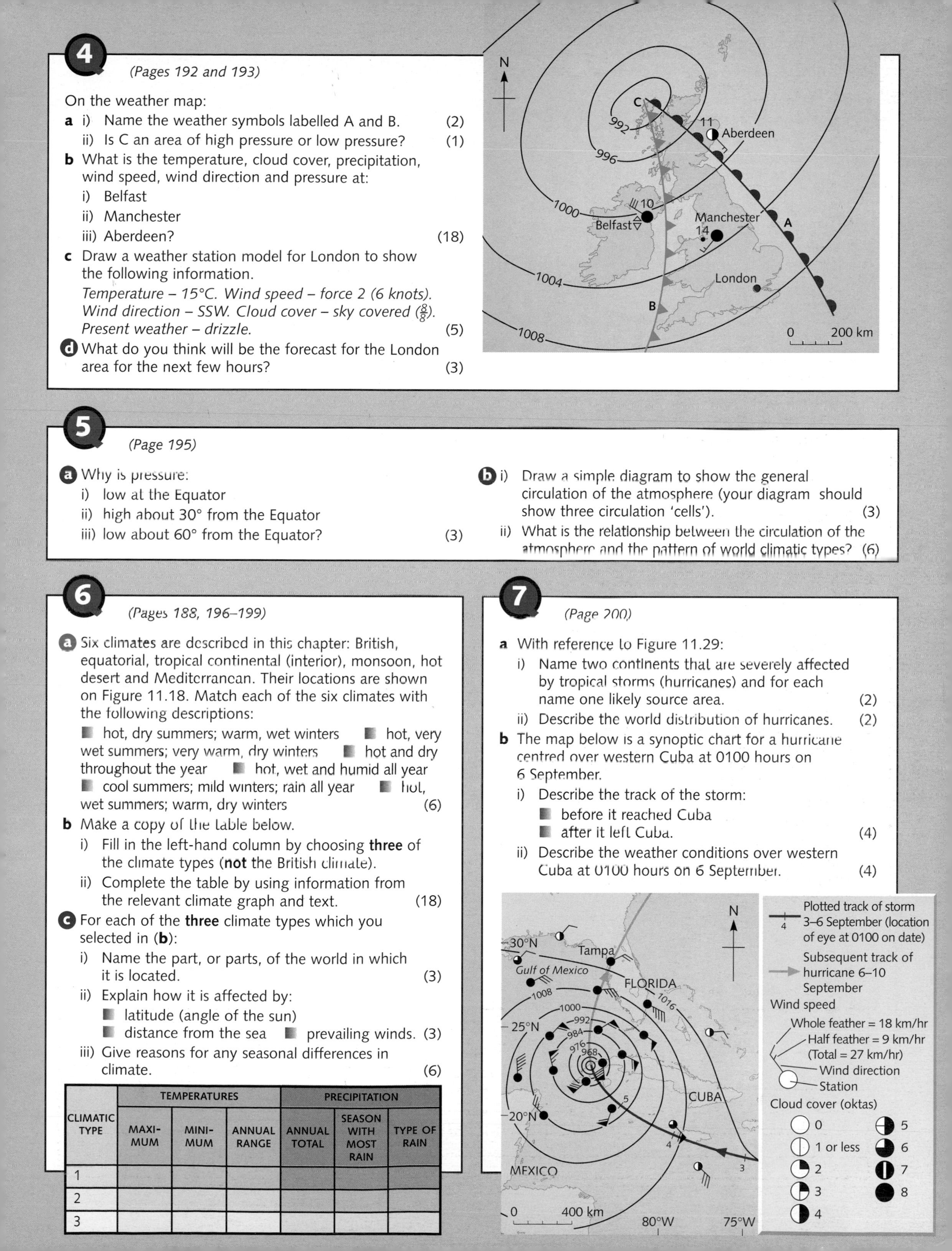

4

(Pages 192 and 193)

On the weather map:

a i) Name the weather symbols labelled A and B. (2)
ii) Is C an area of high pressure or low pressure? (1)

b What is the temperature, cloud cover, precipitation, wind speed, wind direction and pressure at:
i) Belfast
ii) Manchester
iii) Aberdeen? (18)

c Draw a weather station model for London to show the following information.
Temperature – 15°C. Wind speed – force 2 (6 knots). Wind direction – SSW. Cloud cover – sky covered ($\frac{8}{8}$). Present weather – drizzle. (5)

d What do you think will be the forecast for the London area for the next few hours? (3)

5

(Page 195)

a Why is pressure:
i) low at the Equator
ii) high about 30° from the Equator
iii) low about 60° from the Equator? (3)

b i) Draw a simple diagram to show the general circulation of the atmosphere (your diagram should show three circulation 'cells'). (3)
ii) What is the relationship between the circulation of the atmosphere and the pattern of world climatic types? (6)

6

(Pages 188, 196–199)

a Six climates are described in this chapter: British, equatorial, tropical continental (interior), monsoon, hot desert and Mediterranean. Their locations are shown on Figure 11.18. Match each of the six climates with the following descriptions:
- hot, dry summers; warm, wet winters
- hot, very wet summers; very warm, dry winters
- hot and dry throughout the year
- hot, wet and humid all year
- cool summers; mild winters; rain all year
- hot, wet summers; warm, dry winters (6)

b Make a copy of the table below.
i) Fill in the left-hand column by choosing **three** of the climate types (**not** the British climate).
ii) Complete the table by using information from the relevant climate graph and text. (18)

c For each of the **three** climate types which you selected in (**b**):
i) Name the part, or parts, of the world in which it is located. (3)
ii) Explain how it is affected by:
- latitude (angle of the sun)
- distance from the sea
- prevailing winds. (3)

iii) Give reasons for any seasonal differences in climate. (6)

CLIMATIC TYPE	TEMPERATURES			PRECIPITATION		
	MAXI-MUM	MINI-MUM	ANNUAL RANGE	ANNUAL TOTAL	SEASON WITH MOST RAIN	TYPE OF RAIN
1						
2						
3						

7

(Page 200)

a With reference to Figure 11.29:
i) Name two continents that are severely affected by tropical storms (hurricanes) and for each name one likely source area. (2)
ii) Describe the world distribution of hurricanes. (2)

b The map below is a synoptic chart for a hurricane centred over western Cuba at 0100 hours on 6 September.
i) Describe the track of the storm:
- before it reached Cuba
- after it left Cuba. (4)

ii) Describe the weather conditions over western Cuba at 0100 hours on 6 September. (4)

8

(Pages 200 and 201)

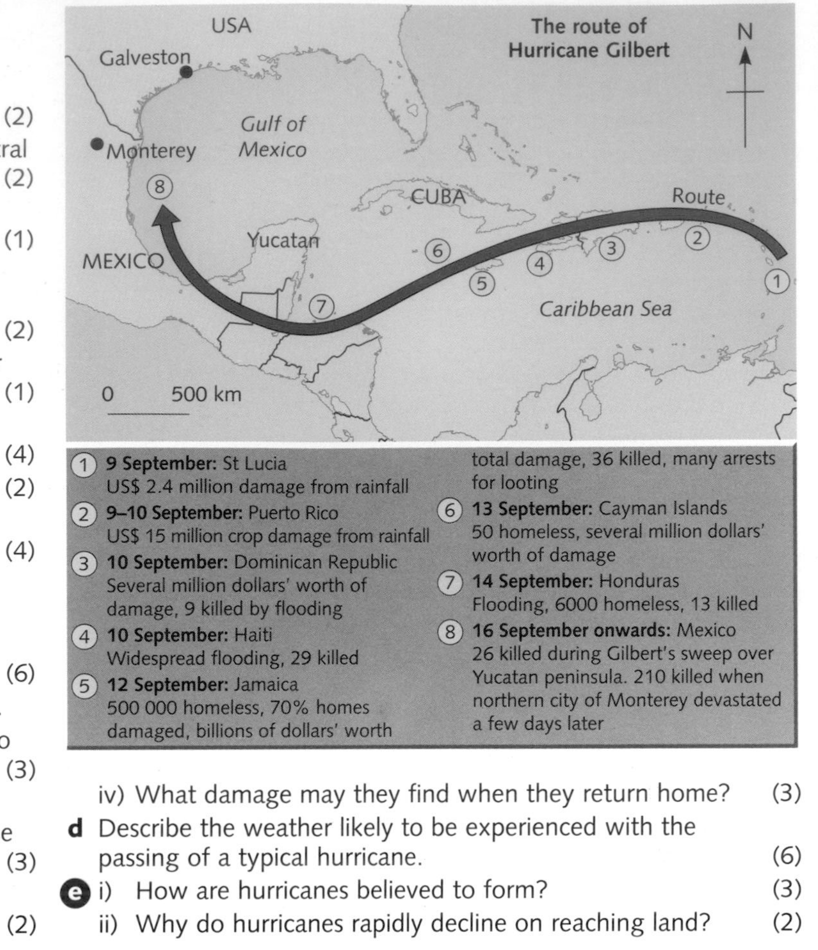

Refer to the map and facts on Hurricane Gilbert.

a i) Name six islands that were affected by the hurricane. (2)
ii) Name two countries on the mainland of Central America that were affected by the hurricane. (2)
iii) Which country had the highest number of deaths? (1)
iv) 70% of homes in Jamaica were damaged by the hurricane. Suggest two ways in which damage would have occurred. (2)
v) Suggest why Galveston was getting ready for the hurricane from 15 September. (1)
vi) What could the government and people of Galveston do to get ready for the hurricane? (4)

b i) Describe the route taken by the hurricane. (2)
ii) Describe how people living in the area were affected by the hurricane. (4)
iii) What measures might people living:
- on an island in the West Indies
- in Galveston (USA)

have taken to reduce the impact of the storm? (6)

c Use the hurricane warning poster (Figure 11.33).
i) List the precautions that people are advised to take when the warning is first given. (3)
ii) Why might people in economically less developed countries either not hear, or choose to ignore, the warning? (3)
iii) Why are people advised to take care immediately after the storm has passed? (2)
iv) What damage may they find when they return home? (3)

d Describe the weather likely to be experienced with the passing of a typical hurricane. (6)

e i) How are hurricanes believed to form? (3)
ii) Why do hurricanes rapidly decline on reaching land? (2)

9

(Pages 202 and 203)

a i) What are the two main causes of acid rain?
ii) Which two chemicals cause rain to be acidic? (4)

b i) Give three ways in which acid rain affects the natural environment.

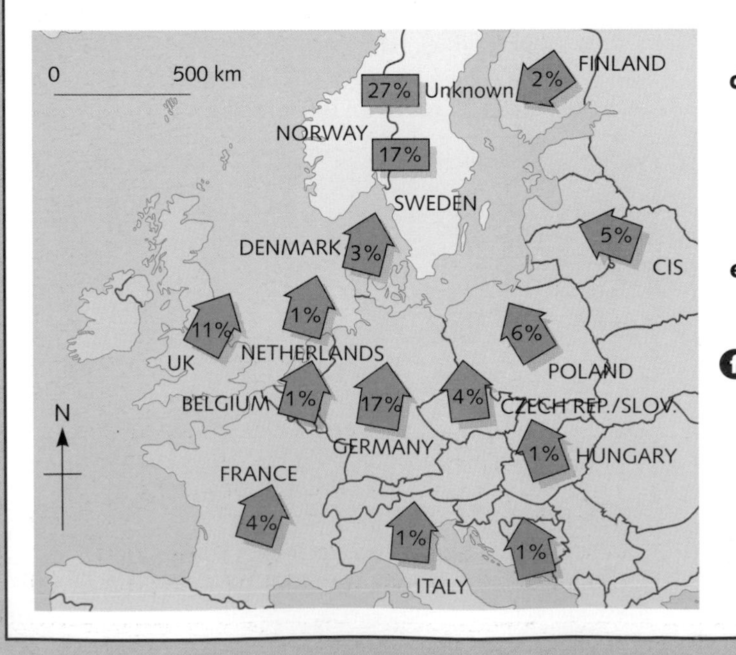

ii) How can acid rain affect buildings and people's health? (5)

c i) How is the level of acidity in rain measured?
ii) Which parts of Europe suffered most from acid rain in 1990?
iii) Why were these areas the worst affected? (4)

d The map shows the origin of the sulphur content of acid rain that poisons lakes in Norway and Sweden.
i) Which two countries were probably the source of most of the sulphur?
ii) Suggest reasons for the production of sulphur in the source countries. (4)

e i) Why is international co-operation needed if the effects of acid rain are to be controlled?
ii) How could the problems of acid rain be reduced? (4)

f It has been estimated that reducing acid rain by eliminating harmful chemicals released by power stations could increase the cost of electricity in Germany and Britain by 20 per cent.
i) What effects would this have upon industry and domestic users?
ii) Would you support such a move? Give reasons for your answer (include the benefits and drawbacks of your choice). (6)

10

(Pages 204 and 205)

a Copy and complete the pie graph by:
i) inserting the names of the four missing greenhouse gases
ii) naming one source for each of the four missing gases. (8)

b i) By how many degrees has the global temperature risen this century?
ii) By how many degrees is it predicted to rise by the end of the next century?
iii) Why are global temperatures rising? (4)

c i) List four advantages to Britain which are likely to result from global warming.
ii) List four problems likely to face Britain as a result of global warming. (8)

d i) How might global warming affect changes in sea-level? (2)
ii) How might these changes affect people living in places like the Nile Delta and Bangladesh? (2)

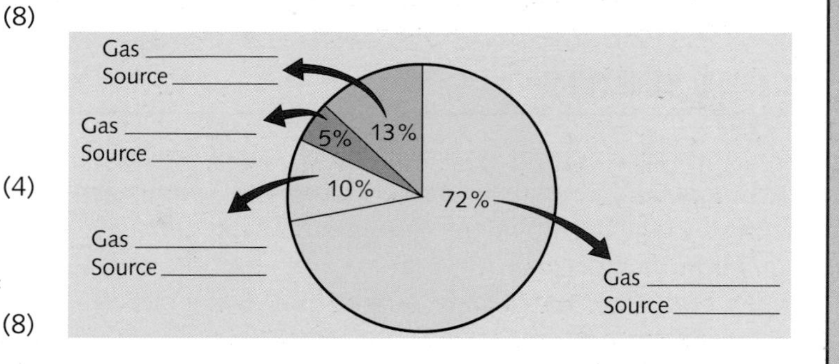

11

(Pages 206 and 207)

a i) Give three reasons why the north and west of Britain usually have a water supply surplus. (3)
ii) Give three reasons why the south and east of Britain often have a water supply deficit. (3)

b i) What caused the 1975–76 drought? (1)
ii) Name three cities that experienced water rationing. (3)

c i) What problems did the drought of 1976 create for each of the following groups of people? (5)

- farmers
- sports enthusiasts
- industrialists
- forestry workers
- householders

ii) Name two other groups of people who might have benefited from the drought. (2)

d i) Using Figure 11.47, give four sources of fresh water supply. (4)
ii) Why, in Essex in October 1997, was 'processed sewage being pumped for the first time into drinking water supplies'? (2)
iii) Name three water authorities likely to be short of water by 2021. (3)

12

(Pages 208 and 209)

a i) Which world climate on Figure 11.18 corresponds with the areas on this map that are always deficient in water? (1)
ii) Which two world climates fit most closely with the areas with a seasonal or sporadic drought? (2)

b i) Describe the location of places where fewest people have access to clean water (Figure 11.51). (3)
ii) Give three reasons why many developing countries are short of reliable, clean water. (3)
iii) Why may urban water supplies be contaminated in developing countries? (2)
iv) Why may rural water supplies be contaminated in developing countries? (2)
v) What problems result from contaminated water? (2)

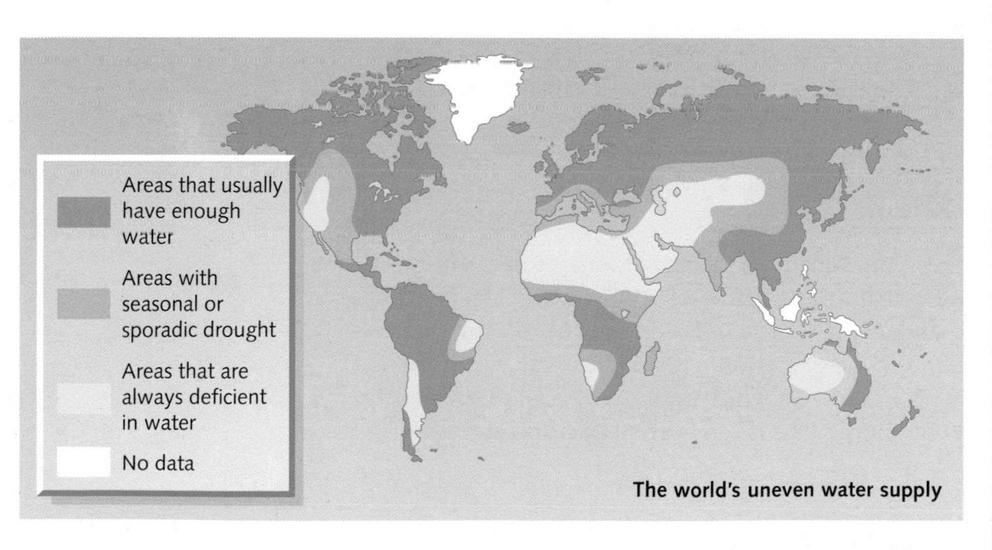

The world's uneven water supply

c i) Why is appropriate technology often more suitable than large schemes financed by overseas countries, in providing a safe, reliable supply of water to people living in developing countries? (4)
ii) Describe how appropriate technology helps local communities to provide water for their people. (3)

12 ECOSYSTEMS

Ecosystems

An *ecosystem* is a natural system in which the life cycles of plants (*flora*) and animals (*fauna*) are closely linked to each other and to the non-living environment. The *non-living environment* includes:

- water – either in the form of rain or from water stored in the soil
- air – which provides oxygen, essential for all forms of life, and carbon dioxide
- solar energy – the sun being the Earth's primary source of energy as well as providing it with heat and light
- rocks – which provide nutrients and which may be permeable (allowing water to pass through them) or impermeable (page 230)
- soils – which vary in depth, acidity (pH), nutrients and fertility.

The *living environment* includes:

- plants, animals, insects and micro-organisms (most of which live in harmony with one another)
- people (who rarely seem to live in harmony with the environment).

LEVEL	EXAMPLES	
MICRO	WATER DROPLET	UNDER A LEAF OR STONE
MESO (MIDDLE)	FRESHWATER POND	WOODLAND
	SAND-DUNES	HEDGEROWS
	SALT MARSH	WETLAND
GLOBAL (BIOME)	TROPICAL RAINFOREST	TROPICAL GRASSLAND
	HOT DESERT	TUNDRA
	CONIFEROUS FOREST	

Figure 12.1
Levels of ecosystems

Ecosystems vary in size from extensive areas of rainforest or desert (known as *biomes* – pages 218–221), to areas of woodland (page 215) and wetland (page 216), down to under a stone or within a droplet of water (Figure 12.1). Any ecosystem depends upon two basic processes: the flow of energy and the recycling of nutrients.

1 Energy flows Each ecosystem is sustained by the flow of energy through it. The main source of energy is sunlight which is absorbed by green plants and converted by the process of *photosynthesis*. Energy is then able to pass through the ecosystem in the *food chain* (Figure 12.2) in which plants are eaten by animals, and some animals consume each other. In other words, each link in the chain feeds on and obtains energy from the link preceding it. In turn it is consumed by and provides energy for the link that follows it. As energy only passes one way, the ecosystem is an *open system*, with inputs, flows, stores and outputs (page 259).

2 Recycling of nutrients Certain nutrients are continually circulated within the ecosystem and so are part of a *closed system*. Each cycle consists of plants taking up nutrients from the soil. The nutrients are then used by plants, or by animals which consume the plants. When the plants or animals die, they decompose and the nutrients are released and returned to the soil ready for future use (Figures 12.3 and 12.5).

Ecosystems can often be very fragile. Some, like the arctic tundra, tropical rainforest and areas of wetland, may take hundreds of years to develop fully. They can, however, be irretrievably damaged in a short time by human activity, e.g. Alaska (Case Study 7), the Amazon forest (page 222) and the Everglades in Florida (page 216).

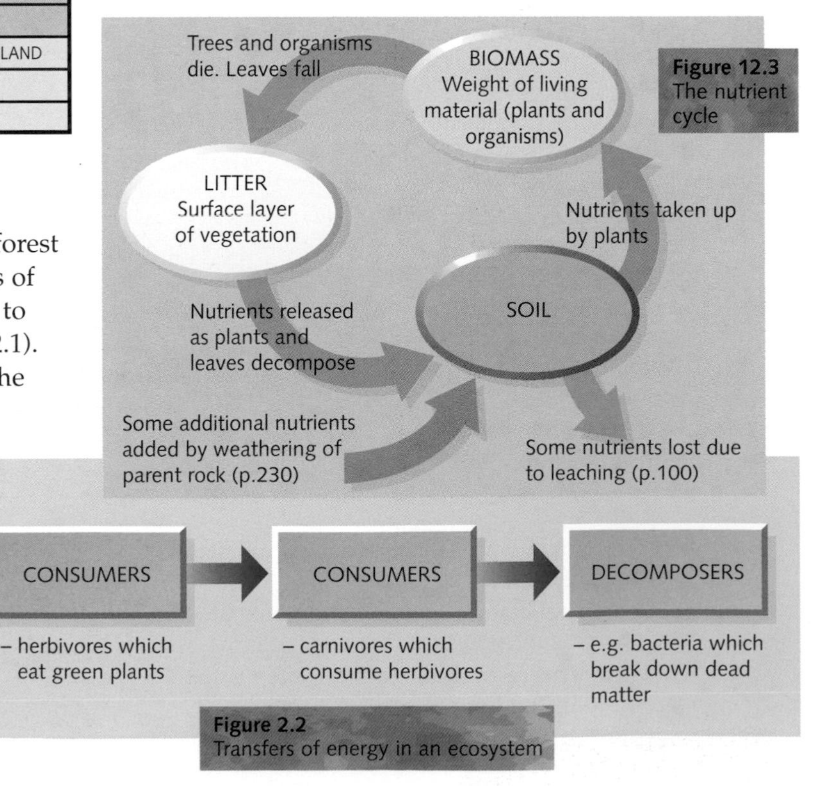

Figure 12.3
The nutrient cycle

NON-LIVING ENVIRONMENT – obtaining solar energy → PRODUCERS – green plants which convert this energy by photosynthesis → CONSUMERS – herbivores which eat green plants → CONSUMERS – carnivores which consume herbivores → DECOMPOSERS – e.g. bacteria which break down dead matter

Figure 2.2
Transfers of energy in an ecosystem

A deciduous woodland ecosystem in Britain

A natural British deciduous woodland (Figure 12.4) consists of three layers of vegetation. The tallest layer, between 30 and 40 metres in height, consists of trees such as oak, ash, birch and beech. The middle layer, ranging from 5 to 15 metres, includes smaller trees and shrubs such as holly and hawthorn. The lowest, or ground, layer may include bracken, brambles, grass and ferns. Figures 12.5, 12.6 and 12.7 show a typical woodland ecosystem.

Figure 12.5 describes the open system of the ecosystem with its inputs (e.g. solar energy, carbon dioxide), flows (e.g. trees absorbing nutrients and water from the soil through their roots), processes (e.g. photosynthesis), stores (e.g. nutrients within the soil) and outputs (e.g. evapotranspiration, oxygen).

Figure 12.6 illustrates how nutrients are recycled in a closed system.

Figure 12.7 shows how energy is transferred through the ecosystem by means of the food chain.

Figure 12.4
A deciduous woodland

This woodland ecosystem, as shown in Figure 12.5, is said to be in balance with nature. Unfortunately, human interference often alters, and even destroys, ecosystems such as this through clearing land for development, e.g. farming and village settlement in earlier times, roads and recreation in more recent times. A vital issue for present generations is: 'How can we gain most benefit from an ecosystem without damaging or destroying it?' – the concept of *sustainable development*.

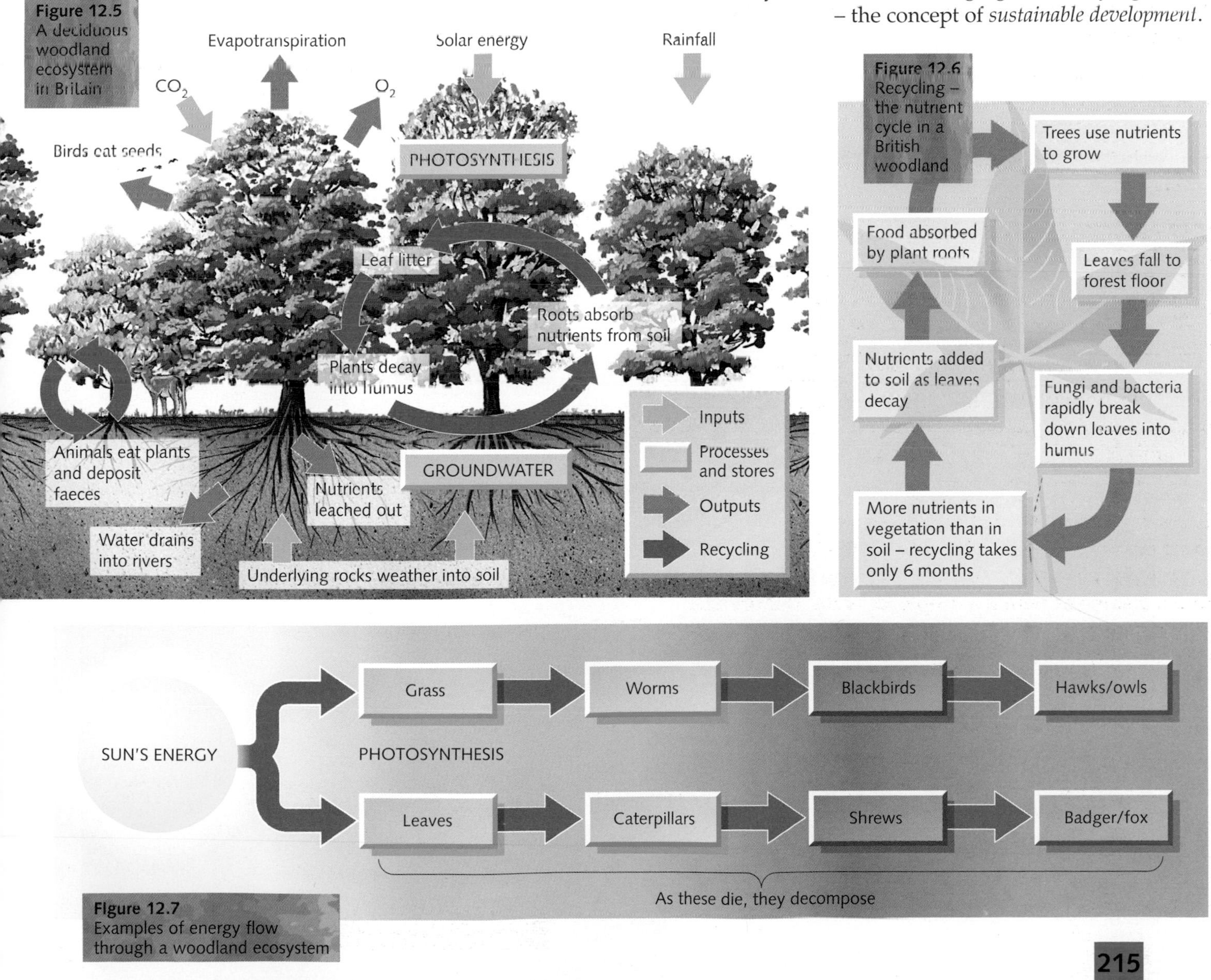

Figure 12.5
A deciduous woodland ecosystem in Britain

Figure 12.6
Recycling – the nutrient cycle in a British woodland

Figure 12.7
Examples of energy flow through a woodland ecosystem

Wetlands – an ecosystem under threat

Wetlands are transition zones between water and dry land where the soil is frequently waterlogged and the water table is at, or near, the surface.

The water may be fresh, brackish or salt, and the land may be inhospitable or highly productive. At all times, however, wetlands produce environments which are havens for special kinds of plants, birds and animals. Wetlands, which may include fen, marsh and swamp, are fragile places whose existence, especially in the last few decades, has become increasingly threatened. This threat was recognised internationally when, in 1971, representatives agreed at a convention held at Ramsar, in Iran, to:

- identify wetlands of international importance
- plan for the sustainable development of wetlands
- establish nature reserves on wetlands
- train people in wetland studies, and
- encourage research on wetland ecosystems.

At present there are 44 Ramsar wetland sites in the UK, all of which have SSSI (Site of Special Scientific Interest) status.

Places

The Everglades, Florida

The Everglades, extending southwards from Lake Okeechobee, form the southern part of the Florida peninsula (Figure 12.8). This low-lying, poorly drained area includes a variety of landscapes and ecosystems, for example:

- mangroves which grow in tidal areas along much of the coast (Figure 12.9) and which provide spawning grounds for fish, a protected hatchery for shellfish, and protect the coast from erosion
- freshwater lakes and marsh away from the coast
- large areas of sawgrass (Figure 12.10), sedges and rushes which occur in wetland areas next to the many small rivers which drain from the lake to the sea
- pine forest (mainly cypress) and tropical trees growing on low hummocks which rise above the wetlands.

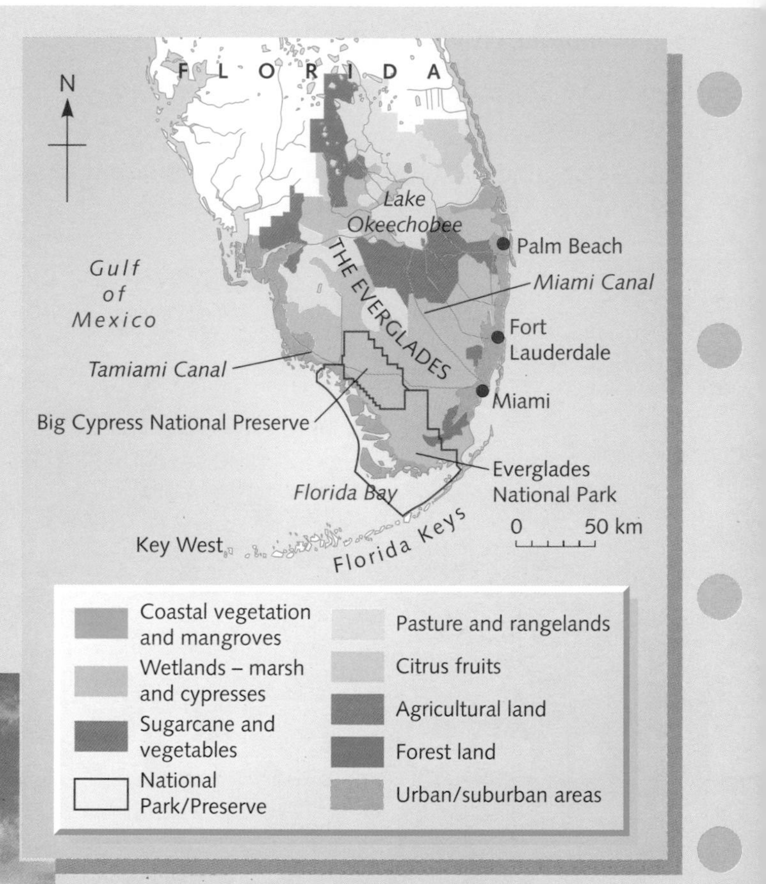

Figure 12.8 The Everglades, Florida

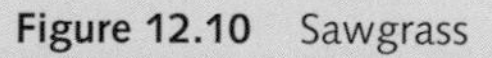

Figure 12.10 Sawgrass

Figure 12.9 Coastal mangroves

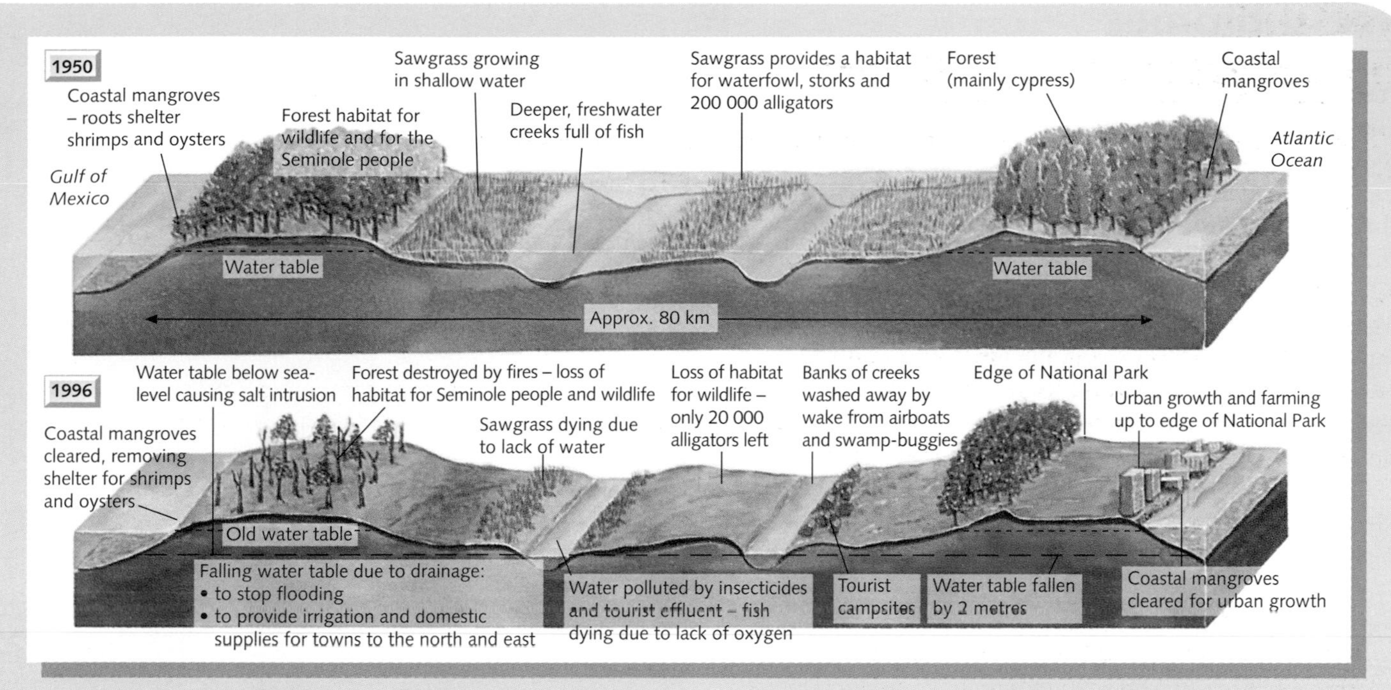

Figure 12.11 How the Everglades have changed since 1950

Figure 12.12 Magazine extract

No sweet solution for the Everglades

The Everglades, one of the largest freshwater marshes in the world, once covered 10 000 km² of southern Florida. Today, less than half remains. The rest has been drained and converted into agricultural or urban use. Now, to add insult to injury, environmentalists are claiming that what remains of these vastly depleted wetlands is slowly being poisoned by agricultural runoff from sugar plantations.

The sugar industry occupies most of the fertile Everglades Agricultural Area, a 223 000 hectare band of drained land in the north that separates the Everglades from Lake Okeechobee. Phosphates, either produced naturally or from artificial fertilisers, are making their way from the fields into the wetland system and disrupting the ecological balance.

Vast stretches of cat-tails, a water grass which thrives on phosphate fertilisation and is much taller and thicker than the indigenous sawgrass, is evidence of this pollution. This plant has encroached on nearly 10 per cent of the whole wetland system, destroying plant and animal habitats. Many animal species, including the Florida panther and the American crocodile, face extinction.

These various ecosystems form habitats for alligators, manatee, otters, ducks and numerous wading birds. Unfortunately, despite the fact that part of the Everglades is a National Park and part is a Nature Reserve, the region is increasingly under threat from human development (Figure 12.11).

- An increase in the use of fresh water for domestic and farming uses means that less reaches the sea. This has resulted in the mangrove swamps becoming too saline (salty) for shellfish which, moving up the food chain, used to provide the main diet for wading birds.
- Sawgrass provides cover for wildfowl and crocodiles. Water abstraction for urban and tourist developments, especially along the east coast, has meant a fall in the water table. As a result the sawgrass has died, and the crocodile is now threatened with extinction (alligators are less threatened because they are 'farmed' for their skin and meat).
- Farming has led to an increase in fertiliser reaching rivers, and urban development means that more effluent is poured into the rivers (Figure 12.12).
- Tourism has brought power boats onto the waterways, which is causing increased erosion of river banks together with noise and oil pollution.
- Pollution of warm springs is threatening the survival of the manatee or sea-cow (Figure 12.13) which needs this warmer water in order to survive the winter.

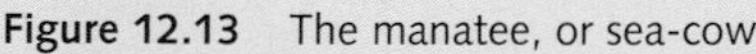

Figure 12.13 The manatee, or sea-cow

World biomes

Tropical rainforest

- Tropical rainforests grow in places that have an equatorial climate. The rainforest is the most luxuriant vegetation system in the world although its trees have had to adapt to the constant high temperatures, the heavy rainfall and the continuous growing season. Over one-third of the world's trees grow here.
- Although the trees are deciduous, the rainforest has an evergreen appearance as the continuous growing season allows trees to shed their leaves at any time.
- Vegetation grows in distinct layers (Figure 12.14). The lowest layer consists of shrubs. Above this is the under canopy, the main canopy and, rising above, the emergents, which can grow to 50 metres in height. Trees have to grow rapidly in order to reach the life-giving sunlight.
- Tree trunks are straight and, in their lower parts, branchless in their efforts to grow tall.
- Large buttress roots stand above the ground to give support to the trees (Figure 12.15).
- Lianas, which are vine-like plants, use the large trees as a support in their efforts to reach the canopy and sunlight.
- As only about 1 per cent of the incoming sunlight reaches the forest floor, there is little undergrowth. Shrubs and other plants which grow here have had to adapt to the lack of light.
- During the wetter months, large areas of land near to the main rivers are flooded.
- Leaves have drip-tips to shed the heavy rainfall.
- Fallen leaves soon decay in the hot, wet climate.
- There are over 1000 different species of tree, including such hardwoods as mahogany, rosewood and greenheart.
- There is dense undergrowth near rivers and in forest clearings where sunlight is able to penetrate the canopy.

Figure 12.14 Vegetation layers in the tropical rainforest

Height in metres
40
30
20
10
0
EMERGENTS
MAIN CANOPY
UNDER CANOPY
SHRUB LAYER
Ground level
Birds, insects, butterflies
Birds, monkeys, parrots
Insects, sloths
Lianas
Buttress roots
Alligators, insects, mosquitoes, tapirs, frogs, piranha

Despite its luxuriant appearance, the rainforest is a fragile environment whose existence relies upon the rapid and unbroken recycling of nutrients (Figure 12.16). Once the forest is cleared (deforestation page 222), then the nutrient cycle is broken. Humus is not replaced and the underlying soils soon become infertile and eroded. Not only is the rainforest unable to re-establish itself, but the land becomes too poor to be used for farming (page 100).

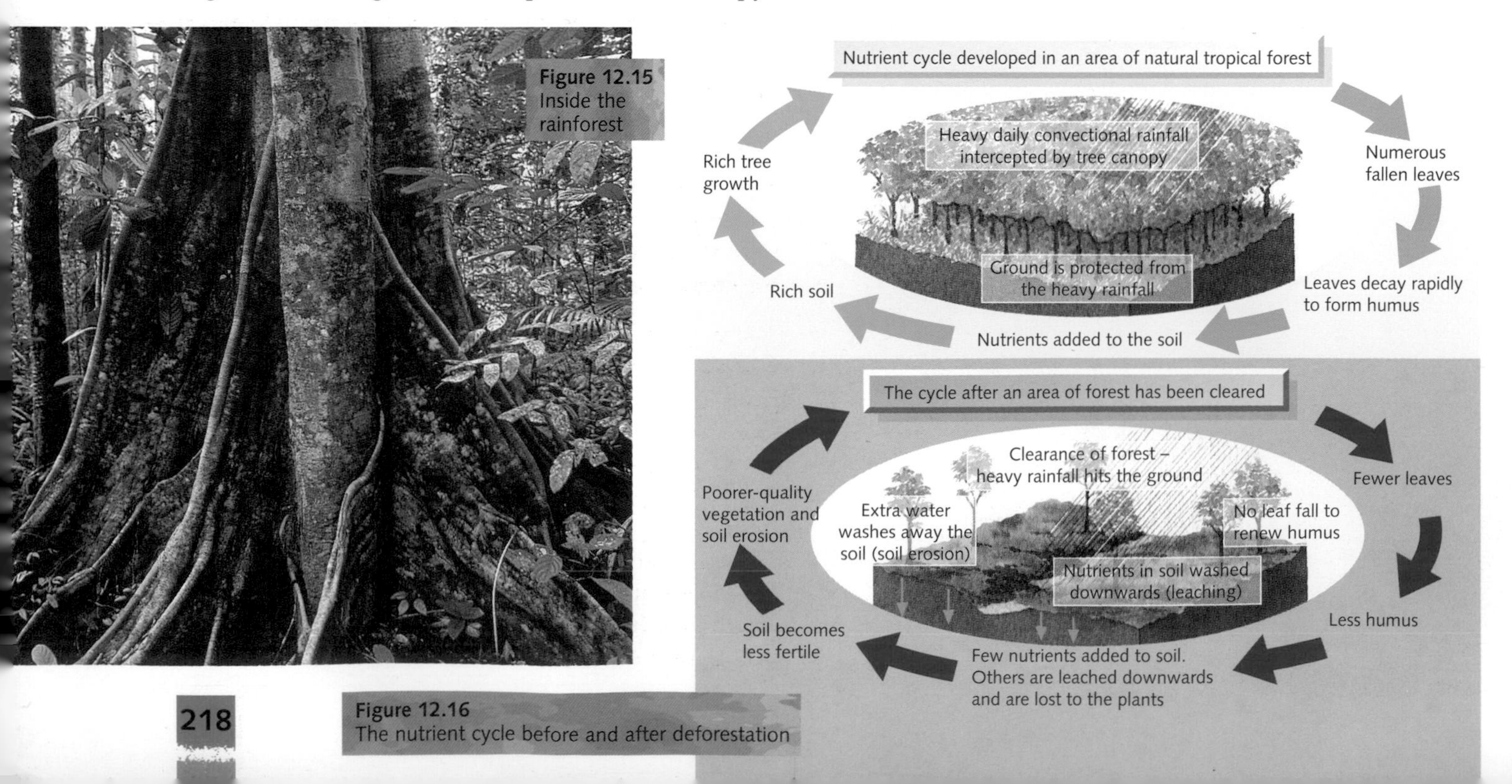

Figure 12.15 Inside the rainforest

Figure 12.16 The nutrient cycle before and after deforestation

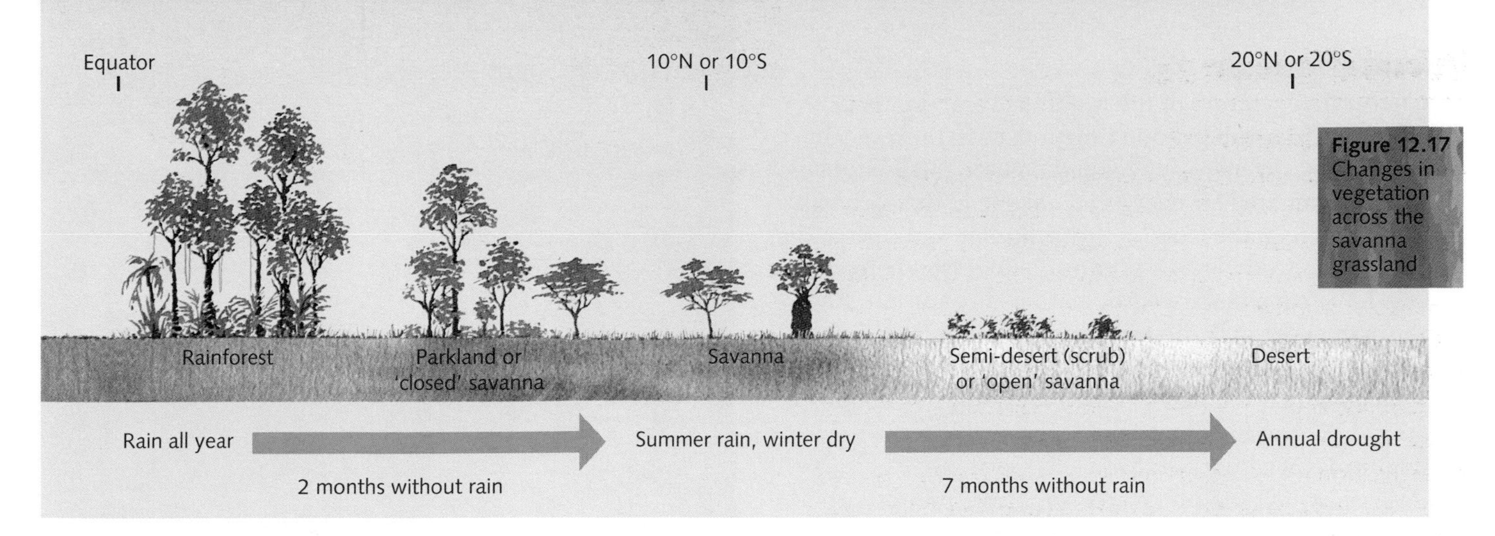

Figure 12.17 Changes in vegetation across the savanna grassland

Tropical savanna grassland

A transect (section) across the savanna grasslands shows how the natural vegetation changes in response to the climate (Figure 12.17). Where the savanna merges with the tropical rainforest (rain all year), the vegetation is dense woodland with patches of tall grass. Moving away from these margins, the vegetation slowly changes to typical savanna grasslands with scattered trees (rain for half the year), and eventually to the drought-resistant bushes and odd clumps of grass on the desert margins (hardly any rain).

Figure 12.18 A baobab tree

The dry season

The scattered deciduous trees lose their leaves, grasses turn yellow and dry up, and the ground assumes a dusty, reddish-brown colour. Some trees shed their leaves while others produce thin, waxy and even thornlike leaves to try to keep transpiration to a minimum. Most plants are *xerophytic* (drought-resistant) with very long roots to tap underground water supplies or with thick bark to store water in the trunk, like the baobab tree (Figure 12.18). Grasses grow in tufts, separated by patches of bare soil. As the dry season progresses, their stalks become stiff, yellow and straw-like and, in time, the plants wither.

The wet season

After the first rains, the grass seeds germinate and trees produce new leaves. Under the hot, wet conditions the grasses grow quickly and can reach a height of 3–4 metres before flowering and producing new seeds. The seemingly endless plains of the Serengeti (Tanzania) and Maasai Mara (Kenya) resemble a vast green sea occasionally interrupted by acacia trees (Figure 12.19). The acacias, with their crowns flattened by the trade winds, provide welcome shelter for wildlife.

The vegetation of these areas has been altered over a period of time by fire, either started deliberately or as a result of electrical storms. More recently, areas nearer the desert margins have experienced desertification (Case Study 13) mainly from pressures resulting from rapid population growth. Trees and shrubs have been removed for fuelwood. As settlements and cultivated areas increase, many nomadic herders, like the Fulani in West Africa and the Maasai in East Africa, find their traditional grazing grounds reduced in size. This leads to overgrazing and soil erosion in the areas to which they are restricted.

Figure 12.19 Savanna grassland with acacia trees

Mediterranean

The natural vegetation of the Mediterranean lands is woodland and scrub. At one time most Mediterranean hillsides were extensively wooded. Where this is still so, they are dominated by evergreen oaks (e.g. the cork oak) and conifers (e.g. the Corsican pine and, in California, the giant sequoia (redwood) – Figure 12.20). Elsewhere, where forests have been destroyed by natural fires or cut down for human needs, a scrub type of vegetation has developed. In Europe there are two major types of scrub:

- *Maquis*, which is a dense, tangled undergrowth more typical on granite and other impermeable rocks (Figure 12.21).
- *Garigue*, which is a much sparser, lower-lying scrub with many aromatic plants such as rosemary and lavender, and which develops on limestone and other permeable rocks (Figure 12.22). In California, a similar type of sagebrush scrub is called *chaparral*.

All these types of vegetation have had to adapt to the summer drought. The vegetation often has:

- either small, waxy, glossy leaves or sharp thorns in order to reduce the amount of moisture lost by transpiration (rosemary can achieve this by curling up its leaves)
- a protective bark which acts as a seal against the heat as well as against transpiration loss
- long tap roots to reach underground water supplies
- a short life cycle which avoids times of drought. Aromatic herbs and other plants germinate during winter rains, flower in spring, and lie dormant (inactive) during the summer drought.

Figure 12.20 Giant sequoias (redwoods) in California

Due to various human activities over many centuries, little of the natural Mediterranean vegetation remains, especially in Europe. Vegetation has been affected in several ways.

- *Deforestation*. Trees have been cut down either to create space for farming or settlement, to use for fuel, or for the construction of ships and buildings. Much of the natural forest was cleared long ago in the time of the Ancient Greek and Roman Empires. Once deforested, the hillsides become vulnerable to soil erosion during the heavy winter rains.
- *Grazing animals*. Herds of sheep and, especially, goats eat leaves of young trees before the plants have time to re-establish themselves.
- *Fire*. Forest fires, sometimes started deliberately, have added to the destruction of vegetation.

Figure 12.21 Maquis vegetation

Figure 12.22 Garigue vegetation

Hot desert

- Plants such as cacti have thick, waxy skins to reduce transpiration, and fleshy stems in which to store water (Figure 12.23).
- Many plants have thin, spiky or glossy leaves, also to reduce transpiration.
- Some plants have long roots to tap underground supplies of water, others have shallow roots to make full use of any rain that falls.
- Seeds can lie dormant for years. After a heavy shower they germinate quickly and the desert literally 'blooms' with flowers that can complete their life cycle within two or three weeks (Figure 12.24).

Figure 12.23 Saguaro cactus

Figure 12.24 The desert in bloom

Coniferous forest

Vegetation consists of vast stands of spruce and trees such as the Scots pine (in Britain) and Douglas fir (in North America) which have had to adapt to the severe climatic conditions and, often, to poor soils (Figure 12.25). Often, in contrast to the vegetation in tropical and deciduous forests, there may be only a single species growing in a given area. Coniferous trees are softwoods, and are valuable for timber, as well as for pulp and paper.

Figure 12.25 Adaptation of conifers to cold climates

Spruce 30 metres

Whirls

Distance between whirls indicates 1 year's growth

Evergreen – no need to renew leaves for the short growing season

Compact conical shape helps stability in the wind

Needles to reduce moisture loss

Trunk is usually straight and tall in attempt to reach the sunlight

Thick resinous bark acts as a protection against cold winds. Thin girth (trunk) due to rapid upward growth

Cones protect the seeds during very cold winters

Downward-sloping and springy branches allow snow to slide off

Very little undergrowth as trees are closely spaced and branches cut out sunlight

Covering of dead pine needles as the cold climate discourages decay

Shallow roots because either:
- soils are thin, or
- subsoil is frozen for much of the year, or
- cold boulder clay soil discourages deep root growth.

Long roots for anchorage against strong winds

The Amazon Basin

Although one-third of the world's trees still grow in the Brazilian rainforest, their numbers are being rapidly reduced due to *deforestation*.

Why is Brazil's rainforest being cleared?

Brazil's rapid population growth since the 1960s has meant that more land was needed for people to live on, more farmland to produce food for the extra numbers, more jobs required for people to earn a living and more resources needed if people's standard of living was to improve. There was also the need to reduce the country's huge national debt (Figure 10.28a). The Brazilian rainforest, at that time largely undeveloped, had the space and the resources.

Farming

Land is cleared for three types of farming.

1 **'Slash and burn'** is the traditional method used by the Amerindians of the rainforest (page 100). Although this is the most sustainable of the three types, it nevertheless causes considerable areas to be cleared, even if only temporarily, each year (Figure 12.26).
2 **Subsistence farming** has increased as a result of the government providing land to some of Brazil's 25 million landless people. In places, 10 km strips of land were cleared alongside highways, and settlers were brought in from places that were even poorer, like the drought areas in the north-east (e.g. Raimundo Jose, page 101).
3 **Commercial cattle ranching** is run by large transnational companies which sell beef mainly to fast-food chains in developed countries. These companies burn the forest, replacing trees with grass (Figure 12.27).

Transport

Over 12 000 km of new roads have been built across the rainforest, the largest being the 5300 km Trans-Amazonian highway (Figure 12.28). These roads were built to develop the region and to transport timber, minerals, farm produce and people. A 900 km railway has been built from Carajas to the coast (Figure 12.31) and numerous small airstrips have been constructed.

Figure 12.26 Destruction of the rainforest

Resources

The main types of resource are:

1 **Timber**, mainly hardwoods, is obtained by logging companies which fell trees for markets in developed countries. While timber is a valuable source of income for Brazil, little attempt has been made to replant deforested areas.
2 **Minerals** provide the region with a vast natural resource. They include iron ore (Figure 12.30), bauxite, manganese, diamonds, gold and silver.
3 **Hydro-electricity** is an important renewable source of energy but the building of dams and the creation of large lakes has led to large areas of forest being flooded.

Settlement

The development of Amazonia has led to an increase in population from 2 million (1960) to almost 30 million (1996). Large tracts of forest have been cleared for the development of such new settlements as Maraba (150 000) and Carajas (Figure 12.31).

Figure 12.27 Commercial cattle ranching in the former rainforest

Figure 12.28 The Trans-Amazonian Highway

Figure 12.29 A member of the Kayapo tribe

Rates of forest clearance

Estimates vary as to how much of the Amazon rainforest has been deforested since clearances began in the early 1960s. The World Bank suggest 15 per cent. In contrast, some environmental groups claim the figure is up to 40 per cent which, if this is accurate, would mean that some 15 hectares (about 15 football pitches) have been cleared *every minute*. In October 1997, an American environmental organisation claimed that there were 24 546 fires burning in the forest – an increase of 28 per cent on 1996.

Effects of the clearances

- Of 30 million known species on the Earth, 28 million are found in the rainforest (99 per cent being insects). A typical patch of 10 km^2 could contain as many as 1500 species of flowering plant, 750 species of tree, 400 types of bird, 150 varieties of butterfly, 100 different reptiles and 60 types of amphibian. Many others have still to be identified and studied. Deforestation has destroyed the habitats of many of these species, some of which may have proven to be of considerable value. (We already get over half of our medicines from the rainforest – one of these, a recently discovered periwinkle, has reduced deaths from child leukaemia from 80 to 20 per cent.)
- There has been a huge reduction in the number of Amerindians (from 6 million when Europeans arrived to the present number of 200 000) and destruction of their traditional culture and way of life. Those remaining, such as the Kayapo, are often forced to live on reservations (Figure 12.29).
- The clearance of trees means that there is no canopy to protect the soil from the heavy afternoon rain, or roots to bind it together. The result is less interception and infiltration and more surface runoff and soil erosion (Figure 15.4). Deforestation also breaks the humus cycle (Figure 12.16) and existing nutrients are rapidly washed (leached) out of the soil leaving it infertile (page 100). This loss in fertility has already caused some of the new subsistence farms and the larger cattle ranches to be abandoned.
- Many rivers have been polluted due to mining operations.
- Deforestation is causing climatic change in two ways. With fewer trees there is less evapotranspiration and, therefore, less water vapour in the air (about one-quarter of the world's fresh water is, at present, stored in the Amazon Basin). With less moisture in the hydrological cycle, there is already evidence of reduced rainfall totals together with the threat of a possible increase in local droughts. At the same time, the burning of the forest is accelerating global warming by releasing huge amounts of carbon dioxide, the main greenhouse gas (page 204).
- It is possible that there are already changes in the composition of the atmosphere. Scientists claim that over one-third of the world's fresh oxygen supply comes from the tropical rainforest. This would be lost if the region is totally deforested.

Figure 12.30 Carajas iron ore mine

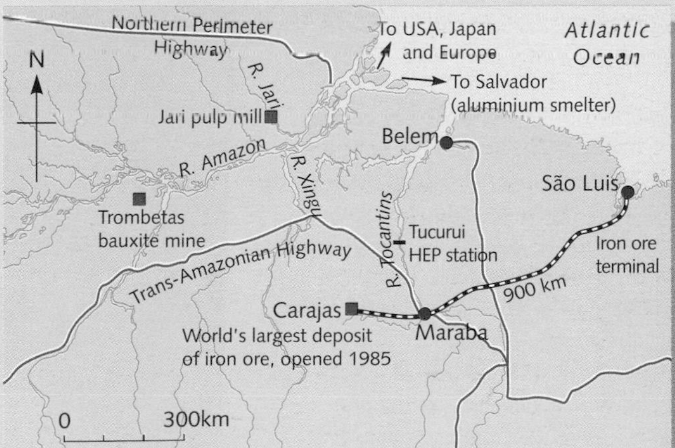

Chance discovery (1966) of one of world's largest iron ore deposits (high-quality 66% iron content). Grand Plan covers $\frac{1}{10}$ of Brazil, but only 1.6% of this area has been cleared of forest. 7000 jobs created. Mine uses HEP from Tucurui. Local steelworks at Maraba. Rest of iron ore taken by rail (each train has 200 wagons) to port of São Luis for transport to steelworks in south-east Brazil, Japan, Korea and Germany.

Figure 12.31
The Grand Carajas scheme

CASE STUDY 12

Rainforests in peninsular Malaysia

Appearance

The natural vegetation of Malaysia is tropical rainforest, which is characteristically evergreen and rich in species (Figure 12.32). Malaysia has 8000 species of flowering plants (including 1000 orchids), 3000 species of tree (of which 300 are palm trees), 500 species of fern and 60 species of bamboo. The most luxuriant rainforest is found in the highlands which form the backbone to the Malay peninsula (Figure 12.33). Here the forest has a closed canopy, at between 35 and 45 metres above the ground, consisting of the crowns of large trees closely fitted together and supported on stout, straight, pillar-like trunks (Figures 12.14 and 12.15). Occasionally a larger tree emerges above the canopy. One of these emergent species is the tualang which, as it can reach 80 metres, is the world's third tallest.

Under the canopy are smaller trees, shrubs and herbs. Smaller plants cling onto the trees. Most of these plants, especially orchids and ferns, are epiphytes. Others, including members of the mistletoe family, are parasitic. On the forest floor, fungi, aided by termites and other insects, rapidly break down any dead trees and fallen litter (leaves), in nature's recycling process. The real opportunists of the rainforest are the lianas which climb over other plants until they can bask in the full sun on the canopy. The most useful lianas belong to the palm family. Their slender, flexible and tough stems are used to make rattan furniture and Melaka (Malacca) cane. The main economic product of Malaysia's forest is timber, with the most important timber-producing family being the Dipterocarps.

Source: Adapted from the *Malaysia Official Year Book*

Extent

Approximately 71 per cent of the Malay peninsula is still covered in trees. This area can be subdivided into 59 per cent rainforest (of which only 10 per cent may be virgin forest, the rest being secondary growth) and 12 per cent of tree crops (mainly oil palm and rubber). Estimates suggest that the rainforest area could decrease by more than half (down to 25 per cent) by the year 2020 at present rates of clearance (1.2 per cent per year).

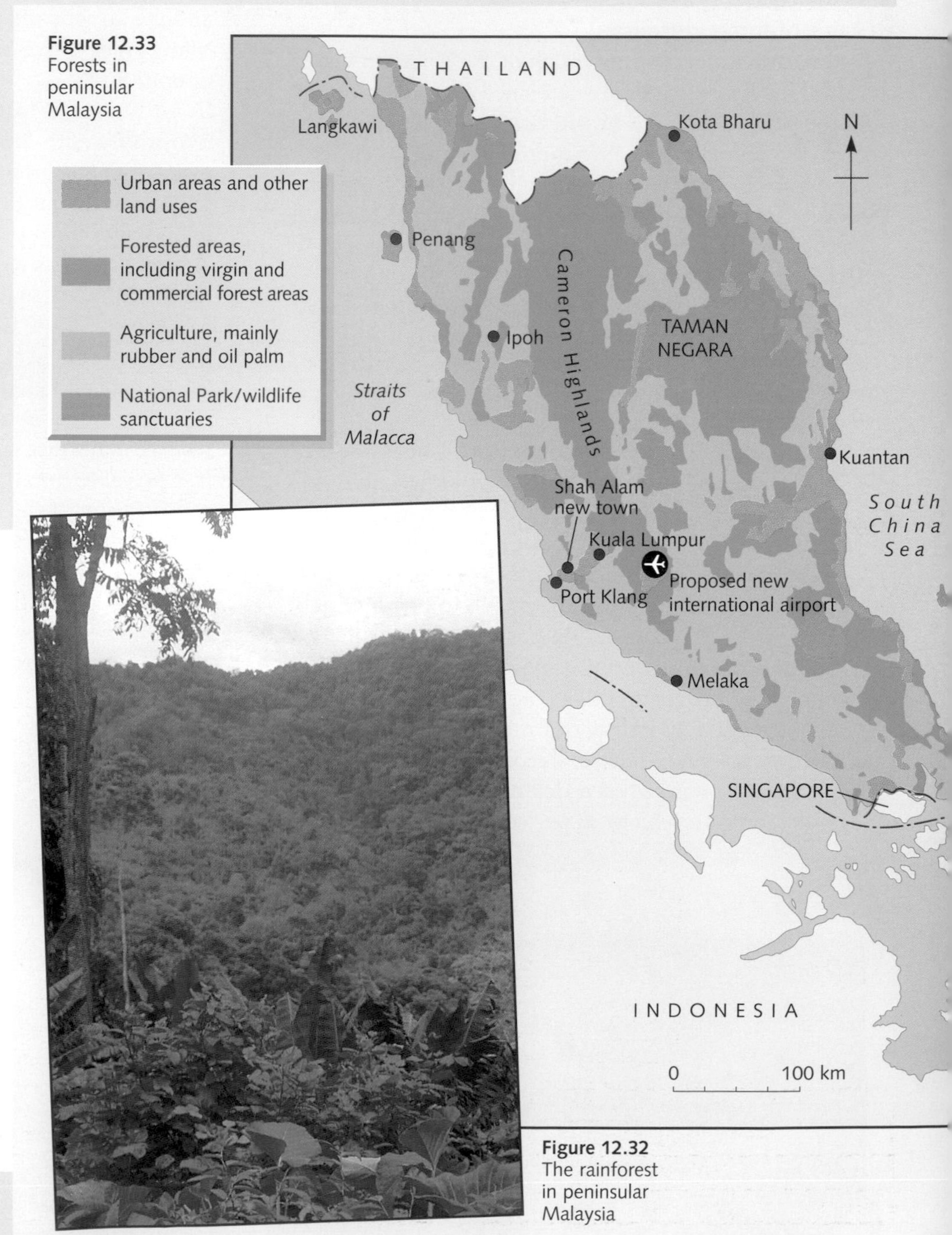

Figure 12.33 Forests in peninsular Malaysia

Figure 12.32 The rainforest in peninsular Malaysia

Conversion to tree crops

During colonial times, extensive areas were cleared of natural forest and replanted with rubber trees. Rubber was grown on large plantations owned and operated by large transnational corporations such as Dunlop and Guthries (Figure 12.34). At one time Malaya, as it was previously known, was the world's leading exporter of rubber. There have been several changes since then.

- There has been a sharp decline in the demand for natural rubber due to the use of synthetic rubber.
- The Malaysian government has taken over the largest plantations from the transnational corporations. While one-third of the former plantation areas are now state-run, the remainder have been divided into smallholdings of 5 hectares and allocated to individual farmers.
- There has been a rapid increase in world demand for palm oil. Since the 1970s, extensive areas of lowland forest have been cleared for oil palm plantations (Figure 12.35), planted under huge government rural development and resettlement schemes. Malaysia has become the world's largest exporter of palm oil (Figure 12.36).

Figure 12.34
A rubber plantation

Figure 12.35
An oil palm plantation

	THOUSAND TONNES		WORLD RANK
	1900	1993	1993
RUBBER	722	1210	3
PALM OIL	49	7400	1

Figure 12.36
Exports of rubber and palm oil

Deforestation

Before the 1970s, there was little deforestation in peninsular Malaysia. Most of the population lived in rural *kampongs* (villages) which were small and virtually self-contained. There was little urban or industrial growth, and often the only forest clearance was for the extraction of minerals, notably tin.

Since then, Malaysia has become one of the newly industrialised countries of the Pacific Rim (page 129), with a government working towards the country becoming 'fully developed' by the year 2020. This means that Malaysia needs more land for its growing population and more wealth for its continued development.

- Deforestation therefore occurs where there is rapid urbanisation (page 28) and a resultant demand for housing, jobs and services. Most deforestation has taken place near the new town of Shah Alam (where most of Malaysia's new industry is located) and Kuala Lumpur (the capital city). Land has also been cleared for motorways and for a new international airport (Subuya).
- Forest is also cut down when money is needed to repay the National Debt and to finance new housing, transport, services and industrial projects. Malaysia, like other developing countries, has had to export primary products (tin, timber, rubber and palm oil) to more economically developed countries (page 176). It is now faced with the option of either conserving its forests or using forest products to earn money (sawn wood is Malaysia's third most valuable export).

Methods of logging

Several different methods of logging have been used in Malaysia (Figure 12.37). Clearfelling was widely used before the 1980s, especially if the logging companies were non-Malaysian. This method of felling often resulted in the total destruction of the rainforest ecosystem, with the trees removed, wildlife habitats lost and the lifestyles of indigenous forest dwellers ruined – all for a rapid economic gain and to satisfy markets in developed countries (Figure 12.38). Although a clearfelled area can regenerate close to its former state in 30 years, it takes 60 years for a replacement Dipterocarp to reach the forest canopy.

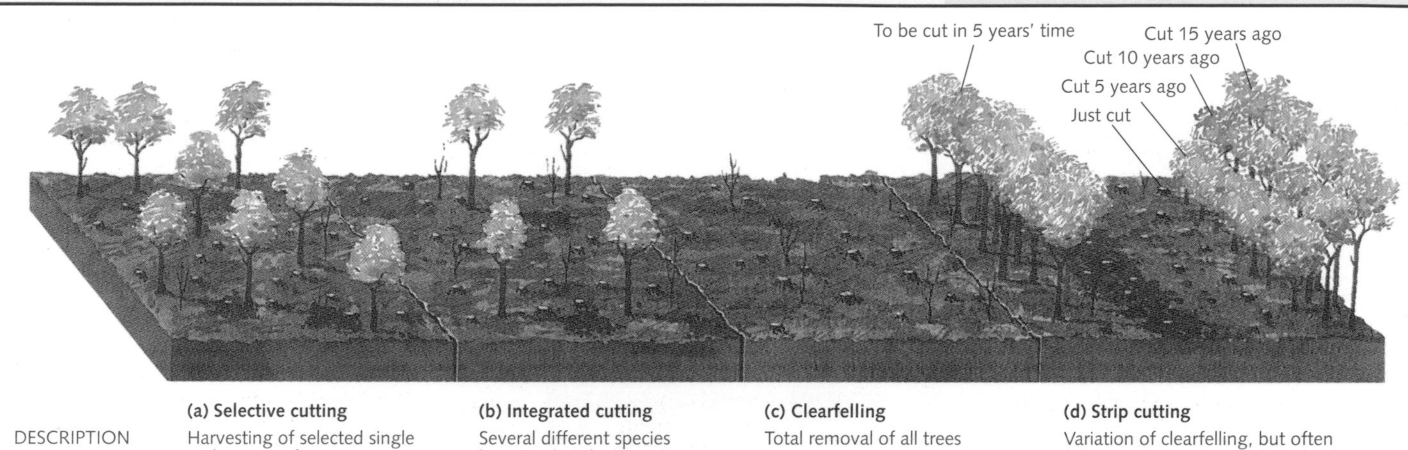

	(a) Selective cutting	(b) Integrated cutting	(c) Clearfelling	(d) Strip cutting
DESCRIPTION	Harvesting of selected single and groups of trees	Several different species harvested at the same time	Total removal of all trees	Variation of clearfelling, but often following the contours of the land
ADVANTAGES	Reduces crowding, promotes growth of younger trees, allows a more natural regeneration. More environmentally friendly	More economical than selective cutting, as trees for different uses (furniture, logs) are harvested at the same time.	Allows total afforestation, ideally with improved plants. Higher yields and profits	Clearing narrow strips avoids huge scars and allows easier regeneration. Constant income and replacement
DISADVANTAGES	Expensive. Removal of some trees by machines usually damages others, especially if Dipterocarps are being harvested	Often poorer-quality trees left. Quality may degenerate over time	Leaves ugly scars. Destroys ecosystem and habitats. Encourages soil erosion and silting of rivers (see diagram on p.229)	Minimises erosion but still destroys sections of the ecosystem

Figure 12.37 Methods of logging

Figure 12.38 Clearfelling in northern Malaysia

Towards more sustainable forestry

Although Western environmental groups remain critical of Malaysia's logging policies, there is little doubt that forest management has improved dramatically since the pre-1980s. As the Malaysian Minister for Primary Products recently stated:

'The West has already become rich by destroying its forests. They have now adopted hypocritical principles which they did not apply to themselves but which they insist on imposing upon us. Yet we are doing a better job of preserving our forests than they ever did.'

Some of the ways by which the Malaysian government and the timber industry have tried to make forestry more sustainable include the following:

- Ensuring that logging companies only use selective cutting methods, do not fell trees under a minimum circumference and do not exceed the maximum number of trees per hectare that they are allowed to cut. Companies employing clearfelling methods do not get their licence renewed.

Figure 12.39 The destructive effects of heavy machinery in the rainforest

'Haze' caused by forest fires in Indonesia is thought to have played a part in the plane crash which killed 234 people in Sumatra last week. The smog was also blamed for a collision between two ships in the Straits of Malacca in which 29 people died. The forest fires have been raging out of control through 1.5 million acres of Sumatra and Kalimantan, enveloping the entire region in dense pollution. Environmentalists are warning of a 'planetary disaster' unless the fires are extinguished.

The Week, September 1997

Why forest fires rage out of control

Even during the hottest part of the day, large areas of Indonesia are shrouded in darkness. Excessive deforestation of the country's rainforests has resulted in four massive forest fires, which are raging out of control.

In 1966, 82% of Indonesia was covered in forest. In 1982 that had shrunk to 68%. Now it's less than 55%. Forest clearings have destroyed the country's protective canopy, allowing the sun's rays to reach the ground and dry tender plants to a crisp. And the Indonesian government does not have the equipment nor the expertise to fight these fires, which are now consuming thousands of acres a day. A large part of Sumatra and Borneo is covered by thick, black smoke. Visibility is so bad that no plane has been able to land for two weeks. Worst of all, the islands are drying up. The lack of moisture in the air is affecting the weather. This is meant to be the rainy season, yet there has been no rain. The Indonesians are terrified, and rightly so. But the government has done little to address their fears because it is even more terrified of offending the ten consortia which own most of the forests.

The Sydney Morning Herald, September 1997

Figure 12.40 The 1997 rainforest fires in Indonesia

- Restricting the use of bulldozers and heavy destructive machinery (Figure 12.39) and keeping their entry tracks (known as skid trails) one kilometre apart.
- Spending money improving the Taman Negara National Park and other protected forest areas, together with increasing the number of recreational parks (there were 68 in 1993). An area amounting to 14 per cent of peninsular Malaysia's forest is protected from 'development'.
- Testing the use of helicopters – although 'helilogging' is a very expensive method of removing logs from the forest, it is far less destructive to the rainforest ecosystem.
- Implementing, through the Community Forestry Development scheme, three particular projects:
  - village forestry, which encourages the planting of traditional fruit trees to ensure a sustained food supply for local people
  - urban forestry, where trees are planted alongside main roads and in open spaces to stabilise temperatures, create shade and reduce soil erosion
  - forest recreation.

Stop press • September 1997

Malaysia was badly affected by smoke from forest fires burning out of control in nearby Indonesia (Figure 12.40). The smoke-laden skies prevented the sun breaking through, affected international and domestic air flights and greatly increased the number of respiratory diseases. At one stage the Malaysian government even considered evacuating children from the worst-affected parts of the country. It was only after a period of heavy rainfall that smoke particles settled and life returned to normal – normal at least, presumably, until the event is repeated during the 1998 Indonesian 'burning season'.

QUESTIONS

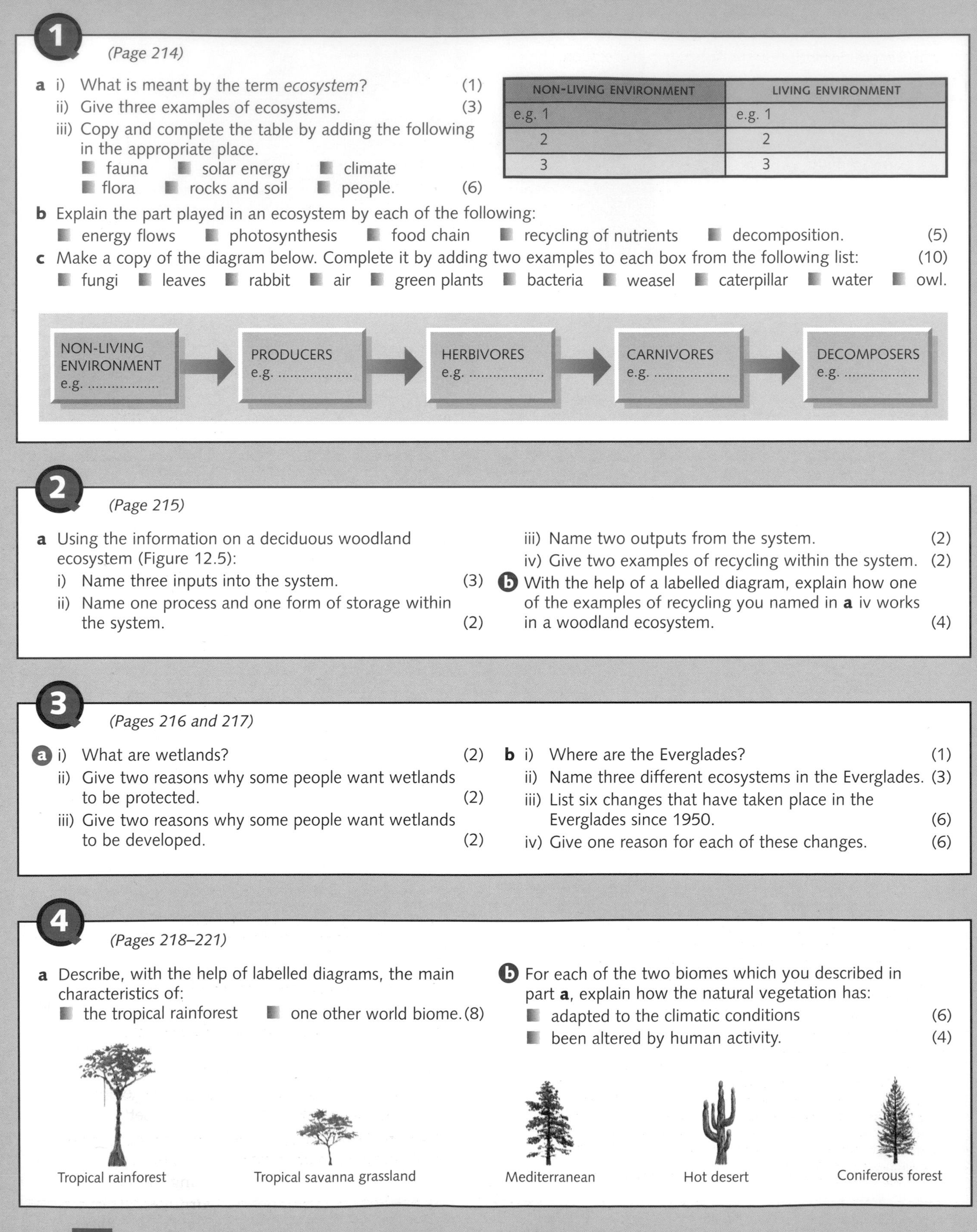

1

(Page 214)

a i) What is meant by the term *ecosystem*? (1)
ii) Give three examples of ecosystems. (3)
iii) Copy and complete the table by adding the following in the appropriate place.
- fauna
- solar energy
- climate
- flora
- rocks and soil
- people. (6)

NON-LIVING ENVIRONMENT	LIVING ENVIRONMENT
e.g. 1	e.g. 1
2	2
3	3

b Explain the part played in an ecosystem by each of the following:
- energy flows
- photosynthesis
- food chain
- recycling of nutrients
- decomposition. (5)

c Make a copy of the diagram below. Complete it by adding two examples to each box from the following list: (10)
- fungi
- leaves
- rabbit
- air
- green plants
- bacteria
- weasel
- caterpillar
- water
- owl.

2

(Page 215)

a Using the information on a deciduous woodland ecosystem (Figure 12.5):
i) Name three inputs into the system. (3)
ii) Name one process and one form of storage within the system. (2)
iii) Name two outputs from the system. (2)
iv) Give two examples of recycling within the system. (2)

b With the help of a labelled diagram, explain how one of the examples of recycling you named in **a** iv works in a woodland ecosystem. (4)

3

(Pages 216 and 217)

a i) What are wetlands? (2)
ii) Give two reasons why some people want wetlands to be protected. (2)
iii) Give two reasons why some people want wetlands to be developed. (2)

b i) Where are the Everglades? (1)
ii) Name three different ecosystems in the Everglades. (3)
iii) List six changes that have taken place in the Everglades since 1950. (6)
iv) Give one reason for each of these changes. (6)

4

(Pages 218–221)

a Describe, with the help of labelled diagrams, the main characteristics of:
- the tropical rainforest
- one other world biome. (8)

b For each of the two biomes which you described in part **a**, explain how the natural vegetation has:
- adapted to the climatic conditions (6)
- been altered by human activity. (4)

5

(Pages 218, 222 and 223)

a How does the tropical rainforest help to protect:
- the soil
- fresh water supplies
- wildlife
- the traditional Amerindian way of life? (8)

b Draw labelled diagrams to show the humus (nutrient) cycle):
- before deforestation
- after deforestation. (4)

c i) Give six reasons why parts of the Brazilian rainforest are being deforested. (6)

ii) Describe how deforestation is affecting:
- wildlife
- plant life
- the Amerindians
- the soil
- water supplies
- the local climate
- the world's climate
- oxygen supplies. (16)

d With the help of a labelled sketch map, describe the main features of the Grand Carajas Plan. (5)

e Brazilians often express different views on deforestation to those of people living in economically more developed countries. Read the extracts below, and study the diagram.

i) Why do some Brazilians feel the developed world has over-reacted? (2)

ii) Why do Brazilians resent being told by developed countries that they should protect the rainforest? (2)

iii) Can you suggest a compromise that might benefit Brazil and protect the rainforest? (2)

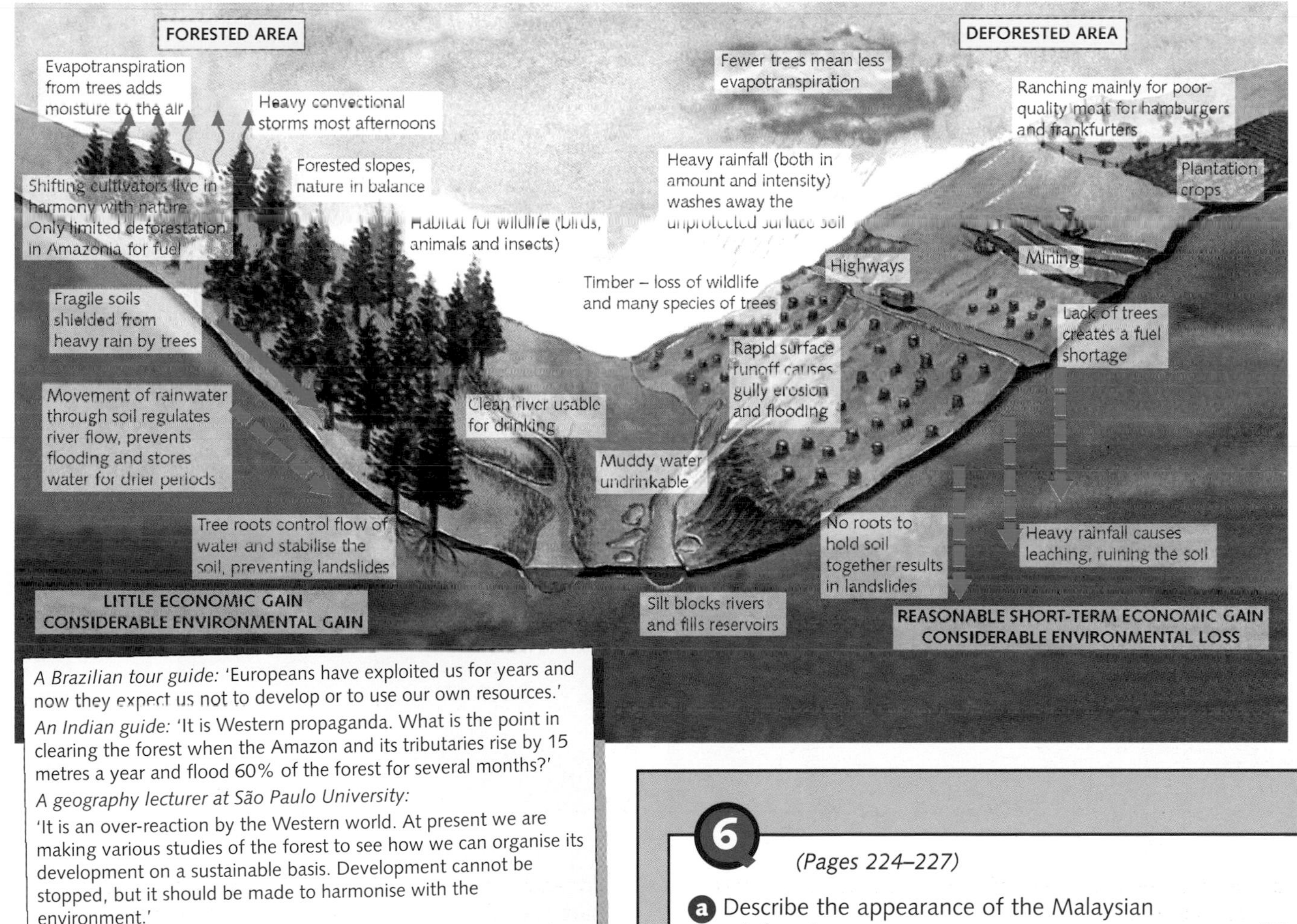

A Brazilian tour guide: 'Europeans have exploited us for years and now they expect us not to develop or to use our own resources.'

An Indian guide: 'It is Western propaganda. What is the point in clearing the forest when the Amazon and its tributaries rise by 15 metres a year and flood 60% of the forest for several months?'

A geography lecturer at São Paulo University:
'It is an over-reaction by the Western world. At present we are making various studies of the forest to see how we can organise its development on a sustainable basis. Development cannot be stopped, but it should be made to harmonise with the environment.'

A Brazilian doctor: 'Conservation groups have to exaggerate if they are to raise funds.'

These views contrast with those of conservation groups.

Friends of the Earth: 'Every minute, an area of rainforest the size of 15 football pitches is being destroyed. At the present rate of destruction, in 40 years' time there will be virtually no more rainforests. That means the Earth's climate will change. We will lose a precious source of medicine. We will be deprived of an important source of food and industrial products. And we will lose over half the world's animal and plant species.'

6

(Pages 224–227)

a Describe the appearance of the Malaysian rainforest. (5)

b Why are tree crops an acceptable alternative to the rainforest? (2)

c Why is Malaysia's rainforest under threat? (3)

d Describe the advantages and disadvantages of:
- clearfelling
- selective felling. (4)

e Describe how Malaysia is trying to develop a more sustainable form of forestry. (4)

13 ROCKS AND SOILS

Rocks – types, formation and uses

Rock types

The Earth's crust consists of many different types of rock. It is usual to group these rocks into three main types. This simple classification is based upon how each type of rock was formed (Figure 13.6).

1 **Igneous rocks** result from volcanic activity. They consist of crystals which formed as the volcanic rock cooled down, e.g. granite (Figure 13.1) and basalt.
2 **Sedimentary rocks** have been laid down in layers, e.g. sandstone (Figure 13.2), coal, limestone and chalk.
3 **Metamorphic rocks** are those that have been altered by extremes of heat and pressure, e.g. marble (Figure 13.3) and slate.

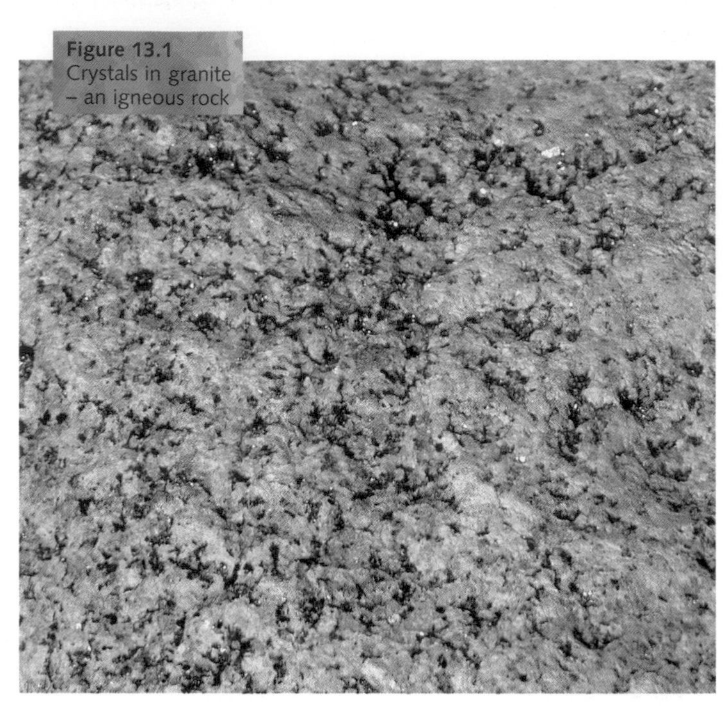

Figure 13.1 Crystals in granite – an igneous rock

Rock structure

The structure of a rock can, among other things, affect its *resistance* to erosion and its *permeability* to water.

Figure 13.2 Sandstone – a sedimentary rock

Figure 13.3 Marble – a metamorphic rock

Resistance

Rocks have different strengths and so produce different landforms. For example:

- The harder a rock is, the more resistant it is likely to be to erosion. Harder rocks are therefore usually found as hills and mountains. The softer and less compact the rock, the more likely it is to be either broken up or worn away (see 'Weathering' page 234). Valleys are formed in soft rocks.
- In a drainage basin (page 258), the more resistant the rock, the steeper the valley sides. Where resistant rock crosses a river's course, it is likely to create waterfalls and rapids (page 263).
- On coasts, resistant rocks form steep cliffs and stand out as headlands (page 287) whereas softer rocks form bays.

Figure 13.4
Hound Tor, Dartmoor

Permeability

An *impermeable* rock is one that does not let water pass through it, in contrast to a *permeable* rock which does allow water to pass through it. Permeable rocks may either:

- consist of tiny pores through which water can pass – such rocks, which include chalk, are said to be *porous*; or
- contain areas of weakness, such as bedding planes, along which water can flow. Horizontal bedding planes, which separate individual layers of rock, can be seen in Figures 13.2 (sandstone) and 13.7 (Carboniferous limestone).

Rock structure, therefore, affects the landforms of an area and can produce distinctive types of scenery, e.g. granite (Figure 13.4), chalk (Figure 13.5) and limestone (pages 232 and 233).

Figure 13.5
Devil's Dyke, a dry valley in chalk, South Downs

Use of rocks

Many rocks have an important economic value. Such rocks are extracted, usually by mining or quarrying, for specific purposes. Some examples of the economic importance of rocks are given in Figure 13.6.

Figure 13.6
Type, formation and uses of rocks

	ROCK	FORMATION	USES
IGNEOUS	GRANITE	MAGMA RISES FROM THE EARTH'S MANTLE (FIGURE 14.3) AND SLOWLY COOLS WITHIN THE EARTH'S CRUST. THE SLOW RATE OF COOLING PRODUCES LARGE CRYSTALS.	BUILDING (ABERDEEN IS KNOWN AS 'THE GRANITE CITY'). POTTERY (FROM KAOLIN, OR CHINA CLAY). SITES FOR RESERVOIRS. GROUSE MOORS (GRANITE GIVES VERY POOR SOILS). SOME TOURISM (TORS – FIGURE 13.4).
	BASALT	MAGMA REACHES THE EARTH'S SURFACE AS LAVA FROM A VOLCANO, AND COOLS VERY QUICKLY. THE RAPID COOLING CREATES SMALL CRYSTALS.	WEATHERS INTO A FERTILE SOIL (PAGE 236). FOUNDATION MATERIAL FOR ROADS. SOME TOURISM (GIANT'S CAUSEWAY AND FINGAL'S CAVE).
SEDIMENTARY	COAL	FOSSILISED REMAINS OF TREES AND PLANTS THAT GREW UNDER VERY HOT, WET CONDITIONS.	THERMAL ENERGY – POWER STATIONS, INDUSTRY, DOMESTIC USE.
	SANDSTONE	RESULTS FROM GRAINS OF SAND BEING COMPRESSED AND CEMENTED TOGETHER.	BUILDING MATERIAL – SITES FOR MANY SETTLEMENTS.
	LIMESTONE	REMAINS OF SHELLS AND SKELETONS OF SMALL MARINE ORGANISMS, e.g. CORAL, WHICH LIVED IN WARM, CLEAR SEAS.	SHEEP PASTURE (SOILS THIN AND POROUS), QUARRIED FOR CEMENT AND LIME. STONE WALLS. TOURISM (CAVES, PAGE 232).
	CHALK	A FORM OF LIMESTONE.	THIN SOILS SUITED TO CEREALS (WHEAT AND BARLEY). ALSO CEMENT AND LIME. SPRING LINE IS IDEAL FOR SETTLEMENT (FIGURE 2.2).
METAMORPHIC	MARBLE	LIMESTONE CHANGES BY HEAT AND PRESSURE.	MONUMENTS (e.g. ITALY).
	SLATE	SHALES AND CLAYS CHANGED BY PRESSURE.	BUILDING MATERIAL (ROOFS).

Limestone

Limestone consists mainly of calcium carbonate. There are several types of limestone including chalk, and Jurassic and Carboniferous limestone. Carboniferous limestone contains many fossils, including coral, indicating that it was formed on the bed of warm, clear seas.

Carboniferous limestone

Since its emergence from the sea, Carboniferous limestone has developed its own distinctive type of scenery, known as *karst*. The development of karst landforms is greatly influenced by three factors: the rock's structure, its permeability and its vulnerability to chemical weathering.

Structure Carboniferous limestone is a hard, grey sedimentary rock which was laid down in layers on the sea-bed. The horizontal junctions between the layers are called *bedding planes*. *Joints* are lines of weakness at right-angles to the bedding planes (Figure 13.7).

Permeability *Permeability* is the rate at which water can either be stored in a rock or is able to pass through it. Chalk, which consists of many pore spaces, can store water and is an example of a *porous rock*. Carboniferous limestone, which lacks pore spaces, allows water to flow along the bedding planes and down the joints, and is an example of *pervious rock*.

Vulnerability to chemical weathering Rainwater contains carbonic acid which is carbon dioxide in solution. Carbonic acid, although weak, reacts with calcium carbonate. The limestone is slowly dissolved, by chemical weathering, and is then removed in solution by running water. Chemical weathering, therefore, widens weaknesses in the rock such as bedding planes and joints.

Underground landforms

Carboniferous limestone areas are characterised by a lack of surface drainage. A river that has its source and headwaters on nearby impermeable rock will flow over the surface until it reaches an area of limestone (Figure 13.7). Various acids in the water, including carbonic acid derived from rainfall, begin to dissolve and widen surface joints to form *swallow holes*, or *sinks* (Figure 13.8). The river will, in time, disappear down one of these swallow holes. Once underground, the river will continue to widen joints and bedding planes through solution. Where solution is more active, underground caverns may form. The river will abandon these caverns as it tries to find a lower level. Should the river meet an underlying impermeable rock, it will have to flow over this rock until it reaches the surface as a *spring*, or *resurgence* (Figure 13.9).

Figure 13.7 Limestone (karst) landforms

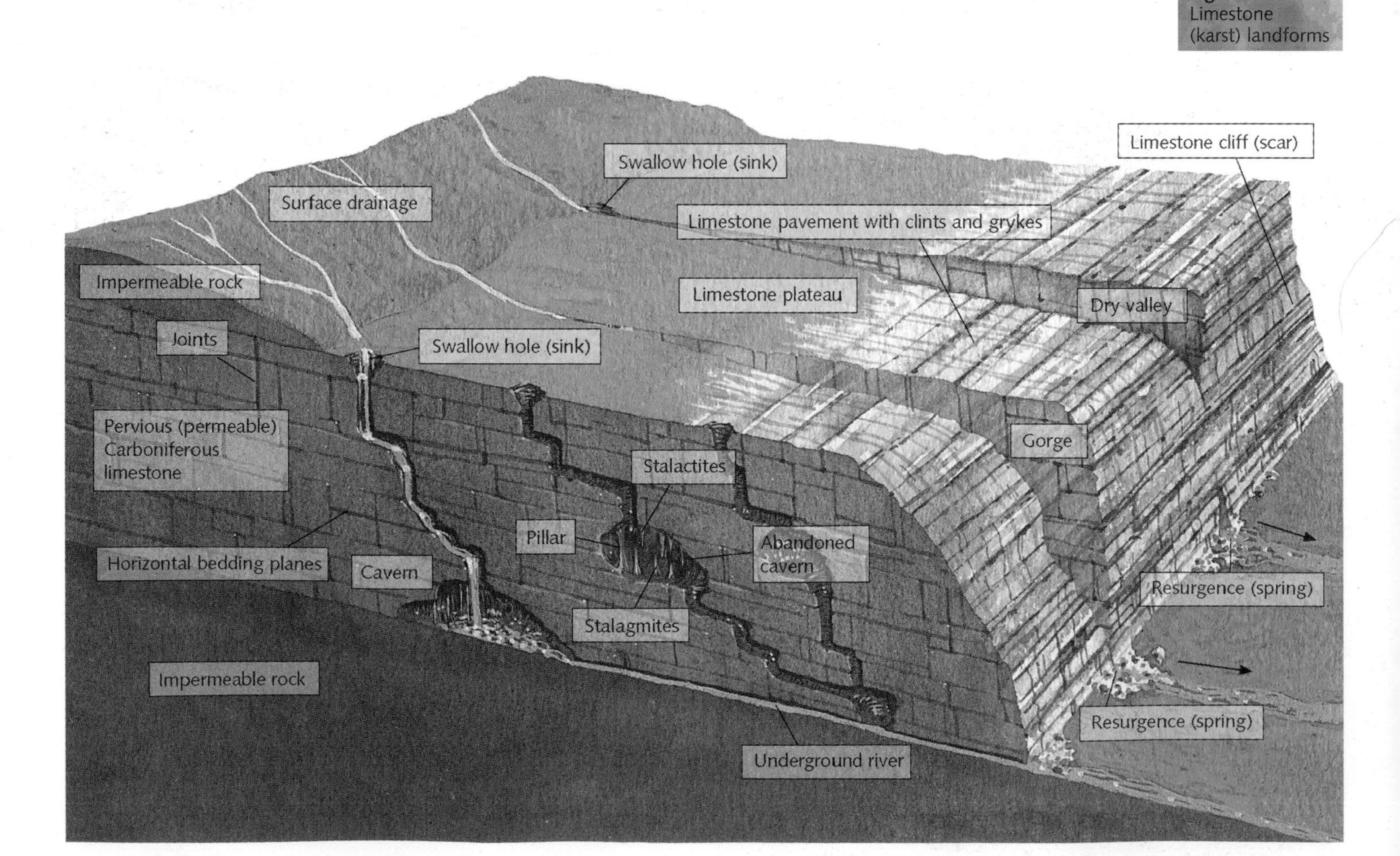

Surface landforms

Dry valleys are evidence that rivers once flowed on top of limestone (Figure 13.10). This might have occurred during the Ice Age when the ground was frozen and acted as an impermeable rock. The dry valleys are usually very steep-sided. Limestone areas often have a flat, plateau-like appearance. The flatness is due to the underlying horizontal bedding planes. Where there is no soil, the top bedding plane will be exposed as a limestone pavement (Figure 13.11). Many joints reach the surface along this pavement. They are widened and deepened by solution to form grooves known as *grykes*. The flat-topped blocks between grykes are called *clints*. Other surface landforms, often more developed in areas outside of Britain, result from limestone having collapsed. Where limestone collapses over an underground river it creates a gorge. If it collapses over a small cave then it forms a small depression called a *doline*; if over a series of caves, it produces a much larger depression known as a *polje*.

Deposition landforms

Water, containing calcium carbonate in solution, continually drips from the ceilings of underground caves. Although it is cold in these caves, some evaporation does take place allowing the formation of icicle-shaped *stalactites* (Figure 13.12). In caves in northern England stalactites only grow about 7.5 mm a year. As water drips onto the floor beneath the stalactite, further deposits of calcium carbonate produce the more rounded *stalagmites*. *Pillars* are the result of stalactites and stalagmites joining together.

Figure 13.12 Stalactites and stalagmites

Figure 13.8 Swallow hole, Malham

Figure 13.9 Resurgence, Malham Cove

Figure 13.10 Watlowes dry valley, Malham

Figure 13.11 Limestone pavement, Malham

Weathering and mass movement

Weathering

Rocks that are exposed on the Earth's surface become vulnerable to weathering. Weathering is the disintegration (breaking up) and decomposition (decay) of rocks *in situ* – that is, in their place of origin. Weathering, unlike erosion, need not involve the movement of material.

There are two main types of weathering:

1 **Physical weathering** is the disintegration of rock into smaller pieces by physical processes without any change in the chemical composition of the rock. It is most likely to occur in areas of bare rock where there is no vegetation to protect the rock from extremes of weather. Two examples of physical weathering are *freeze–thaw* and *exfoliation*, to which a third, *biological weathering* by tree roots, can be added.

2 **Chemical weathering** is the decomposition of rocks caused by a chemical change within the rock. It is more likely to occur in warm, moist climates, as these encourage chemical reactions to take place. An example of chemical weathering is *limestone solution*.

Freeze–thaw, or *frost shattering* as it is sometimes called, occurs in cold climates when temperatures are around freezing point and where exposed rock contains many cracks. Water enters the cracks during the warmer day and freezes during the colder night. As the water turns into ice it expands and exerts pressure on the surrounding rock. When temperatures rise, the ice melts and pressure is released. Repeated freezing and thawing widens the cracks and causes pieces of rock to break off. Where broken-off rock collects at the foot of a cliff it is called *scree* (Figure 13.13).

Exfoliation, or *onion weathering*, occurs in very warm climates when exposed, non-vegetated rock is repeatedly heated and cooled. The surface layers heat up and expand more rapidly during the day and cool and contract more rapidly at night than do the inner layers. This sets up stresses within the rock which cause the surface layers to peel off, like the layers of an onion, to leave rounded rocks and hills (Figures 13.14 and 8.35).

Biological weathering is when tree roots penetrate and widen bedding planes and other weaknesses in the rock until blocks of rock become detached (Figure 13.15).

Limestone solution is caused by carbonic acid (i.e. carbon dioxide in solution) which occurs naturally in rainwater. Although it is only a weak solution, it reacts chemically with rocks such as limestone which contain calcium carbonate. As the limestone slowly dissolves, it is removed in solution by running water. Solution widens bedding planes and joints to create the distinctive landforms described on pages 232 and 233.

Figure 13.13
Screes beside Moraine Lake in the Canadian Rockies

Figure 13.14
Exfoliation: Ayers Rock, Australia

Figure 13.15
Biological weathering, Cumbria

Figure 13.16
Terracettes on a hillside

Mass movement

Mass movement describes the downhill movement of weathered material under the force of gravity. The movement, which occurs on all slopes and is an almost continuous process, can cause the transport of soil, stones and rock. However, the speed of movement can vary considerably, from soil creep, where the movement is barely noticeable, to mudflows and landslides, where the movement becomes increasingly more rapid.

Soil creep is the slowest of downhill movements, occurring on very gentle and well-vegetated slopes. Although material may move by less than 1 cm a year, its results can be seen as step-like *terracettes* on banks and hillsides (Figure 13.16). Other effects of soil creep are shown on Figure 13.17.

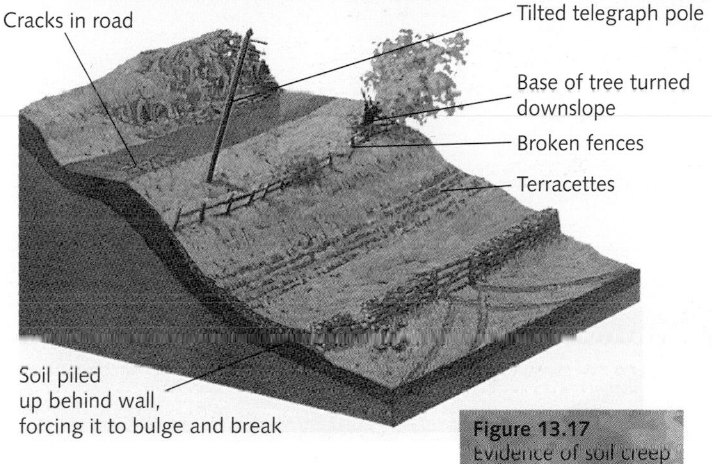

Figure 13.17
Evidence of soil creep

Mudflows and *landslides* are rapid movements which often take place with little warning. They are most likely to occur on steep slopes.

- They may occur after periods of heavy rain when loose surface material becomes saturated and the extra weight causes material to move downhill. This can occur on steep valley sides as at Aberfan in South Wales (1966), on mountainsides as in Rio de Janeiro's favelas (Figure 4.34), or along coasts, especially where soil and softer rock overlie harder rock (Figure 16.29).
- They may follow Earth movements such as earthquakes, as in Peru (1970), or after volcanic eruptions as in the case of Nevado del Ruiz in Colombia (1985) and Pinatubo in the Philippines (1991).

Armero, Colombia

The South American volcano of Nevado del Ruiz is situated in the Andes in Colombia. As it had not erupted since 1595, little notice was taken when gas and steam were emitted in 1985. A further warning went unheeded when melting snow and ice, resulting from hot magma rising towards the surface, caused a small mudflow 20 metres in height and 27 km in length. The volcano eventually erupted on 13 November. As lava, ash and hot rocks were ejected, the remaining snow melted and a tremendous volume of meltwater was released. The meltwater, swelled by torrential rain (usually associated with volcanic eruptions), rushed down the Lagunillas Valley. As it did so it became increasingly muddy as it picked up ash deposited from previous eruptions. The front wave of mud was probably 30 metres high and travelled at over 80 km per hour. Some 50 km from the summit, the mudflow (or *lahar* to use its proper name) emerged on more open ground, upon which was situated the town of Armero. It was 11 p.m. and most of the town's 22 000 inhabitants had already gone to bed when the mudflow struck. By morning, a layer of mud up to 8 metres deep covered Armero (Figure 13.18). The death toll was put at 21 000, though few bodies were ever recovered.

Figure 13.18
Mudflows near Armero, Colombia

Places

Soils

Soil forms a thin upper layer over most parts of the Earth's land surface. It provides the foundation for plant, and consequently animal, life on land.

Formation

There are two main stages in soil formation:

1 Parent, or underlying, rock is weathered to give a layer of loose, broken material known as *regolith*. Regolith may also result from the deposition of material by water, wind, ice and volcanic activity.
2 The addition of water, gases (air), living organisms (biota) and decayed organic matter (humus).

HORIZON
Surface layers — Leaf litter – decaying leaves and vegetation
A — Often dark brown/black due to high humus content. Much organic activity
Lighter colour due to downward washing (leaching) of nutrients and iron
B — Brighter colour due to inwashing (accumulation) of nutrients and iron. Less humus
C — Weathered parent material
Parent rock

Figure 13.20 Soil profile model

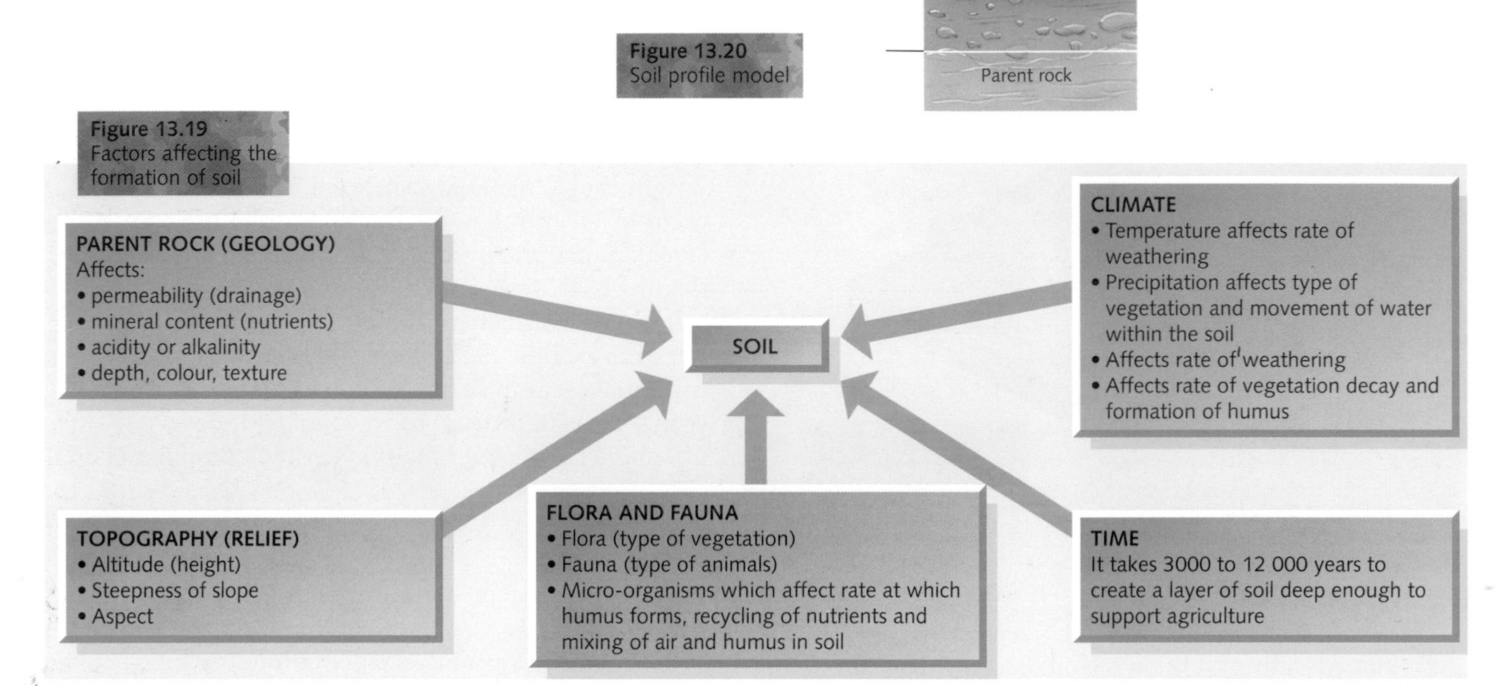

Figure 13.19 Factors affecting the formation of soil

Soil formation is also dependent upon five main factors, all of which are closely interconnected and interdependent. These factors are summarised on Figure 13.19. Increasingly, however, a sixth factor, that of human interference, is rapidly altering many of the world's natural soils, with the result that in many parts of the world soil, which is normally a sustainable resource (page 106), is being eroded and lost.

The soil profile

The soil profile is a vertical section that shows different layers, known as *horizons* (Figure 13.20). Apart from the surface layer, which consists mainly of fallen leaves and partly decayed vegetation, there are three main horizons, referred to as the A, B and C horizons. However, not all soils have the three horizons, nor the clear-cut boundaries between them that are shown on the model profile in Figure 13.20.

World soils

On a global scale, soils often develop a profile that reflects the climate and vegetation of the area in which they form (Figure 13.21). Of these, three have been taken as examples – two because they can be found in Britain, the third because it has been referred to previously in Case Studies 6 and 12.

Figure 13.21 Relationship between the world's climate, vegetation and soil types

CLIMATE (BIOME)	VEGETATION	SOIL TYPE
ARCTIC	TUNDRA	TUNDRA
COLD	CONIFEROUS FOREST	PODSOL
BRITISH TYPE	DECIDUOUS FOREST	BROWN EARTH
TEMPERATE CONTINENTAL	TEMPERATE GRASSLAND	CHERNOZEM
MEDITERRANEAN	MEDITERRANEAN	MEDITERRANEAN
DESERT	DESERT	RED-YELLOW DESERT
TROPICAL CONTINENTAL	TROPICAL GRASSLAND	FERRUGINOUS
EQUATORIAL	TROPICAL RAINFOREST	TROPICAL RED EARTH (FERRALITIC)

Brown earths develop on lowland areas in southern Britain, especially where the natural vegetation is deciduous woodland and where there is a thick undergrowth. The heavy leaf fall and rapid decay of plants in autumn gives a thick litter layer. This is rapidly broken down, as temperatures are warm enough for the presence of earthworms and other living organisms. This decay, together with an equally rapid recycling of nutrients, forms a dark brown upper horizon. As precipitation slightly exceeds evapotranspiration, there is a slow downward movement of water which causes some leaching. Leaching, together with the earthworms acting as mixing agents, means that there is no clear-cut boundary between the A and B horizons. As some minerals, including iron, are leached downwards, the lower layer tends to develop a reddish-brown tint. Both chemical and biological weathering (page 234) cause a rapid breakdown of parent rock to give a relatively deep and fertile soil.

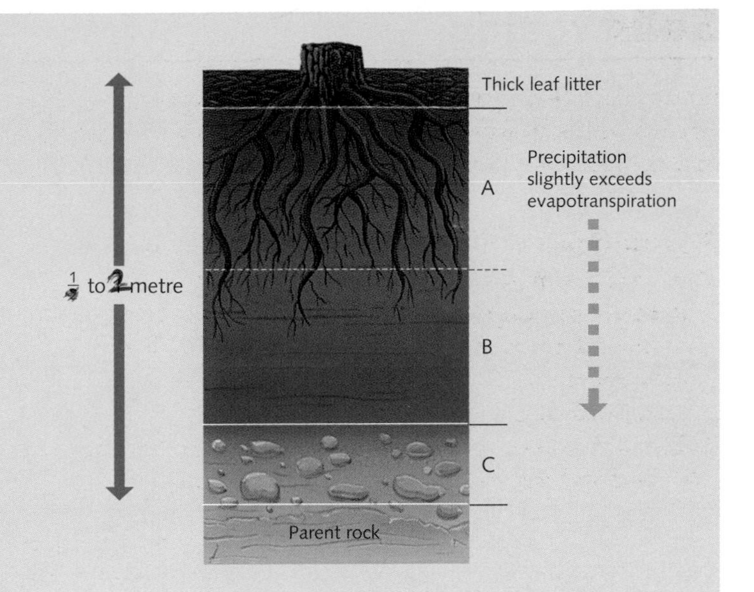

Figure 13.22
Brown earths

Podsols tend to occur in higher, wetter areas of northern and western Britain, especially where the vegetation is coniferous forest. Pine needles form a thin, acid layer and, due to lower temperatures, take a long time to decay. The cold climate also discourages earthworms and other living organisms. This creates a marked boundary between the A and B horizons and slows down the rate at which nutrients are recycled. Although precipitation is fairly light, the low temperatures limit evapotranspiration. This results in minerals, especially iron, being leached downwards to leave the A horizon an ash-grey colour. Where the iron is deposited, it forms a rust-coloured hard pan which impedes drainage. The slow rate of weathering of the parent rock gives a shallow soil which, due to its acidity and lack of humus, is usually infertile.

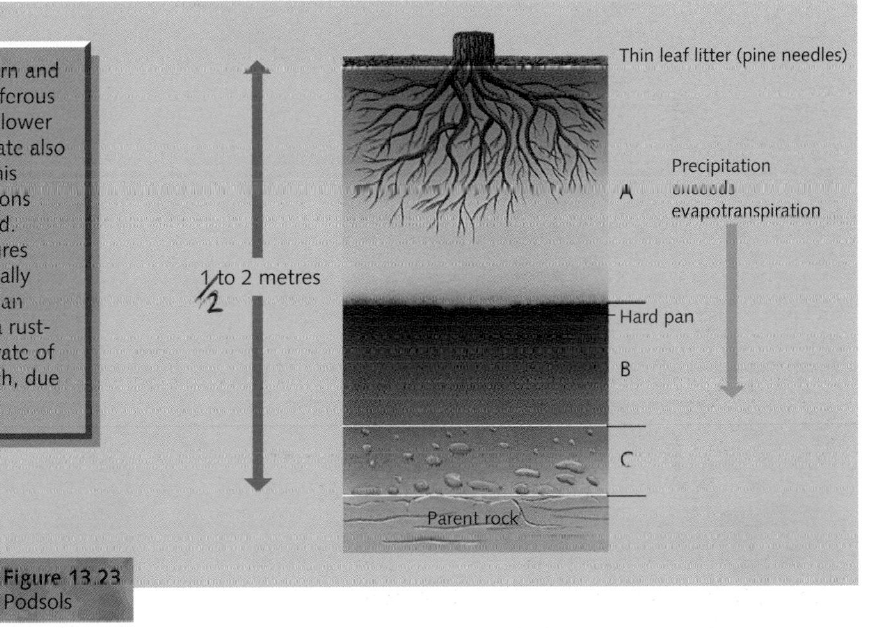

Figure 13.23
Podsols

Figure 13.24
Tropical red earths

Tropical red earths are associated with tropical rainforests. The heavy, continuous leaf fall gives a thick litter layer. This layer is rapidly broken down, partly due to the hot, wet climate and partly due to the presence of large numbers of soil organisms. As precipitation exceeds evapotranspiration, there is a steady leaching of minerals which include recycled nutrients and iron. Whereas leaching often transports nutrients to rivers (where they are lost to the system), iron is deposited in the lower layers to give the soil its characteristic red colour. Leaching and mixing agents (living organisms) also prevent the formation of well-defined horizon boundaries. The climatic conditions are ideal for chemical weathering, and the resultant rapid decomposition of the bedrock (page 234) gives a deep soil.

Under normal conditions, the tropical red earths are relatively fertile soil, but once the nutrient cycle is broken (by farming or deforestation), leaching increases and the soil rapidly loses its fertility (Case Studies 6 and 12).

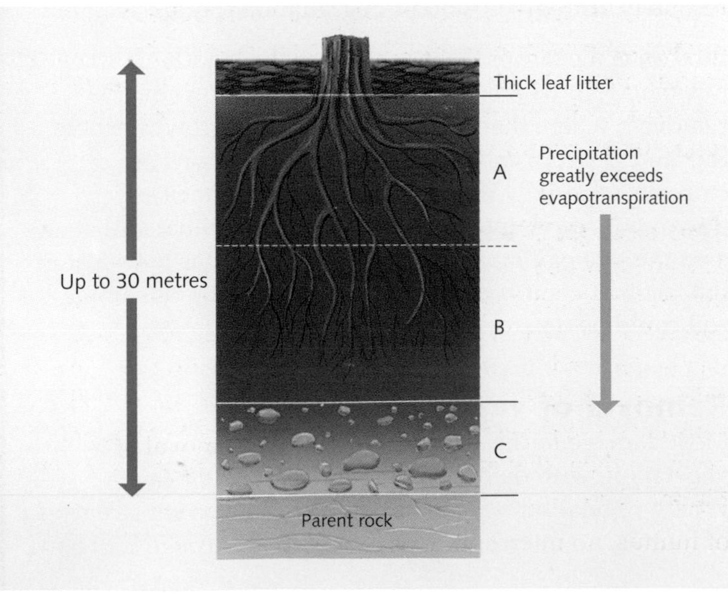

Soil erosion

Cutting down trees and removing hedges leaves soil exposed to wind and increases water erosion, which forms gullies
In tropical areas, deforestation increases leaching and surface runoff
Mining ruins large areas
Wind blows away soil on exposed areas
Overcultivation impoverishes the soil
1 million hectares of arable land lost every year in USA to highways, urbanisation and industry (land is lost at an even greater rate near fast-growing cities in Latin America)
Ploughing up and down hillsides increases surface runoff
Overgrazing exposes land to erosion by wind and water
Heavy machinery compacts the ground
Agribusiness has little regard for the soil
Irrigation without adequate drainage can cause salinity and waterlogging
Overcropping and monoculture impoverish the soil

Figure 13.25
Some causes and effects of soil erosion

It can take up to 400 years for 1 cm of soil to form, and between 3000 and 12 000 years to produce a sufficient depth for farming. At present, only 11 per cent of the Earth's land surface is classified as prime agricultural land.

This land is needed to feed an ever growing world population. So far the increase in population has been matched by increases in food production but the capacity of the soil to produce enough food is being stretched to the limit (page 98). The problem is aggravated where human development is actually ruining this essential resource, through erosion or degradation (Figure 13.25).

Erosion is most rapid in areas where the land is mismanaged; where the protective vegetation cover is removed; where there is rapid population growth; where the land is steep; and where climatic conditions are extreme, especially if rainfall is seasonal, comes as downpours or is unreliable. The UN has estimated that an area the size of China and India combined has been lost in the last half-century, and that one-third of the remaining soil could be destroyed by the year 2030.

Removal of vegetation

Most damage to the soil results from the removal of vegetation, with deforestation being the major cause. Where vegetation is removed there will be no replacement of humus, no interception of rain by plants (page 259) and no roots to bind the soil together, so the surface will be left exposed to rain and wind. If the rainfall occurs as heavy thunderstorms and in areas of steep slopes, the soil will be washed downhill and removed by rivers to leave deep, unusable gullies (Figure 13.26). If heavy rain falls on more gentle slopes or if the climate is dry and windy (Figure 13.27), loose material will be removed to leave areas of bare rock. The removal of hedgerows (page 97), the ploughing of grassland and the collection of fuelwood (page 109) all increase the risk of erosion by water and wind.

Overcultivation and overgrazing

Overcultivation occurs when crops are grown on the same piece of land year after year. *Overgrazing* is where there are too many animals for the amount of grass available. Overcultivation and overgrazing tend to occur mainly in developing countries where the increase in population means that the land is in constant use. Local farmers neither have the money to buy fertiliser for their the land, nor the time to allow a resting (fallow) period for the soil to recover naturally.

Figure 13.26
Gully erosion

Agribusiness

In many of the economically more developed countries, agriculture has become an agribusiness – that is, the land is farmed using mass production methods. In most cases, this means that the care of the soil comes a poor second to maximising output and profit (pages 84 and 85). The application of fertiliser and animal slurry may increase yields, but harms water supplies; heavy machinery reduces laborious manual tasks but compacts the soil, reducing its capacity to absorb water and air and impeding plant roots; irrigation allows new areas to be farmed, but causes salinisation; while pesticides eliminate pests, but can also kill useful insects.

The Dust Bowl, USA

The American Mid-West suffers from fluctuations in the amount of its annual precipitation. John Steinbeck's book, *The Grapes of Wrath*, gives a dramatic picture of the effect of a drought in the 1930s on the land and its people. The book begins with a description of how the soil was blown away to form a dust bowl. He describes how, by May, clouds had disappeared to allow the sun to beat down daily upon the corn (maize). By June, the vegetation had turned pale through lack of moisture, and teams of horses drawing carts caused clouds of dust. With the passing of each day this dust seemed to rise higher into the sky and to take longer to settle. The rain clouds of mid-May came and went without giving rain, and were replaced by a wind which grew daily in strength. The topsoil was picked up and carried away. The sun became an increasingly dim red ring in a darkening sky. People put handkerchiefs around their noses and mouths, and goggles over their eyes before going out, and unsuccessfully tried to stop the dust coming in through the doors and windows of their homes. Suddenly, one night, the wind dropped. The people waited for the morning.

'They knew it would take a long time for the dust to settle out of the air. In the morning the dust hung like fog, and the sun was as red as ripe new blood. All day the dust sifted down from the sky, and the next day it sifted down. An even blanket covered the earth. It settled on the corn, piled up on the tops of the fence posts, piled up on the wires; it settled on the roofs, blanketed the weeds and trees'.

John Steinbeck, *The Grapes of Wrath*

Some of the dust was carried by the wind as far as Washington DC, 2000 km to the east, while in the worst-affected states such as Oklahoma, many farms were left abandoned. Those farmers who remained were to suffer extreme poverty for many years.

During a similar drought in 1989, dust from the American Mid-West again settled on cities on the east coast of the USA.

Figure 13.27
Soil erosion caused by the wind

Figure 13.28 A derelict farm in the Dust Bowl

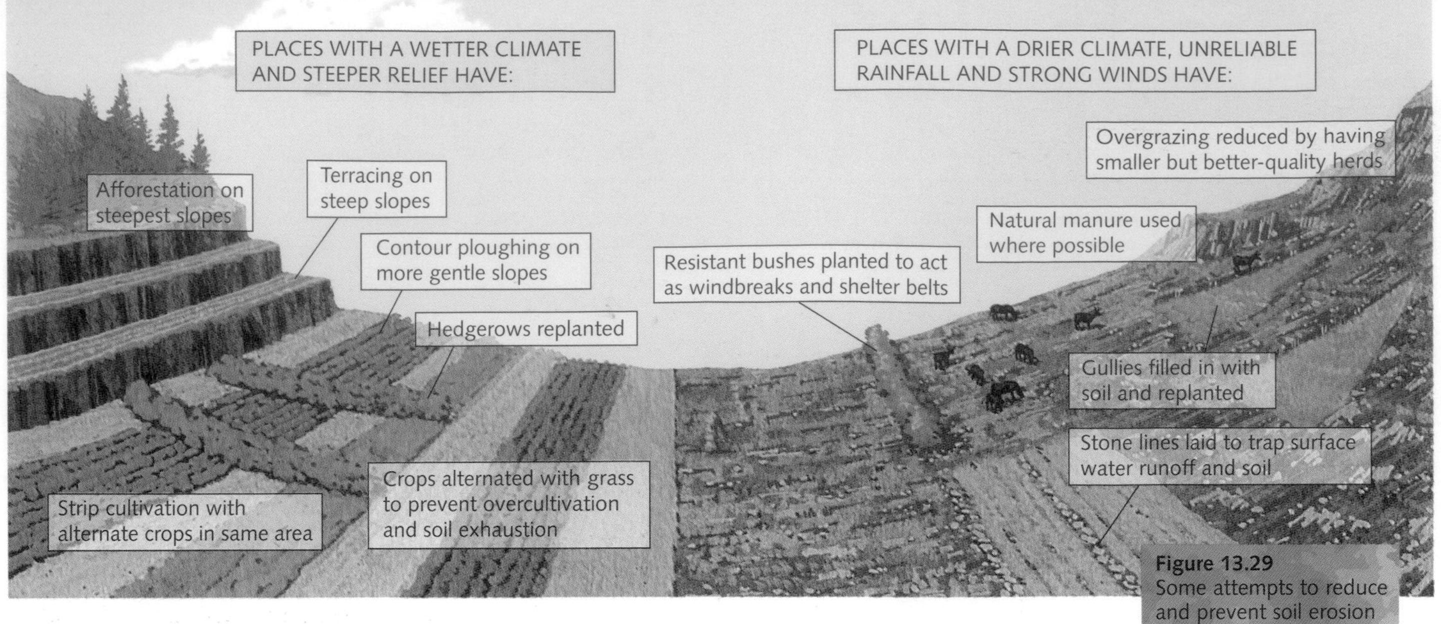

Figure 13.29 Some attempts to reduce and prevent soil erosion

Soil management

Soil is a sustainable resource, but only if it is carefully managed. Without careful management, soil, which may have taken centuries to form, can be lost within days and, under extreme conditions, sometimes even in minutes. Some suggested methods of preventing, or at least limiting, soil erosion while at the same time sustaining its productivity are shown on Figure 13.29.

Various techniques – such as terracing (Figure 13.30); the replanting of trees, grass and hedges; contour ploughing and strip cultivation (Figure 13.31) – can reduce soil erosion by over 50 per cent. Also, as well as reducing erosion, the addition of mulch and manure increases soil fertility.

The best protection against soil erosion is to prevent it from being exposed to wind and rain. Trees, bushes and grass can act as windbreaks, improve water retention and bind the soil together. *Terracing* is when artificial 'steps' are cut into steep hillsides and the front of each flat terrace is edged with mud or stone walls known as *bunds*. The bunds trap water and soil. *Contour farming* is when crops are planted around the hillside rather than up and down the slope. *Strip farming* is when two or more crops are grown in the same field. Sometimes one crop may grow under the shelter of a taller crop – a technique that has long been used in Mediterranean areas. In India, vetier grass, with its deep roots to bind the soil, is being planted in contour strips across hillsides. In drier parts of Africa, stone lines (Figure 13.39) are used to trap surface water runoff and soil. Mulching, ideally using the waste from harvested crops, keeps the ground moist by protecting it from evaporation.

Figure 13.30 Bananas being grown on terraces in Madeira

Figure 13.31 Contour farming

CASE STUDY 13

Desertification in the Sahel

There are more than a hundred definitions of the term *desertification*. Taken literally it means 'turning the land into desert'. It is a process of land degradation, mainly in arid and semi-arid lands where the rainfall is unreliable, caused by human mismanagement of a fragile environment. The causes of desertification are complex, but it is now widely accepted that it results from a combination of physical processes and human activity.

Over 900 million people are affected by desertification. They live in countries that are amongst the world's poorest and have to rely upon the sustainable use of the land for their food, income and employment. While it is the Sahel region of Africa that has received most attention to date, the problem of desertification is far more widespread, as shown in Figure 13.32. In late 1996, over 60 countries ratified the Convention to Combat Desertification (CCD) whose main objective was to 'mitigate the effects of drought and/or desertification'.

The Sahel

The effects of desertification are greatest in the Sahel. The Sahel is a narrow belt of semi-arid land which lies immediately to the south of the Sahara Desert and which extends across most of Africa. (The map in Figure 13.32 was drawn up at the 1977 Nairobi Conference on Desertification.) It has a climate that is a transition between the tropical continental (page 197) and the hot desert (page 199). Rainfall is confined to just one or two months of the year, but both the total amount of rain and the length of the 'rainy season' are unreliable. In some years the total rainfall may come in several downpours when the water is immediately lost through surface runoff. In other years the rains may fail altogether. Since the late 1960s, there have been several lengthy droughts including those in Ethiopia (1983), the Sudan (three between 1984 and 1991) and Somalia (early 1990s). Each drought means less grazing land and fewer crops in an area where several countries have some of the highest birth and population growth rates in the world. In other words, there are fewer available resources to be shared by an increasing number of people (Figure 13.33).

Figure 13.33 The effects of drought on people

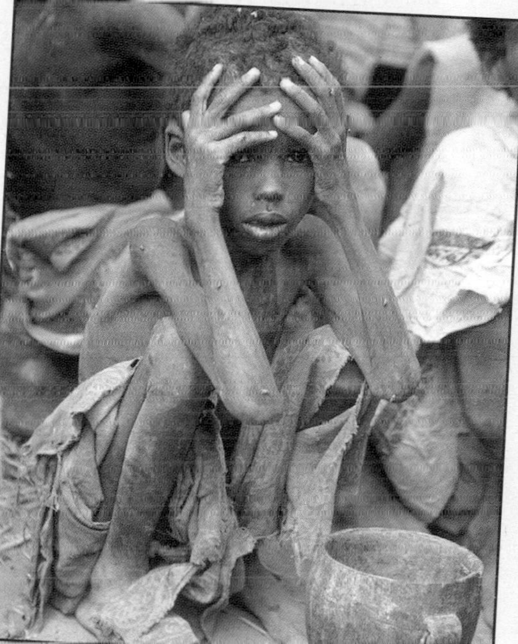

Figure 13.32 Areas at risk from desertification

Worst areas	% total population at risk
1 Ethiopia	18% at risk
2 Sudan	23% at risk
3 Chad	30% at risk
4 Niger	42% at risk
5 Somalia	26% at risk

Causes of desertification

It is suggested that desertification results from a combination of:

- climatic change – a decrease in rainfall and, possibly, the effects of global warming
- population growth – an increase in numbers of animals and people.

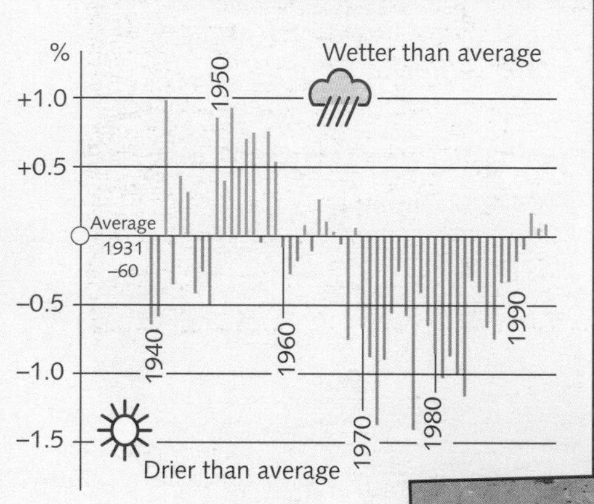

Figure 13.34 Annual rainfall for 14 weather stations in the Sahel, 1931–95

Climatic change

Even at the best of times, water supply is always a problem in these semi-arid lands. During the 1940s and early 1960s there were years when rainfall was above average and years when it was below average (Figure 13.34). After that, and for the next 20 years, rainfall totals remained well below average. The resultant drought caused seasonal rivers and water holes to dry up, the water table to fall and non-drought-resistant vegetation, notably grass, to die. More recently, global warming has been suggested as a contributory cause of desertification. This is because higher temperatures mean increased evapotranspiration (the warmer the air the more vapour it can hold), reduced condensation and, therefore, less rainfall.

Overgrazing

The basic economy of many Sahel communities is animal grazing – cattle, camels, goats and sheep. In some areas, livestock numbers increased by up to 40 per cent during the wetter years of the 1950s. When the drought began, it meant that there were too many animals for the amount of grass available and any protective grass covering was soon stripped away (Figure 13.35). Animals had to be taken to wells and water holes (the only source of available water) where the herds were too large for the amount of grass available. The result was the loss of vegetation at an increasing distance from the source of water, and the turning of the land into a desert (Figure 13.36).

Figure 13.35 Overgrazed land in Kenya

Figure 13.36 The results of overgrazing

BEFORE
Traditional farming practices: bare ground created as a result of cutting wood for fuel, and grazing around wells.

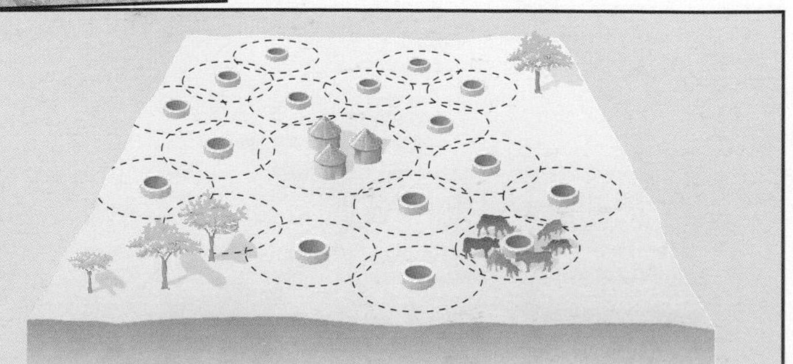

AFTER
As the population grew, more wells were bored, more trees cleared and the number of animals increased. There has been no change in the traditional herding practices, and more land has been left overgrazed and bare.

Population growth

The population of most Sahel countries, which has risen rapidly since 1950 (Figure 13.37), is predicted to increase at an annual rate of between 3.4 and 3.9 per cent until at least the turn of the century. This growth has mainly been as a result of high birth rates within each country and, in several cases, immigration resulting from civil wars in neighbouring states. As the population has grown, so too has the pressure upon the land. Farmers have been forced to grow crops on marginal land which is more vulnerable to erosion, and to grow the same crop year after year on the same piece of land which had previously been allowed 'rest', or fallow, periods. This has led to overcultivation, the exhaustion of the soil and, finally, to soil erosion. The increase in population has also led to a greater need for fuelwood for cooking. Villagers, who have to walk further each year in their search for wood, cause deforestation and accelerate desertification (Figure 7.8).

The results of desertification

Whether it is climatic change, overgrazing or population growth, in all cases the soil loses its protective vegetation cover, contains less humus and holds less moisture. As it becomes exposed to the wind and the occasional heavy downpour, it becomes increasingly at risk from erosion. These events in turn increase the vulnerability of local people to drought, food shortages and the longer-term risk to their traditional way of life.

Is the desert advancing?

In the 1970s there was concern that the Sahara Desert was advancing southwards by between 5 and 10 km a year. Attempts were made to try to halt its progress, such as planting trees and shrubs to create a 'green belt' along the southern fringe (Figure 13.38) or constructing stone lines to try to capture what moisture became available (Figure 13.39). Recent satellite photographs have shown, contrary to earlier belief, that the Sahara has not made any significant advance southwards. Instead it is now thought that the southern fringes contract and extend according to periods of rain and drought (remember that the 1970s coincided with a time of drought). While this suggests that the threat of desertification may be less than was previously feared, it does not mean that the risk has disappeared. Indeed, it should have increased our awareness that the semi-arid lands are a very fragile environment that needs careful management.

MAP NUMBER (FIGURE 13.32)		POPULATION (MILLIONS)			GROWTH PER DECADE 1950–90	ESTIMATED GROWTH FOR DECADE 1990–2000
		1950	1995	EST.2000		
1	ETHIOPIA	19.6	51.6	66.4	26%	39%
2	SUDAN	9.2	30.0	33.6	29%	35%
3	CHAD	2.7	6.3	7.3	21%	21%
4	NIGER	2.4	9.1	10.8	34%	34%
5	SOMALIA	2.4	9.2	9.7	33%	38%
	(UK	50.6	58.3	58.4	3%	4%)

Figure 13.37
Comparative population change in the Sahel area and the UK

Figure 13.38
A green belt along the southern edge of the Sahara Desert

Figure 13.39
Stone lines in Burkina Faso

QUESTIONS

1

(Pages 230 and 231)

Make a large copy of the following table and complete it by filling in the missing sections. (14)

ROCK TYPE	TWO EXAMPLES	TWO USES	HOW THE ROCK WAS FORMED
IGNEOUS	1 GRANITE 2	1 2	
	1 2	1 2	
METAMORPHIC	1 2	1 2 ROOFING MATERIAL	

2

(Pages 232–235)

a What is meant by the term *weathering*? (2)

b What is the difference between physical and chemical weathering? (2)

c For each of the following types of weathering:
- freeze–thaw
- exfoliation
- biological
- solution

i) describe the weathering process (4 x 3)

ii) link it, using the following list, to the conditions under which it is most likely to occur:
- a hot, rocky desert
- a limestone (karst) area
- a cold, mountainous area
- a woodland on a rocky slope. (4)

d i) Name an area of karst (limestone) scenery you have studied. (1)

ii) From the area you have studied, describe fully the characteristics and the formation of:
- one surface landform
- one underground landform. (6)

iii) Describe the course of a river as it crosses the area shown in Figure 13.7. (4)

3

(Pages 236 and 237)

a Briefly describe two stages in the formation of soil. (2)

b The pie graph shows the components of a typical soil.

i) What proportion is made up of mineral matter, organic material and biota? (1)

ii) Match the empty boxes numbered 1 to 4 with the following:
- sand
- earthworms
- stones
- humus. (4)

c How does each of the following affect the formation of a soil?
- Parent rock
- Climate
- Topography
- Flora and fauna (8)

d On a copy of the soil profile, clearly label the following features:
- humus
- parent rock
- leaf litter
- A horizon
- B horizon
- direction of leaching
- weathered parent rock. (7)

e Draw fully labelled diagrams to show the soil profiles of a podsol and any one other world soil type that you have studied. (8)

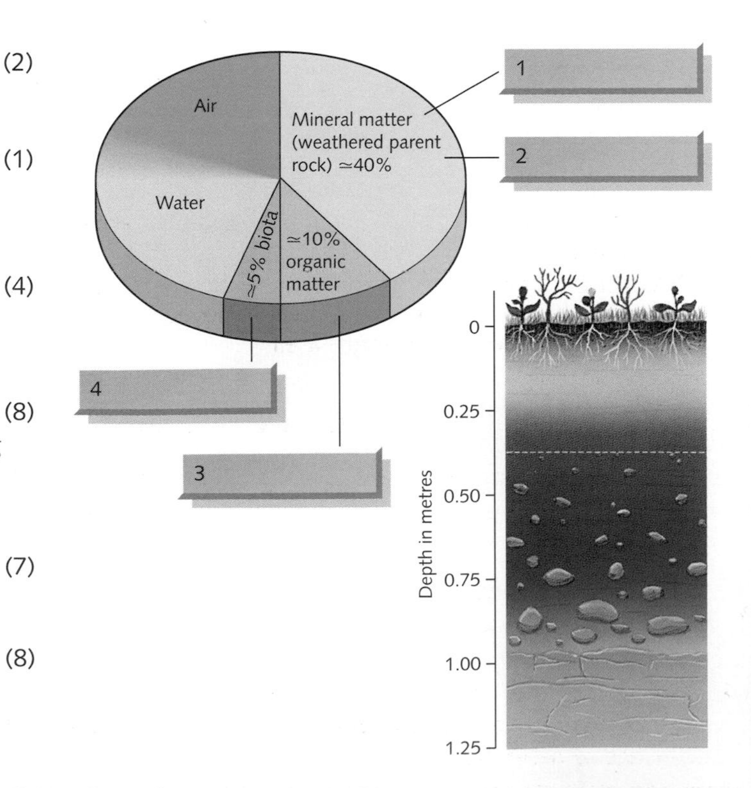

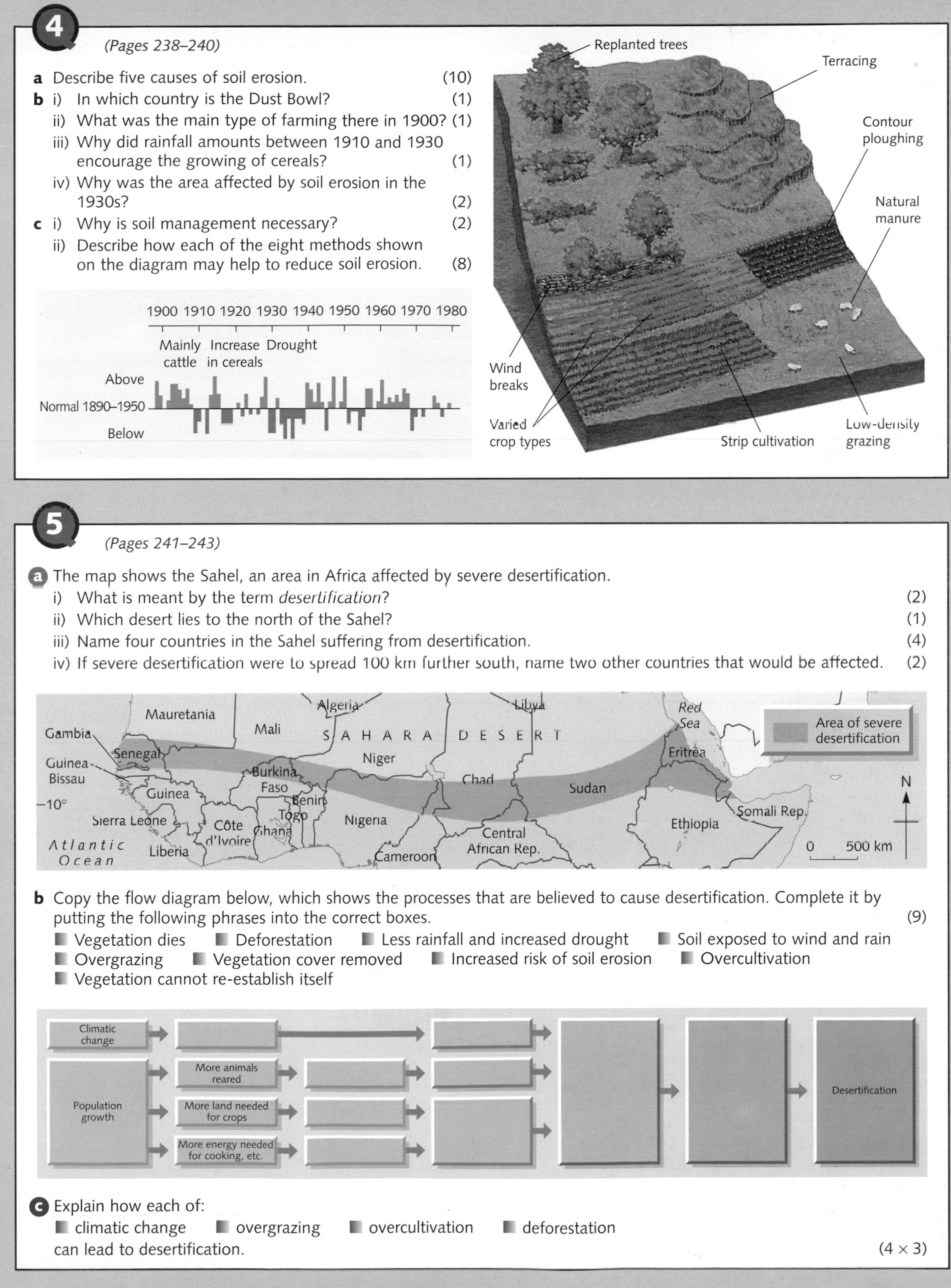

4

(Pages 238–240)

a Describe five causes of soil erosion. (10)

b i) In which country is the Dust Bowl? (1)

ii) What was the main type of farming there in 1900? (1)

iii) Why did rainfall amounts between 1910 and 1930 encourage the growing of cereals? (1)

iv) Why was the area affected by soil erosion in the 1930s? (2)

c i) Why is soil management necessary? (2)

ii) Describe how each of the eight methods shown on the diagram may help to reduce soil erosion. (8)

5

(Pages 241–243)

a The map shows the Sahel, an area in Africa affected by severe desertification.

i) What is meant by the term *desertification*? (2)

ii) Which desert lies to the north of the Sahel? (1)

iii) Name four countries in the Sahel suffering from desertification. (4)

iv) If severe desertification were to spread 100 km further south, name two other countries that would be affected. (2)

b Copy the flow diagram below, which shows the processes that are believed to cause desertification. Complete it by putting the following phrases into the correct boxes. (9)

- Vegetation dies
- Deforestation
- Less rainfall and increased drought
- Soil exposed to wind and rain
- Overgrazing
- Vegetation cover removed
- Increased risk of soil erosion
- Overcultivation
- Vegetation cannot re-establish itself

c Explain how each of:

- climatic change
- overgrazing
- overcultivation
- deforestation

can lead to desertification. (4 × 3)

14 PLATE TECTONICS

Tectonic activity

Earthquakes and volcanic eruptions are caused by movements within the Earth. While there are many thousand gentle Earth movements each year, occasionally one is sufficiently violent to cause severe damage to property, to disrupt human activity, and to result in loss of life. Earth movements cannot, as yet, be predicted. Scientists do, however, know which parts of the world are most likely to be affected by these movements, even if they cannot say when, or with what severity, they will occur.

Earthquakes

Figure 14.1 identifies the location of places where earthquake activity is most frequent. It also locates some of the more recent major earthquakes. The map clearly shows that there is a well-defined distribution pattern, with most earthquakes occurring in long, narrow belts. These belts include those that:

- encircle the whole of the Pacific Ocean
- extend down the entire length of the mid-Atlantic Ocean
- stretch across southern Europe and Asia, linking the Atlantic and the Pacific Oceans.

Figure 14.1
Earthquake activity

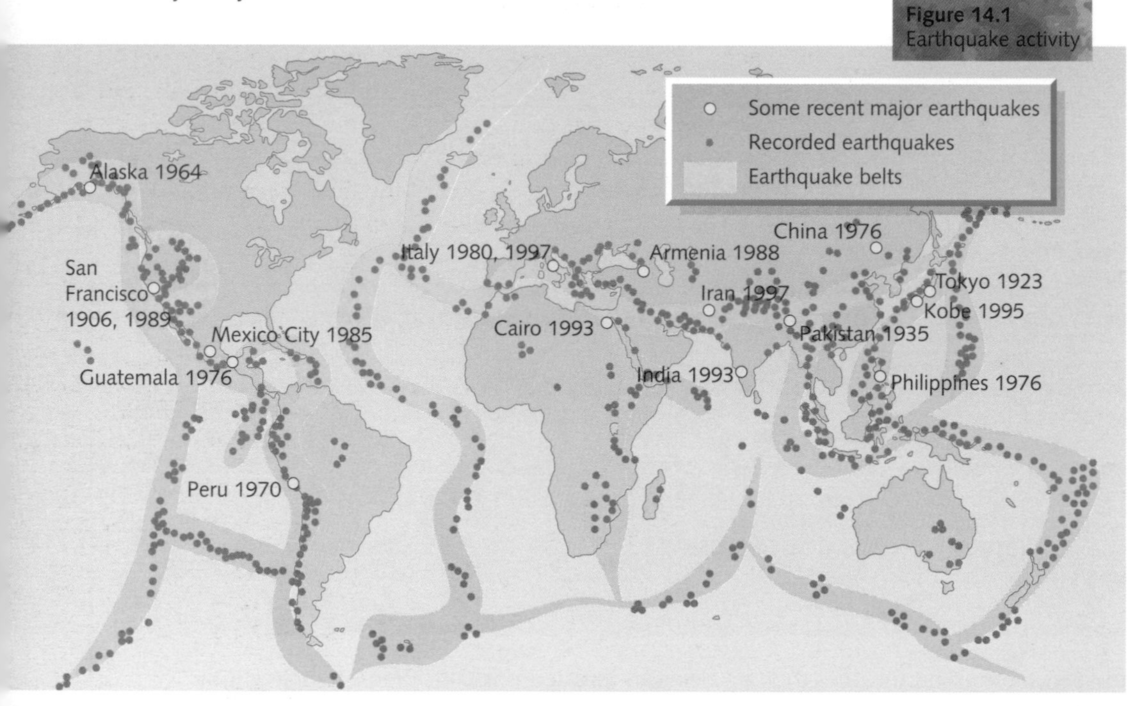

Volcanoes

The location of the world's major centres of volcanic activity, including some of the most recent major volcanic eruptions, is shown on Figure 14.2. This map shows that volcanoes also occur in long, narrow belts. These belts include:

- the so-called 'Pacific Ring of Fire' which encircles the whole of the Pacific Ocean
- the one that extends down the entire length of the mid-Atlantic Ocean
- smaller areas in southern Europe, the Caribbean, east Africa and the mid-Pacific Ocean.

Figure 14.2
Volcanic activity

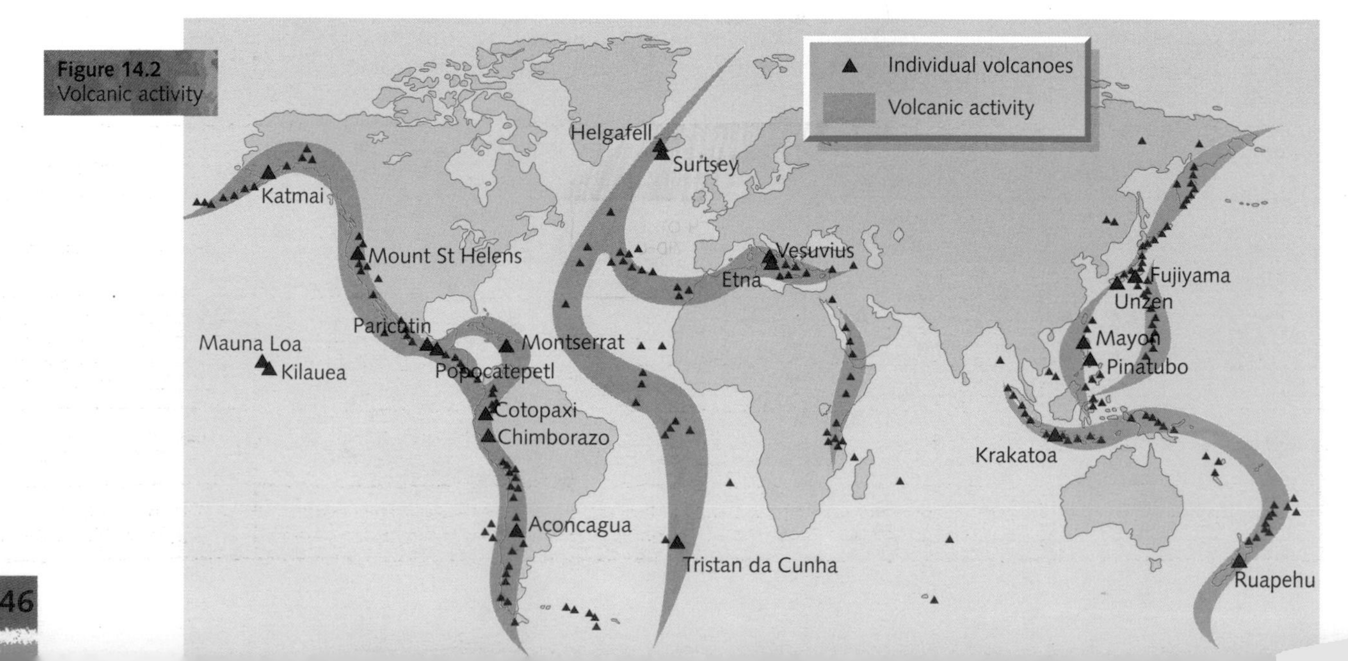

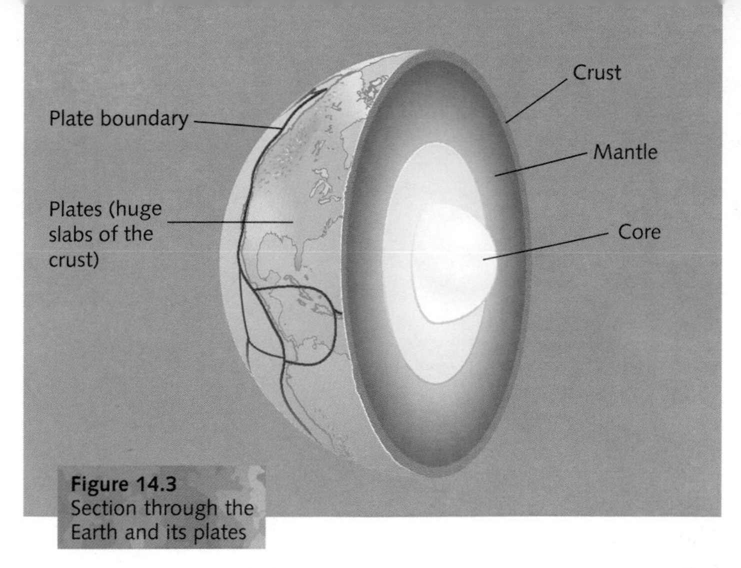

Figure 14.3 Section through the Earth and its plates

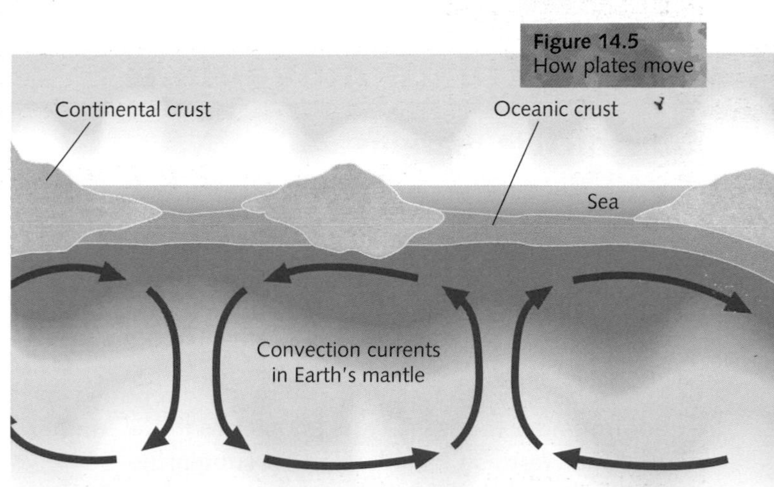

Figure 14.5 How plates move

Figures 14.1 and 14.2 show that earthquakes and volcanic activity often occur at similar places on the Earth's crust, and in narrow *zones of activity*. If the Earth were the size of an apple, its *crust* would be no thicker than the apple's skin. Underneath the crust is the *mantle* (Figure 14.3) where temperatures are so high that rock exists in a semi-molten state. The crust is broken into several large, and other smaller, segments known as *plates* which float, like rafts, on the mantle (Figure 14.4). Heat from within the Earth creates convection currents which cause the plates to move, perhaps by a few centimetres a year (Figure 14.5). Plates may either move away from, towards, or sideways past, neighbouring plates. Plates meet at *plate boundaries* and it is at these boundaries that most of the world's earthquakes and volcanic eruptions occur, and where high mountain ranges are located (compare Figures 14.1 and 14.2). Very little activity takes place in the rigid centre of plates.

Plates consist of two types of crust: continental and oceanic. Continental crust is older, lighter, cannot sink and is permanent. Oceanic crust is younger, heavier, can sink and is constantly being destroyed and replaced. It is these differences in crust that account for the variation in processes and landforms and the level of activity at plate boundaries (Figure 14.6).

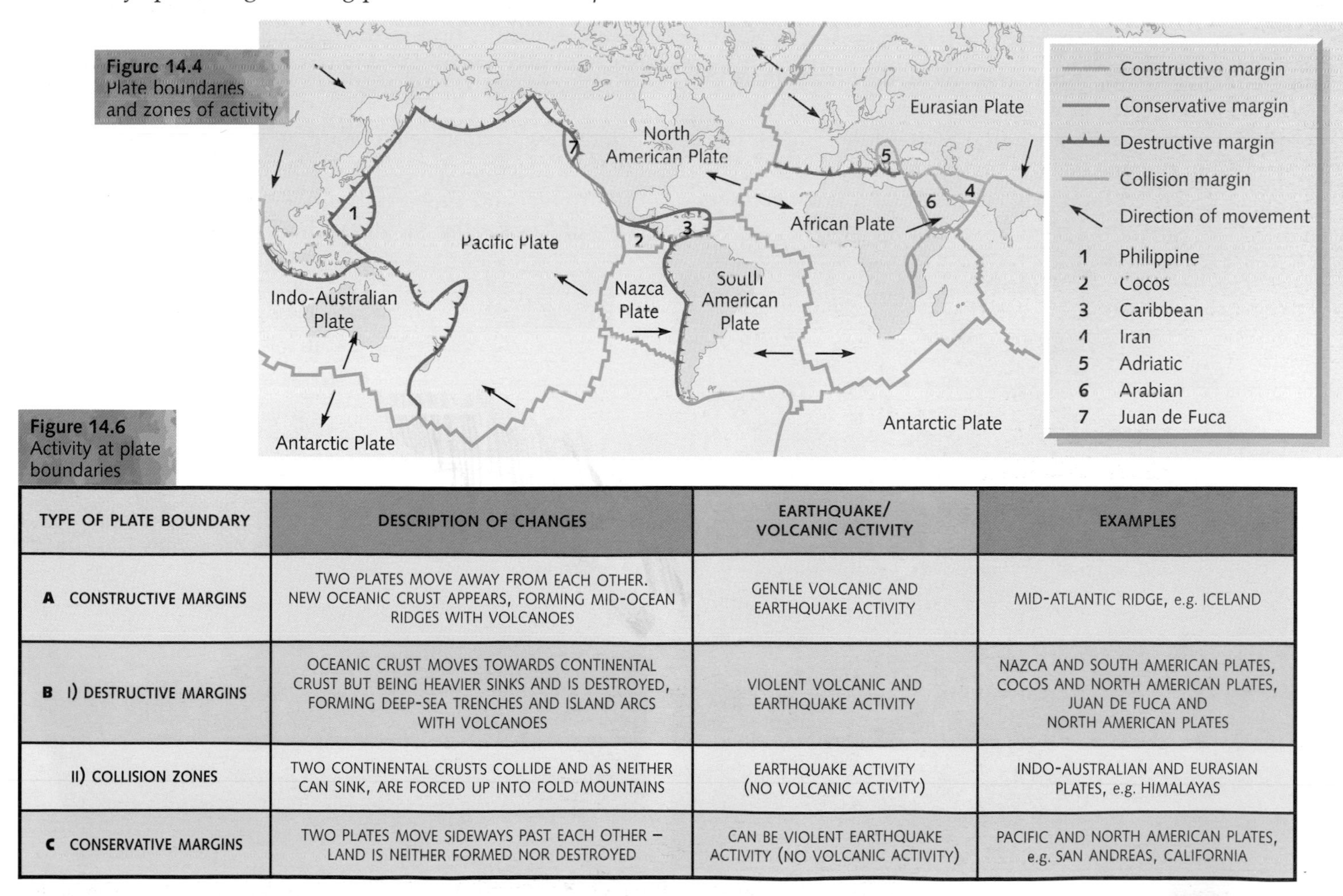

Figure 14.4 Plate boundaries and zones of activity

Figure 14.6 Activity at plate boundaries

TYPE OF PLATE BOUNDARY	DESCRIPTION OF CHANGES	EARTHQUAKE/ VOLCANIC ACTIVITY	EXAMPLES
A CONSTRUCTIVE MARGINS	TWO PLATES MOVE AWAY FROM EACH OTHER. NEW OCEANIC CRUST APPEARS, FORMING MID-OCEAN RIDGES WITH VOLCANOES	GENTLE VOLCANIC AND EARTHQUAKE ACTIVITY	MID-ATLANTIC RIDGE, e.g. ICELAND
B I) DESTRUCTIVE MARGINS	OCEANIC CRUST MOVES TOWARDS CONTINENTAL CRUST BUT BEING HEAVIER SINKS AND IS DESTROYED, FORMING DEEP-SEA TRENCHES AND ISLAND ARCS WITH VOLCANOES	VIOLENT VOLCANIC AND EARTHQUAKE ACTIVITY	NAZCA AND SOUTH AMERICAN PLATES, COCOS AND NORTH AMERICAN PLATES, JUAN DE FUCA AND NORTH AMERICAN PLATES
II) COLLISION ZONES	TWO CONTINENTAL CRUSTS COLLIDE AND AS NEITHER CAN SINK, ARE FORCED UP INTO FOLD MOUNTAINS	EARTHQUAKE ACTIVITY (NO VOLCANIC ACTIVITY)	INDO-AUSTRALIAN AND EURASIAN PLATES, e.g. HIMALAYAS
C CONSERVATIVE MARGINS	TWO PLATES MOVE SIDEWAYS PAST EACH OTHER – LAND IS NEITHER FORMED NOR DESTROYED	CAN BE VIOLENT EARTHQUAKE ACTIVITY (NO VOLCANIC ACTIVITY)	PACIFIC AND NORTH AMERICAN PLATES, e.g. SAN ANDREAS, CALIFORNIA

Types of plate movement

Constructive margins

At constructive margins, such as the Mid-Atlantic Ridge (Figure 14.7), two plates move away from each other. Molten rock, or magma, immediately rises to fill any possible gap and forms new oceanic crust. The Atlantic Ocean is widening by about 3 cm a year which means that the Americas are moving away from Eurasia and Africa.

In November 1963 an Icelandic fishing crew reported an explosion under the sea several kilometres to the south-west of the Westman Islands. Further submarine explosions, accompanied by ejections of steam and ash, gave birth to the island of Surtsey. The permanence of the new island was guaranteed when, in April 1964, lava began to flow from the central vent. When the eruptions ended, in 1967, the island measured 2.8 km^2 and reached 178 metres above sea-level. Within a month plants and insects had begun colonisation.

Later eruptions elsewhere in Iceland caused the temporary evacuation of Heimaey in 1973 and the melting of part of the Vatnajökull ice cap in 1996.

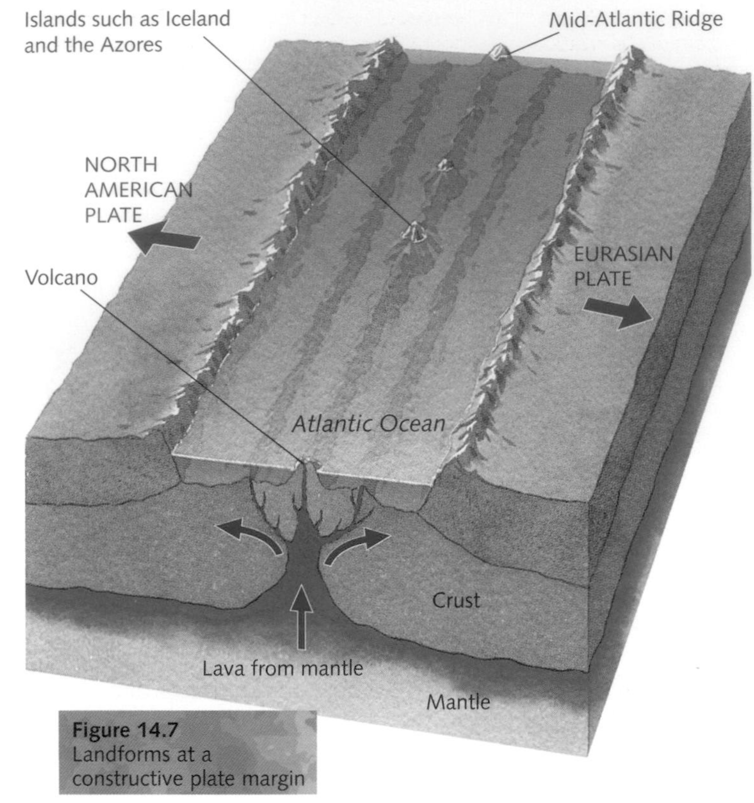

Figure 14.7
Landforms at a constructive plate margin

Destructive margins

Destructive margins occur where plates consisting of oceanic crust move towards plates of continental crust. To the west of South America (Figure 14.8), the Nazca Plate (oceanic crust) is moving towards the American Plate (continental crust). Where they meet, the Nazca Plate is forced downwards to form a *subduction zone* and an associated deep-sea trench (the Peru–Chile trench). The increase in pressure, as the plate is forced downwards, can trigger severe earthquakes. As the oceanic crust continues to descend, it melts, partly due to heat resulting from friction caused by contact with the American Plate and partly due to the increase in temperature as it re-enters the mantle. Some of the newly formed magma, being lighter than the mantle, rises to the surface to form volcanoes (e.g. Chimborazo and Cotopaxi) and a long chain of fold mountains (the Andes).

Sometimes, at destructive margins, the magma rises offshore to form *island arcs* such as the West Indies (Figure 14.9 and Case Study 14), Japan and, off the south coast of Alaska, the Aleutian Islands.

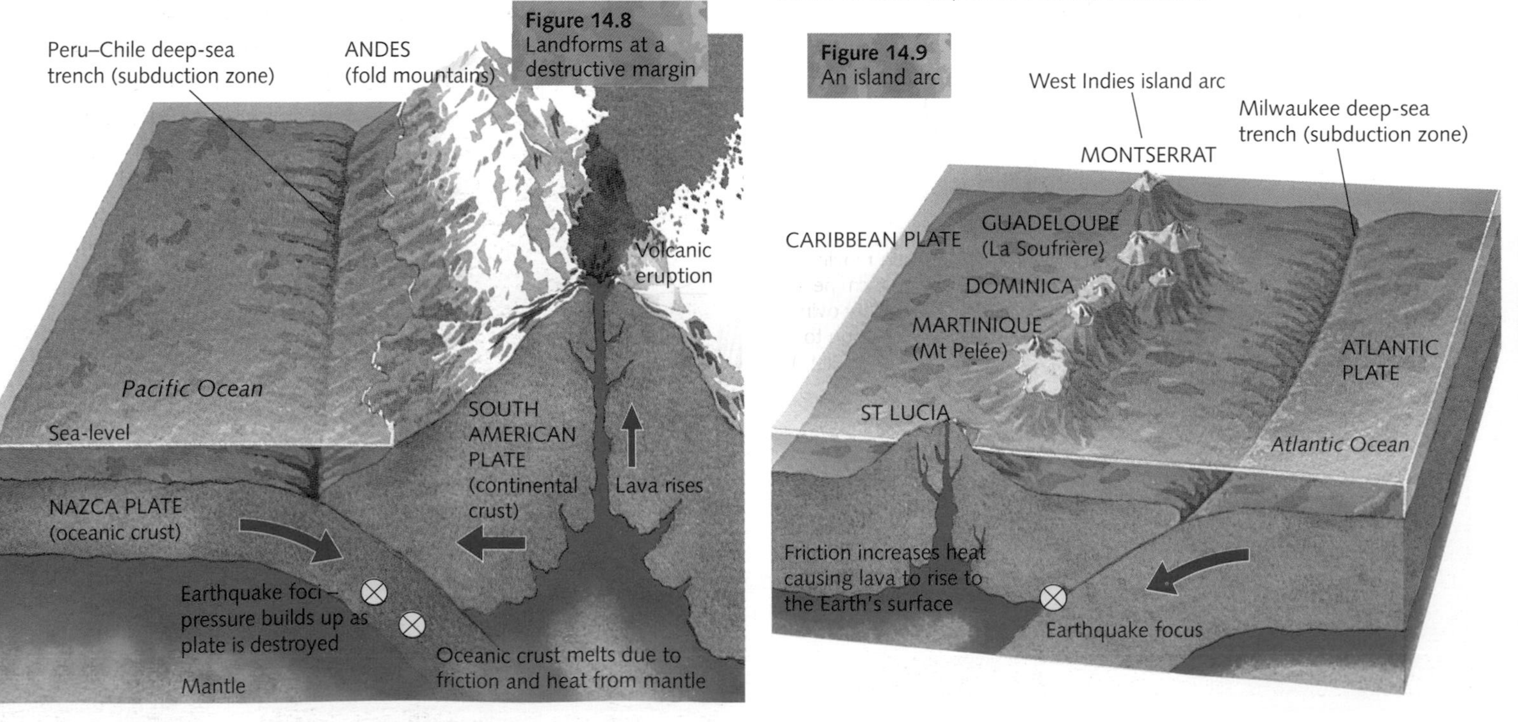

Figure 14.8
Landforms at a destructive margin

Figure 14.9
An island arc

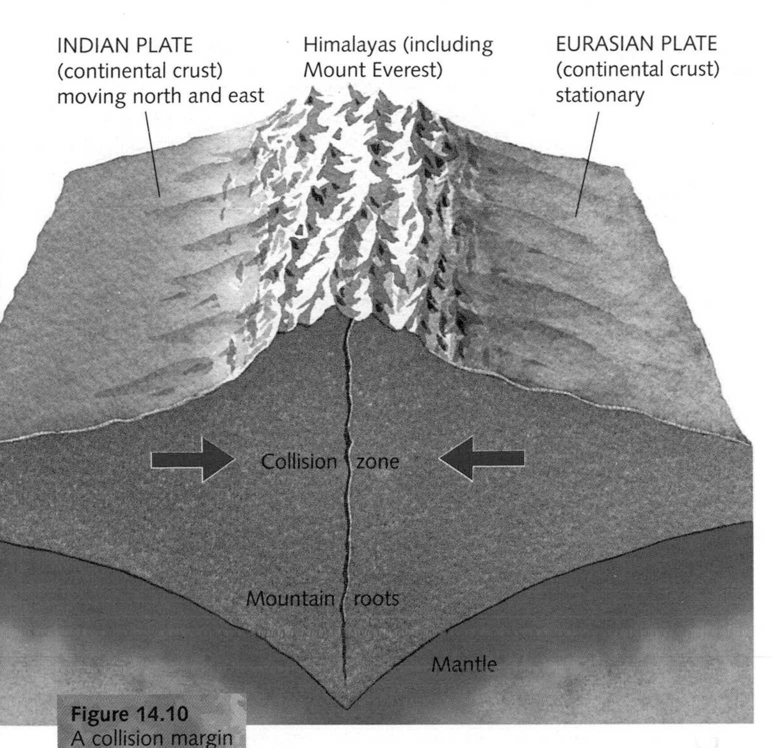

Figure 14.10
A collision margin

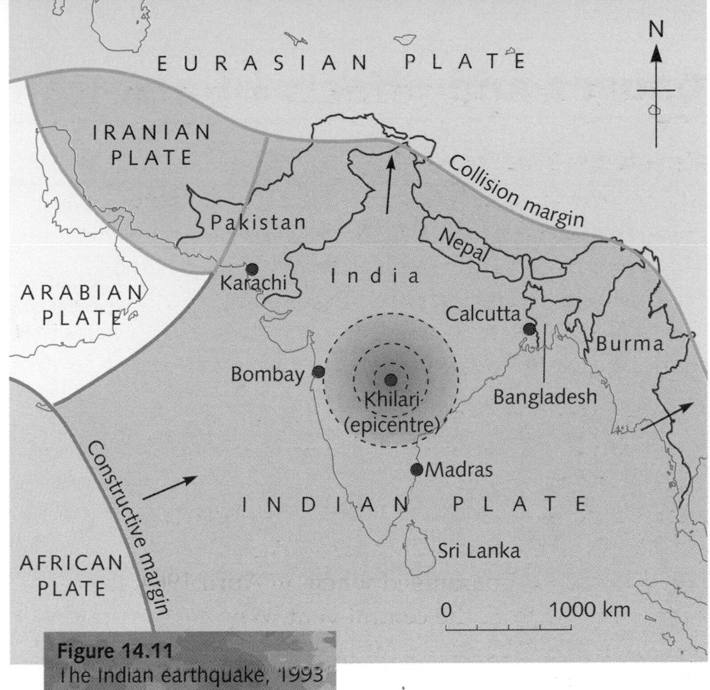

Figure 14.11
The Indian earthquake, 1993

Collision margins

Collision margins are where two plates consisting of continental crust, move together. As continental crust can neither sink nor be destroyed, the rocks between them are forced upwards. The Indian Plate is moving into the Eurasian Plate at a rate of 5 cm a year (Figure 14.10). The land between them, which was once the bed of the now non-existent Tethys Sea, has been buckled and pushed upwards to form the Himalayas – indeed evidence suggests that Mount Everest is increasing in height. This movement, which is still taking place, accounts for major earthquakes such as that which caused the deaths of 10 000 people and left another 150 000 homeless in central India in 1993 (Figure 14.11).

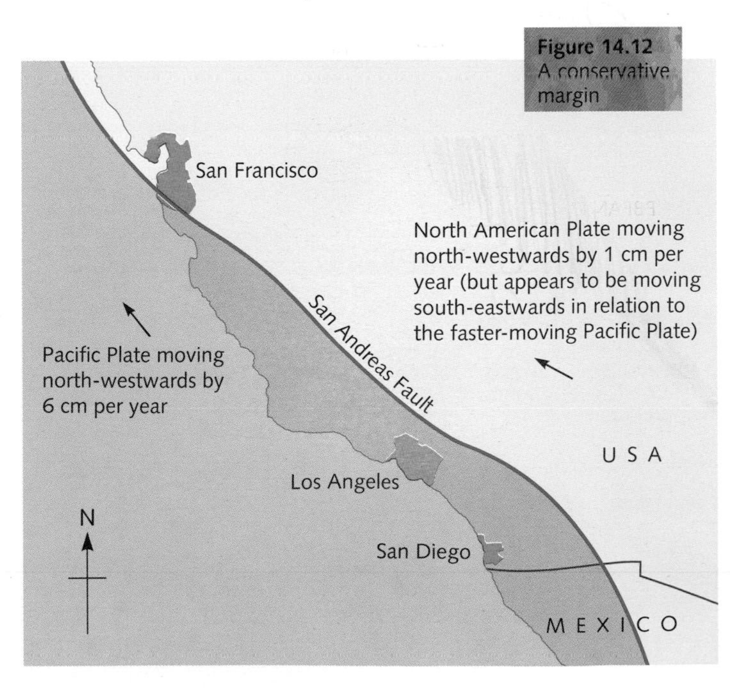

Figure 14.12
A conservative margin

Conservative margins

Conservative margins are found where two plates slide past one another. As crust is neither being formed nor destroyed at this plate boundary, new landforms are not created and there is no volcanic activity. However, earthquakes can occur if the two plates 'stick'. This is the situation in California, where the San Andreas Fault marks the junction of the Pacific and North American Plates (Figure 14.12). The American Plate moves more slowly than, and at a slight angle into, the Pacific Plate. Instead of the plates slipping evenly past each other, they tend to stick – like a machine without oil. When sufficient pressure builds up, one plate is jerked forwards sending shockwaves to the surface. These shockwaves caused an earthquake in San Francisco in 1906, when the ground moved by 6 metres, more than 450 people were killed and 28 000 buildings destroyed (Figure 14.13). Similar shockwaves caused another earthquake in 1989. Short-term predictions suggest that a further major earthquake could occur at any time in the San Francisco area, and long-term predictions suggest that in the distant future Los Angeles (on the faster-moving Pacific Plate) could be further north than San Francisco (on the North American Plate).

Figure 14.13
Destruction of San Francisco in 1906

Causes and effects of a volcanic eruption

Mount St Helens, USA

Figure 14.14 shows the causes of the Mount St Helens eruption. The Juan de Fuca Plate (oceanic crust) moves eastwards towards the North American Plate (continental crust), and is forced downwards. This movement creates friction which produces earthquakes and, due to an increase in temperature, destroys the oceanic crust. Volcanic eruptions take place where, and when, the magma rises to the Earth's surface. Through the centuries, a series of volcanic eruptions had formed the Cascades mountain range. Mount St Helens (2950 metres high) is one peak within this range. As, by 1980, it had been inactive for over 120 years, most people living near to it did not accept that one day it might erupt again.

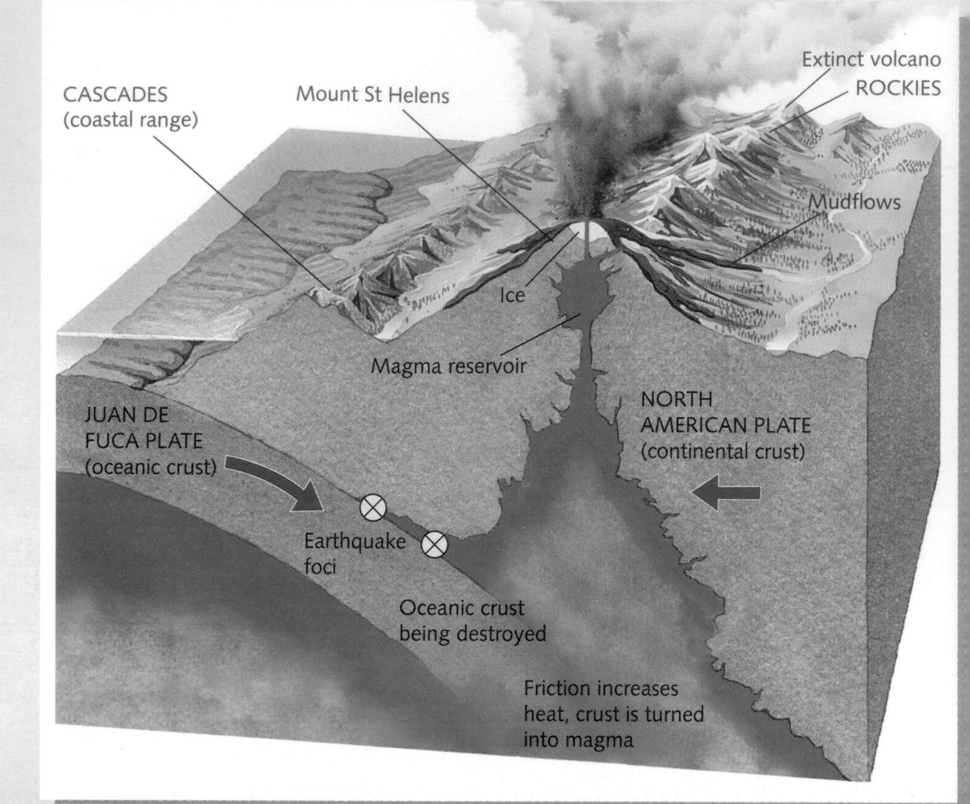

Figure 14.14 Causes of the Mount St Helens eruption

Timetable of events

Spring 1980

On 20 March there was a minor earthquake which measured 4.1 on the Richter scale. There were many more tremors over the next few days until, on 27 March, there was a small eruption of steam and ash. Minor eruptions occurred daily, attracting an increasing number of tourists. By early May the north side of the mountain began bulging by 1.5 m a day, indicating a build-up of magma and an increase in pressure. This reached a peak on the morning of 18 May.

18 May

0830 hours Ash and steam erupted.

0832 hours An earthquake (magnitude 5 on the Richter scale) caused the bulge to move forwards and downwards (Figure 14.15a). The released material formed a landslide of rock, glacier ice and soil. This raced downhill to fill in Spirit Lake and then, reinforced by the water displaced from this lake, moved rapidly down the northern fork of the Toutle Valley (Figure 14.16). The mudflow reached Baker Camp, but floodwater continued down the valley and sediment blocked the port of Portland on the Columbia River.

0833 hours The exposed magma exploded sideways sending out blast waves of volcanic gas, steam and dust (called a *nuée ardente*) which moved northwards for 25 km. Within this range every form of life – plant and animal – was destroyed (Figure 14.17).

Rest of morning A series of eruptions ejected gas, ash and volcanic 'bombs' (rocks). The thicker ash rose 20 km and drifted eastwards before settling. Inhabitants at Yakima, 120 km away, could only go out if they wore face masks.

Three days later The volcanic 'plume' (cloud) of fine ash reached the east coast of the USA.

Several days later The ash had completely encircled the world.

Figure 14.15
The eruption of Mount St Helens
(a) Early 1980

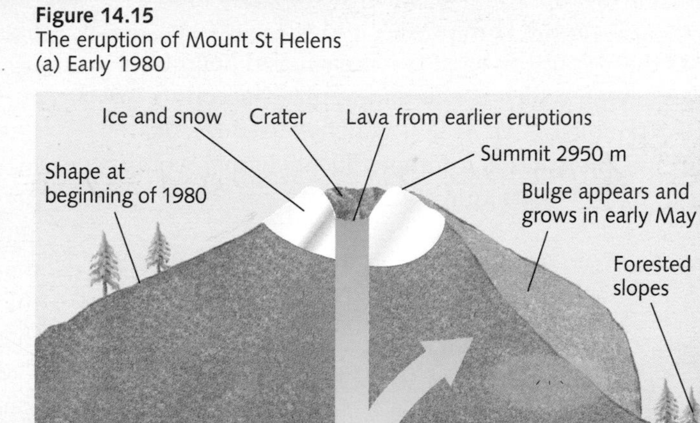

(b) 0833 hours, 18 May 1980

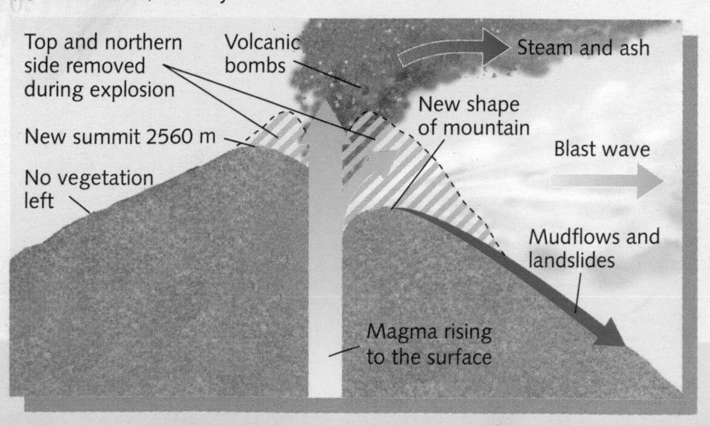

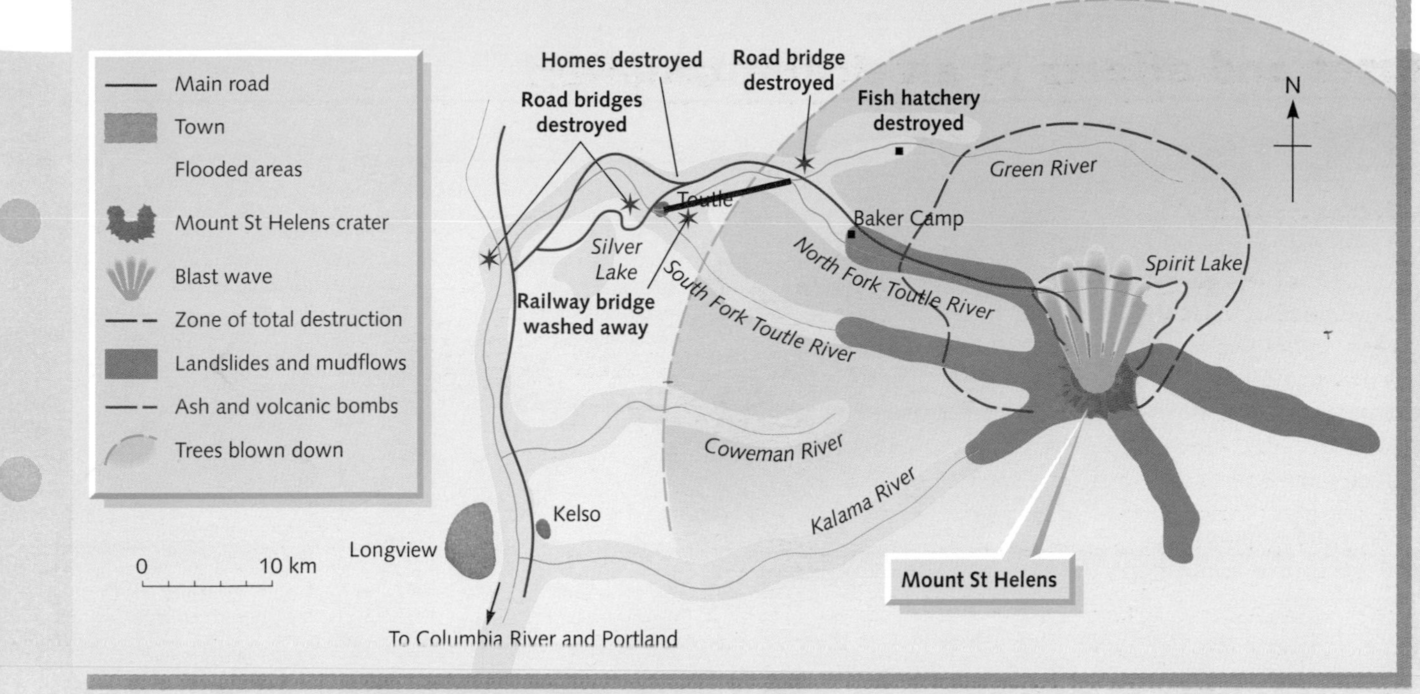

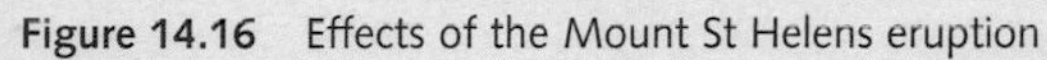

Figure 14.16 Effects of the Mount St Helens eruption

Figure 14.17
Destruction resulting from the Mount St Helens eruption

Figure 14.18
View of Mount St Helens crater following the eruption

Consequences of the eruption (Figure 14.16)

Mount St Helens itself had been reduced by 390 m to 2560 m. A crater (more like an amphitheatre in appearance) 3 km long and 0.5 km deep had been created on the north-facing slope (Figure 14.18).

Human life Sixty-one deaths were reported, most of them caused by the release of poisonous gases which accompanied the blast waves.

Settlements Several logging camps were destroyed – luckily, as it was a Sunday, no one was working or living there.

Rivers and lakes Ash which fell into rivers and lakes raised the water temperature, while sediment and mud also choked channels. The combined effect was the death of all fish, including those in a hatchery, and the loss of 250 km of former top-class salmon and trout rivers. Spirit Lake was filled in.

Communications Floodwaters washed away several road and railway bridges. Falling ash hindered the smooth running of car engines in three states.

Forestry Every tree in the 250 km^2 forest and lying within the 25 km blast zone north of the volcano was totally flattened and destroyed (Figure 14.17). Trees, carried down by rivers in flood, caused a log jam 60 km away. Some 10 million trees had to be replanted.

Services Electricity supplies were interrupted and telephone wires cut.

Wildlife As with the trees, nothing survived within the blast zone.

Farming Estimates suggested that 12 per cent of the total crop was ruined by settling dust. Fruit and alfalfa were hardest hit. Crops and livestock on valley floors were lost due to flooding.

Causes and effects of an earthquake

Kobe, Japan

Earthquakes are measured on the *Richter scale* (Figure 14.19). Each level of magnitude on the scale is 10 times greater than the level below it. This means that the San Francisco earthquake of 1906 (measuring 8.2) was 10 times stronger than the one that affected Kobe in 1995 (which measured 7.2) and well over 100 times stronger than the one that hit Assisi in 1997 (which was 5.7 – Case Study 14).

Why did it happen?

Kobe is on a minor fault, the Nojima Fault, which lies above a destructive plate margin (Figure 14.20). It is here that the Philippine Plate (oceanic crust) is forced downwards on contact with the Eurasian Plate (continental crust). Over the centuries, volcanic activity at this plate boundary has created the Japanese island arc (page 248). However, associated Earth movements mean that Japan is under constant threat from severe earthquakes (e.g. the 1923 Kanto earthquake which killed 140 000 people in Tokyo and Yokomaha).

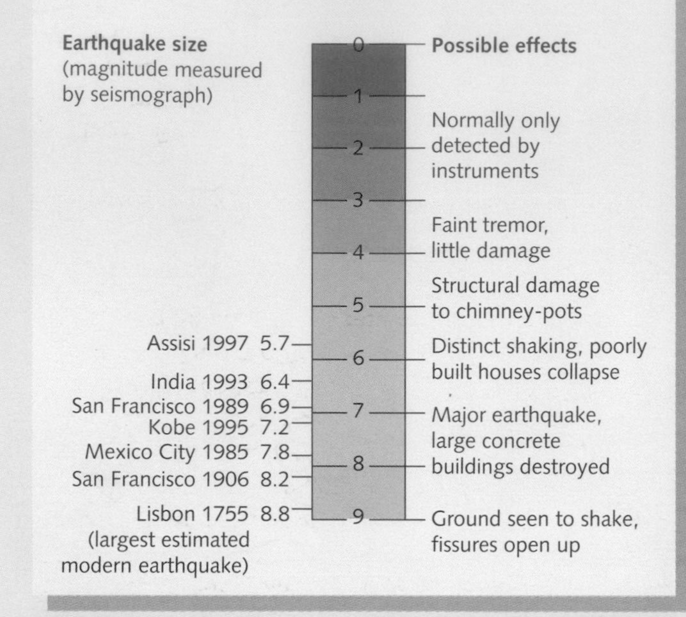

Figure 14.19
The Richter scale, a measurement of earthquakes

When did it happen?

The earthquake occurred at 5.46 on the morning of 17 January 1995. It recorded 7.2 on the Richter scale and lasted for 20 seconds. During this time the ground moved 18 cm horizontally and 12 cm vertically. The widespread devastation was due to the earthquake's focus being so near to the surface (Figure 14.21) and its epicentre being so close to Kobe.

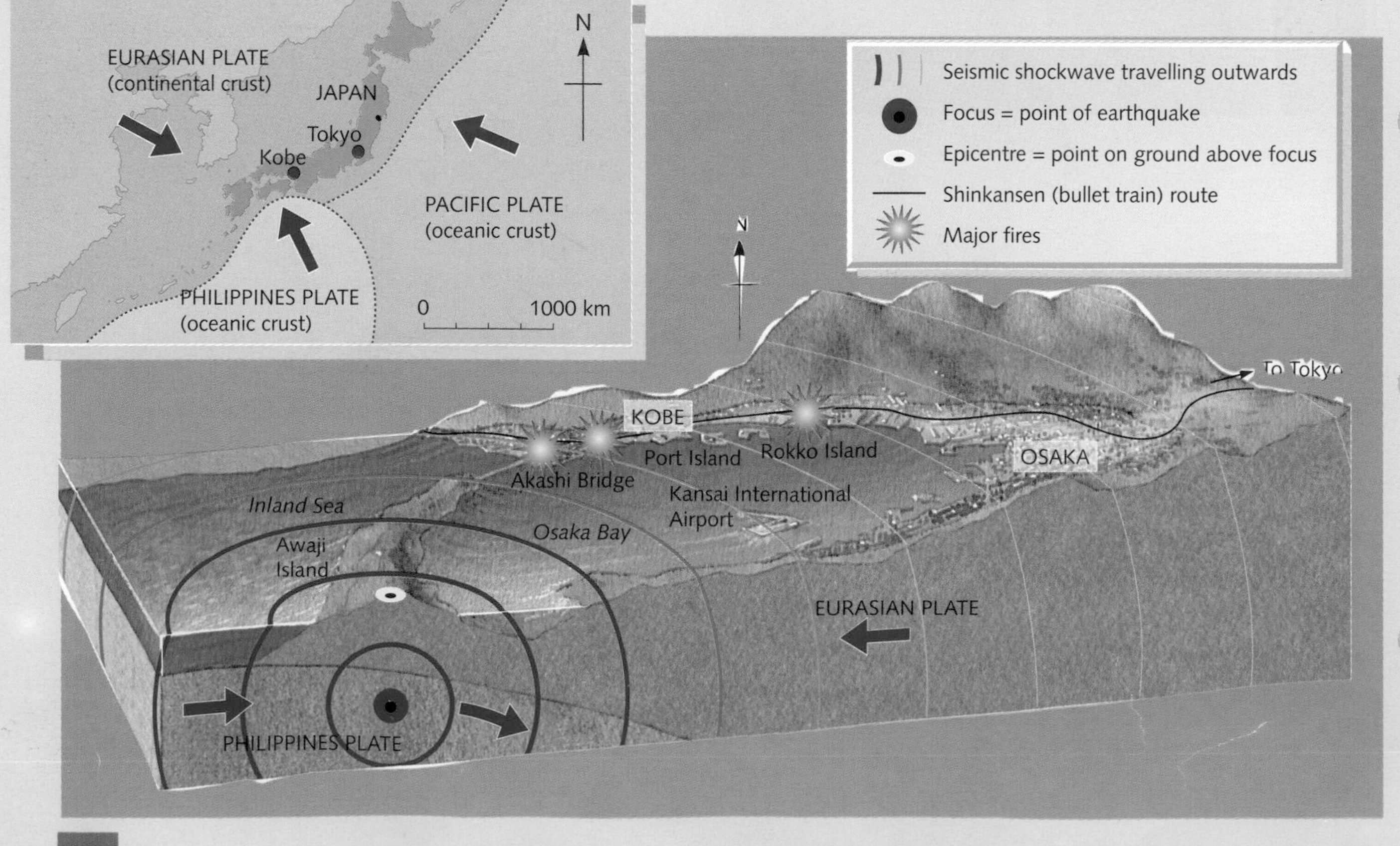

Figure 14.20 Location of Kobe

Figure 14.21 The 1995 Kobe earthquake

What were its effects?

The effects can be divided into primary and secondary types.

Primary effects

- Nearly 200 000 buildings collapsed, and also a 1 km stretch of the elevated Hanshin Expressway (Figure 14.22) and numerous bridges along a 130 km section of the bullet train route. Several trains on minor lines were derailed.
- 120 of the 150 quays in the port of Kobe were destroyed.

Secondary effects

- Electricity, gas and water supplies were disrupted.
- Fires (Figure 14.23), caused by broken gas pipes and ruptured electricity mains, raged for several days, destroying a further 7500 houses (many of which were made of wood). At one time, the wall of fire extended for over 400 metres.
- Roads were at gridlock, delaying ambulances and fire-engines.
- An estimated 230 000 people were made homeless and had to live in temporary shelters (unheated school gyms or in open parks) at a time when night-time temperatures dropped to –2°C. There was a short-term shortage of blankets, clean water and food.
- People were afraid to return home as the 716 recorded aftershocks lasted several days after the main event (74 were strong enough to be felt by humans).
- Industries, including Mitsubishi and Panasonic (page 139), were forced to close.

Final figures put the death toll at 5500, injuries at 40 000 and houses totally destroyed at 180 000. However, the newly opened Kansai International Airport and Akashi Bridge were both undamaged – presumably due to their high-tech construction aimed at withstanding earthquakes (page 49).

What happened in the months after the earthquake?

- Kobe's infrastructure, including water, electricity, gas and telephone services, was fully operational by July.
- The area worst affected by fire had been cleared of rubble but little rebuilding had taken place. Most commercial buildings in central areas had been repaired.
- All rail services were back to normal by August.
- One year later the port of Kobe was 80 per cent functional but the Hanshin Expressway remained closed.
- Replacement buildings had to meet stronger earthquake-resistance standards. High-rise buildings had to have flexible steel frames, smaller buildings had to have concrete frames with reinforcing bars to absorb shockwaves, houses were not to be built just from brick (which shakes loose) or wood (which burns too easily) but with fire-resistant materials. New buildings had to be built on solid rock, not clay, as water rises to ground-level during an earthquake, causing clay to 'liquefy' into mud. This results in the collapse of buildings.
- There was an increase in the number of seismic instruments to record earth movements in the region.

Figure 14.22 The collapsed Hanshin Expressway

Figure 14.23 Fires following the earthquake

Montserrat and Assisi

Montserrat – a volcanic eruption

Montserrat was formed by volcanic action. It is located on a destructive plate margin where the Atlantic and Caribbean Plates, as they move towards each other, have formed an island arc (Figure 14.9). In the early 1990s, the National Trust described the island as 'one of the few perfect ecosystems left in the world', to which the Insight Guide to the Caribbean added: 'The centrepiece of this magnificent ecosystem is Soufrière Hills, an inactive volcano and sulphur spring'. That was before July 1995.

Soufrière Hills volcano had been dormant for 400 years (the last big eruption was 10 000 years ago). The soils surrounding it, consisting of weathered volcanic material, were extremely fertile and, together with the hot, wet climate, had attracted many farmers to the area. In 1991 the island's population was given as 11 000.

1995 During July, people living in the capital of Plymouth became aware of an increase in sulphur fumes, but it was not until steam was seen coming from the summit that they made the link with the former volcano (Figure 14.24). When the first explosions began several days later, nearby farmers hurriedly collected what harvest they could. Further eruptions caused ash to fall on Plymouth and, with scientists predicting a major explosion, the area's inhabitants were evacuated to a 'safe zone' in the north. Although the south of the island became increasingly covered in ash, there was no big eruption.

1996 Several times during the year, magma pushed upwards to create a dome. Each time this occurred, pressure built up under the dome until it was forcibly removed by an eruption. The eruptions sent *pyroclastic flows* (rivers of hot gas, ash, mud and rock) down the mountain in all directions (Figure 14.26), some reaching the sea to increase the island's size.

1997 People who had been evacuated to the 'safe zone' became increasingly impatient at not being allowed to return home. In fact, against advice given by both government and scientists, some did go back. The biggest explosion to date took place in June. It caused large pyroclastic flows which, for the first time, over-ran several small settlements and caused the death of 19 people. Ash also fell in the safe zone. An eruption on 8 August created a pyroclastic flow which virtually destroyed what was left of Plymouth (Figure 14.25). This resulted in the danger zone being extended, and the 1000 inhabitants of Salem were ordered to evacuate to the already overcrowded northern part of the island – a decision which led to a mini-riot. By the end of the year, 7000 people had actually left the island.

1998 . . . It is now up to *you* to listen to what happens to Montserrat.

- Will the predicted big explosion come, causing the possible destruction of an island that owes its existence to volcanic activity?
- Will volcanic activity decrease, allowing those remaining on the island to return to their homes and re-establish their lives?
- Will the periodic eruptions of 1996–97 continue, leaving the remaining islanders with an uncertain future?

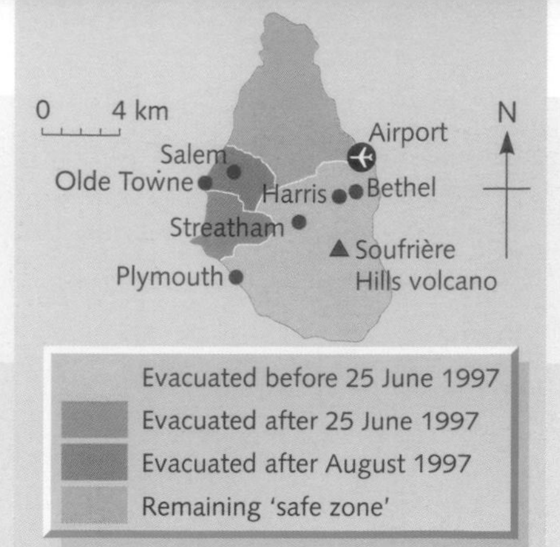

Figure 14.24 Montserrat, 1997

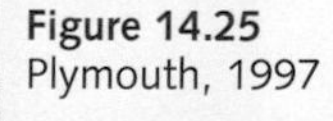

Figure 14.25 Plymouth, 1997

Figure 14.26 Pyroclastic flows near Plymouth, 1997

Figure 14.27
The Basilica of St Francis

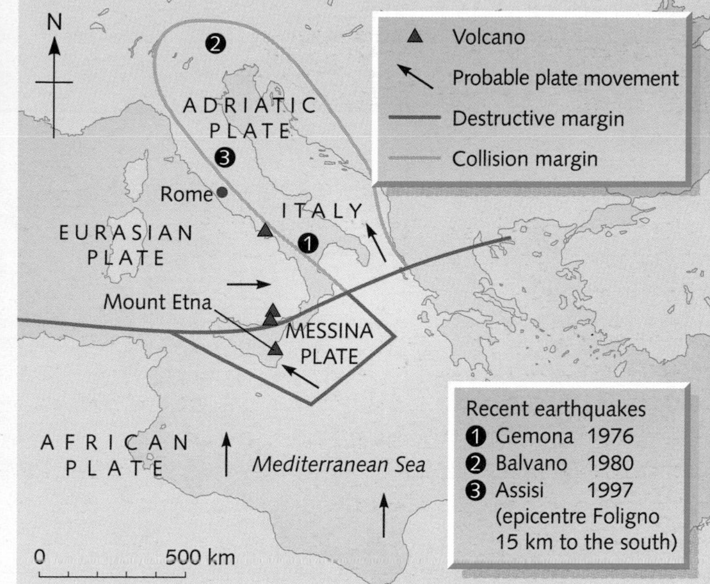

Figure 14.28
Recent earthquakes in Italy

Assisi – an earthquake

Assisi is one of many hilltop towns in central Italy. It is best known as the birthplace (1182) of St Francis, the patron saint of animals and the founder of the Roman Catholic Church's largest monastic order – the Franciscans. During his lifetime, two churches (the Upper and Lower Basilicas) were built (Figure 14.27). His life was recorded in the Upper Basilica in a famous series of 28 frescoes (wall-paintings) painted by Giotto (1296), one of the earliest of Italian Renaissance painters.

Assisi is also situated near to a complex plate boundary where the small Adriatic Plate is squeezed in between the two larger Eurasian and African Plates (Figure 14.28). To the south is a destructive margin with volcanic activity (Etna and Vesuvius) and earthquakes (one in 1980 killed 3000 people). To the north is a collision margin with fold mountains (Alps) and earthquakes (one in 1976 caused over 1000 deaths).

19 September 1997 An earthquake struck Assisi and the surrounding region in the early hours of 19 September. In mid-morning a second earthquake, measuring 5.7 on the Richter scale (Figure 14.19), caused considerable damage and some loss of life (Figure 14.29). To make matters worse, strong aftershocks, one measuring 4.9, were still being experienced two weeks after the initial event. Each major shockwave caused panic among the 50 000 townspeople, who were afraid to return home and were sleeping in tents and public buildings. These later shockwaves caused further damage to the Basilica and other buildings.

Figure 14.29
Newspaper reports on the Assisi earthquake

The day after the event
In Assisi, six people were killed and 80 per cent of housing was damaged (five more died in the surrounding area). Part of the roof of the thirteenth-century Upper Basilica of St Francis of Assisi collapsed during the second earthquake. Two Franciscan monks and two surveyors who were in the Basilica assessing damage caused by the first 'quake, were killed by falling masonry. Later, it was reported that two frescoes had been seriously though not irreparably damaged. Nearby, the twelfth-century cathedral was also made unstable.

A week after the event
Early reports suggest that most of Giotto's 28 frescoes appear to have survived – although restorers will have a mammoth task in piecing together the thousands of fragments strewn across the damaged buildings. The tomb of St Francis, under the Lower Basilica, is said to be safe. However, recriminations have already begun. Some 30 years ago, wooden beams and joints (which had supported the basilica for over 700 years and had withstood several earthquakes) had been replaced by reinforced concrete.

The mayor's attempts to close the town to visitors failed when local businesses objected (80 per cent of Assisi's revenue comes from tourism). Thousands of people, many elderly, were still spending the chilly nights in temporary shelters. Many complained that the world seemed more concerned about damaged frescoes than with homeless people.

QUESTIONS

1

(Pages 246–249)

a The map shows some of the major plate boundaries.

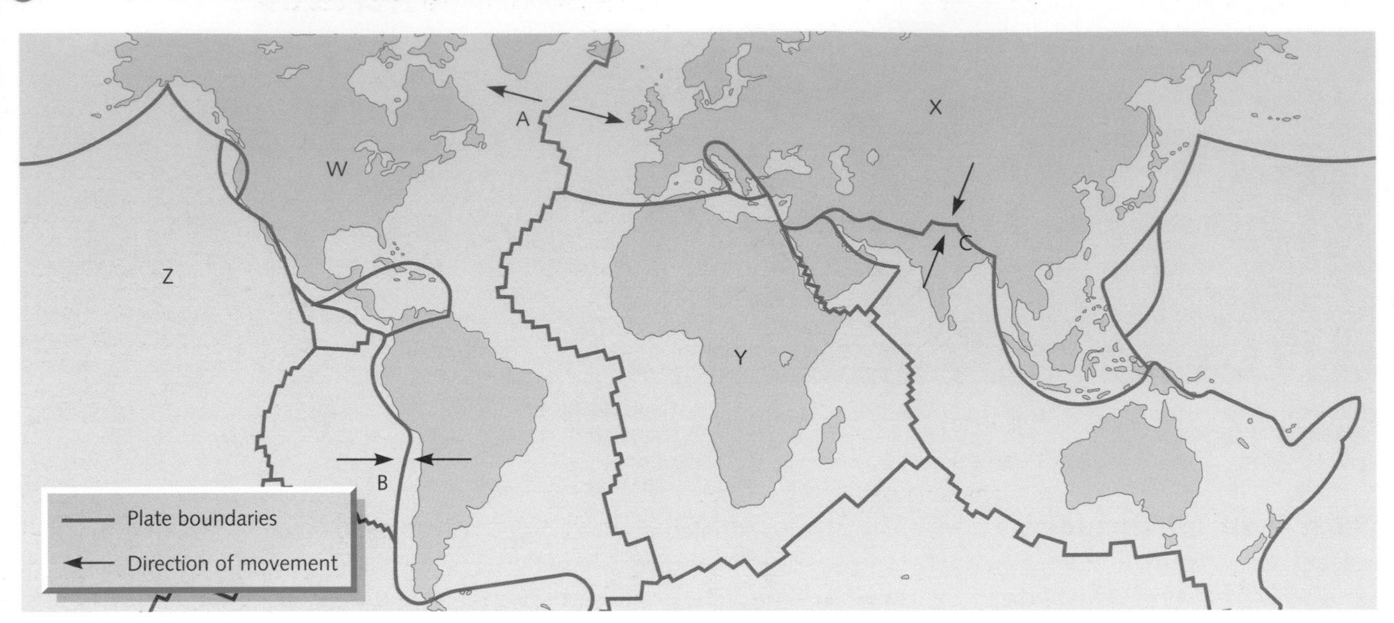

i) Name the plates labelled W, X, Y and Z. (4)
ii) What is happening to the plates at A, B and C? (3)
iii) Name the type of plate boundary at A, B and C. (3)

b i) Name one volcano in each of North America, South America, Europe and Asia. (4)
ii) Name one recent earthquake in each of North America, Europe and Asia. (3)
iii) Using Figure 14.1, describe where most earthquakes are likely to occur. (3)
iv) Using Figure 14.2, describe where most volcanic eruptions are likely to occur. (3)

c i) What are plates? What causes them to move? (2)
ii) Give three differences between continental and oceanic crust. (3)
iii) Copy and complete the following table by:
- naming the two types of crust involved (continental or oceanic)
- giving the direction of movement (away from, towards or sideways past)
- giving an actual example or location
- putting one tick in the earthquake column
- putting one tick in the volcano column. (5 x 4)

PLATE MARGIN	TWO TYPES OF CRUST INVOLVED	DIRECTION OF PLATE MOVEMENT	EXAMPLE/ LOCATION	EARTHQUAKE			VOLCANO		
				VIOLENT	LESS VIOLENT	RARE	VIOLENT	LESS VIOLENT	RARE
CONSTRUCTIVE									
DESTRUCTIVE									
COLLISION									
CONSERVATIVE									

d With the aid of a labelled diagram, explain why:
i) volcanoes occur at constructive plate margins (3)
ii) volcanoes and earthquakes occur at destructive plate margins (4)
iii) fold mountains and earthquakes occur at collision plate margins (4)
iv) earthquakes occur at conservative plate margins. (3)

e Why do people continue to live in places where there is a risk of either an earthquake or a volcanic eruption? (2)

2

(Pages 250 and 251)

a Why did Mount St Helens erupt in 1980? (2)

b Copy the flow diagram and complete it by adding the following information to the correct boxes. (8)

- Landslides, mudflows and floodwaters affect surrounding area
- Ash and steam erupts
- Bulge develops on north side of mountain
- Earthquake causes bulge to move outwards
- Blast wave from main explosion kills everything within 25 km to north
- Ash completely encircles the world
- Gas, ash and volcanic bombs continually ejected
- Ash plume reaches east coast of USA

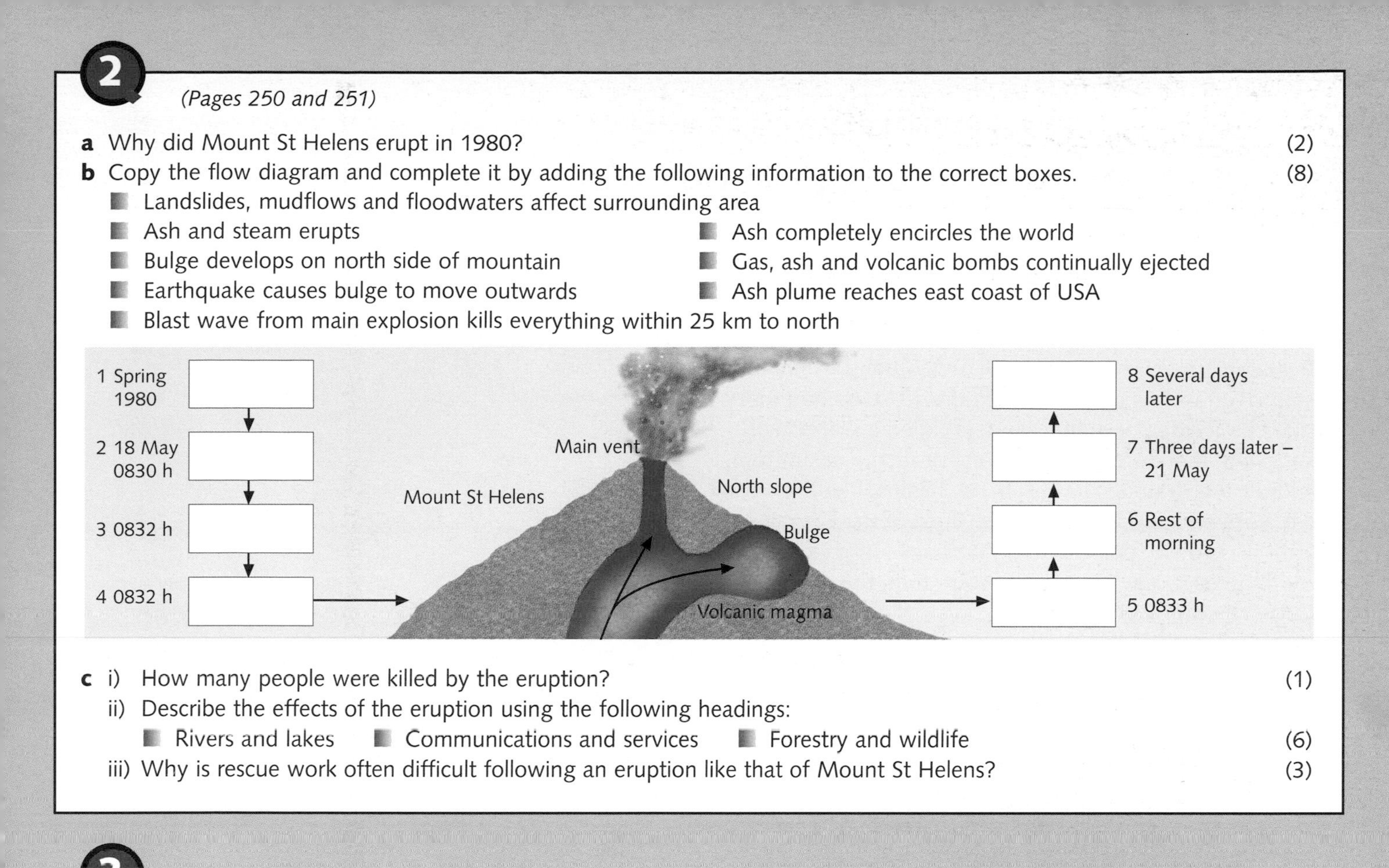

c i) How many people were killed by the eruption? (1)

ii) Describe the effects of the eruption using the following headings:

- Rivers and lakes
- Communications and services
- Forestry and wildlife (6)

iii) Why is rescue work often difficult following an eruption like that of Mount St Helens? (3)

3

(Pages 252 and 253)

a i) When did the Kobe earthquake occur? (1)

ii) Why did an earthquake occur at Kobe? (3)

b The effects of an earthquake may be divided into primary and secondary.

i) What is the difference between primary and secondary effects? (2)

ii) What were the main:
- primary effects
- secondary effects of the Kobe earthquake? (6)

c Using the diagram and page 253, describe what the authorities in Kobe might do to try to reduce the effects of a possible future earthquake. (4)

d Why do many Japanese continue to live in places like Kobe and Tokyo which they know are at risk from future earthquakes? (3)

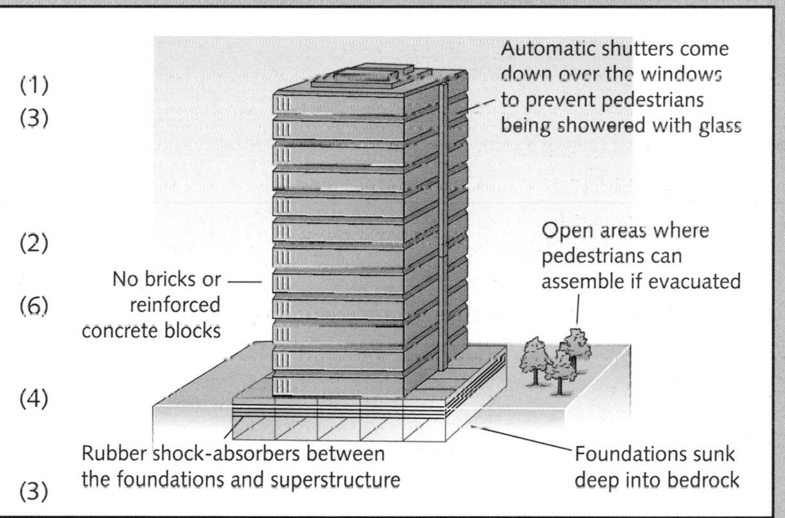

4

(Pages 254 and 255)

a i) Where is Montserrat? How did the island form? (3)

ii) Why did people live near to the Soufrière Hills volcano? (2)

b Imagine you were born in, and lived in, the capital of Plymouth.

i) Write a diary to describe the events that have taken place since the volcano erupted in July 1995. (6)

ii) Explain why you had to leave Plymouth in late 1995. (2)

iii) Since then you have lived in temporary shelter in the north of the island. Given a choice, would you prefer to continue to live there, return to the south of the island, or leave Montserrat altogether? Give reasons for your answer. (3)

c i) Where is Assisi? Why is it at risk from earthquakes? (3)

ii) Why do many people live in Assisi knowing that it is in an earthquake risk area? (2)

d Imagine you were born in Assisi and work for the local newspaper. Write two accounts for the newspaper:

i) the first to describe the scene the day after the earthquake (4)

ii) the second to update the situation one week after the event. (3)

15 DRAINAGE BASINS AND RIVERS

Drainage basins

A *drainage basin* or *river basin* is an area of land drained by a main river and its tributaries (Figure 15.1). Its boundary, marked by a ridge of higher land, is called a watershed. A *watershed*, therefore, separates one drainage basin from neighbouring drainage basins. Some basins, like the Mississippi which drains over one-third of the USA, are enormous. Others, possibly that of your local river, can be small. However, size is less important than the drainage density. The drainage density is:

$$\frac{\text{the total length of all the streams in the drainage basin}}{\text{the total area of the drainage basin}}$$

The drainage basin of the River Exe (Figure 15.2) has a much higher density than the basin in Figure 15.1. The density is highest on impermeable rocks and clays, and lowest on permeable rocks and sands. The higher the density the greater the risk of flooding, especially as a result of a flash flood (page 266).

Drainage basins can store rainwater, either within the river channel itself, or in lakes and in the ground. Excess water is carried back to the sea by rivers. Rivers form part of the hydrological (water) cycle (Figure 15.3). The seas and oceans contain 97 per cent of the world's water, but being salty it is not suitable for use by terrestrial plants, animals and people.

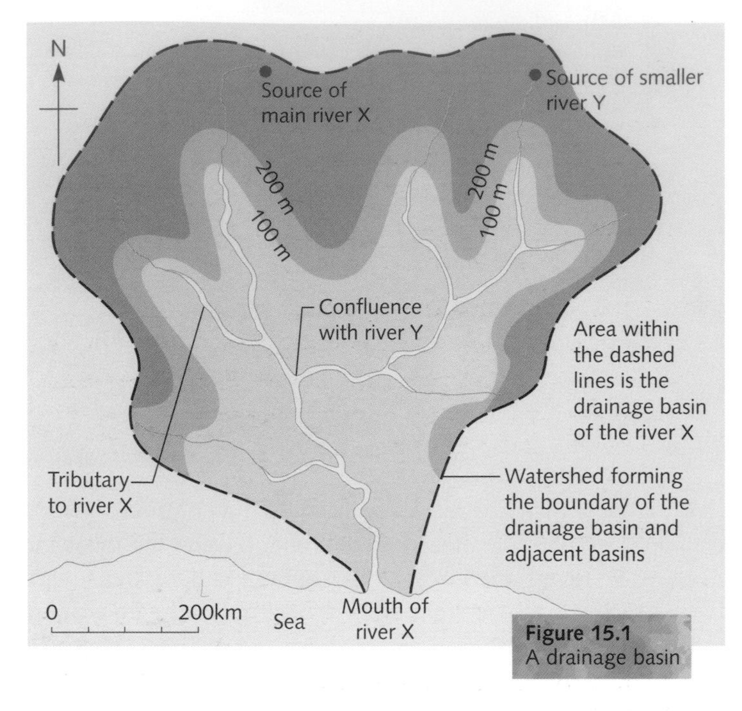

Figure 15.1
A drainage basin

Two per cent of the world's water is stored as ice and snow in arctic and alpine areas. That leaves 1 per cent which is either fresh water on land or water vapour in the atmosphere. As the amount of fresh water and vapour is limited, it has to be recycled over and over again. It is this constant recycling of water between the sea, air and land which is known as the *hydrological cycle*. As no water is added to or lost from the hydrological cycle, it is said to be a *closed system*.

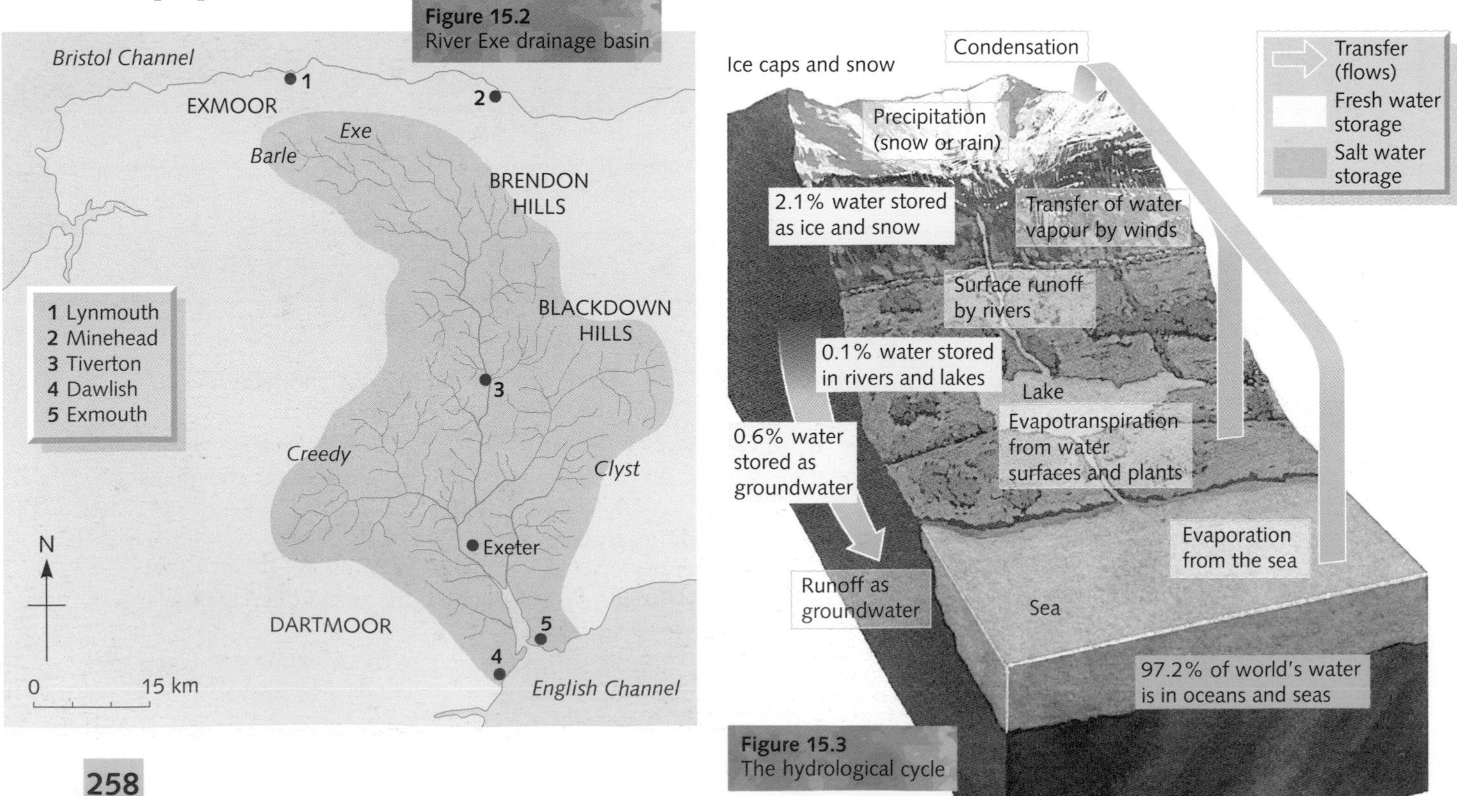

Figure 15.2
River Exe drainage basin

Figure 15.3
The hydrological cycle

The drainage basin system

A drainage basin forms part of the hydrological cycle but, unlike the hydrological cycle, it is an *open system*.

INPUTS → FLOWS → STORES → FLOWS → STORES → FLOWS → OUTPUTS

It is an open system because it has:

- **Inputs** where water enters the system through precipitation (rain and snow).
- **Outputs** where water is lost to the system either by rivers carrying it to the sea or through evapotranspiration. *Evapotranspiration* is the loss of moisture directly from rivers or lakes (evaporation) or from vegetation (transpiration).

Within the system are *stores* and *transfers* (flows).

- **Stores** are places where water is held, e.g. in pools and lakes on the surface or in soil and rocks underground.
- **Transfers** are processes by which water flows, or moves, through the system, e.g. infiltration, surface runoff, throughflow.

A typical drainage basin system is shown in Figure 15.4.

When it rains, most water droplets are intercepted by trees and plants. Interception is greatest in summer. If the rain falls as a short, light shower then little water will reach the ground. It will be stored on leaves and then lost to the system through evaporation.

When the rain is heavier and lasts longer, water will drip from the vegetation onto the ground. At first it may form pools (*surface storage*) but as the ground becomes increasingly wet and soft, it will begin to infiltrate. *Infiltration* is the downward movement of water through tiny pores in the soil. This downward transfer will be greatest in porous rock or soil such as chalk or sand, and least in impermeable rock or soil like granite or clay. The water will then either be stored in the soil or slowly transferred sideways or downwards. The movement of water sideways is called *throughflow* and it is likely to form, eventually, a spring on a valley side. When the movement of water is downwards it is called *percolation*. Percolation forms *groundwater*, which is water stored at a depth in rocks. Groundwater flow is the slowest form of water transfer. The fastest process of water movement is surface runoff. *Surface runoff*, sometimes referred to as *overland flow*, occurs when either the storm is too heavy for water to infiltrate into the soil, where the soil is impermeable, or when the soil has become saturated. The level of *saturation*, i.e. when all the pores have been filled with water, is known as the *water table*. Although some rain may fall directly into the river, most water reaches it by a combination of surface runoff, throughflow and groundwater flow. Rivers carry water to the sea where it is lost to the system.

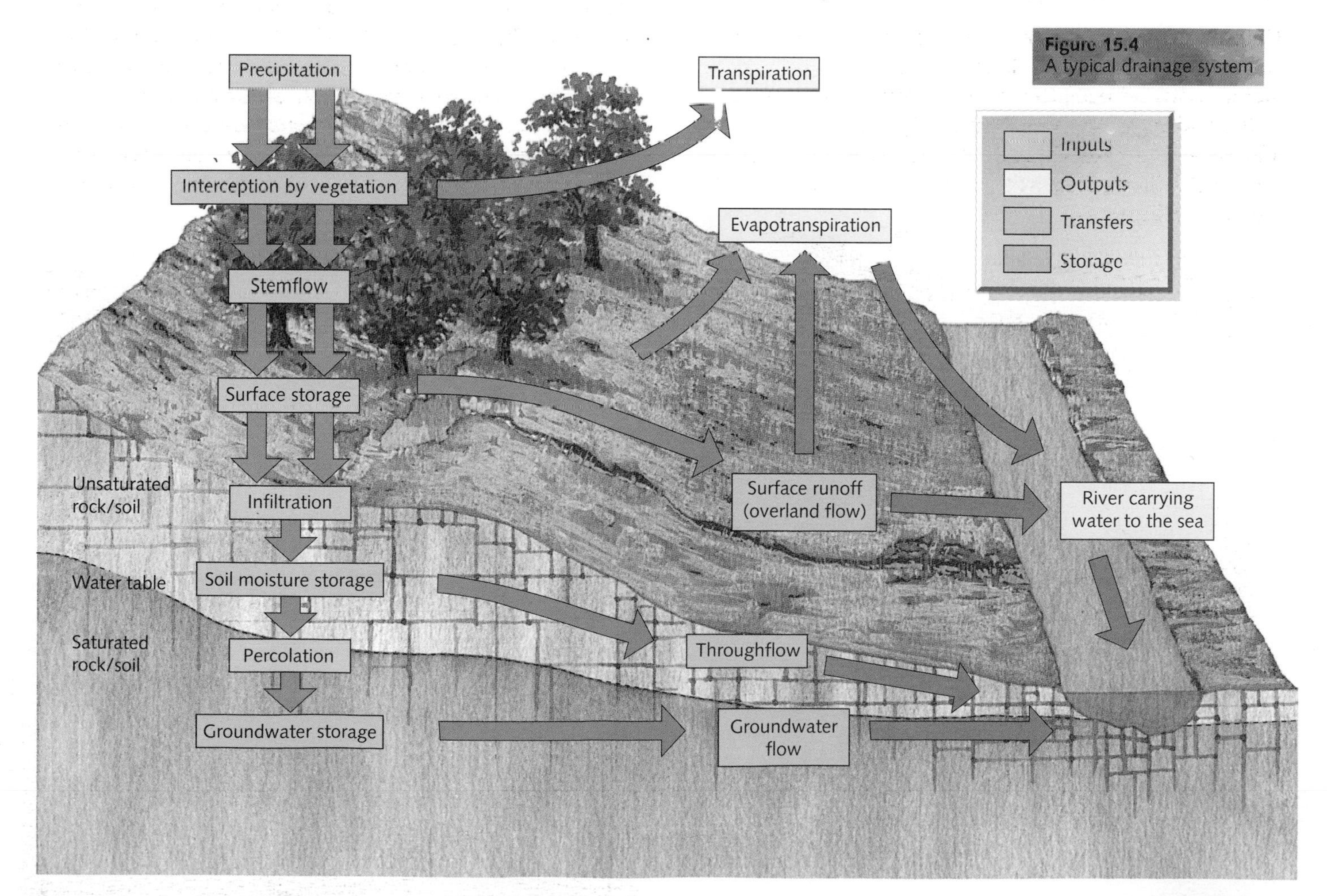

Figure 15.4 A typical drainage system

River discharge and flood hydrographs

Discharge depends upon the river's velocity and volume.

Velocity is the speed of the river. It is measured in metres per second.

Volume is the amount of water in the river system. It is the cross-sectional area of the river's channel measured in square metres.

Discharge is the velocity of the river times its volume. It is the amount of water in the river passing a given point at a given time, measured in cumecs (cubic metres per second).

In some drainage basins, discharge and river levels rise very quickly after a storm. This can cause frequent, and occasionally serious, flooding. Following a storm in these basins, both discharge and river levels fall almost as rapidly, and after dry spells, become very low. In other drainage basins, rivers seem to maintain a more even flow.

The flood (storm) hydrograph

A *hydrograph* is a graph showing the discharge of a river at a given point (a gauging station) over a period of time. A flood or storm hydrograph shows how a river responds to one particular storm (Figure 15.5). When a storm begins, discharge does not increase immediately as only a little of the rain will fall directly into the channel. The first water to reach the river will come from surface runoff, and this will later be supplemented by water from throughflow. The increase in discharge is shown by the rising limb.

Figure 15.5 A flood, or storm, hydrograph

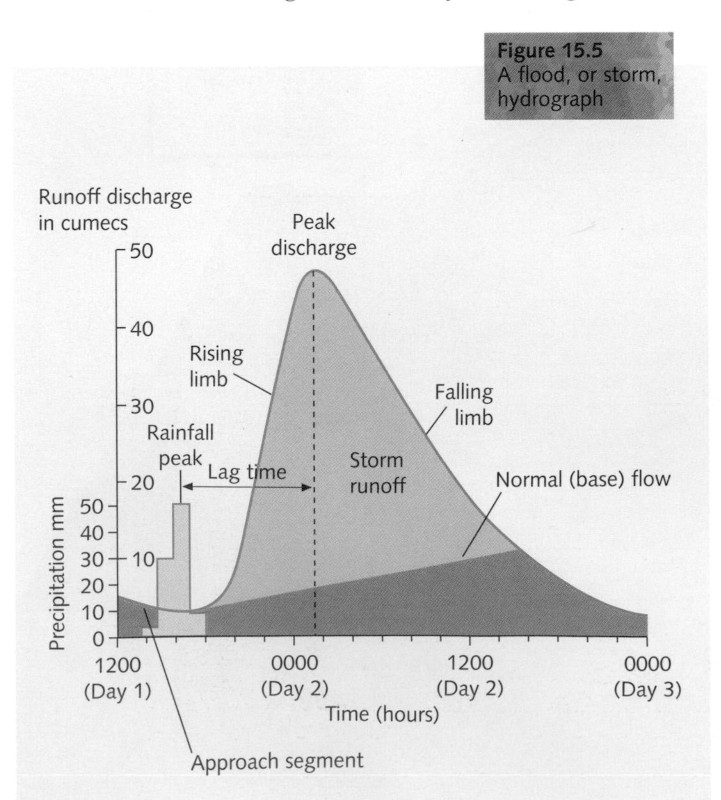

Figure 15.6 Differences in flood hydrographs between two adjacent drainage basins

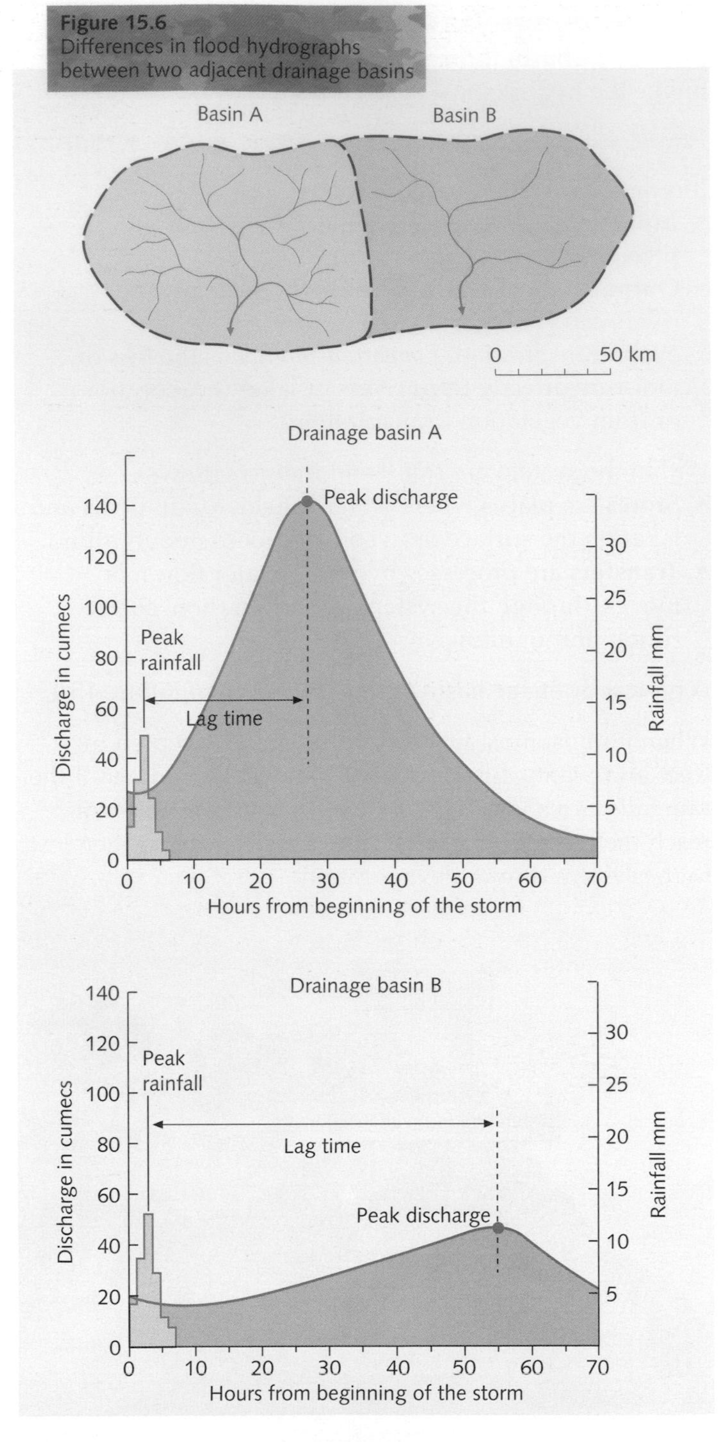

The gap between the time of peak (maximum) rainfall and peak discharge (highest river level) is called *lag time*. A river with a short lag time and a high discharge is more likely to flood than a river with a lengthy lag time and a low discharge.

Factors affecting the shape of the flood hydrograph

It is possible that two drainage basins, located side by side, can receive approximately the same amount of rainfall and yet have very different hydrograph shapes (Figure 15.6). The river in basin A is likely to flood regularly, the river in basin B probably never. The reasons for this difference may be due to one factor, or a combination of factors (Figure 15.7).

FACTOR	DRAINAGE BASIN A	DRAINAGE BASIN B
RELIEF	FASTER RUNOFF ON STEEP SLOPES	SLOWER RUNOFF ON MORE GENTLE SLOPES
ROCK TYPE	SURFACE RUNOFF ON IMPERMEABLE ROCK	THROUGHFLOW AND GROUNDWATER FLOW AS RAINFALL INFILTRATES INTO PERMEABLE (POROUS) ROCK
SOIL	VERY THIN SOIL; LESS INFILTRATION	DEEPER SOIL; MORE INFILTRATION
NATURAL VEGETATION	THIN GRASS AND MOORLAND; LESS INTERCEPTION	FOREST; MOST INTERCEPTION. ROOTS DELAY THROUGHFLOW AND ABSORB MOISTURE. EVAPOTRANSPIRATION REDUCES CHANCES OF WATER REACHING RIVER
LAND USE	URBANISATION; INCREASED TARMAC (IMPERMEABLE LAYER) AND DRAINS (INCREASED RUNOFF). ARABLE LAND EXPOSES MORE SOIL	RURAL AREA WITH LITTLE TARMAC, CONCRETE OR DRAINS. TREE CROPS AND ARABLE FARMING INCREASE INTERCEPTION
USE OF RIVER	LIMITED USE	WATER EXTRACTED FOR INDUSTRY, DOMESTIC USE AND IRRIGATION. DAM BUILT TO STORE WATER
DRAINAGE DENSITY	HIGHER DENSITY MEANS MORE STREAMS TO COLLECT WATER QUICKLY	LOWER DENSITY; FEWER STREAMS TO COLLECT WATER

Figure 15.7
Reasons for differences in flood hydrographs for two drainage basins

Extreme weather conditions are often the major cause of a river flooding. A torrential thunderstorm, continuous rainfall for several days, or a heavy snowfall melting while it rains all increase the discharge of a river. Although high summer temperatures increase evapotranspiration and reduce the amount of water available to reach a river, they can make the ground hard, reducing infiltration when it does rain. Freezing conditions in winter can make the ground impermeable.

Hydrographs can also be drawn for longer periods of time. The Environment Agency, formerly the National Rivers Authority (NRA), produces hydrographs to show, among other things, if there is a relationship between the two variables of rainfall and discharge. Figure 15.8, which covers a very wet month in late 1992, was provided by the then NRA for the River Torridge in North Devon.

Notice that:

- discharge is dependent upon rainfall, but rainfall does not depend upon discharge
- most of the discharge peaks are a half or full day after the rainfall peak (lag time)
- the highest discharge peak came after several very wet days during which river levels had no time to drop, rather than after the wettest day which followed a relatively dry spell when the river level had fallen
- the River Torridge responds very quickly to rainfall and so would appear to pose a flood risk.

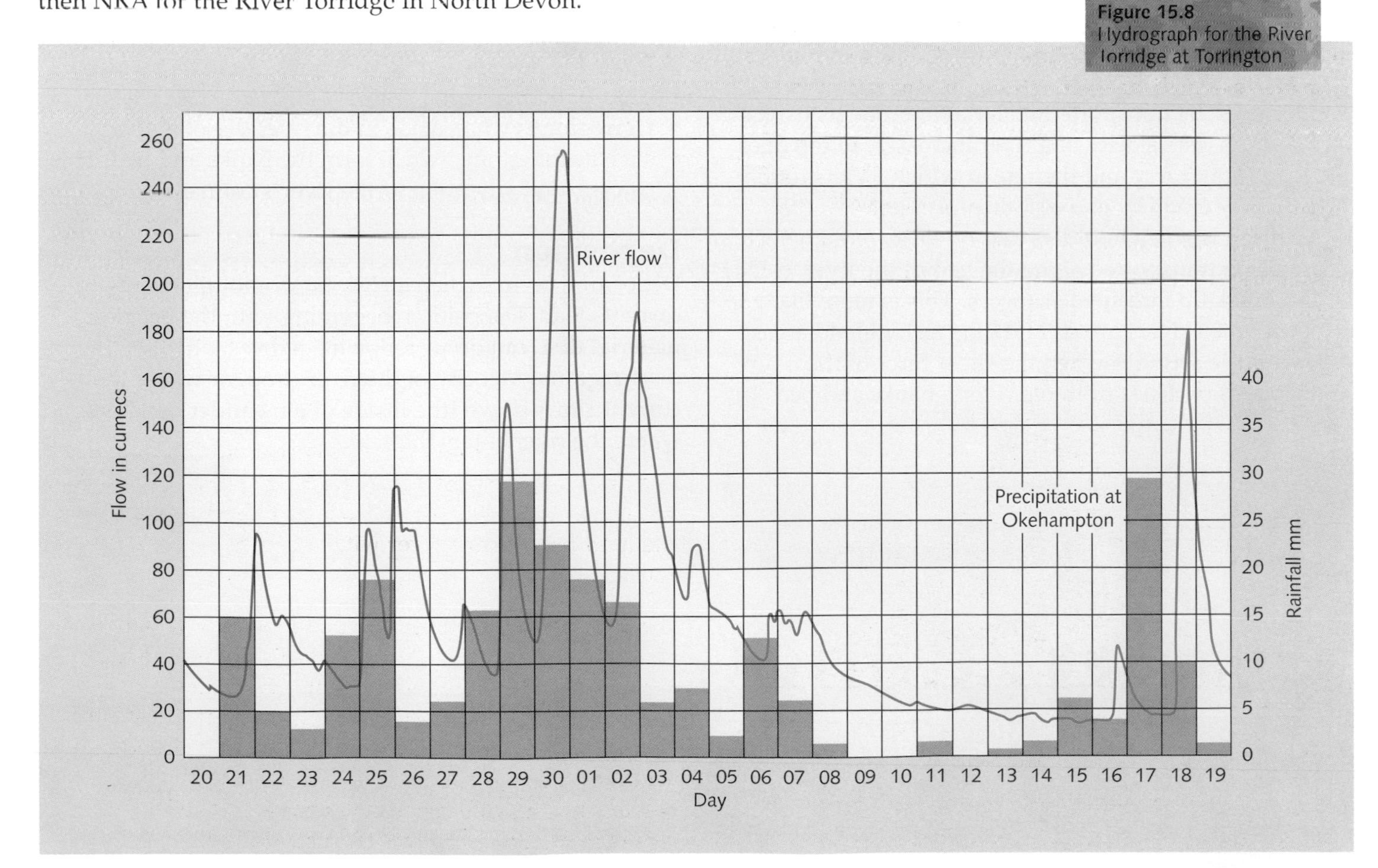

Figure 15.8
Hydrograph for the River Torridge at Torrington

River processes

Energy is needed in any system, not just the drainage basin, for transfers to take place (page 214).

In the case of a river most of this energy, an estimated 95 per cent under normal conditions, is needed to overcome friction. Most friction occurs at the *wetted perimeter*, i.e. where the water comes into contact with the river's banks and bed. The channel of a mountain stream, often filled with boulders, creates much friction (Figure 15.9). As a result, water flows less quickly here than in the lowlands where the channel becomes wider and deeper (Figure 15.10).

Following a period of heavy rain, or after the confluence with a major tributary, the volume of the river will increase. As less water will be in contact with the wetted perimeter, friction will be reduced and the river will increase its velocity. The surplus energy, resulting from the decrease in friction, can now be used to pick up and transport material. The greater the velocity of a river the greater the amount of material, both in quantity and size, that can be carried. The material that is transported by a river is called its *load*.

Figure 15.9
Rapids at Aberglaslyn, North Wales

Figure 15.10
Velocity and discharge in the upper and lower course of a highland stream

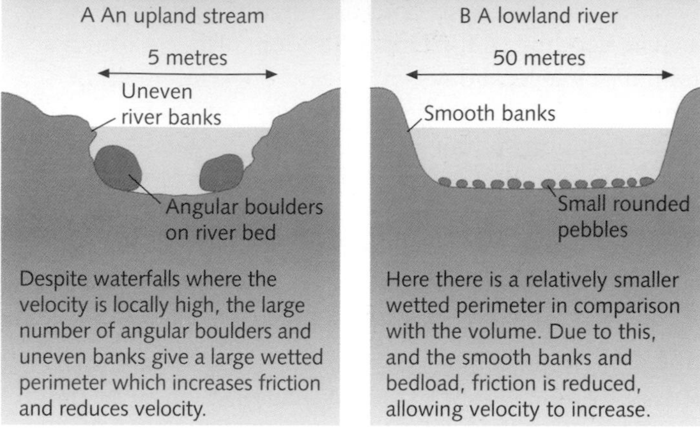

Transportation

A river can transport its load by one of four processes: traction and saltation, along its bed; and suspension and solution, within the river itself (Figure 15.11).

Erosion

A river uses the transported material to erode its banks and bed. As the velocity of a river increases, so too does the load it can carry and the rate at which it can erode. A river may erode by one of four processes:

- **Attrition** is when boulders and other material, which are being transported along the bed of the river, collide and break up into smaller pieces. This is more likely to occur when rivers are still flowing in highland areas.
- **Hydraulic action** is when the sheer force of the river dislodges particles from the river's banks and bed.
- **Corrasion** occurs when smaller material, carried in suspension, rubs against the banks of the river. This process is more likely in lowland areas by which time material will have been broken up small enough to be carried in suspension. River banks are worn away by a sand-papering action which is also called *abrasion*.
- **Corrosion** is when acids in the river dissolve rocks, such as limestone, which form the banks and bed. This can occur at any point of the river's course.

Deposition

Deposition occurs when a river lacks enough energy to carry its load. Deposition, beginning with the heaviest material first, can occur following a dry spell when the discharge and velocity of the river drop, or where the current slows down (the inside of a meander bend or where the river enters the sea).

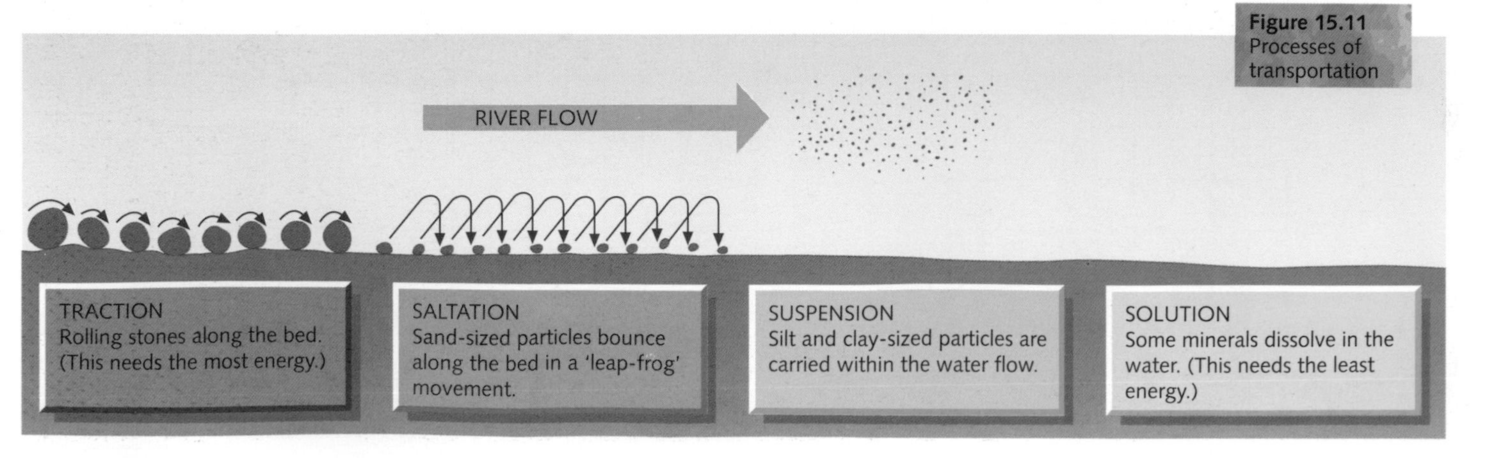

Figure 15.11
Processes of transportation

River landforms in a highland area

Figure 15.12
V-shaped valley and interlocking spurs, in the Peak District

V-shaped valleys and interlocking spurs

Any spare energy possessed by a river near to its source will be used to transport large boulders along its bed. This results in the river cutting rapidly downwards, a process called *vertical erosion*. Vertical erosion leads to the development of steep-sided, narrow valleys shaped like the letter V (Figure 15.12). The valley sides are steep due to soil and loose rock being washed downhill following periods of heavy rainfall. The material is then added to the load of the river. The river itself is forced to wind its way around protruding hillsides. These hillsides, known as *interlocking spurs*, restrict the view up or down the valley.

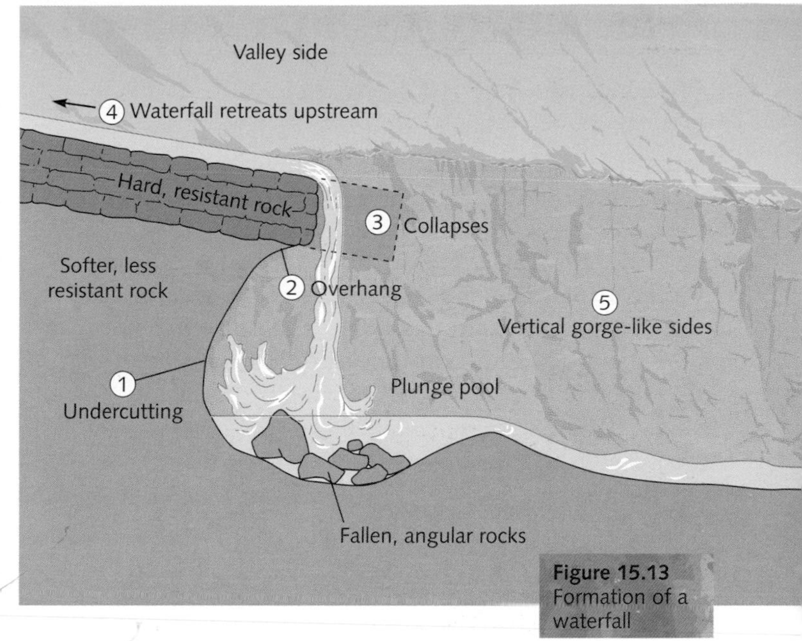

Figure 15.13
Formation of a waterfall

Waterfalls and rapids

Waterfalls form when there is a sudden interruption in the course of a river. They may result from erosion by ice (Figure 16.9), changes in sea-level (Figure 16.26), and Earth movements. However, many waterfalls form when rivers meet a band of softer, less resistant rock after flowing over a relatively hard, resistant rock (Figure 15.13). The underlying softer rock is worn away more quickly, and the harder rock is undercut. In time the overlying harder rock will become unsupported and will collapse. After its collapse, some of the rock will be swirled around by the river, especially during times of high discharge, to form a deep plunge pool. This process is likely to be repeated many times, causing the waterfall to retreat upstream and leave a steep-sided gorge (Figure 15.14). The Niagara Falls are retreating by one metre a year. *Rapids* occur where the layers of hard and soft rock are very thin, and so no obvious break of slope develops as in a waterfall .

Figure 15.14
Niagara Falls

River landforms in a lowland area

Meanders and ox-bow lakes

As a river approaches its mouth it usually flows over flatter land and develops increasingly large bends known as *meanders* (Figure 15.15). Meanders constantly change their shape and position. When a river reaches a meander most water is directed towards the outside of the bend (Figure 15.16). This reduces friction and increases the velocity of the river at this point. The river therefore has more energy to transport material in suspension. This material will erode the outside bank by corrasion. The bank will be undercut, collapse and retreat to leave a small river cliff. The river is now eroding through *lateral*, not vertical, *erosion*.

Meanwhile, as there is less water on the inside of the bend, there is also an increase in friction and a decrease in velocity. As the river loses energy it begins to deposit some of its load. The deposited material builds up to form a gently sloping slip-off slope (Figure 15.17).

Continual erosion on the outside bends results in the neck of the meander getting narrower until, usually at a time of flood, the river cuts through the neck and shortens its course. The fastest current will now be flowing in the centre of the channel and deposition is more likely next to the banks. The original meander will be blocked off to leave a crescent-shaped *ox-bow lake*. This lake will slowly dry up, except during periods of heavy rain.

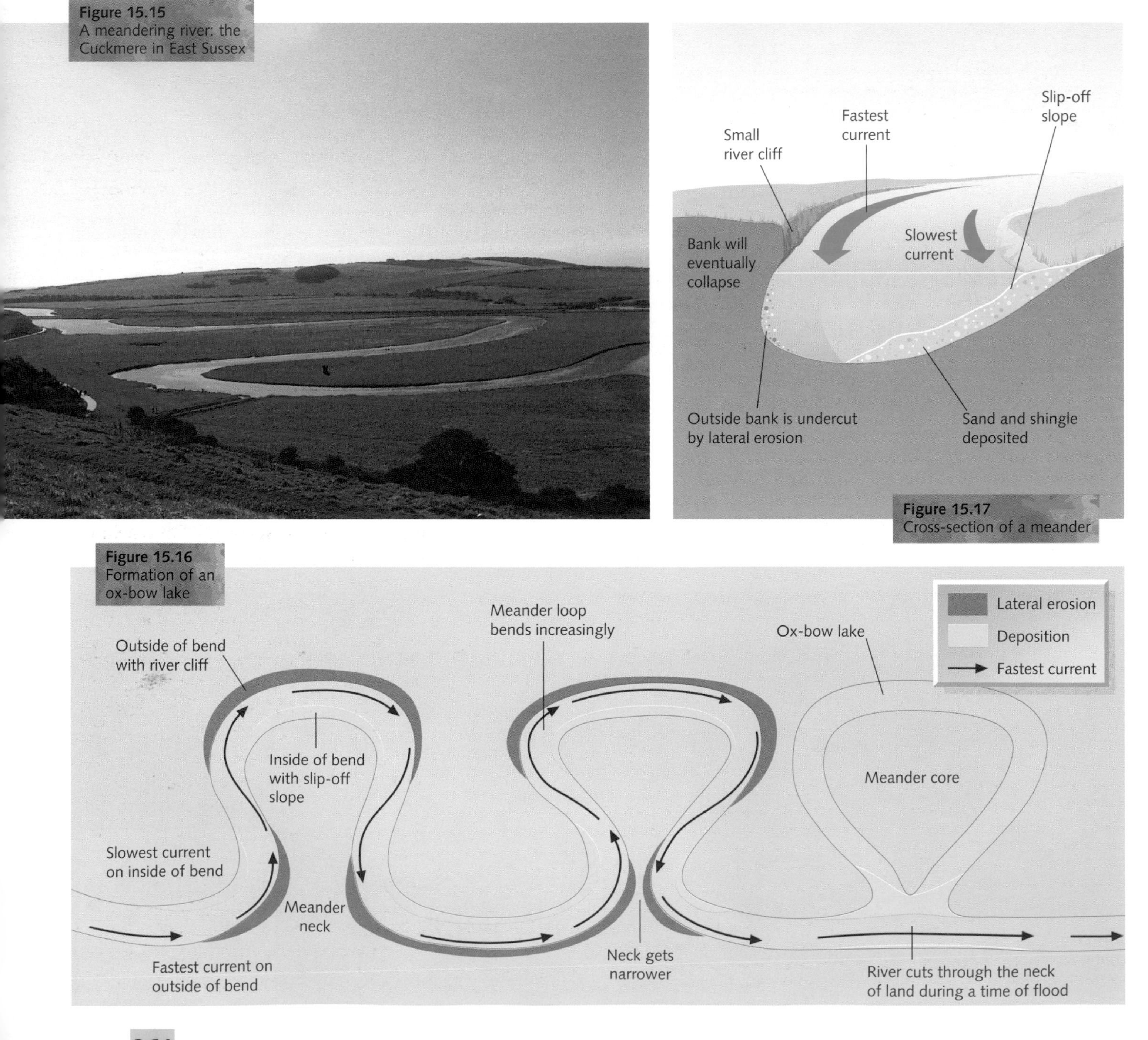

Figure 15.15
A meandering river: the Cuckmere in East Sussex

Figure 15.17
Cross-section of a meander

Figure 15.16
Formation of an ox-bow lake

Floodplain and levées

The river widens its valley by lateral erosion. At times of high discharge, the river has considerable amounts of energy which it uses to transport large amounts of material in suspension. When the river overflows its banks it will spread out across any surrounding flat land. The sudden increase in friction will reduce the water's velocity and fine silt will be deposited. Each time the river floods another layer of silt is added and a flat *floodplain* is formed (Figure 15.18). The coarsest material will be dropped first and this can form a natural embankment, called a *levée*, next to the river. Sometimes levées are artificially strengthened to act as flood banks. Some rivers, like the Mississippi, flow between levées at a height well above their floodplain. If, during a later flood, the river breaks through its levées, then widespread flooding may occur.

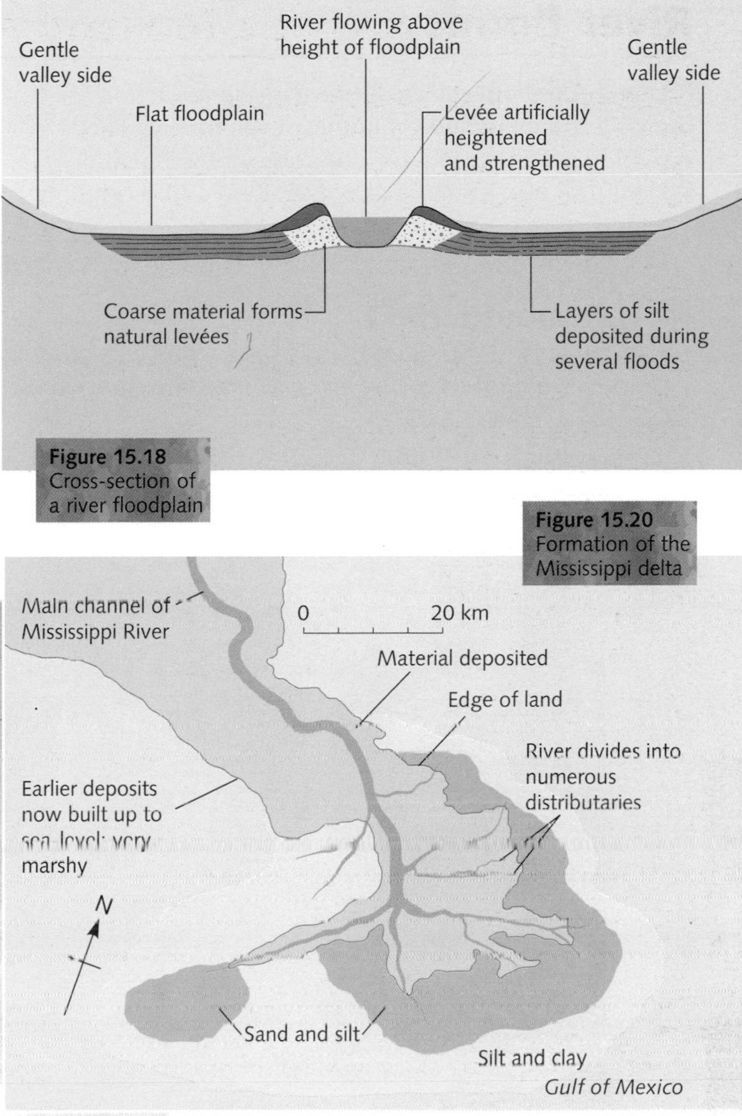

Figure 15.18 Cross-section of a river floodplain

Figure 15.20 Formation of the Mississippi delta

Figure 15.19 The Mississippi delta

Deltas

As large rivers approach the sea, they have the energy to carry large amounts of fine material in suspension. On reaching the sea, the river current may suddenly be reduced and the material deposited (Figure 15.19). Sometimes deposition occurs in the main channel, and blocks it. The river has then to divide into a series of channels, called *distributaries*, in order to find its way into the sea. Over a period of time the deposited materials of sand and silt may build upwards and outwards to form a *delta* (Figure 15.20). Deltas are only likely to form where the amount of material brought down by a river is too great for sea currents to remove it (e.g. Mississippi and Ganges) or in seas that are virtually tideless (the Nile and the Rhône in the Mediterranean Sea). Deltas can also form when a river flows into the gentle waters of a lake.

River floods

Rivers throughout the world provide an attraction for human settlement and economic development. They provide a water supply for domestic, industrial and agricultural use; a source of power, food and recreation; and a means of transport. However, under extreme climatic conditions, and increasingly due to human mismanagement, rivers can flood and cause death and widespread damage.

Places

Lynmouth 1952

One of the worst floods in living memory in Britain devastated the North Devon village of Lynmouth in 1952 (Figure 15.21).

Figure 15.21
Devastation in Lynmouth, 1952

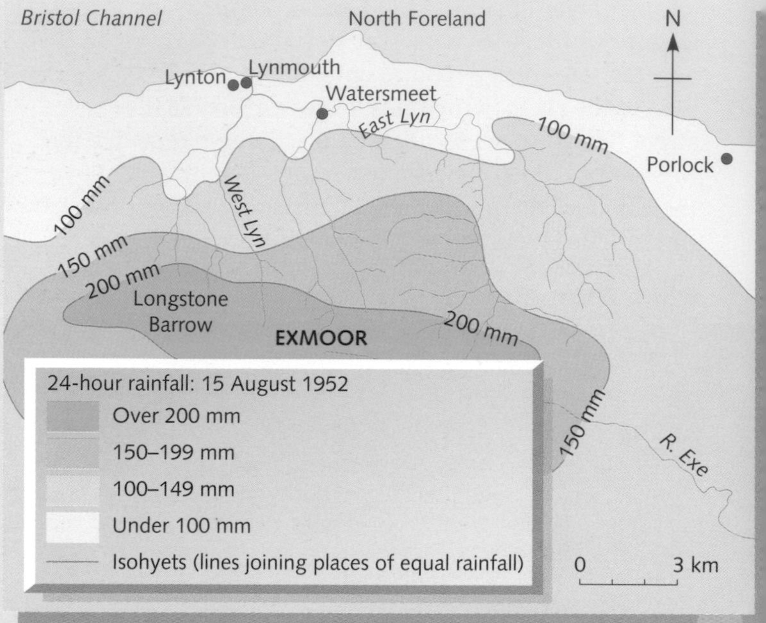

Figure 15.22
Drainage basin of East and West Lyn Rivers

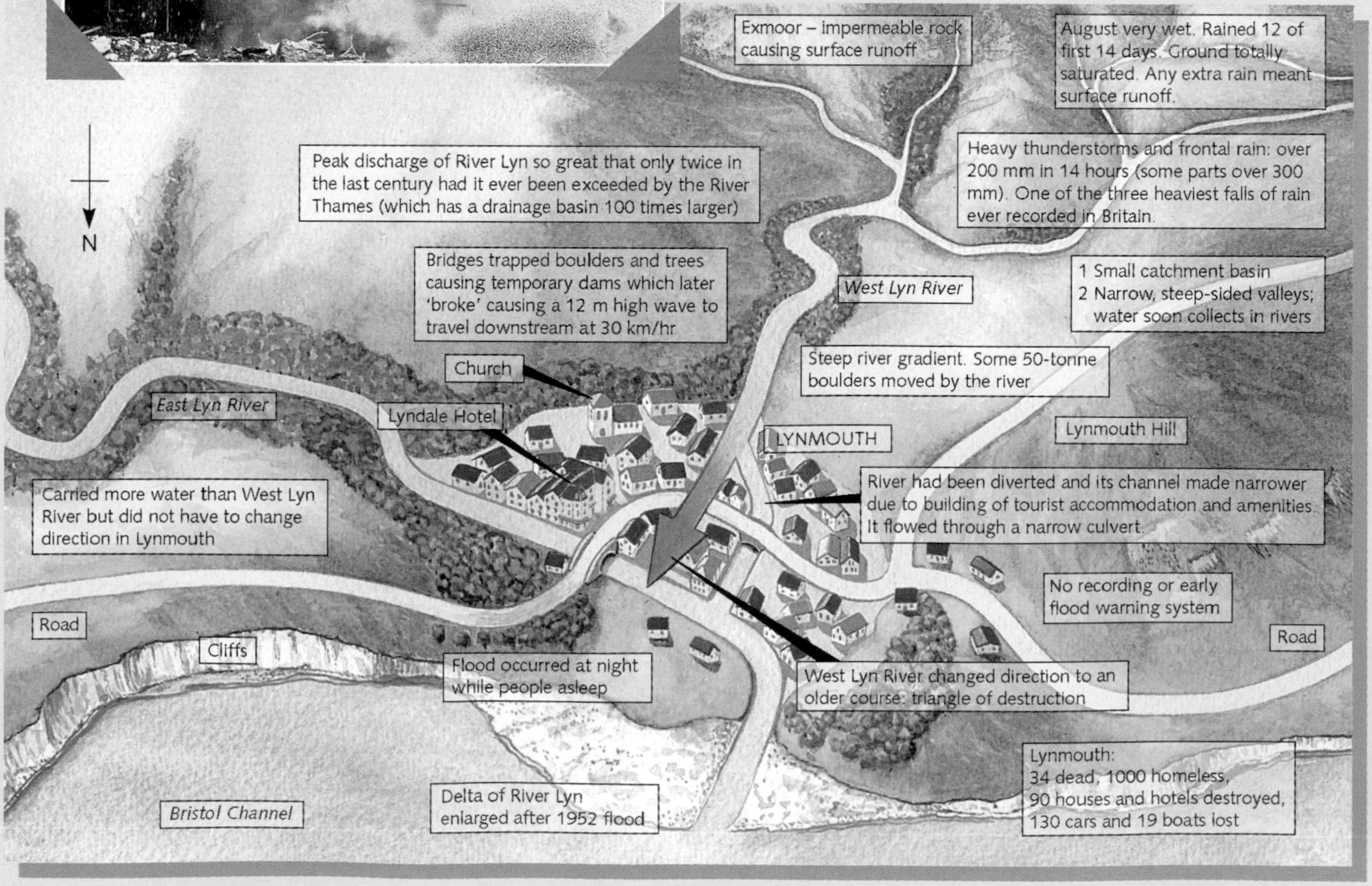

Figure 15.23 Causes and effects of the Lyn flood, 15 August 1952

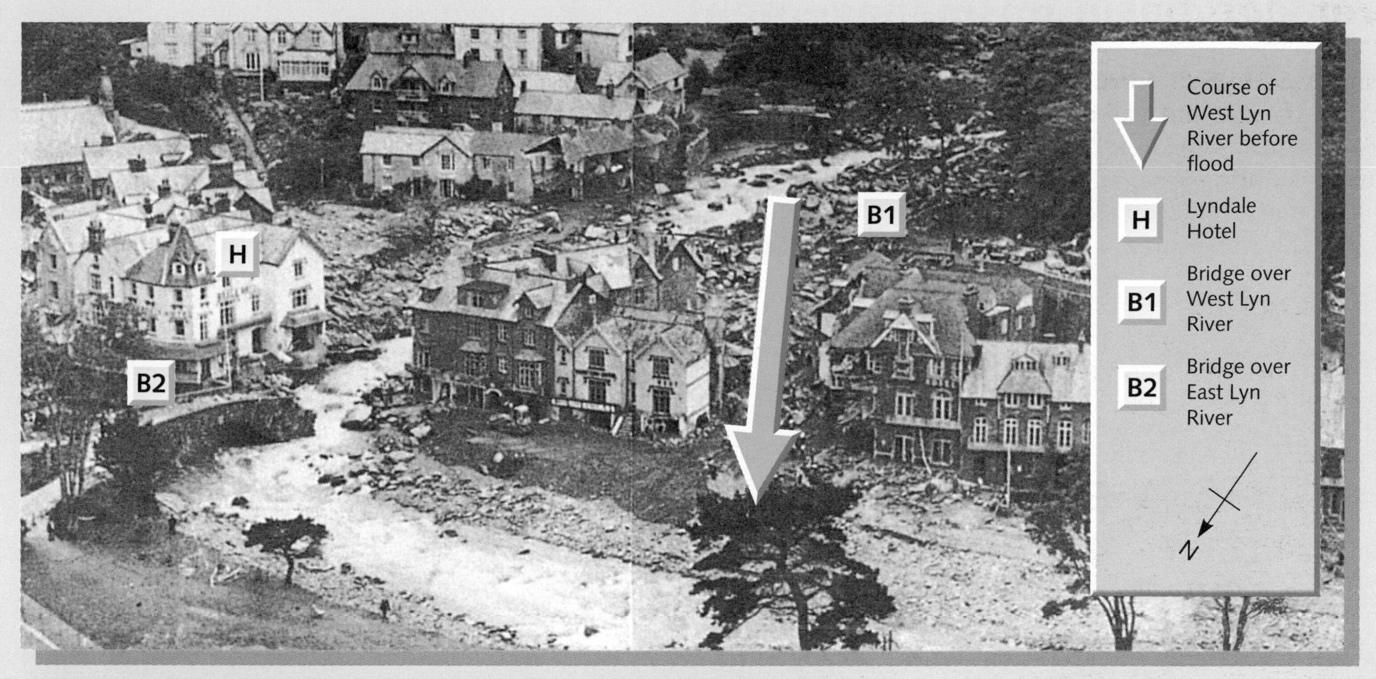

Figure 15.25
River Lyn changing course while in flood

Cause of the 1952 flood

The first fortnight of August 1952 was exceptionally wet throughout south-west England. This meant that if any more rain fell on the already saturated soils of Exmoor it would very quickly reach the various tributaries of the East and West Lyn and, within a few hours, would be rushing through Lynmouth since this village lay at the junction of the two rivers. Tragically for the people of Lynmouth an unprecedented deluge of rain fell on 15 August – one of the three heaviest falls of rain ever recorded in 24 hours in the British Isles. Thunderstorms associated with a small frontal depression produced torrential rain as the Exmoor plateau squeezed every drop of rainfall from the overlying moist atmosphere: 9 inches [229 mm] was measured near Longstone Barrow about 5 miles [8 km] south of Lynmouth, and some localities may have suffered as much as 12 inches [305 mm]. Over 3000 million gallons of water fell into the drainage area of the two Lyn rivers. The river channels could not cope with such an overwhelming amount of water and devastating floods were inevitable.

Figure 15.24 The River Lyn flood, 1952

- A West Lyn allowed to follow its natural route (the one taken during the flood)
- B Floodplain left open to take excess water
- C Larger bridges with much wider spans
- D Land where hotel was is now left open (car park).
- E East Lyn channel has been straightened to allow flood water to flow off more quickly

Figure 15.26 Lynmouth in the 1990s

A flood of this size may occur only once in every 100 or even 200 years. There were several physical reasons why the situation of Lynmouth was a flood risk. Figure 15.22 shows a small river basin with a high drainage density (page 258) in an area that can expect heavy rainfall. The rocks in the basin are impermeable, and the valley sides and the gradient of the rivers are both steep (Figure 15.23). The flood risk had, however, been increased by human activity (Figure 15.23). Even so, no one was prepared for the events of 15 August 1952 (Figure 15.24). The real cause of the damage was the huge volume of water which enabled the West Lyn to move gigantic boulders and to transport vast amounts of material. The river also changed its course so that it flowed through the centre of the village (Figure 15.25).

The flood has had a lasting impact upon the village, both physically and emotionally. Lynmouth has a different appearance today, as it was rebuilt to try to ensure the safety of its inhabitants rather than to try to recapture all of its former character (Figure 15.26). The flood management plan has, to date, managed to cope with any excess water in the Lyn drainage basin despite exceptionally heavy rainfall on three occasions since it was implemented.

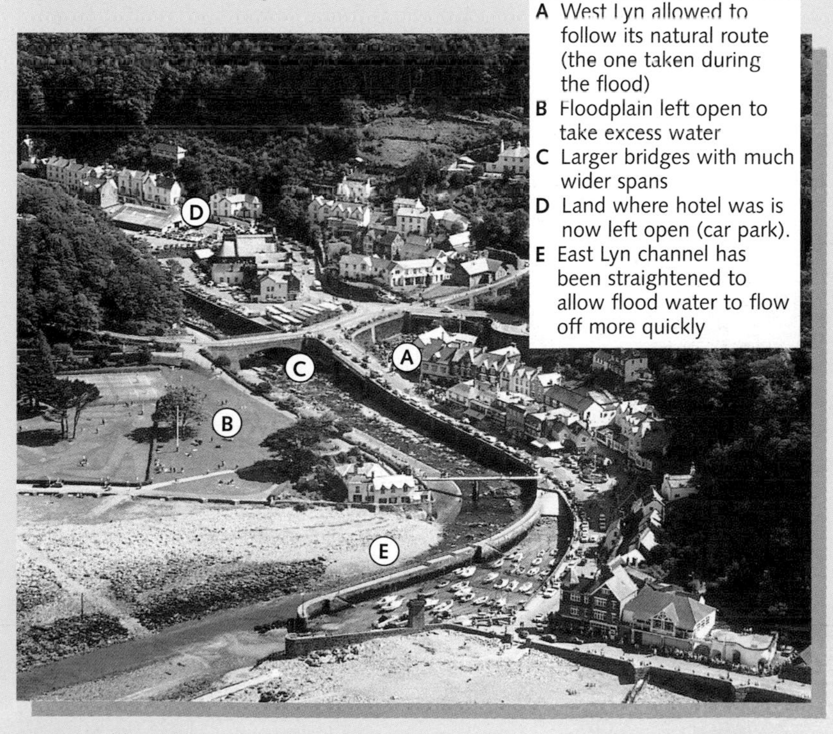

Drainage basin mismanagement

For many centuries, people across the world have perceived rivers to be a cheap and convenient method of removing their waste, whether it was domestic, agricultural or, more recently, industrial. In England and Wales, the Environment Agency categorises incidents of water pollution by both source and type (Figure 15.27). In 1995:

- **By source**, the sewage and water industry accounted for 30 per cent of incidents (Figure 15.27b), with surface water outfall and combined sewer overflows accounting for the largest proportion. Although the total number of substantiated incidents increased slightly between 1991 and 1995, mainly in agriculture (e.g. fertiliser and farm slurry – page 96), the number of major incidents decreased considerably. This decrease in major incidents has been credited partly to the efforts of the Environment Agency and partly to increased awareness by the general public in reporting incidents. Of concern, however, has been the increase in major incidents relating to road accidents (Figure 15.27b).
- **By type**, sewage and oil accounted for over half of the substantiated cases (Figure 15.27c), with oil and chemicals being the cause of most major incidents.

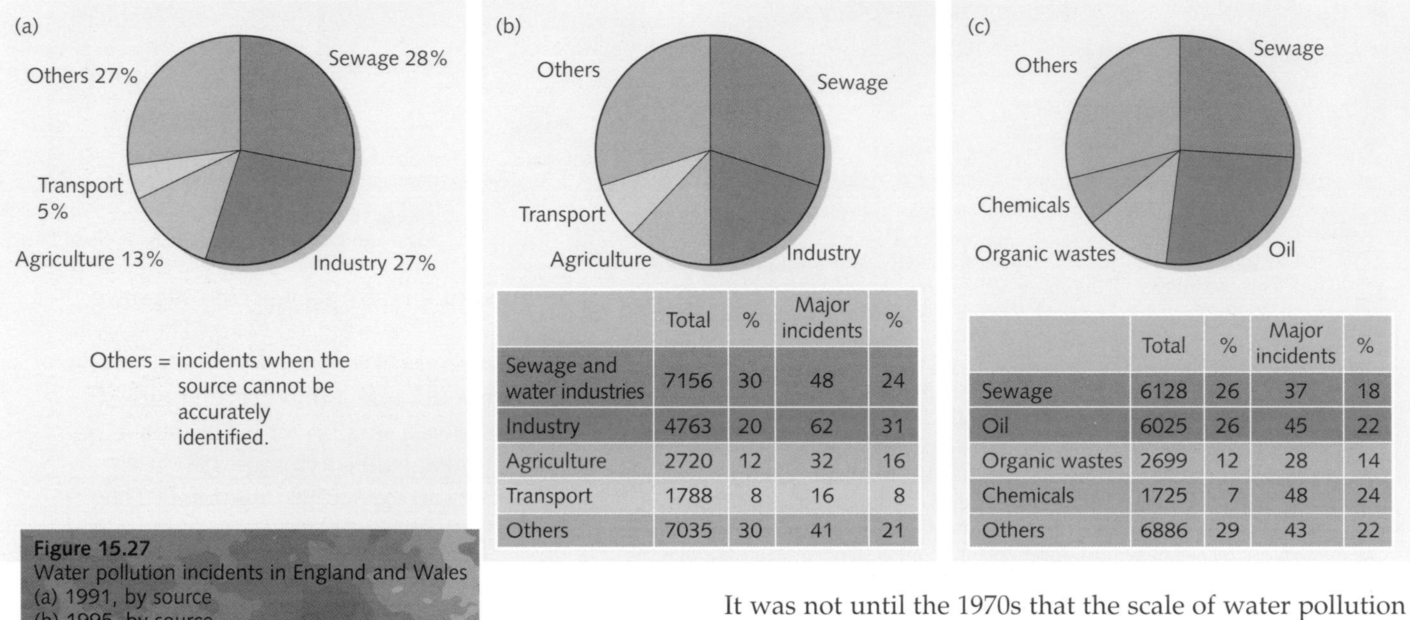

	Total	%	Major incidents	%
Sewage and water industries	7156	30	48	24
Industry	4763	20	62	31
Agriculture	2720	12	32	16
Transport	1788	8	16	8
Others	7035	30	41	21

	Total	%	Major incidents	%
Sewage	6128	26	37	18
Oil	6025	26	45	22
Organic wastes	2699	12	28	14
Chemicals	1725	7	48	24
Others	6886	29	43	22

Figure 15.27
Water pollution incidents in England and Wales
(a) 1991, by source
(b) 1995, by source
(c) 1995, by type

It was not until the 1970s that the scale of water pollution was first recognised in the USA. As shown in Figure 15.28, the main causes were factories and cities releasing untreated waste into lakes and rivers, and runoff from farms, construction sites, cities and mines.

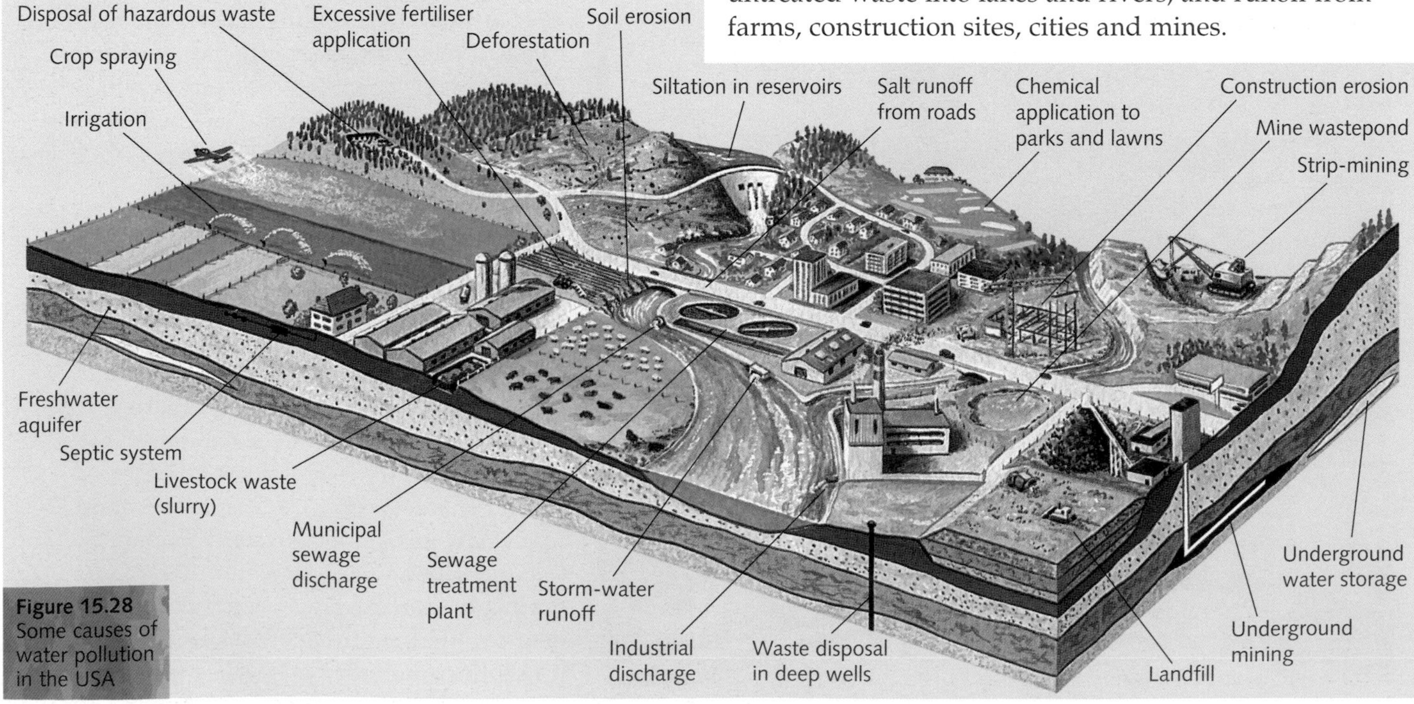

Figure 15.28
Some causes of water pollution in the USA

Water pollution in Japan before 1980

In its attempts to industrialise, Japan initially paid little attention to the effects this would have on its environment. Water resources, both freshwater (rivers and lakes) and saltwater (the sea), became severely polluted, causing major health problems in humans and adversely affecting the habitats of plants and animals. Some of the main causes of water pollution are described here.

Farming As Japan became more industrialised and wealthy, and as its population continued to grow rapidly in number, it could afford to use an increasing amount of fertiliser and pesticides needed to increase food supply. Unfortunately, much of the fertiliser was washed downwards through the soil (leaching) by the heavy rainfall, and into padi-fields, rivers, lakes and, eventually, the sea. Nitrate fertiliser encouraged the growth of algae (Figure 9.33) and other water plants which used up oxygen and left insufficient for fish to survive. Figure 15.29 shows how increases in the amount of ammonia and nitrate in the water reduce oxygen and plant life. The use of pesticides often killed bees and other useful insects as well as undesirable pests.

Sewage Many rural areas were too isolated to have mains sewerage, while urbanisation was often too rapid for the authorities to keep pace with the demand for extra drains. In rural areas untreated sewage was liable to find its way into water intended for either drinking purposes or for rice-fields. In urban areas, many of which were near to the coast, it was often allowed to escape into the sea. The worst-affected areas were the Inland Sea and Tokyo Bay (Figure 15.30).

Industry Industry also released large amounts of waste into rivers, lakes and the sea. Water in lakes such as Lake Biwa (Figure 15.30) became dull, lifeless, and unsuitable for domestic use or plant and animal life. The larger industries of steel, shipbuilding, chemicals, cars and engineering, were located along the coast or at mouths of rivers where they could easily discharge their waste into estuaries or the sea, killing fish and other marine life. In one extreme case during the 1950s, mercury waste was released into Minamata Bay. It was converted by bacteria into a substance which, having been absorbed by fish, entered the food chain. This led to the death of many birds and cats (which ate the fish) and eventually, as mercury accumulated in human bodies, over 100 people died. Children were born with mental and physical defects (e.g. blindness, deformed limbs). In 1973, the government advised pregnant women to avoid all sea-food, and recommended that the remainder of the population did not consume more than the equivalent of six prawns or 0.5 kg of tuna fish per week.

Power stations Thermal power stations ejected hot water into rivers and the sea, raising the temperature beyond that usually tolerated by plants and fish, and reducing the oxygen content. The construction of nuclear power stations (Figure 7.7) increased the risk of radioactive leaks into adjacent rivers and coastal waters.

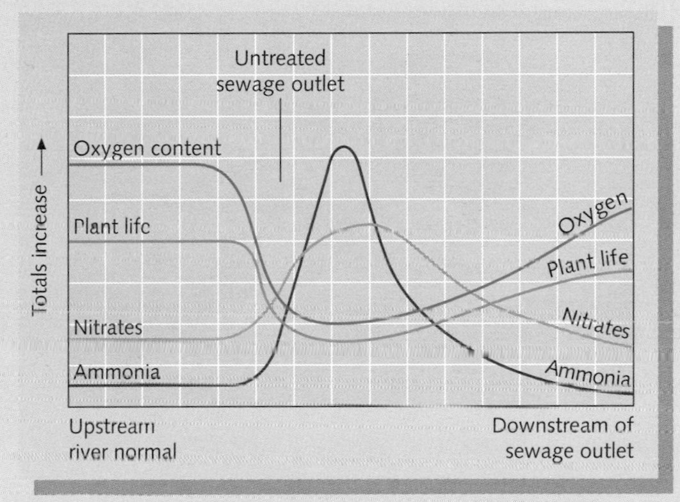

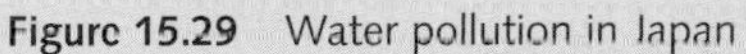
Figure 15.29 Water pollution in Japan

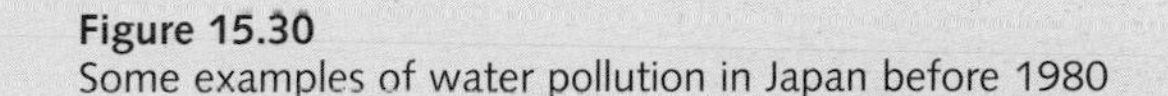
Figure 15.30
Some examples of water pollution in Japan before 1980

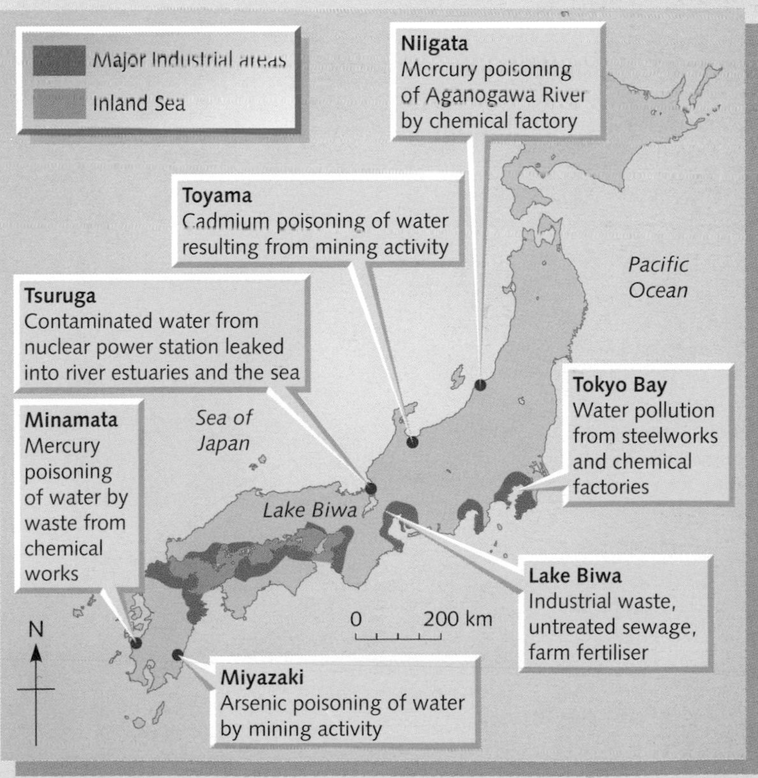

Drainage basin management

Should rivers be allowed to flood?

Some rivers, like the Mississippi in the USA, the Po in Italy and the Huang He in China, have built up their beds so that they flow above the level of their floodplains. Such rivers can only be prevented from flooding by building dams, diversion channels and artificial embankments (i.e. levées – Figure 15.18). The problem arises when a serious flood risk occurs. If the levées break, or if they have not been built high enough, then large areas of land are flooded and numerous lives are put at risk. Following the Mississippi floods of 1993, scientists and engineers began asking whether the dams, diversion channels and levées had actually aggravated that and other floods. There are two schools of thought. One accepts that as floods are part of a river's natural cycle and the drainage basin's ecosystem, then flooding should be allowed as a natural event (Figure 15.31). The other argues for better flood defences and a more effective control of rivers (Figure 15.32).

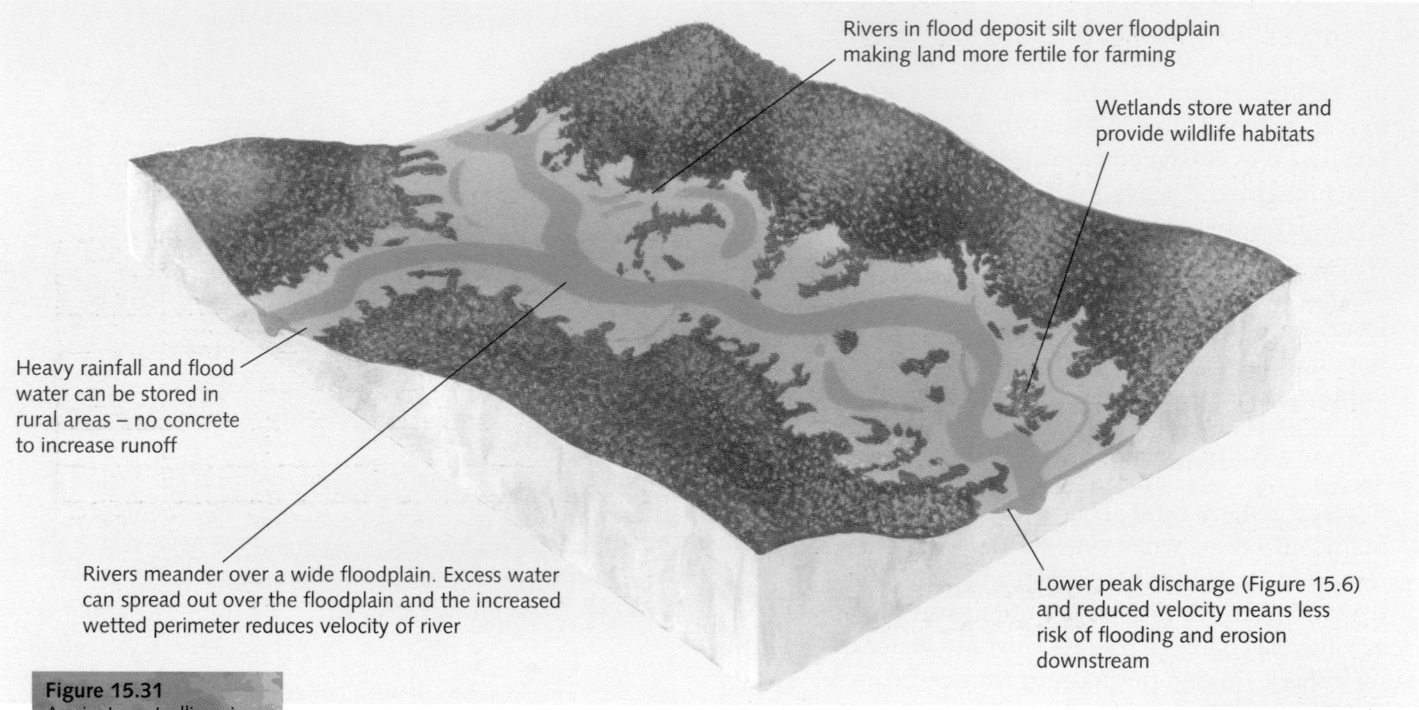

Figure 15.31
Against controlling rivers

Figure 15.32
For controlling rivers

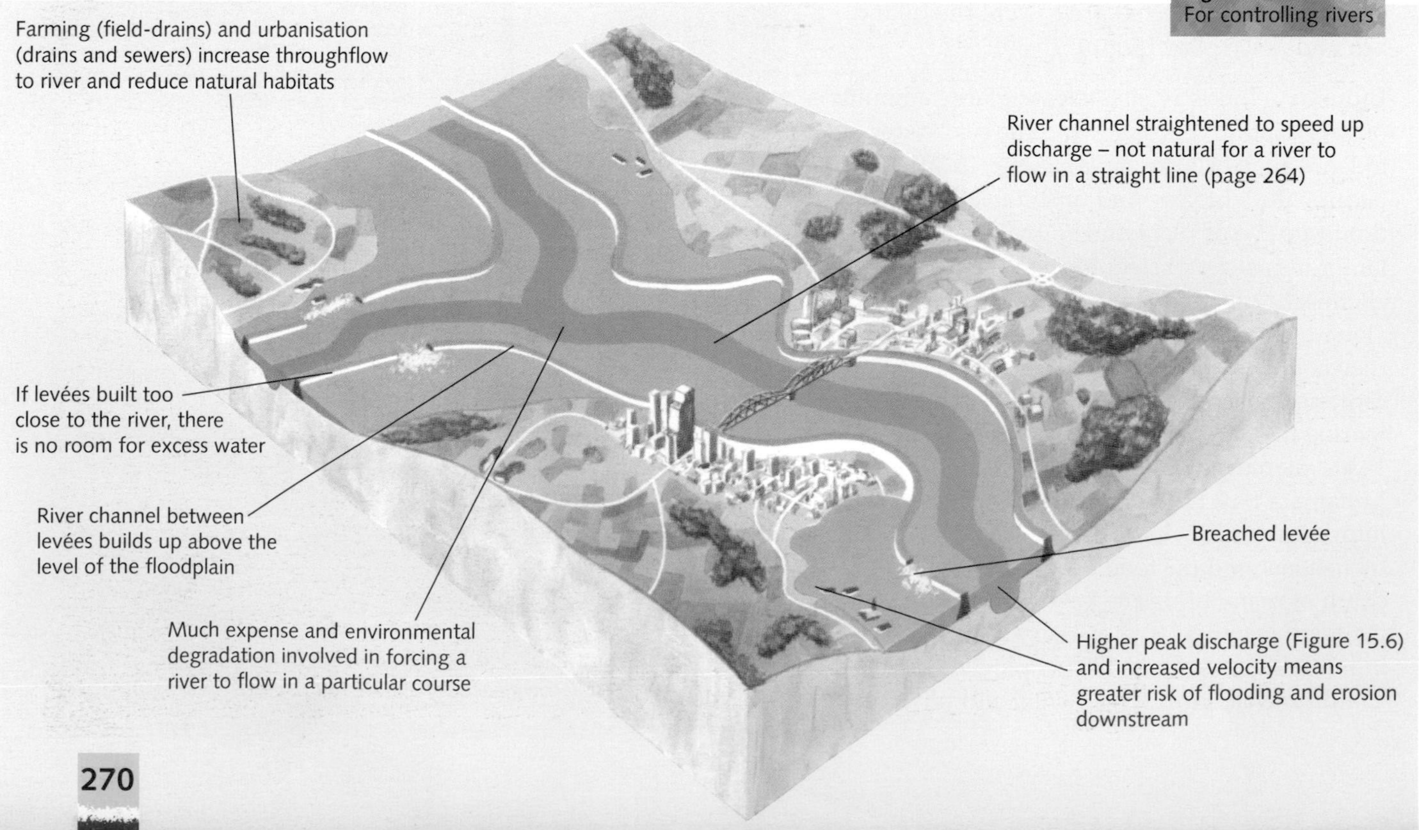

Clean-up schemes in the UK

The Environment Agency, which took over the role of the National Rivers Authority in 1996, hopes by the turn of the century to:

- improve water quality by
  - improving water treatment works and so reducing the discharge of sewage
  - reducing the amount of contaminated land and the number of derelict industrial sites
  - reducing and controlling effluent from existing industries and discharge from waste disposal sites
  - pumping water from abandoned coal mines
  - reducing agricultural pollution in order to protect supplies of drinking water and wildlife habitats
- regulate water abstraction so that water supplies and river levels can both be maintained during times of drought
- improve physical habitats to allow the development of sustainable fish and other wildlife populations
- continue to improve flood defences where deemed viable and economic (Figures 15.31 and 15.32), and maintain early warning systems
- improve visitor access (e.g. riverside trails) and areas for recreation (e.g. for anglers, birdwatchers and canoeists).

Water quality in Japan since 1980

During the 1970s the Japanese government produced standards designed to improve the quality of their environment. High on the list were standards designed to control the levels of water pollution (Places, page 269). Fortunately for Japan, the country had both the technology and the capital needed to tackle its environmental problems.

Within a few years there was a marked improvement in water quality in rivers, lakes and the surrounding sea areas. Levels of river pollutants, such as nitrate, phosphorus, lead, cadmium, ammonia and mercury, all fell significantly. This was especially so in the case of rivers flowing into Japan's two most industrialised bays at Tokyo and Osaka (Figure 15.33).

RIVERS	1980	1985	1990	1995
YODO (OSAKA)	3.3	3.4	2.5	2.1
TAMA (TOKYO BAY)	6.7	4.7	4.6	3.5
ISHIKARI	1.5	1.5	1.7	1.0
CHIKUKO	1.9	2.2	1.7	0.9
LAKES				
LAKE BIWA	27	22	13	2

POLLUTANTS INCLUDE NITRATES, PHOSPHATES, AMMONIA, LEAD AND CADMIUM

Figure 15.33
Total concentrations (%) of pollutants in selected Japanese rivers and lakes

There has also been an improvement in the quality of lake water. Lake Biwa is now a major tourist area with numerous newly created 'swimming resorts', and its southern portion is a designated Quasi National Park (Figure 15.34). Fish have returned to rivers, previously too polluted for their survival, and to the Inland Sea, where now there are fish farms and oyster beds.

Figure 15.34
Lake Biwa

The Three Gorges Dam, China

The Yangtze (Chang Jiang) River and the Three Gorges

The Yangtze, or Chang Jiang – the 'Long River' – is the longest river in Asia. Its source, near to the border with Tibet, is 5600 km from where it enters the Yellow Sea near Shanghai (Figure 15.35). Its drainage basin, which covers 20 per cent of China, is home to one in thirteen of the world's population.

On a 190 km stretch of the river in west-central China, lie the famed Three Gorges (Figure 15.36). The westernmost gorge, Qutang Gorge, is 13 km long and walled by nearly vertical cliffs. The next, Wu Gorge, is 43 km long and steeped in mythology. The last, Xiling Gorge, is 75 km long and filled with dangerous shoals and hidden rocks.

Figure 15.35
The Three Gorges Dam project

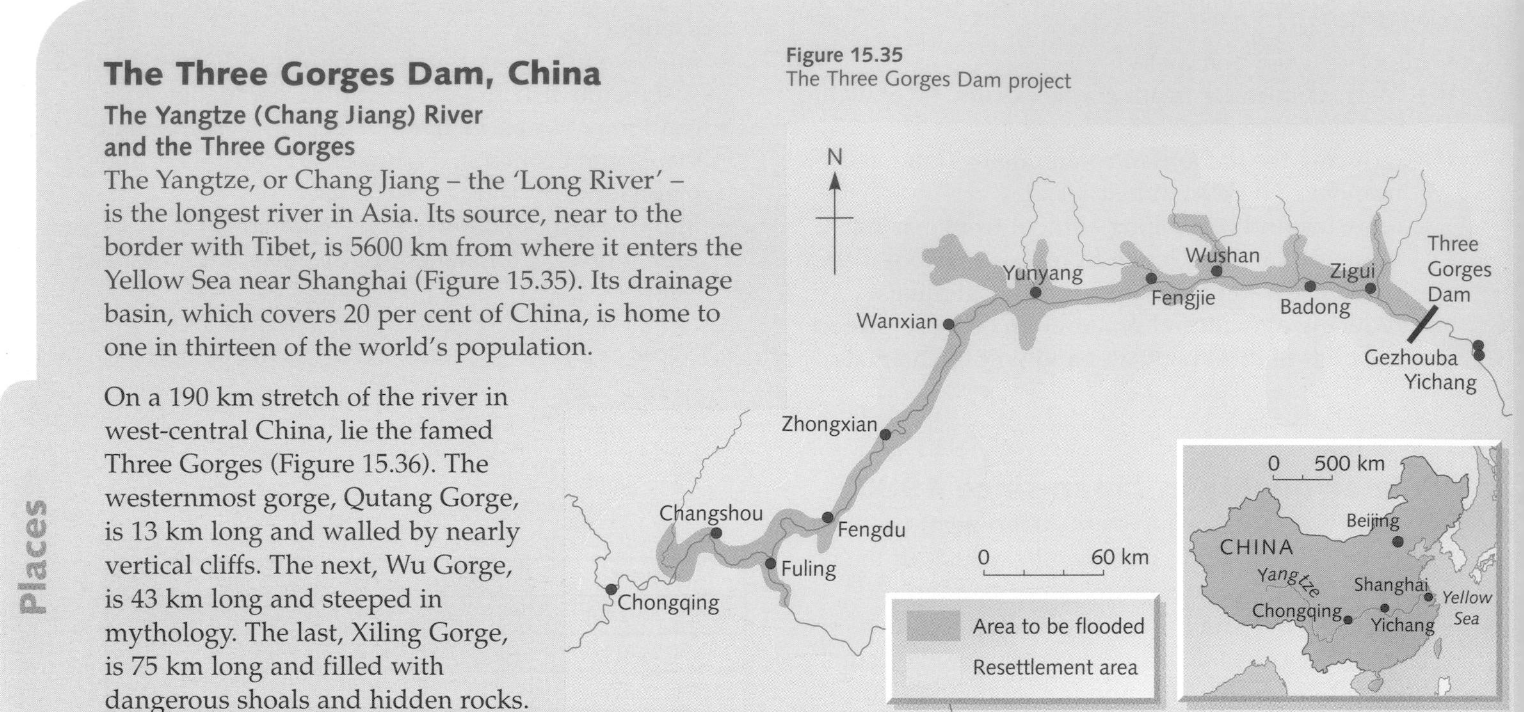

Figure 15.36 In one of the Three Gorges before work began on the project

In each of the past 20 centuries there has been at least one catastrophic flood. Already this century there have been three: in 1931, 1934 and 1954, and over 300 000 people have been drowned. For much of the river's middle section, some of the most fertile land in the country and many of its people are protected only by vulnerable levées. Each year, summer rains bring the prospect of a good harvest and the fear of the river bursting its banks.

The Three Gorges Dam project

The Three Gorges Dam will be the world's largest (Figure 15.37). It is meant to control flooding, improve navigation and to provide energy-inefficient China with cheap hydro-electricity. The project will cost at least $20 billion, it will take 18 years to complete, force 1.2 million people to move home and will flood vast tracts of farmland. Great doubts remain on the project's long-term economic, social and ecological costs.

The dam	Built across the Yangtze at Sandouping (Figure 15.35).1983 metres long, 185 metres high, 18 years to complete (in 2010).
Water levels	175 metres at dam site. Between 215 and 225 metres in the Qutang and Wu Gorges.
Lake/Reservoir	660 km long. Storage capacity 39.3 billion cubic metres.
Power plant	Will produce 17 680 MW (compared with 12 600 MW at the world's present largest at Itaipù – page 110).
Navigation	At present only a few vessels can negotiate the Three Gorges, and only at certain times of year. In future 10 000 vessels will be able to reach Chungking (Chongqing). The volume of trade is expected to increase by 400 times.

Figure 15.37
Fact File

Work on the project

Figure 15.38, adapted from an article in the *Sydney Morning Herald*, describes the scene at the dam in early 1996. The first major stage, the diversion of the Yangtze to allow the construction of the dam to begin, was completed in October 1997.

Figure 15.38
Work on the Three Gorges project, 1996

This is the biggest building site in the world. Senior engineer Li Junlin pointed to where there had once been an island, and then to the huge man-made earthworks which will be used to divert the mighty Yangtze River. As he spoke, hundreds of workers with bulldozers and cement machines toiled away below him shifting millions of tonnes of soil and rock. Li then pointed out two markers high on opposite sides of the river. Between these points, a 160 metres high concrete curtain will divide the valley, flooding land equivalent in area to the island of Singapore. When the dam is completed, the famous Three Gorges scenery will be half drowned by the world's largest reservoir.

Construction of the dam and hydro-electric plant probably represents the largest earth-moving exercise in history. By the scheduled completion date in 2010, some 95 million cubic metres of soil and rock will have been excavated. Most controversially, 1.2 million people, 13 cities, 140 towns, 4500 villages and 1600 industrial enterprises will have been relocated. Engineer Li gives a well-rehearsed argument in favour of the dam – flood control, all-year-round navigation, 10 per cent of China's present electricity requirements and, by replacing coal-burning power stations with hydro-electric stations, dramatically reducing emissions of sulphur dioxide (Figure 11.42b). He claims that the authorities have taken into account the problems of the reservoir rapidly filling with silt and the threats of landslides and earthquakes (the dam lies near an earthquake fault line). Li has little to say on such problems as the rehousing of people, the movement of settlements, the loss of 7000 hectares of forest, the drowning of 23 000 hectares of fertile farmland, the loss of fertile silt to farms downriver from the dam, or the risk of disaster on an unprecedented world scale should the dam fail (outside experts predict such an event could have a death toll of several million). The authorities have forbidden free debate and access to information that might allow their own people, as well as the outside world, to judge for themselves.

Adapted from an article by Teresa Poole, *The Independent Newspaper*, Australia

See question 14 on page 281.

Figure 15.39
The Three Gorges dam under construction

CASE STUDY 15

The Nile Valley

Although the Nile is the world's longest river (6995 km), its drainage basin covers an area less than half that of the Amazon (just over 3000 km^2) and its mean annual discharge is seven times less. Almost 85 per cent of the Nile's water which reaches Egypt originates in the Ethiopian Highlands, the remainder coming from the equatorial areas of East Africa (Figure 15.40). Unlike other large rivers, nearly all the Nile's water comes from just three tributaries – the White Nile, the Blue Nile and the Atbara – and for the last 2700 km of its journey to the Mediterranean Sea it flows through desert without receiving any additional tributary.

- The tributaries of the White Nile, which drain mountainous areas near to the Equator, flow into Lake Victoria. Here, the equatorial climate, with rain spread throughout the year, and Lake Victoria, which acts as a natural reservoir releasing water at an even rate, combine to give the White Nile a fairly constant discharge (Figure 15.41).
- In contrast, the Blue Nile and Atbara Rivers rise in the Ethiopian Highlands where rainfall, and consequently the flow of the rivers, is seasonal (Figure 15.41). It was, however, the six months of heavy rain in Ethiopia that was responsible for the Nile's annual flood.

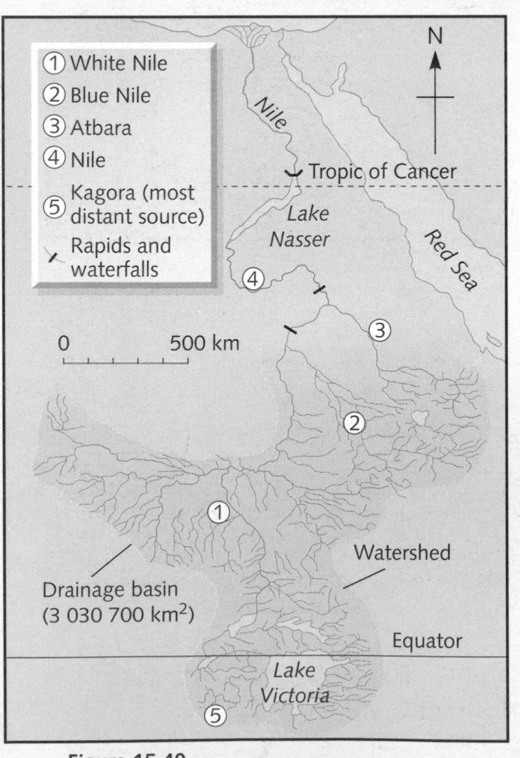

Figure 15.40
The drainage basin of the River Nile

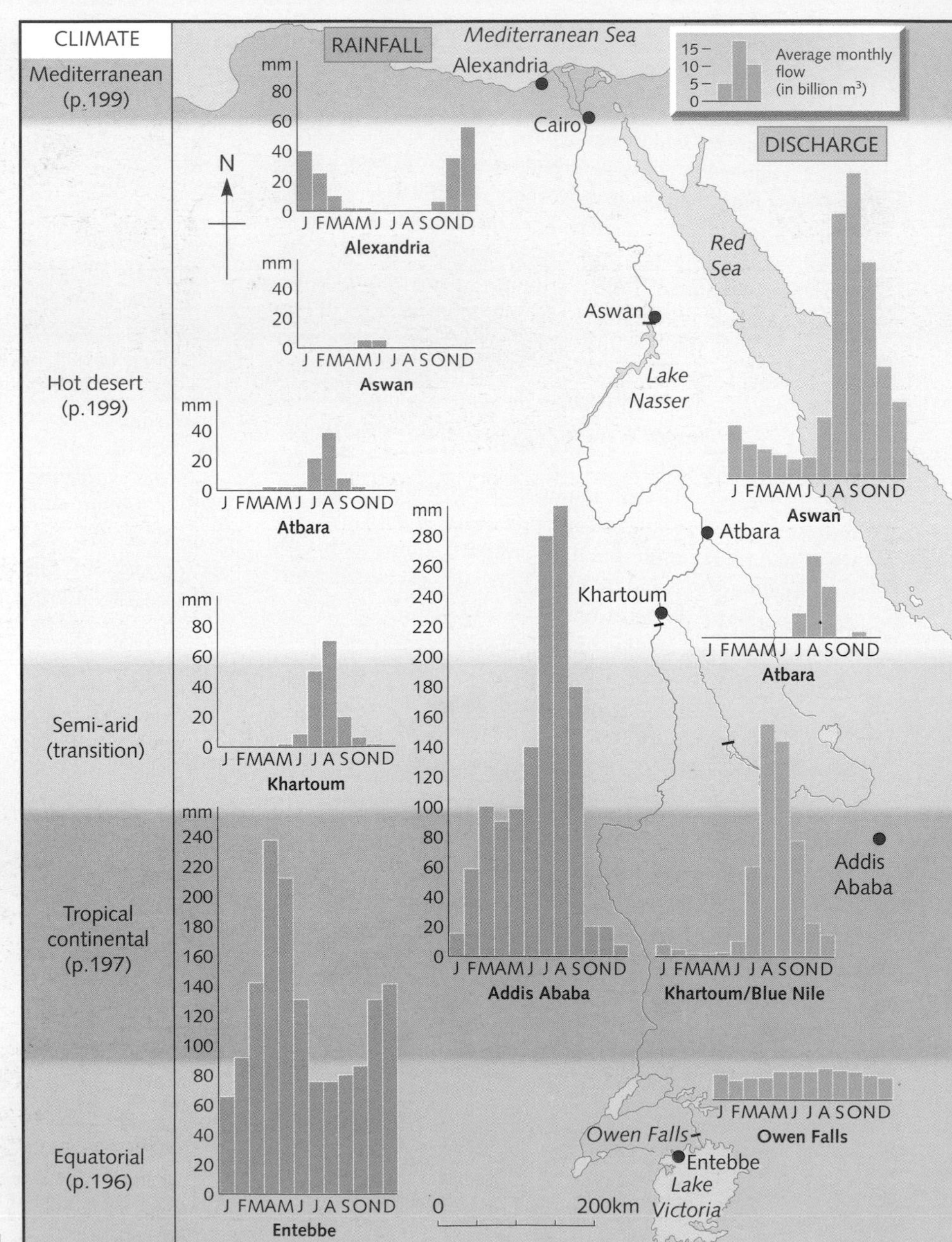

Figure 15.41
Rainfall and river discharge at places in the Nile basin

River landforms

The White Nile flows out of Lake Victoria at the Owen Falls (although the Falls are now drowned by a hydro-electric dam). After dropping into the Rift Valley, the White Nile flows through Lakes Kyoga and Albert before entering a huge flat, swampy area known as The Sudd. As it meanders across this flat area, the river can lose up to 75 per cent of its water through evaporation. Meanwhile the Blue Nile and Atbara Rivers descend from the Ethiopian Highlands by a series of waterfalls (in the rainy season) and through steep-sided gorges. The confluence of the White Nile and Blue Nile is at Khartoum. North of here, the Nile flows over the first of six major cataracts (waterfalls) caused by outcrops of resistant rock crossing the valley. At Aswan, in Egypt, the river crosses the last cataract and flows through a valley fringed by cliffs up to 500 metres high (Figure 15.42). Later, as the valley and floodplain widen, silt has been deposited to a depth of 10 metres. Sediment has also built up at the mouth of the river to form, north of Cairo, the Nile delta (Figure 15.43).

A multipurpose river

Some of the many traditional uses of the Nile are summarised on 15.44.

An international river

The first Nile Waters Agreement, between Egypt and Sudan (1929), partitioned the annual flow of the Nile between the two countries. It tried to reconcile Egyptian and Sudanese growing demands for water, especially for irrigation. Under the second Nile Agreement (1959), from an average annual flow of 74 billion m^3, Egypt and Sudan receive 75 and 25 per cent respectively.

Figure 15.42
The Nile at Aswan – steep valley sides, rocks forming cataracts

Figure 15.43
Satellite image of the lower Nile Valley and delta

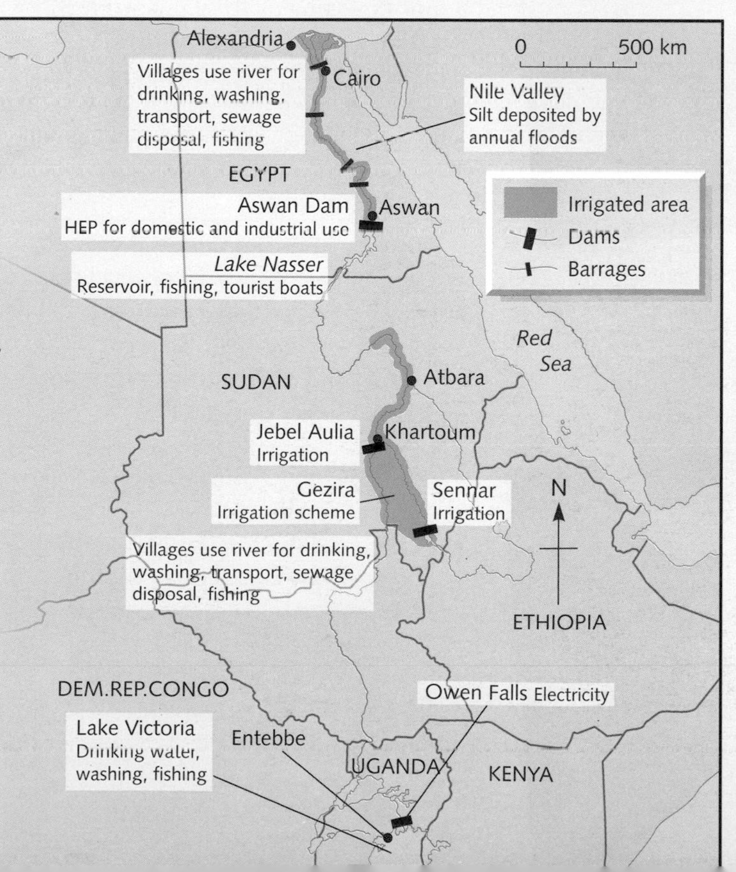

Figure 15.44
The multipurpose function of the Nile before 1960

However, Ethiopia, the source of some 85 per cent of the Nile's water entering the Sudan, is not party to any agreement. Should Ethiopia develop hydro-electric power or encourage irrigation, reducing the Nile's flow by perhaps 10 per cent, this could have serious repercussions in both Sudan and Egypt. Six other countries lie in the upper basin of the White Nile – Burundi, Kenya, Rwanda, Tanzania, Uganda and the Democratic Republic of the Congo. All have a high birth rate and, therefore, potentially large future demands on the headwaters. In order to promote economic and political co-operation among eight of the nine Nile basin countries (Ethiopia was not invited), Egypt co-sponsored the formation of the *Undugu* group (Swahili for 'fraternity').

Egypt, the gift of the Nile

For thousands of years, the Nile basin has supported an agricultural population. The river supplied Egypt's water needs, the annual floods deposited nutrient-rich silt both next to the river and in the delta, and irrigation made farming possible in an otherwise desert area. Water for irrigation was obtained by two methods.

1 Each autumn, the annual flood water was allowed to cover the land, where it remained trapped behind small bunds until it deposited its silt.
2 During the rest of the year when river levels were low, water was lifted one or two metres by a *shaduf, saquia* (*sakia*) *wheel* or *Archimedes screw*.

However, even since the time of the Pharaohs, the Egyptians had wanted to control the Nile so that its level would stay constant throughout the year. Although barrages of increasing size had been built during the early twentieth century (Figure 15.44), they had not been able to store sufficient water to meet the needs of the country's rapidly growing population and increasing demand for food. The problem was how to increase the area of cultivated land (Figure 15.45) and, at the same time, produce electricity.

Figure 15.45
The narrow strip of irrigated land alongside the Nile

Figure 15.46
Fact File – the Aswan High Dam

Figure 15.47
The Aswan High Dam

Figure 15.48
Lake Nasser from the High Dam

The Aswan High Dam

One perceived answer was the building of the Aswan High Dam. The dam and associated hydro-electric power station, which took 11 years to construct, was opened in 1971 (Figures 15.46 and 15.47). Lake Nasser, which formed behind the dam, is 550 km in length – about the same as England from north to south (Figure 15.48). The dam is a multipurpose scheme. It was built:

- to stop, by storing water in Lake Nasser, serious flooding in the lower Nile Valley and, by releasing it throughout the year, to maintain a constant river level (for transport as well as irrigation)
- to provide water all the year round (and with sufficient in reserve to survive several possible years of drought), for domestic, agricultural and industrial use
- to increase the area of cultivation and to allow two, and sometimes three, crops to be grown on the same piece of land each year
- to provide hydro-electric power (about one-quarter of the country's needs).

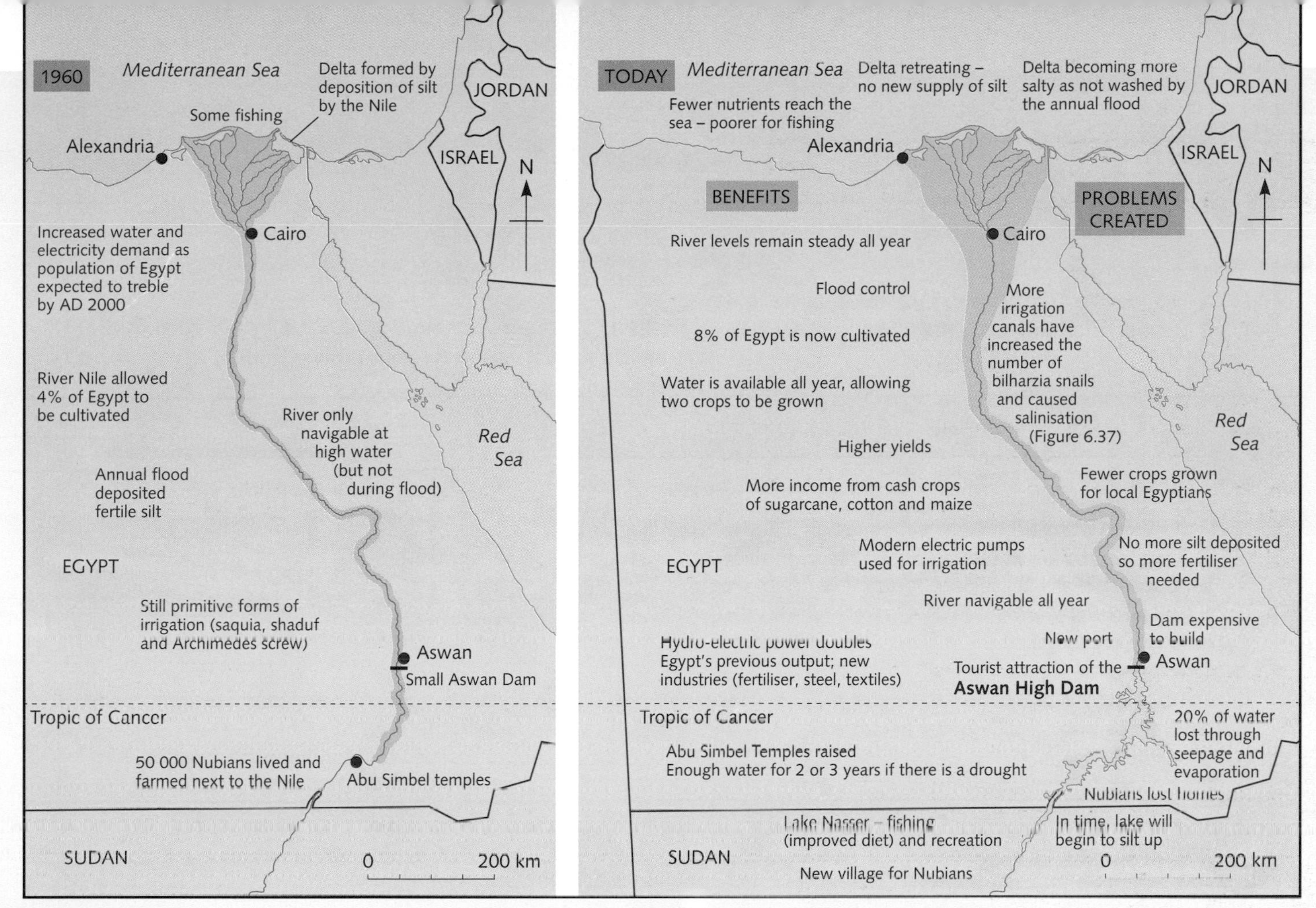

Figure 15.49 The Nile in Egypt (a) 1960 and (b) today

While the scheme has indeed brought many benefits to Egypt (mainly economic and social), it has also created many problems (often environmental and health), some of which were unforeseen (Figure 15.49).

The present and the future

Egypt's population continues to grow at such a rapid rate that it has doubled since the opening of the Aswan High Dam (33 million in 1971, 67 million in 1997, an estimated 85 million by 2010). With an increasing number of mouths to feed, urban sprawl devouring 1 per cent of the richest land in the delta each year, and every cultivated hectare having to support 28 people, it is not surprising that Egyptians believe President Mubarak when he claims: 'Greening the desert is less a dream than a necessity'. Twenty years ago the 200 km road from Cairo to Alexandria ran through desert. Now it is bordered by factories and drip-irrigation fields.

In October 1997, the President opened the Al-Salam, or Peace, Canal. This canal takes water from the Nile into the north of the Sinai peninsula (Figure 15.50). It is hoped that by early next century:

- water from the canal will irrigate 240 000 hectares of desert
- people will migrate here rather than, as at present, to Cairo (pages 62–63)
- less cultivated land in the Nile Valley and delta will need to be built upon.

The major concern is that as the flow of the Nile is depleted, salinisation (Figure 6.37) and coastal erosion may increase.

A more ambitious scheme, the New Delta project, would send water from Lake Nasser on a 500 km journey to link a string of desert oases (Figure 15.50). If successful, up to 600 000 hectares could be used to produce food.

Figure 15.50
Recent and proposed irrigation schemes

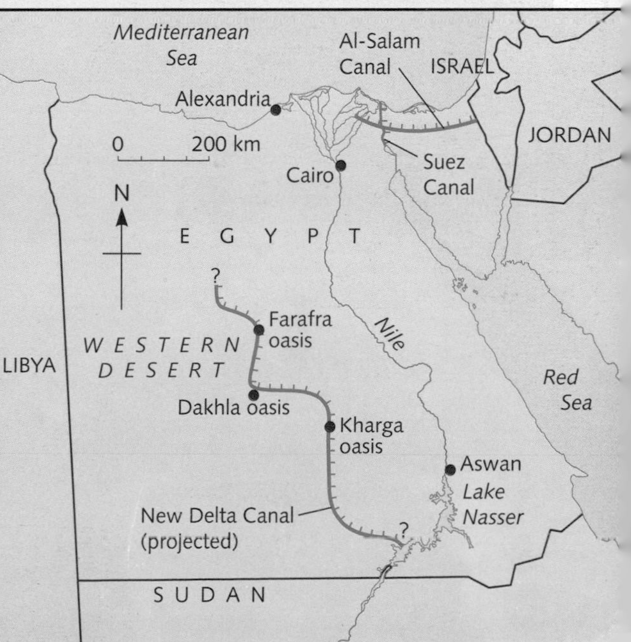

QUESTIONS

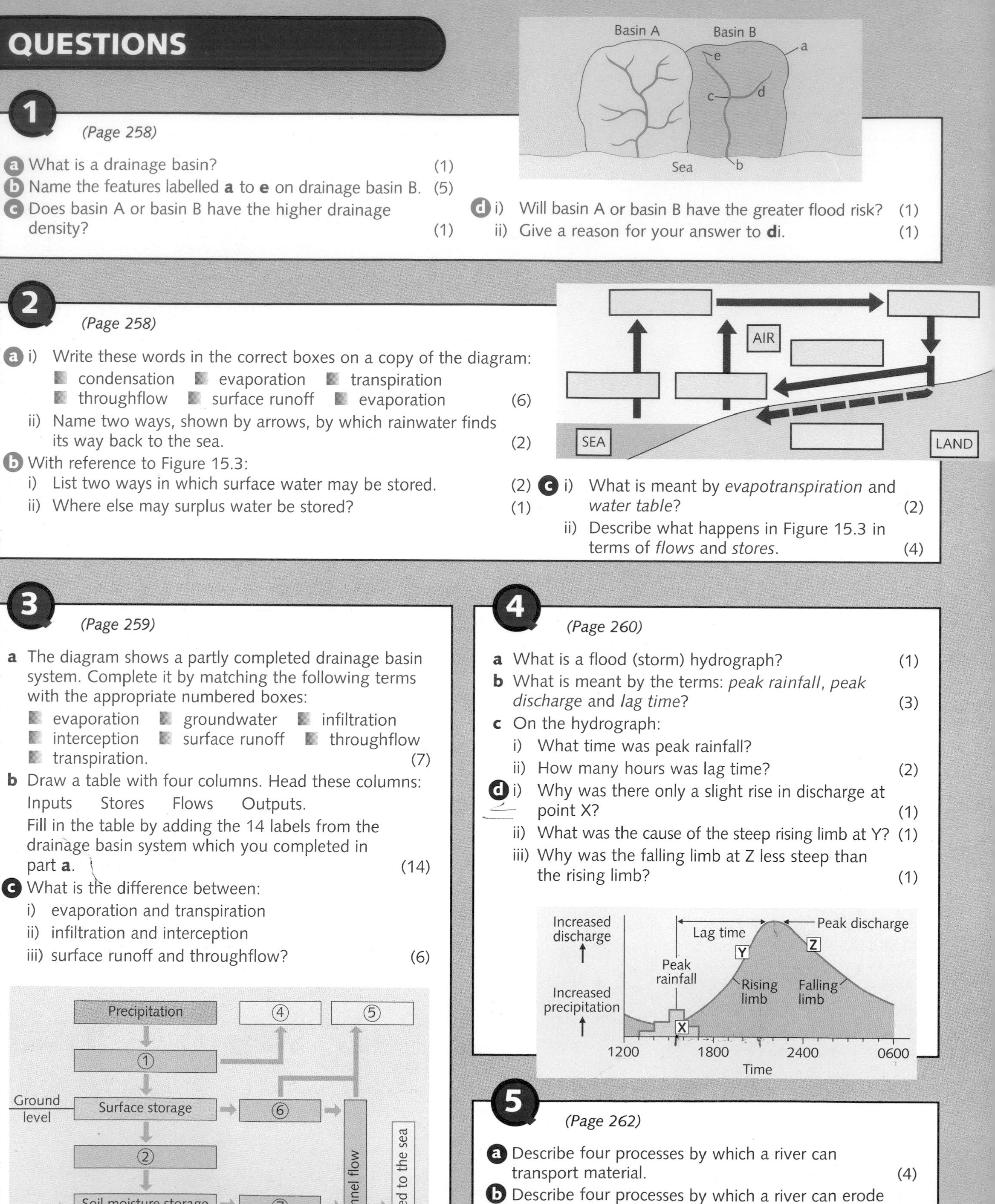

1

(Page 258)

a What is a drainage basin? (1)

b Name the features labelled **a** to **e** on drainage basin B. (5)

c Does basin A or basin B have the higher drainage density? (1)

d i) Will basin A or basin B have the greater flood risk? (1)
ii) Give a reason for your answer to **d**i. (1)

2

(Page 258)

a i) Write these words in the correct boxes on a copy of the diagram:
- condensation
- evaporation
- transpiration
- throughflow
- surface runoff
- evaporation (6)

ii) Name two ways, shown by arrows, by which rainwater finds its way back to the sea. (2)

b With reference to Figure 15.3:
i) List two ways in which surface water may be stored. (2)
ii) Where else may surplus water be stored? (1)

c i) What is meant by *evapotranspiration* and *water table*? (2)
ii) Describe what happens in Figure 15.3 in terms of *flows* and *stores*. (4)

3

(Page 259)

a The diagram shows a partly completed drainage basin system. Complete it by matching the following terms with the appropriate numbered boxes:
- evaporation
- groundwater
- infiltration
- interception
- surface runoff
- throughflow
- transpiration. (7)

b Draw a table with four columns. Head these columns:
Inputs Stores Flows Outputs.
Fill in the table by adding the 14 labels from the drainage basin system which you completed in part **a**. (14)

c What is the difference between:
i) evaporation and transpiration
ii) infiltration and interception
iii) surface runoff and throughflow? (6)

4

(Page 260)

a What is a flood (storm) hydrograph? (1)

b What is meant by the terms: *peak rainfall*, *peak discharge* and *lag time*? (3)

c On the hydrograph:
i) What time was peak rainfall?
ii) How many hours was lag time? (2)

d i) Why was there only a slight rise in discharge at point X? (1)
ii) What was the cause of the steep rising limb at Y? (1)
iii) Why was the falling limb at Z less steep than the rising limb? (1)

5

(Page 262)

a Describe four processes by which a river can transport material. (4)

b Describe four processes by which a river can erode its banks and bed. (4)

c What is the difference between vertical and lateral erosion? (2)

6

(Pages 260 and 261)

The hydrograph for 5 July gives information about rainfall and discharge of a stream flowing over an area of impermeable rock.

a i) What is meant by the terms *impermeable rock* and *discharge*? (2)

ii) What is the opposite term to *impermeable*? (1)

b i) Copy and complete the hydrograph by using the figures in the table. (2)

ii) When did storm 1 start? (1)

iii) What was the lag time between the start of storm 1 and the first peak in stream discharge? (1)

iv) Suggest two reasons why there was a lag time. (2)

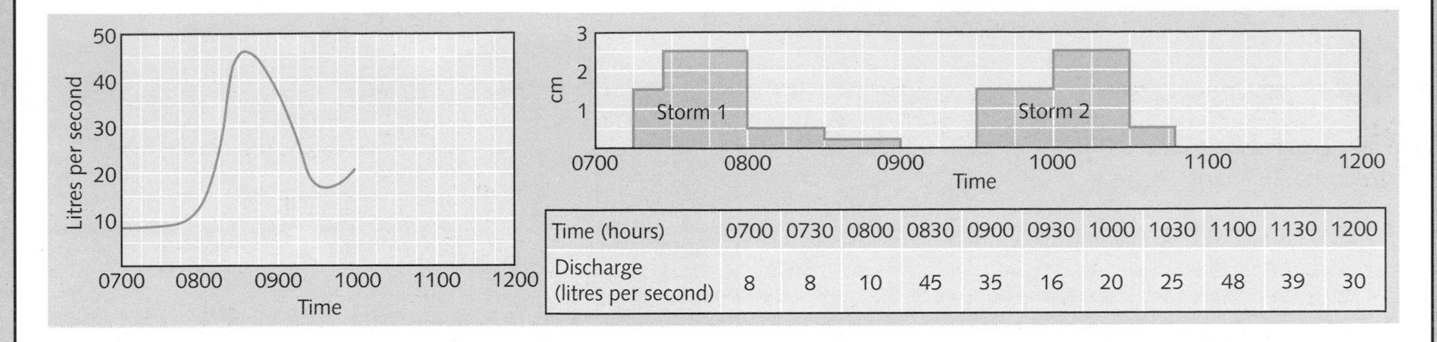

Time (hours)	0700	0730	0800	0830	0900	0930	1000	1030	1100	1130	1200
Discharge (litres per second)	8	8	10	45	35	16	20	25	48	39	30

7

(Page 261)

The diagram shows two hydrographs.

a Complete the table by saying which hydrograph, A or B, is the more likely to correspond to each of the following pairs

- a long period of gentle rain and a short, heavy thunderstorm
- a basin with steep valley sides and a basin with gentle sides
- an area of impermeable rock and an area of permeable (porous) rock
- an area of forest and an area of bare rock
- a mainly urbanised basin and a mainly rural basin
- a river with a dam built across it and a river with no dam
- a basin with a high drainage density and a basin with a low drainage density (7)

b Give one reason for each of your answers in part **a**. (7)

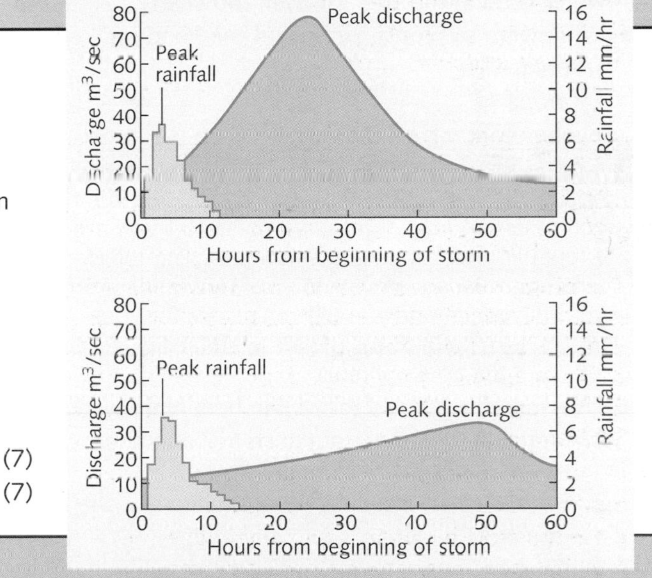

8

(Pages 263 and 264)

a Fieldsketch A shows a waterfall and gorge.

i) On a copy of the sketch, mark on and label: an area of hard rock, an area of soft rock, a plunge pool, the position of the gorge, and a place where rock is likely to collapse. (5)

ii) Describe and explain the main features of the waterfall. (3)

b Explain how a gorge forms. (3)

c Fieldsketch B shows a meander.

i) On a copy of the sketch, mark on and name areas of erosion, areas of deposition and the position of the fastest current. (3)

ii) Why is erosion and deposition taking place where you have shown? (6)

d On a cross-section of a meander, label a river cliff and slip-off slope. (3)

e Explain how an ox-bow lake forms. (3)

9

(Pages 262–265)

a Fifteen river features have been labelled **a** to **o** on the fieldsketch. Match up the correct letter with the feature named in the following list:
delta, floodplain, gorge, interlocking spur, levées, meander, mouth, ox-bow, plunge pool, rapids, river cliff, slip-off slope, source, V-shaped valley, waterfall. (15)

b The sketch has also divided the river valley into two sections labelled **A** and **B**. For each pair of terms below, say which is likely to occur in zone A and which in zone B. In each case give a reason for your answer. (9)

- vertical erosion and lateral erosion
- attrition and corrasion
- traction and suspension.

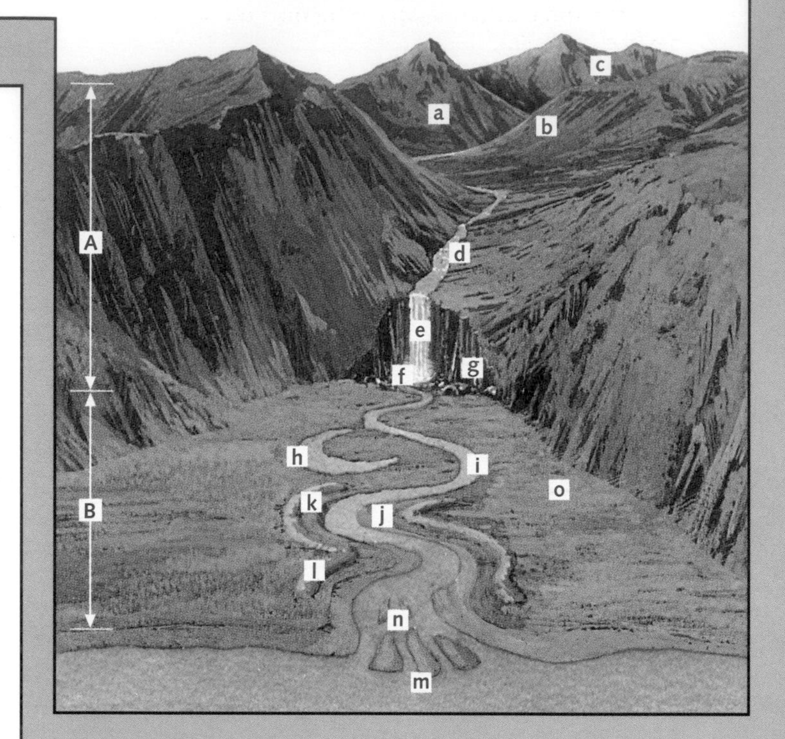

10

(Pages 266 and 267)

One of the worst UK floods in living memory hit Lynmouth in August 1952.

a What natural (physical) features of the River Lyn's drainage basin made Lynmouth vulnerable to a possible flood? (6)

b How had the development of Lynmouth increased the risk of damage from flooding? (3)

c How did extreme weather conditions cause the worst flood in the area for over 200 years? (3)

d What were the consequences of the flood on
- human life
- property
- the economy of the town? (6)

e A flood the size of that which hit Lynmouth may only occur once in every 100 or 200 years. After the flood, you were put in charge of a scheme to try to prevent a repetition of flooding in Lynmouth. The following ten suggestions were put to you. How appropriate do you think each one was? Give reasons for your answers (think in terms of safety, suitability, costs, jobs and the environment). (10)

1. Widen the river so that it can hold more water.
2. Straighten the river to reduce friction and to carry excess water away more quickly.
3. Build large flood banks (levées) alongside the river.
4. Construct wider arches to any bridges.
5. Plant trees on Exmoor and in the drainage basin to reduce the speed of surface runoff.
6. Install a flood warning system.
7. Divert the course of the river away from Lynmouth.
8. Move the buildings in Lynmouth to a new site away from the river.
9. Build a dam on the river to hold back any flood water.
10. Do nothing and just pray that it will not happen again in your lifetime.

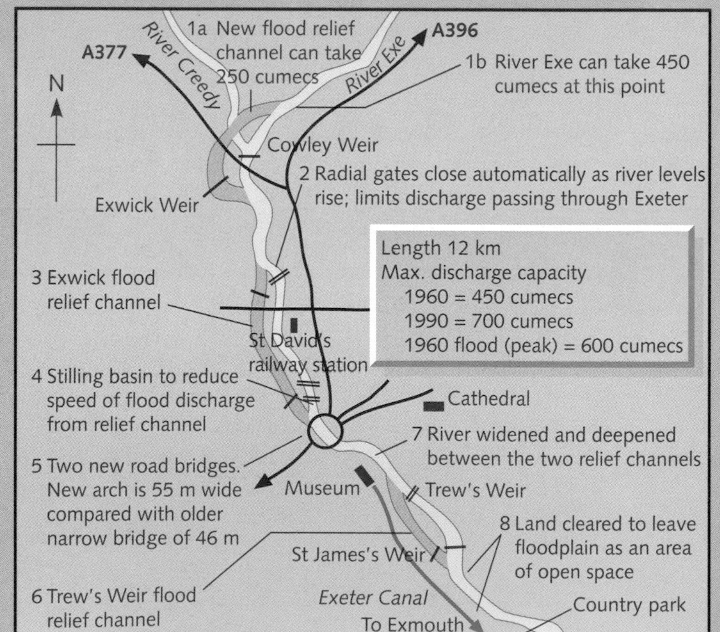

The River Exe, like the Lyn, has its source on Exmoor. Heavy rainfall on Exmoor takes 24 hours to reach Exeter which, over the centuries, has experienced frequent flooding. During 1960, the wettest year this century in the West Country, 1000 homes in Exeter were flooded on 27 October and 1200 houses and business premises on 4 December. Residents demanded action. Once costs had been balanced against risks, the Exeter Flood Defence Scheme was built. (The more the money spent on prevention schemes, the less likely should be the risk of future flooding.) The scheme was completed in 1977. Since then, although the flood relief channels have been needed on average 3 to 4 times a year, no property has been flooded.

11

(Pages 266 and 267)

a Why was a flood prevention scheme thought necessary for Exeter? (4)

b Describe the main points of the flood prevention scheme. (5)

c How successful do you think the scheme has been to date? (2)

12

(Pages 268, 269 and 271)

For any river (or lake) that you have studied which suffers from water pollution:

a Name the river (or lake).

b i) On a copy of the star diagram, list 6 ways by which the river (or lake) may have become polluted. (6)

ii) Describe any **four** of these ways. (4)

c Describe three attempts made by the authorities to clean up the river (or lake). (6)

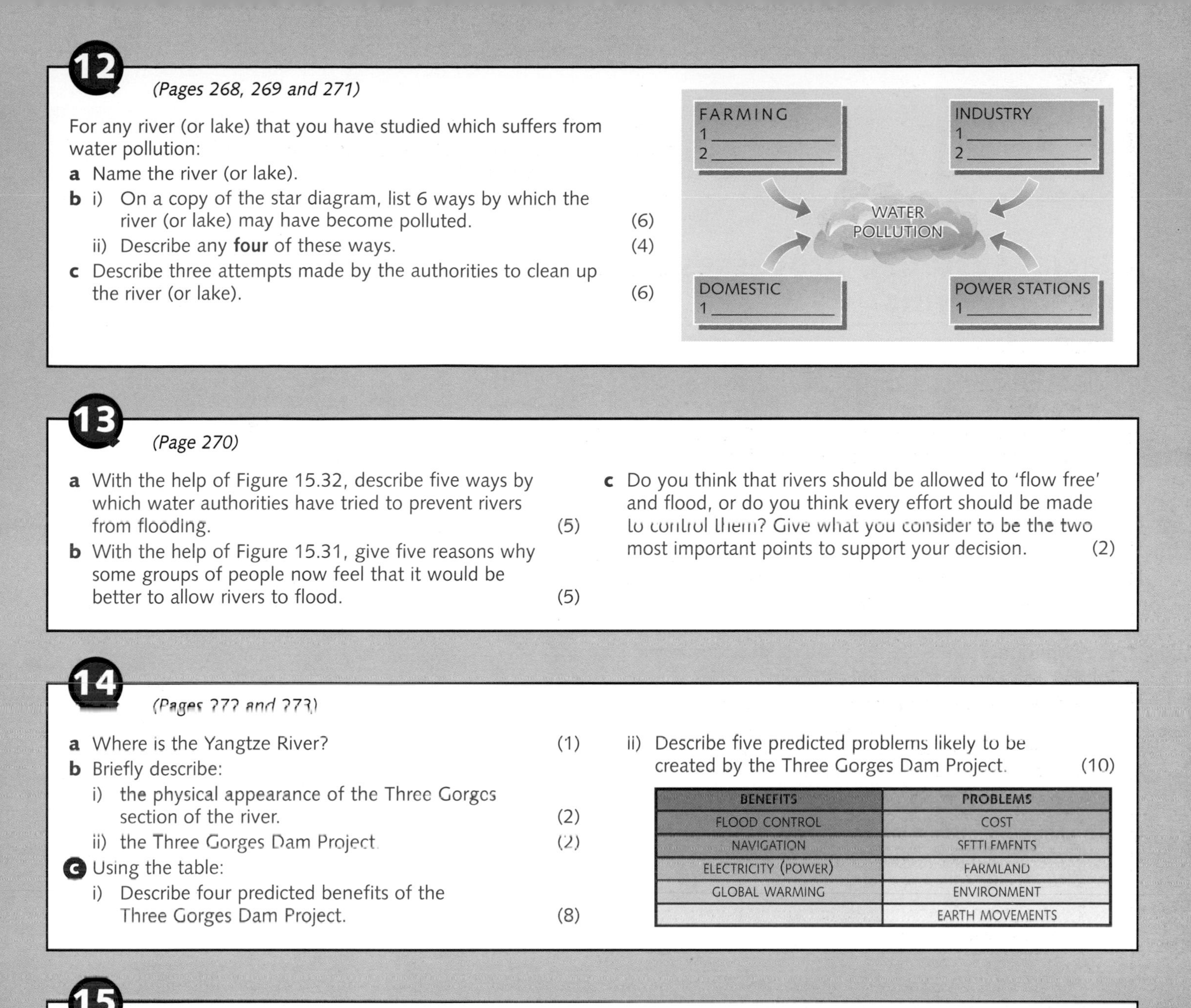

13

(Page 270)

a With the help of Figure 15.32, describe five ways by which water authorities have tried to prevent rivers from flooding. (5)

b With the help of Figure 15.31, give five reasons why some groups of people now feel that it would be better to allow rivers to flood. (5)

c Do you think that rivers should be allowed to 'flow free' and flood, or do you think every effort should be made to control them? Give what you consider to be the two most important points to support your decision. (2)

14

(Pages 272 and 273)

a Where is the Yangtze River? (1)

b Briefly describe:

i) the physical appearance of the Three Gorges section of the river. (2)

ii) the Three Gorges Dam Project. (2)

c Using the table:

i) Describe four predicted benefits of the Three Gorges Dam Project. (8)

ii) Describe five predicted problems likely to be created by the Three Gorges Dam Project. (10)

BENEFITS	PROBLEMS
FLOOD CONTROL	COST
NAVIGATION	SETTLEMENTS
ELECTRICITY (POWER)	FARMLAND
GLOBAL WARMING	ENVIRONMENT
	EARTH MOVEMENTS

15

(Pages 274–277)

a On a copy of the map of the Nile basin, name: countries 1 to 3, rivers 4 to 7, cities 8 to 13, lakes 14 and 15, sea areas 16 and 17, lines of latitude 18 and 19, river feature at 20. (20)

b i) Describe the annual flow of the White Nile. (2)

ii) Describe the annual flow of the Blue Nile. (2)

iii) Give two reasons for the differences in flow. (2)

c i) In which three months was the Nile likely to flood in Egypt? (2)

ii) Give two advantages of the flood. (2)

iii) Give two disadvantages of the flood. (2)

d i) What is an 'international river'? (1)

ii) Why may an international river cause problems? (2)

e i) Why was the Aswan High Dam built? (3)

ii) Give six ways in which the High Dam has benefited Egypt. (6)

iii) Give three economic and three environmental problems caused by the Aswan High Dam. (6)

f Describe one recent or proposed Egyptian water diversion project. Why is it thought to be needed? (4)

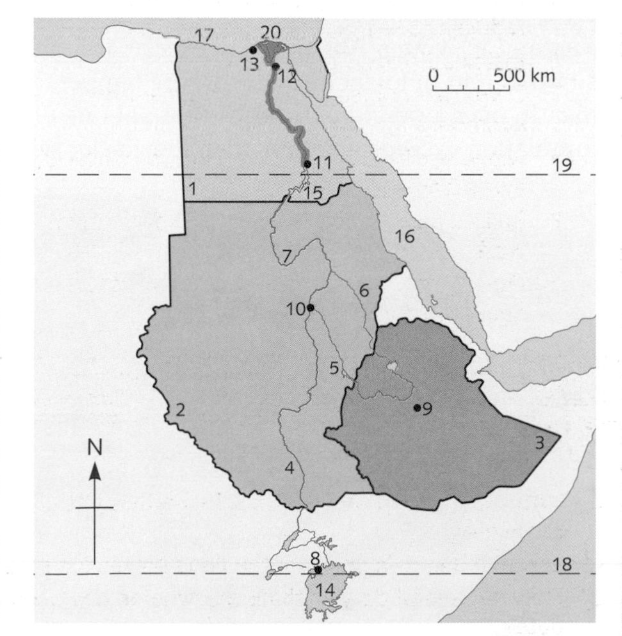

16 GLACIATION AND COASTS

Glacial systems, processes and landforms

Although it is too warm for glaciers to be found in Britain today, much of the spectacular highland scenery of Scotland, Wales and northern England owes its attractiveness to the work of ice in earlier times. At the height of the Ice Age, northern Britain was covered by a large ice sheet. At other times glaciers extended down valleys leading from the higher mountains.

The glacier system

A glacier, like a river, behaves as a system with inputs, stores, flows and outputs (Figure 16.1). Inputs come from precipitation in the form of snow falling directly onto the glacier, or from avalanches along the glacier sides. Inputs mainly occur near to the head of a glacier in the *zone of accumulation* (Figure 16.1). During fresh falls of snow, air is trapped between the flakes. As more snow falls, the underlying layers are compressed, the air is squeezed out and the snow becomes firmer (like making a snowball). As more snow accumulates, the underlying layers are compressed into ice. Ice without any air (oxygen) left in it turns blue. The glacier itself is water held in storage and, as ice, flows (transfers) downhill under the force of gravity.

Outputs from the system are mainly meltwater with a limited amount of evaporation. When a glacier melts it is called *ablation* (Figure 16.1). Inputs (accumulation) are likely to exceed outputs (ablation) near to the head of a glacier and in winter. Ablation will exceed accumulation in summer and in lower altitudes where temperatures are higher. If, over a period of time, the annual rate of accumulation exceeds ablation, then the glacier will advance. If ablation exceeds accumulation, the glacier will retreat. At present most, but not all of the world's remaining glaciers are retreating (Figure 16.2).

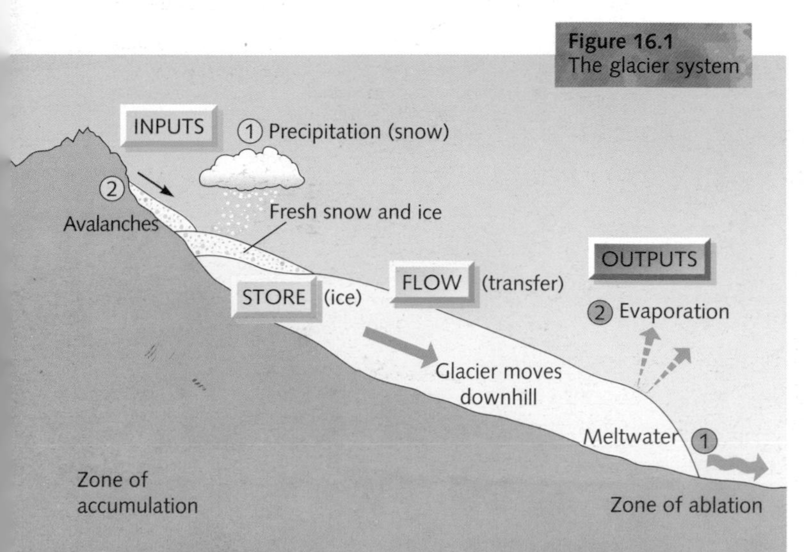

Figure 16.1
The glacier system

Erosion

A glacier can erode much faster than a river, but, like a river, it can only erode if it has a continuous supply of material. The main source of material for a glacier results from the process of freeze–thaw weathering (or frost shattering) (page 234). Freeze–thaw occurs in rocks that have many joints and cracks in them, and where temperatures are frequently around freezing point. Water, which gets into the cracks during the day, freezes at night. As it freezes it expands and puts pressure on the surrounding rock. When the ice melts, pressure is released. Repeated freezing and thawing widens the cracks and causes jagged pieces of rock to break off. The glacier uses this material, called *moraine*, to widen and deepen its valley.

There are two main processes of glacial erosion:

1 **Abrasion** is when the material carried by a glacier rubs against and, like sandpaper, wears away the sides and floor of the valley. It is similar to corrasion by a river, but on a much larger scale.
2 **Plucking** results from glacial ice freezing onto solid rock. As the glacier moves away it pulls with it large pieces of rock.

Figure 16.2
Yerupaja Glacier, Peru

Glacial landforms

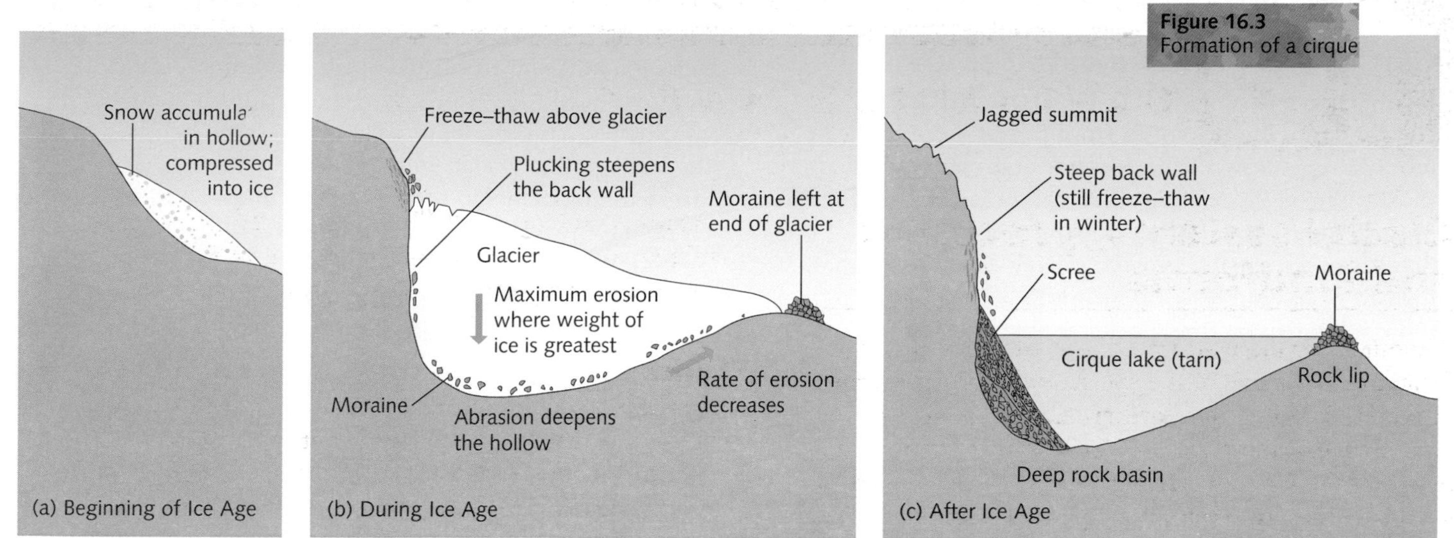

Figure 16.3
Formation of a cirque

Cirques

Cirques, which are also known as *corries* and *cwms*, are deep, rounded hollows with a steep back wall and a rock basin. They began to form at the beginning of the Ice Age when snow accumulated in hollows on hillsides, especially in hollows with a less sunny north- and east-facing aspect (Figure 16.3a). Snow turned into ice, and the ice moved downhill. Freeze–thaw and plucking loosened and removed material from the back of the hollow, creating a steep back wall (Figure 16.3b). Moraine, dragged along the base of the glacier, deepened the floor of the hollow by abrasion, and formed a rock basin. A *rock lip* was left where the rate of erosion decreased. This lip was often heightened by the deposition of moraine. After the Ice Age the rock lip and moraine acted as a natural dam to meltwater, and many rock basins are now occupied by a deep, round *cirque lake* or *tarn* (Figures 16.3c and 16.4).

Figure 16.4
A cirque in Snowdonia, North Wales

Arêtes and pyramidal peaks

When two or more *cirques* develop back to back (or side by side), they erode backwards (or sideways) towards each other. The land between them gets narrower until a knife-edged ridge, called an *arête*, is formed (Figure 6.5). Where three or more cirques cut backwards into the same mountain, a *pyramidal peak*, or *horn*, develops (Figure 16.6). Arêtes radiate from the central peak.

Figure 16.5
A pyramidal peak: Machhappuchare, Nepal

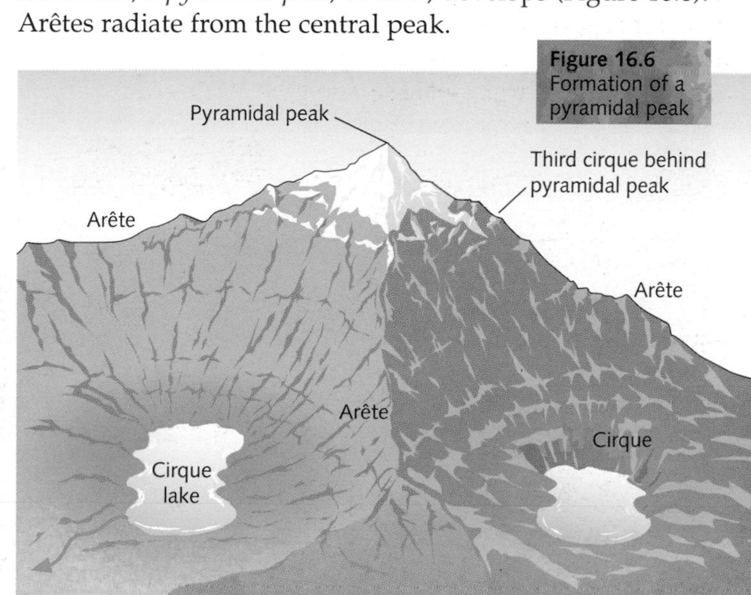

Figure 16.6
Formation of a pyramidal peak

Glacial landforms

Glacial troughs, truncated spurs, hanging valleys and ribbon lakes

Glaciers, moving downhill from their source in the mountains, follow the easiest possible route which, in most cases, is an existing river valley. Unlike a river, however, the glacier often fills the whole valley and this gives it much greater erosive power. This means that, instead of having to wind around obstacles, such as interlocking spurs, the glacier is able, mainly through abrasion, to widen, deepen and straighten its valley (Figure 16.7). The result is that the characteristic V shape of a river valley in a highland area is converted into the equally characteristic U shape of a *glacial trough* (Figure 16.8). As the glacier moves downvalley it removes the ends of interlocking spurs to leave steep, cliff-like, *truncated spurs*.

Between adjacent truncated spurs are *hanging valleys*. Before the Ice Age, tributary rivers would have their confluence with the main river at the same height. During the Ice Age, the glacier in the main valley would be much larger than glaciers in the tributary valleys, and so it could erode downwards much more rapidly. When the ice melted, the tributary valleys were left 'hanging' above the main valley. Each tributary river has now to descend to the main river by a waterfall (Figure 16.9).

Many glacial troughs in highland Britain contain long, narrow, ribbon lakes (Figure 16.8). *Ribbon lakes* are partly the result of erosion when a glacier over-deepens part of its valley, perhaps in an area of softer rock or due to increased erosion after being joined by a tributary glacier. They may also be partly created by deposition of moraine across the main valley.

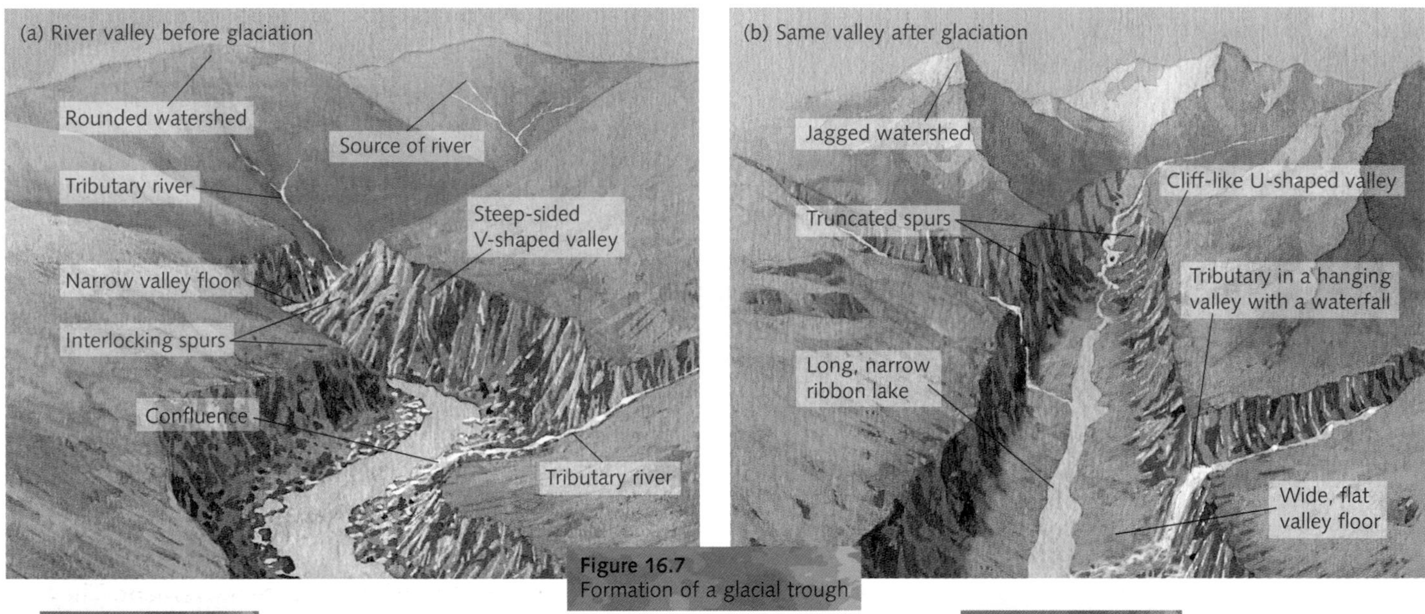

Figure 16.7
Formation of a glacial trough

Figure 16.8
A glacial trough with ribbon lake in Austria

Figure 16.9
Hanging valley with waterfall in California

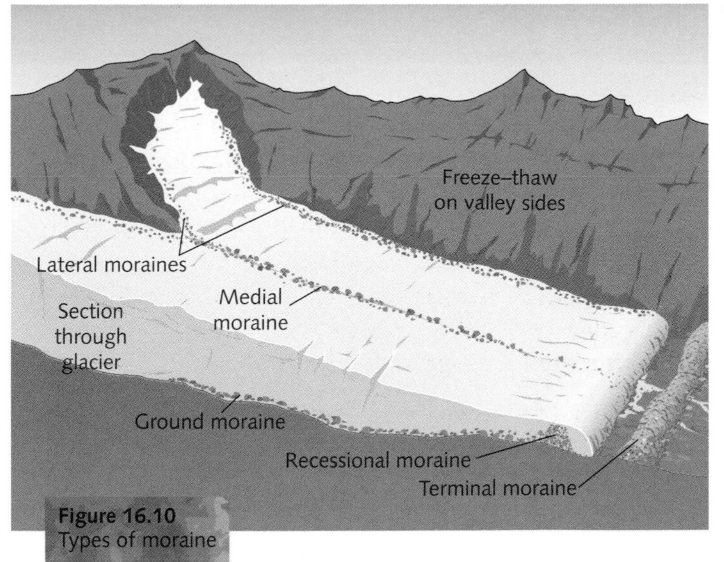

Figure 16.10
Types of moraine

Figure 16.11
Terminal moraine in Greenland

Transportation and deposition

Moraine is material, mainly angular rock, which is transported and later deposited by a glacier. It is deposited when there is a rise in temperature. As the glacier begins to melt, it cannot carry as much material. There are several types of moraine (Figure 16.10):

- *Lateral moraine* is material derived from freeze–thaw weathering of valley sides and which is carried at the sides of a glacier.
- *Medial moraine* is found in the centre of a glacier and results from two lateral moraines joining together.
- *Ground moraine* is material dragged underneath a glacier which, when deposited, forms the flat valley floor. Ground moraine is also referred to as *till* or *boulder clay*.
- *Terminal moraine* marks the maximum advance of a glacier. It is material deposited at the snout, or end, of a glacier (Figure 16.11). If a glacier remains stationary for a lengthy period then a sizeable mound of material, extending across the valley, can build up.
- *Recessional moraines* form behind, and parallel to, the terminal moraine. They mark interruptions in the retreat of a glacier when it remained stationary for long enough for further ridges to develop across the valley. Both terminal and recessional moraines can act as natural dams behind which ribbon lakes can form.

Figure 16.12
An erratic in Greenland

Glaciers can transport material many kilometres. *Erratics* are rocks and boulders carried by the ice and deposited in an area of totally different rock (Figure 16.12). Material from Norway can be found on parts of England's east coast, and Lake District rock on Anglesey.

Drumlins are smooth, elongated mounds of material formed parallel to the direction of ice movement. They often consist of stones and clay, and are believed to result from the load, carried by a glacier, becoming too heavy and being deposited. They owe their streamlined shape to later ice movement (Figure 16.13).

Figure 16.13
Drumlins in Cumbria

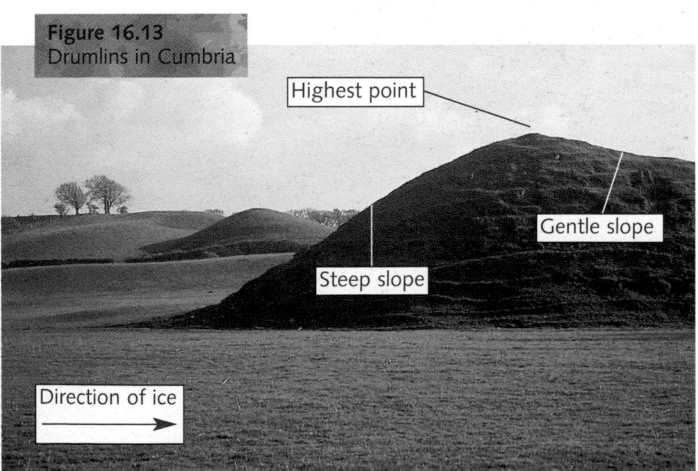

Coastal processes and erosion landforms

The *coast* is a narrow contact zone between land and sea. It is constantly changing due to the effects of land, air and marine processes. On many coastlines the dominant process results from the action of waves. Waves are usually created by the transfer of energy from the wind blowing over the surface of the sea (the exceptions are tsunamis which result from submarine Earth movements). The larger the wave, the more energy it contains. The largest waves are formed when winds are very strong, blow for lengthy periods and cross large expanses of water. The maximum distance of water over which winds can blow is called the *fetch*. In the case of south-west England the fetch is from the south-west. This also coincides with the direction of the prevailing, or most frequent, wind. In eastern England the fetch is generally from the east.

Water particles within a wave move in a circular orbit (Figure 16.14). Each particle, or a floating object, tends to move vertically up and down. It is only the shape of the wave and its energy that is transferred horizontally towards the coast. However, when a wave reaches shallow water, the velocity at its base will be slowed due to friction with the sea-bed, and the circular orbit is changed to one that is more elliptical (Figure 16.14). The top of the wave, unaffected by friction, becomes higher and steeper until it breaks. Only at this point does the remnant of the wave, called the *swash*, actually move forwards. The swash transfers energy up the beach, the *backwash* returns energy down the beach.

There are two types of wave (Figure 16.15).

1 *Constructive waves* have limited energy. Most of this is used by the swash to transport material up the beach.
2 *Destructive waves* have much more energy. Most of this is used by the backwash to transport material back down the beach.

Erosion

Waves, like rivers (page 262), can erode the land by one of four processes.

- *Corrasion* (abrasion) is caused by large waves hurling beach material against a cliff.
- *Attrition* is when waves cause rocks and boulders on the beach to bump into each other and to break up into small particles.
- *Corrosion* (solution) is when salts and other acids in seawater slowly dissolve a cliff.
- *Hydraulic pressure* is the force of waves compressing air in cracks in a cliff.

Headlands and bays

Headlands and *bays* form along coastlines where there are alternating outcrops of resistant (harder) and less resistant (softer) rock (Figure 16.16). Destructive waves erode the areas of softer rock more rapidly to form bays. The waves cannot, however, wear away the resistant rock as quickly and so headlands are left protruding out into the sea. The headlands are now exposed to the full force of the waves, and become more vulnerable to erosion. At the same time they protect the adjacent bays from destructive waves.

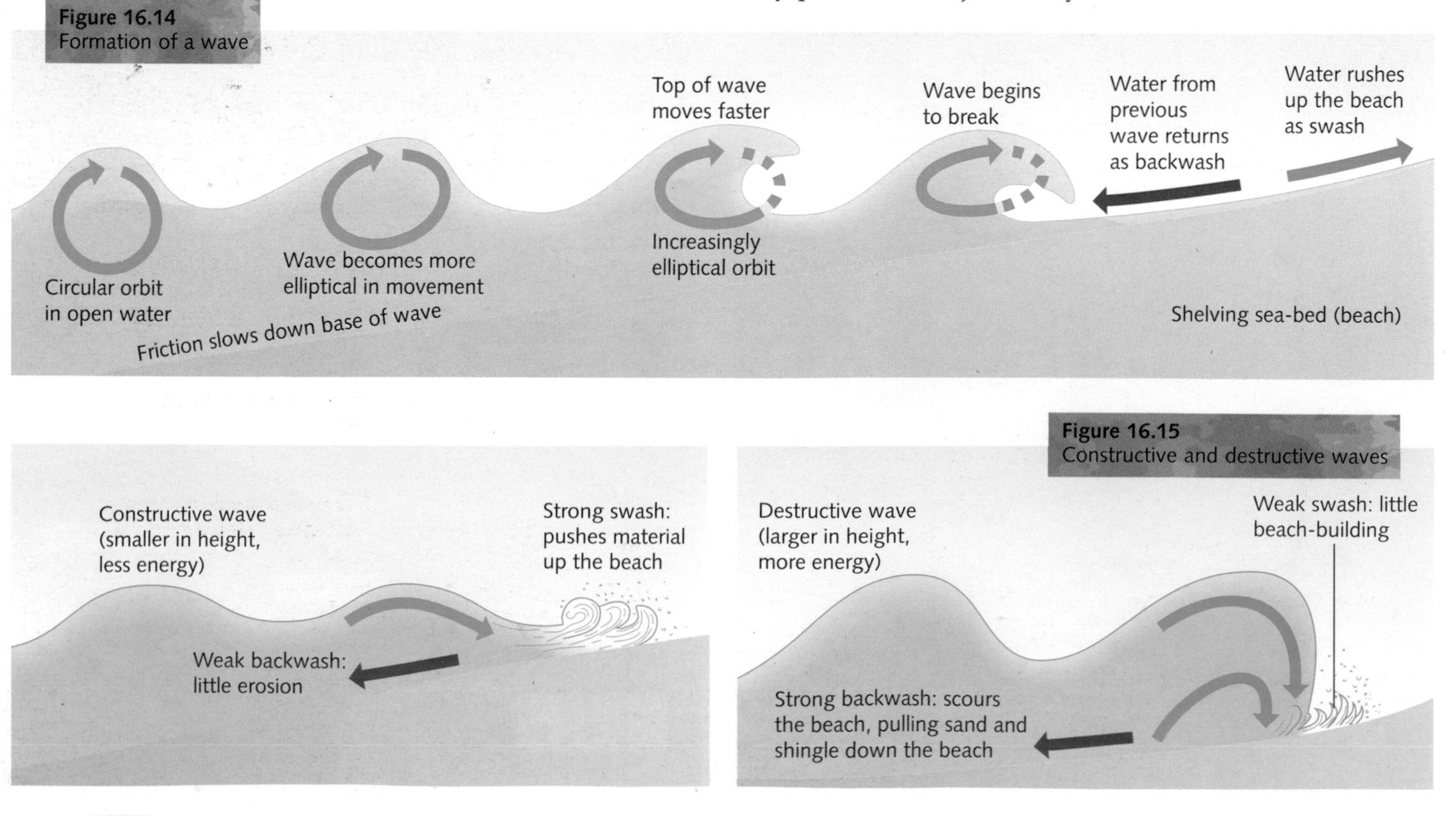

Figure 16.14 Formation of a wave

Figure 16.15 Constructive and destructive waves

Cliffs, wave-cut notches and wave-cut platforms

Wave erosion is greatest when large waves break against the foot of the cliff. With wave energy at its maximum, the waves undercut the foot of the cliff to form a *wave-cut notch* (Figure 16.17). Over a period of time the notch enlarges until the cliff above it, left unsupported, collapses. As this process is repeated, the cliff retreats and, often, increases in height. The gently sloping expanse of rock marking the foot of the retreating cliff is called a *wave-cut platform* (Figure 16.18). Wave-cut platforms are exposed at low tide but covered at high tide.

Caves, arches and stacks

Cliffs are more likely to form where the coastline consists of resistant rock. However, within resistant rocks there are usually places of weakness, such as a joint or a fault (Figure 16.19). Corrasion, corrosion and hydraulic action by the waves will widen any weakness to form, initially, a *cave*. If a cave forms at a headland, the cave might be widened and deepened until the sea cuts through to form a *natural arch* (Figure 16.20). Waves will continue to erode the foot of the arch until its roof becomes too heavy to be supported. When the roof collapses it will leave part of the former cliff isolated as a *stack* (Figure 16.21). In time, further wave action will result in the stack collapsing to leave a *stump*.

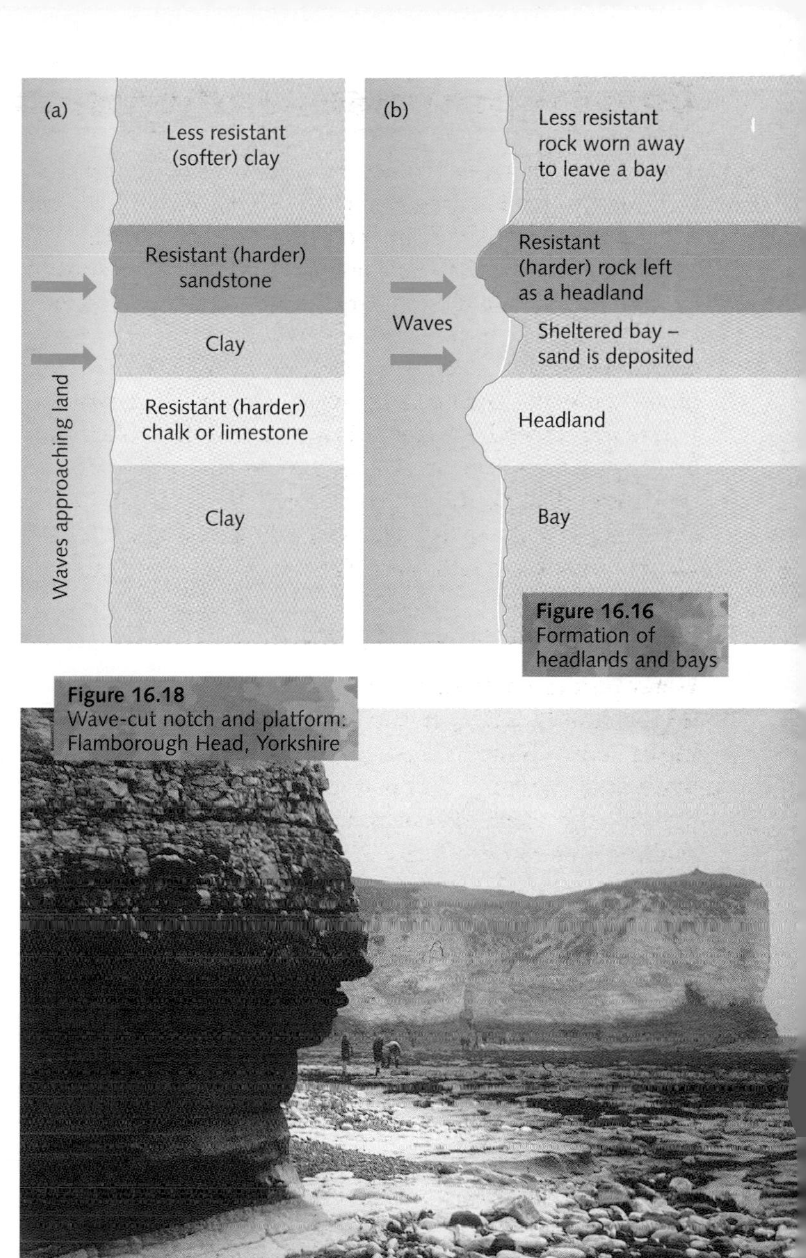

Figure 16.16 Formation of headlands and bays

Figure 16.18 Wave-cut notch and platform: Flamborough Head, Yorkshire

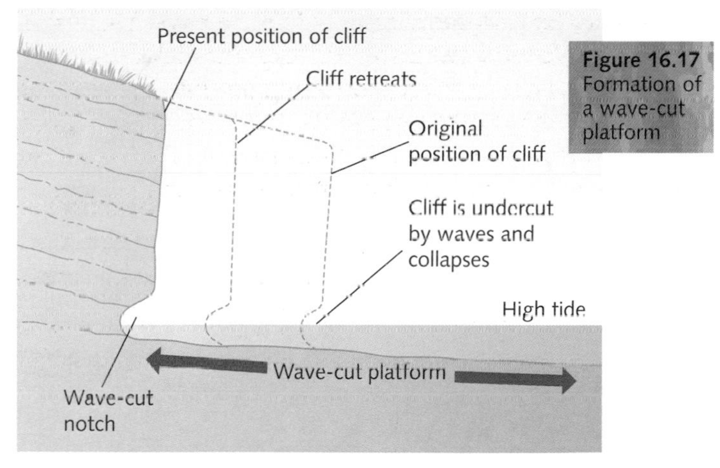

Figure 16.17 Formation of a wave-cut platform

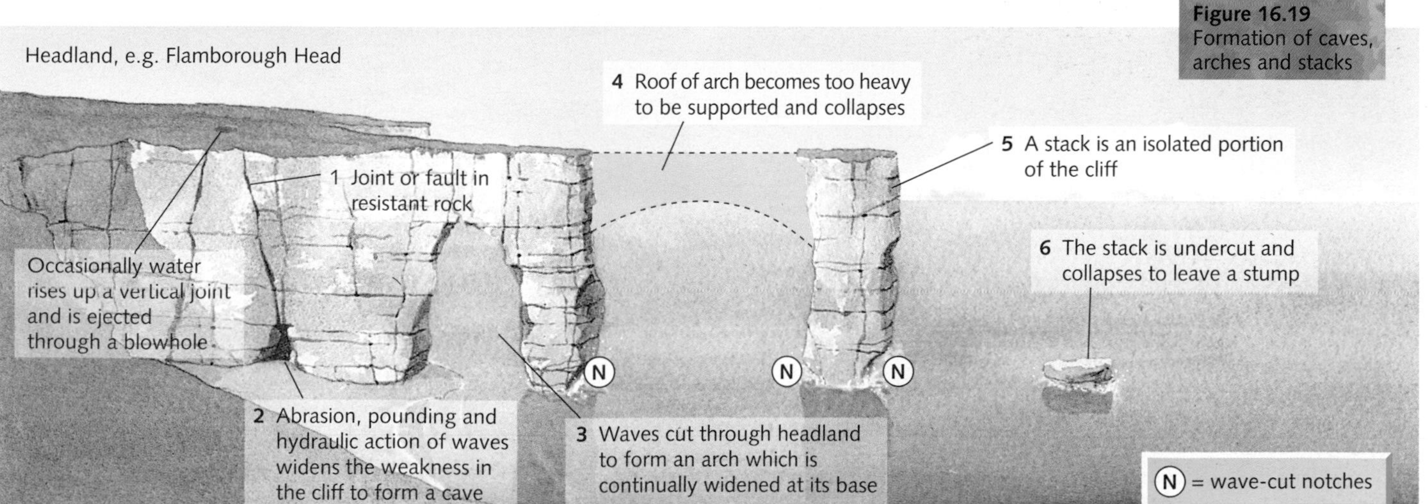

Figure 16.19 Formation of caves, arches and stacks

Figure 16.20
A natural arch and stack, Pembroke in Wales

Transportation

Although waves do carry material up and down a beach, the major movement is along the coast by a process called *longshore drift* (Figure 16.22). Waves rarely approach a beach at right-angles, but rather from a direction similar to that from which the wind is blowing. When a wave breaks, the swash carries material up the beach at the same angle at which the wave approached the shore. As the swash dies away, the backwash returns material straight down the beach, at right-angles to the water, under the influence of gravity. Material is slowly moved along the coast in a zig-zag course. The effect of longshore drift can best be seen when wooden groynes have been built to prevent material from being moved along the beach (Figure 16.23).

Figure 16.21
A stack

Deposition

Sand and shingle being transported along the coast by longshore drift will, in time, reach an area where the water is sheltered and the waves lack energy, e.g. a bay. The material may be temporarily deposited to form a beach. Beaches are not permanent features as their shape can be altered by waves every time the tide comes in and goes out. Shingle beaches have a steeper gradient than sandy beaches.

Figure 16.22
Longshore drift

Figure 16.23
Groynes at Worthing, West Sussex

Figure 16.24 Hurst Spit, Hampshire

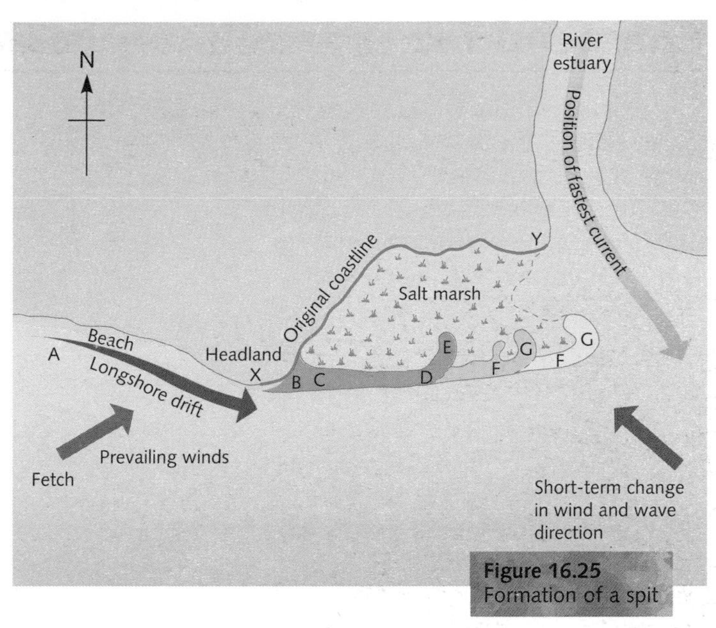

Figure 16.25 Formation of a spit

Spits

A *spit* is a permanent landform resulting from marine deposition. It is a long, narrow accumulation of sand or shingle, with one end attached to the land, and the other projecting at a narrow angle either into the sea or across a river estuary. Many spits have a hooked or curved end (Figure 16.24). They form where longshore drift moves large amounts of sand and shingle along the coast, and where the coastline suddenly changes direction to leave a shallow, sheltered area of water. In Figure 16.25 line 'X' to 'Y' marks the position of the original coastline. As the fetch and prevailing winds are, in this example, from the south-west, material is moved eastwards along the coast by longshore drift. After headland X the direction of the original coastline changes and larger material (shingle) is deposited in water sheltered by the headland (B). Further deposition of finer material (sand) enables the feature to build up slowly to sea-level (C) and to extend length (D). Occasionally the wind changes its direction (e.g. it comes from the east). This in turn causes the waves to alter their direction (e.g. approach from the south-east).

During this time some material at the end of the spit may be pushed inland to form a curved end (E). When the wind returns to its usual direction the spit resumes its growth eastwards (F). Spits become permanent when sand is blown up the beach, by the prevailing wind, to form sand-dunes. *Salt marsh* is likely to develop in the sheltered water behind the spit. The spit is unable to grow across the estuary as the river current carries material out to sea. Should there be no river, the spit may grow across the bay to form a *bar*.

Figure 16.26 Geiranger Fiord, Norway

Sea-level changes

During the Ice Age large amounts of water were held in storage as ice and snow. This interruption in the hydrological cycle (page 258) caused the world's sea-level to fall. After the Ice Age the sea-level rose as the ice and snow melted. Many coastal areas were drowned, creating landforms such as fiords and rias. Both *fiords* and *rias* are drowned valleys. Fiords are found where glaciers overdeepened valleys until they were below sea-level (Figure 16.26) while rias occur in valleys which were formed by rivers (Figure 16.27). Fiords are long, narrow inlets with high, cliff-like sides. They are very deep, apart from a shallow entrance. Rias are more winding with relatively low, gentle sides. The depth of a ria increases towards the sea.

Figure 16.27 A ria: the Fal estuary

The need for coastal management

The coastline in the UK and other parts of the world is dynamic and always changing. The changes may result either from:

- natural (physical) processes of weather, mass movement, waves and the tide, or
- human intervention and activity.

The problems and pressures created by these changes can vary within short distances along any given stretch of coastline. One example is the New Forest coastline in Hampshire which extends from the Dorset border to Southampton Water (Figure 16.28).

Places

The New Forest coastline

Pressures on the coastline

The New Forest District Council which, with other authorities, is responsible for this stretch of coastline, divides it into three main sections.

1 Christchurch Bay (Highcliffe to Hurst Spit)
Towards Highcliffe, the coast is characterised by narrow shingle beaches backed by cliffs which exceed 30 metres in height. The cliffs, which consist mainly of sand and clay, are subject to severe erosion and have, in places, retreated by over 60 metres since 1971 (Figure 16.29). The area between Barton on Sea and Milford on Sea, which is mainly built up, also suffers from coastal erosion, especially after heavy rain when mudflows and landslips are possible. Towards the eastern end of the bay, the cliffs reduce in height and give way to the long shingle bank that forms Hurst Spit (Figure 16.24). The spit is vulnerable to erosion, having been breached during storms several times in recent years.

2 Western Solent (Hurst Spit to Calshot)
This section is distinguished by extensive coastal marshes of considerable wildlife value, such as those at Keyhaven and Lymington (Figure 16.30). The coastal area is low-lying, with the threat of flooding at times of highest tides or during severe storms. The New Forest extends to the shore between the mouths of the Lymington and Beaulieu Rivers. These two river estuaries, together with the estuary at Keyhaven, are centres for recreational sailing and some boat-building and repair (Figure 16.31).

Figure 16.29
Naish Farm Holiday Village

Figure 16.30
Keyhaven marsh

3 Southampton Water (Calshot to Redbridge)
Much of the coastal area of this river estuary is now developed either for major industry, such as the Esso oil refinery and petrochemicals complex (Figure 16.32) and the Fawley power station, or for housing development, as at Hythe. The remaining coastal marshes and intertidal areas are of great nature conservation value, although they are at risk from water pollution caused by domestic sewage and industrial waste.

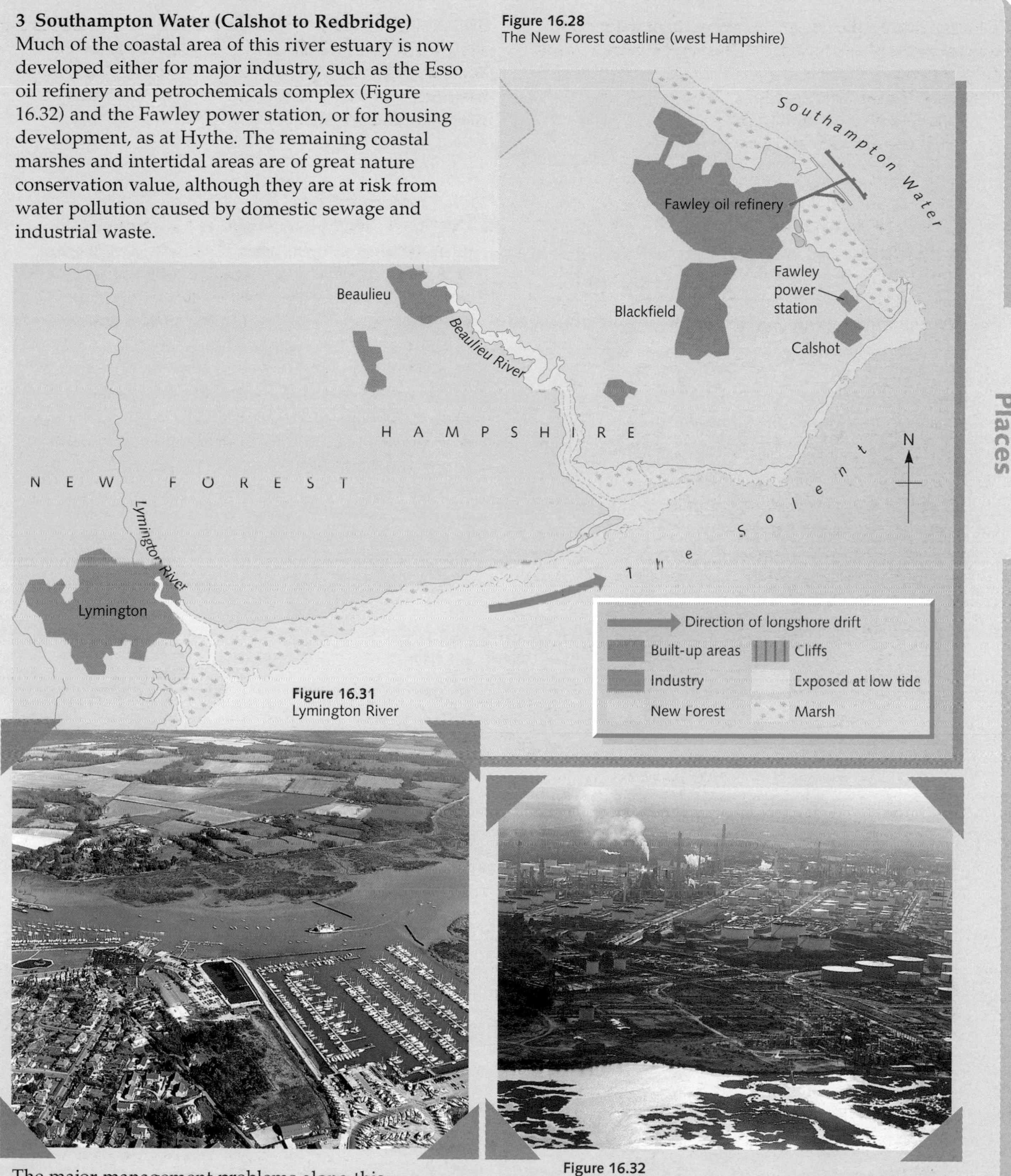

Figure 16.28
The New Forest coastline (west Hampshire)

Figure 16.31
Lymington River

Figure 16.32
Fawley refinery

The major management problems along this coastline are:

- Should the cliffs be allowed to erode, should Hurst Spit be allowed to be breached, and should low-lying areas be left at risk of coastal flooding . . . or should each be protected?
- Should natural landscapes and wildlife habitats be protected or should they be developed for people who want to live and/or work here, and for people who wish to visit the area?

The management of the New Forest coastline

The growing pressures on this stretch of coastline have been described by the New Forest District Council (1997) as follows:

'The coastline is under increasing pressure and threat. The scenery and the recreational potential draw large numbers of visitors, especially in summer, and also provide a very attractive environment in which to live, work and retire. Increasing human impact and influence on the coastline is creating a range of pressures which threaten to change irrevocably the special character, and the natural balance and ecological diversity of one of the District's major assets. In addition natural forces are constantly changing the shape of the coastline through the action of waves, currents, groundwater movements and weathering. These processes have a considerable impact on human activity and the quality of the environment, and must be taken into account when developing management proposals for the coast.'

The New Forest District Council published its strategy document 'New Forest 2000' in 1990. The Council hopes to sustain and improve the quality of life in the area by *'protecting the ecology, archaeology and character of the coast; ensuring that necessary coastal protection works are carried out; and improving recreational enjoyment of the coast where these can be achieved in concert with nature conservation interests'*. The objectives of the Coastal Management Plan are shown in Figure 16.33.

The Council does acknowledge that certain: *'objectives will conflict. For example, recreation uses can be incompatible with nature conservation interests; so can coast protection works, which can also affect access to beaches and the appearance of the coast. Measures to prevent water pollution can involve building installations in sensitive areas.'*

Coastal protection

a Over the years various attempts have been made to protect the cliffs between Highcliffe and Milford on Sea. Traditional defences included the building of concrete sea-walls and the construction of groynes (Figure 16.23). Although their main purpose is to prevent material moving along the beach, they can help to reduce the effect of breaking waves and, by widening the beach, protect the cliff. However, it is now realised that concrete sea-walls, apart from being eyesores and at variance with local habitats, absorb, rather than deflect, wave energy. Without constant maintenance and expense, they can be breached and, as at Barton on Sea, cliff erosion is renewed. The modern approach is to work *with* nature, rather than against it, and to implement schemes that are more cost-effective, which retain wildlife and which enhance the environment, e.g. by rock revetments and stone groynes (Figure 16.34).

Figure 16.33
Objectives of the New Forest District Council's coastal management plan

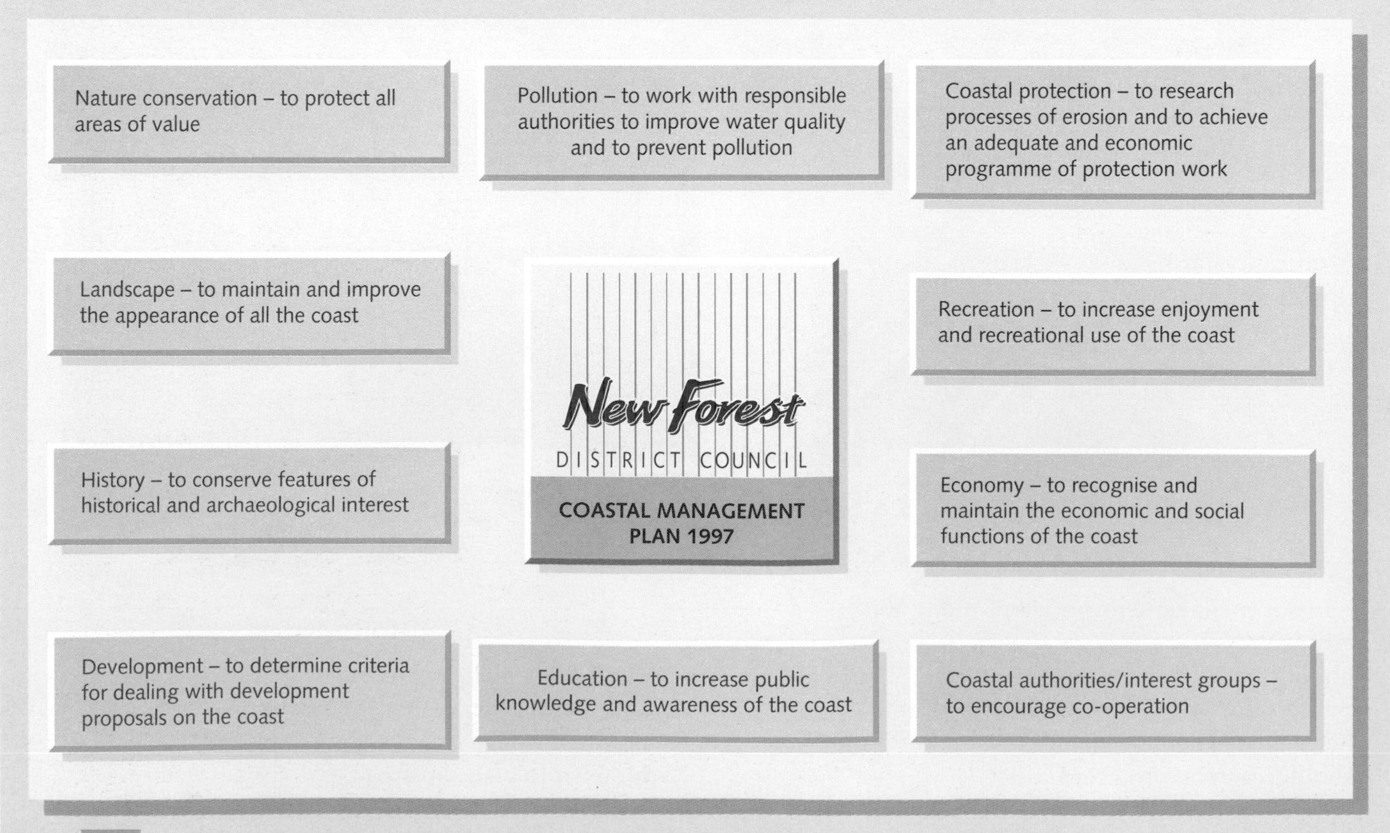

b With Hurst Spit being increasingly breached, the problem (like that at Spurn Head on Humberside) is whether to spend large sums of money in a likely losing battle to stabilise the feature, or to accept that processes which initially formed the landform have changed, and to let nature have its way.

Figure 16.34
Rock protection works at Barton on Sea

c Some low-lying areas, especially near Lymington where there was severe flooding in 1989, have been protected from coastal flooding by dykes and flood walls (Figure 16.35). As with rivers (page 270), there is a growing body of opinion which believes that flooding of coastal areas should be allowed as a natural event (which would preserve natural habitats) rather than protecting them by expensive schemes (even if this means property is put at risk).

d Attempts have been to protect wildlife habitats by creating Nature Reserves (Figure 16.30) and a bird sanctuary, and to encourage sustainable tourism by establishing two country parks and giving the New Forest the status, if not the title, of a National Park (Figure 16.36).

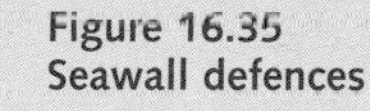

Figure 16.35
Seawall defences

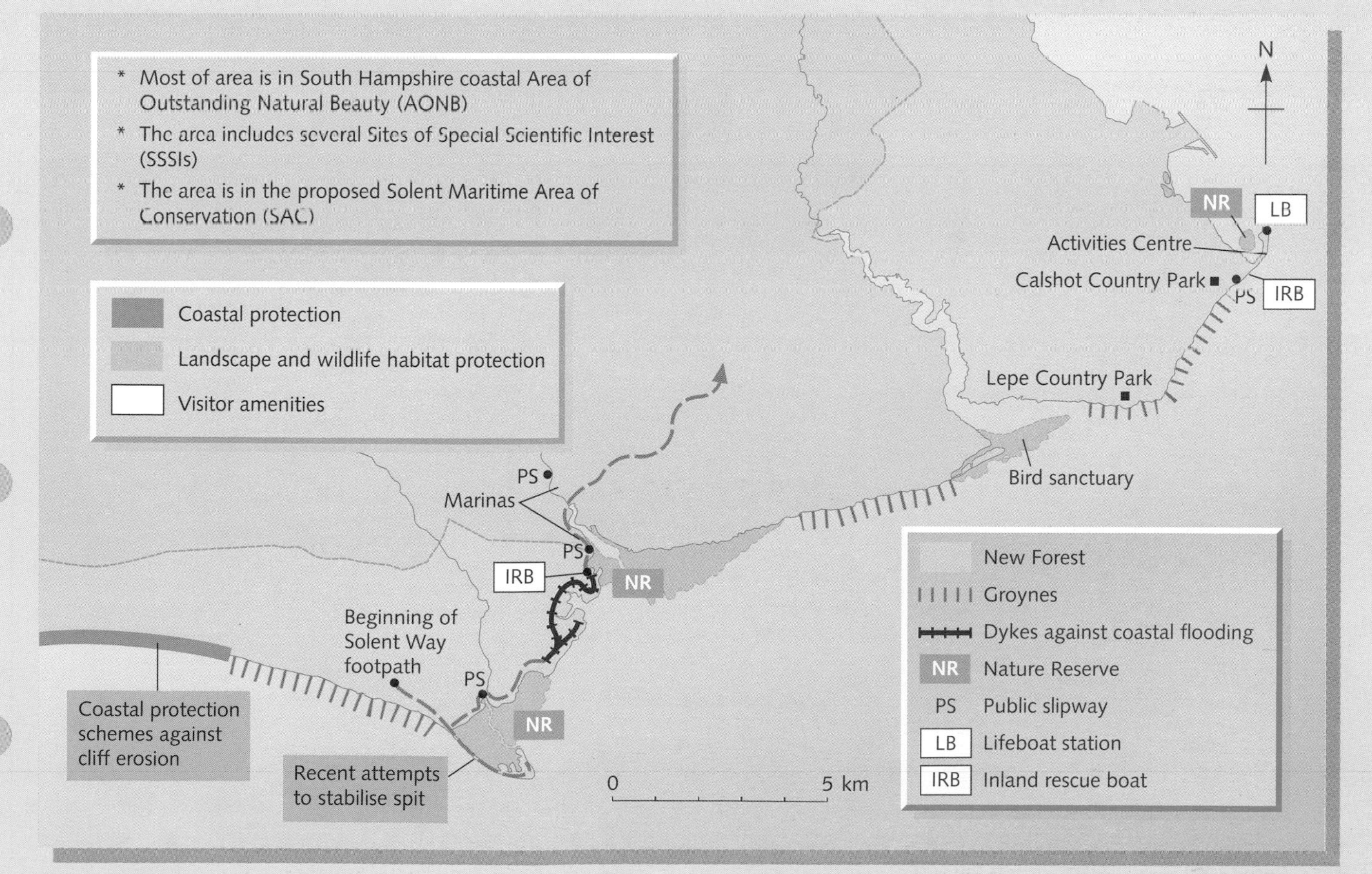

Figure 16.36
Coastal protection, wildlife conservation and visitor amenities

CASE STUDY 16

Flooding and flood prevention in Bangladesh

Most of Bangladesh's 120 million inhabitants live on floodplains which, whether they are river or coastal plains, are seasonally flooded. For most people, the annual flood is essential for their survival and way of life. It brings water in which to grow the main crops of rice and jute, and silt to fertilise the fields. All too frequently, especially in more recent times when the rapid growth in population has meant that people have to live in areas of greater risk, severe floods have caused considerable loss of life, damage to property, and disruption of the local economy (Figures 16.37 and 16.38). The worst-affected areas are often those along the coast. Figure 16.39 describes the effects of three recent floods.

Figure 16.37
Flooding in a rural area

Figure 16.38
Flooding in an urban area

May 1985
Three days after the tropical cyclone hit the coastal islands of Bangladesh, countless bodies are floating in the Bay of Bengal. Hundreds of survivors on bamboo rafts and floating roof-tops, stalked by sharks and crocodiles, are awaiting rescue. The Red Cross suggested that the tidal wave, 9 metres in height and extending 150 km inland, may have claimed the lives of 40 000 people. An official source in Hatia claimed that 6000 people, many in their sleep, were washed out to sea and the only survivors were those who climbed to the top of palm trees and clung on despite the 180 km/hour winds. No links have yet been made with the more remote islands. Already there is the threat of typhoid and cholera, as fresh water has been contaminated. Famine could result as the rice crop has been lost, and it will take next year's monsoon rains to wash the salt out of the soil. Thousands of animals and most of the coastal fishing fleet seem likely to have been lost. The people of Bangladesh, already amongst the world's poorest, will be even more destitute.

September 1988
The heaviest monsoon rains ever recorded left 2400 people dead and 25 million homeless. Within a period of 10 days, rivers rose by 8 metres. Up to 80 per cent of the country was, at one time, at least 1 metre deep in floodwater. With road and rail bridges swept away, stranded villages could not be reached, and with Dhaka airport under water and closed for five days, foreign aid was prevented from arriving. As much of the country's water supply became contaminated, hospitals soon filled with cases of dysentery, diarrhoea and cholera. The government made a world-wide appeal for food (to replace the lost rice crop), medicines, water purification tablets and blankets.

April 1991
The tropical cyclone which hit the coast of Bangladesh in late April brought with it winds of 225 km/hour and waves 7 metres in height. It swept over unprotected offshore islands and the flat delta of the Ganges–Brahmaputra rivers where the land is never more than one or two metres above sea-level. To escape the flood, people climbed trees and onto roof-tops – but the wind was too strong for them to cling on for long. Over 150 000 people and half a million cattle were drowned, entire villages swept away, thousands of hectares of crops lost, electricity supplies cut off, and roads and fishing boats destroyed. When the floods eventually subsided, people were faced with food shortages and disease caused by water supplies contaminated with sewage and dead bodies.

Figure 16.39
Three recent floods

Why are coastal areas prone to flooding?

Bangladesh is trapped between two sets of floods: one caused by tidal surges and a rising sea-level, and the other by rivers (Figure 16.40).

- Silt, deposited at the mouth of the Ganges and Brahmaputra, has formed a large delta (page 265). As the silt accumulated upwards and outwards, it created many flat islands which divide the several rivers into numerous distribu-taries. As the marshy islands are ideal for rice growing, they have attracted large numbers of farmers. Further deposition of silt blocks the main channels and increases the flood risk by raising the beds of the rivers. Flooding is most likely to occur in late sum-mer following the heavy, seasonal monsoon rains (page 198) and snowmelt in the Himalayas. Deforestation in the Himalayas may be a contributory factor.
- As tropical cyclones are funnelled up the Bay of Bengal, the force of the wind increases and water is pushed northwards. Towards Bangladesh, the Bay of Bengal becomes narrower and the sea shallower (due to deposition of silt by the rivers), so that the water builds up to form a *storm surge* (Figure 16.41). The surge may be 4 metres in height and topped by waves reaching another 4 metres. The wall of water sweeps over the flat, defenceless islands of the delta, carrying away the flimsy buildings and any life-form in its path. Local inhabitants, many without telephones or televisions, may not get any advance warning or, even if they do, cannot find land high enough upon which to escape the rising water.
- Although Bangladesh's contribution to global warming is minimal (page 206), the effects of this process upon the country are expected to be considerable. As global temperatures increase and ice caps melt, the predicted rise in the world's sea-level will result in many parts of Bangladesh, including the whole delta region, being totally submerged. For every few centimetres that sea-level rises, the more frequent and serious will flooding be along the coast of Bangladesh.

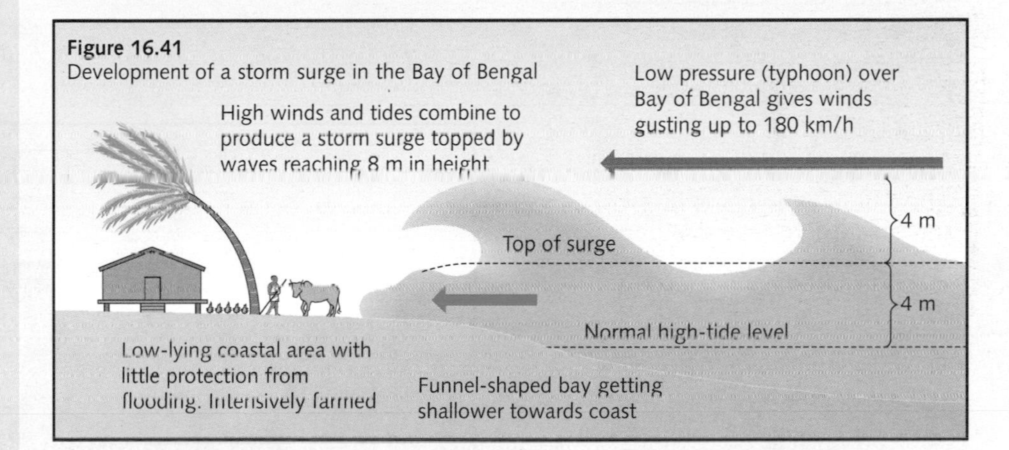

Figure 16.41
Development of a storm surge in the Bay of Bengal

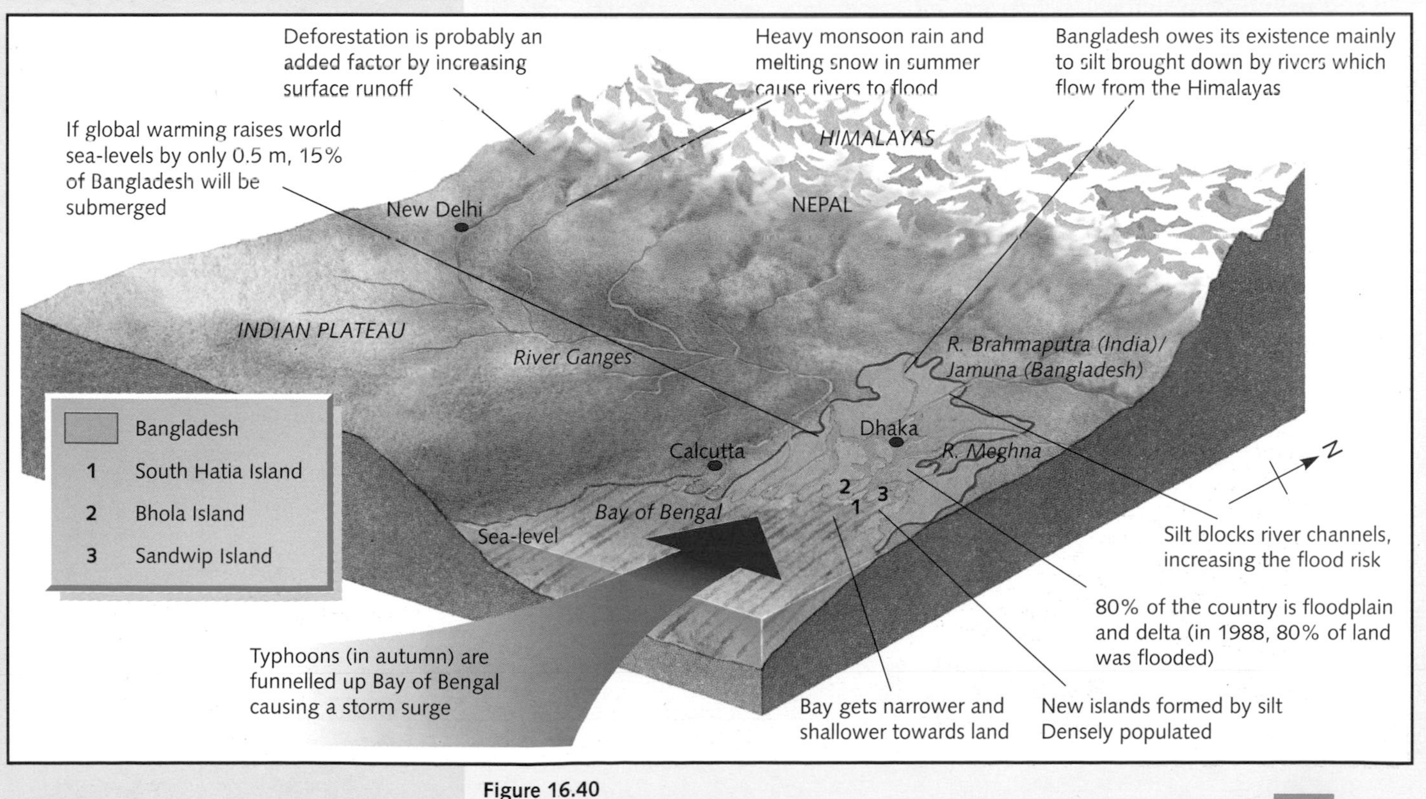

Figure 16.40
Causes of flooding in Bangladesh

What can be done to reduce the flood risk?

In the USA, levées (flood banks) have been built to try to stop the Mississippi River from flooding. In the Netherlands, large dykes, with dams across river mouths, have been constructed to protect the land from flooding by the sea. Both schemes were expensive to implement and are expensive to maintain. Similar schemes have been proposed for Bangladesh but, unlike the USA and the Netherlands, Bangladesh is one of the world's poorest countries. Any flood protection scheme here would need large loans (which would mean that Bangladesh would fall further into debt) and technical help from richer organisations and/or countries. Several very different proposals have been made, including the following:

1 That flooding should be allowed, as it is essential for agriculture and fishing, but that extremes of flooding should be controlled, ideally through appropriate technology (page 136).

2 The Flood Action Plan (FAP – 1990), a massive scheme, supported by several wealthy countries and financed by the World Bank. The FAP would involve setting up regional planning groups to study and monitor local river processes, and the construction of huge flood banks to protect the land from (initially) river flooding and (possibly later) coastal flooding (Figure 16.42). The scheme has, however, provoked considerable opposition from groups that are concerned with its cost and the potential social and environmental implications. Doubt was also cast upon the scheme following the 1993 Mississippi floods where attempts to prevent flooding only led to increased flood damage. The first findings of the FAP (1995) suggested that while flood protection might be economically desirable for large urban areas, the reverse is the case in rural areas which are dependent upon farming and fishing.

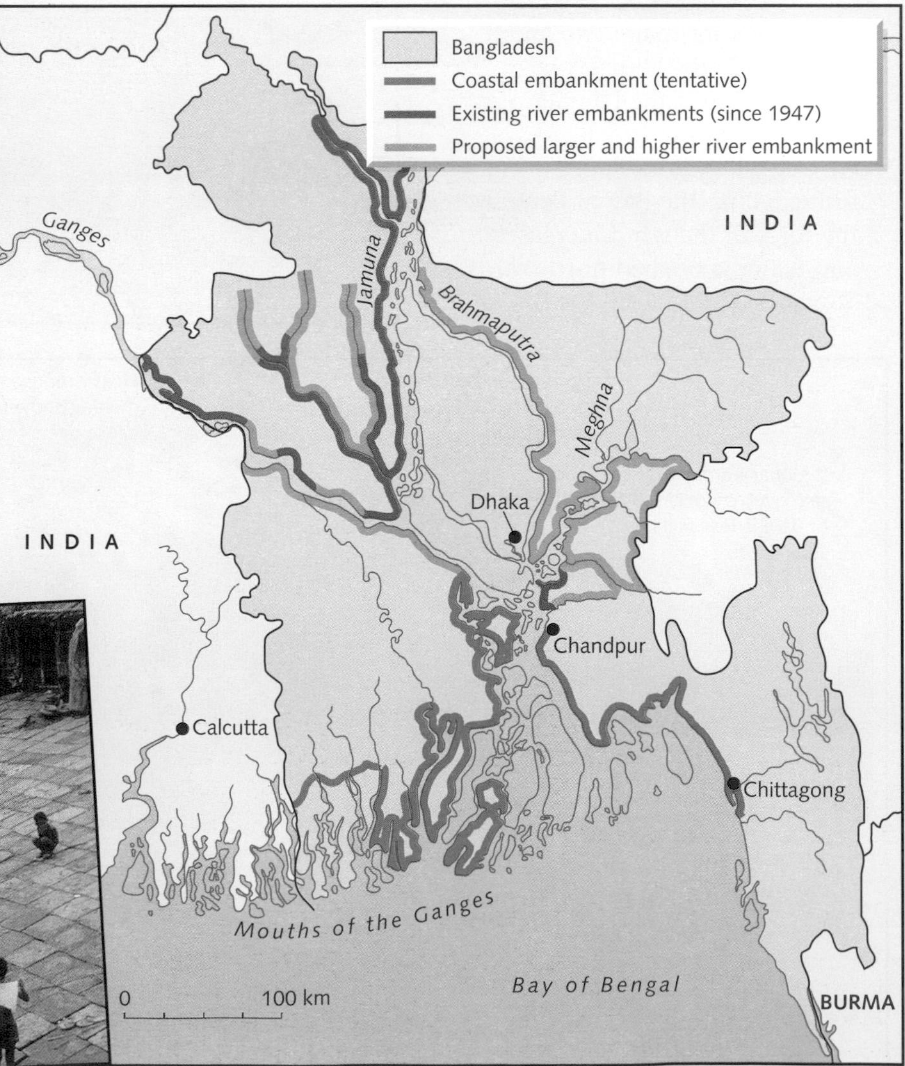

Figure 16.42
Proposed Flood Action Plan, 1990

Figure 16.43
City dwellers wash in the river, next to the Dhaka flood embankment

3 Smaller, more sustainable projects such as:
   - flood embankments to protect key urban areas such as that at Dhaka (Figure 16.43)
   - the building of flood shelters as protection against both river and coastal flooding (Figure 16.44)
   - improved forecasting, using satellite images and early warning systems (Figure 16.45).

A flood shelter is a strange sight. A vast table of earth, big enough to house up to 150 families made homeless by the floods. People can survive on them for many weeks if necessary, until the floods subside. There is space for grain storage and for cattle. There are latrines, a freshwater well and even a school for children. There are boats to rescue people in times of emergency and to ferry them back to the shelter.

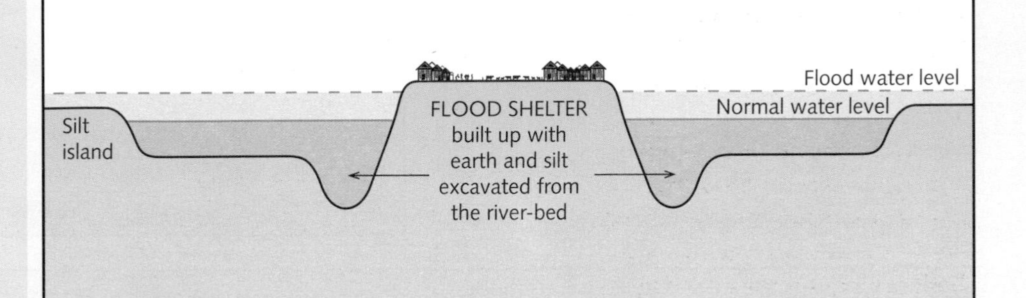

Figure 16.44
Flood shelter – a haven from the floods

Figure 16.45
A cyclone shelter poster

Figure 16.46
The Dhaka flood embankment

1997 – two hopeful signs of success

Dhaka has not experienced flooding since its new embankment was completed in 1994 (Figure 16.46).

In May 1994, Bangladesh was hit by one of the most ferocious cyclones in living memory. Despite this, the death toll was in the lower hundreds rather than, as in earlier storms, the higher thousands (Figure 16.39). The lower death toll was credited to:

- Physical conditions: the cyclone hit during daylight and at low tide (the storm surge was less high).
- Human factors:
  - a new, sophisticated meteorological office gave 36 hours' warning
  - more flood shelters (Figure 16.44) were available
  - government and donor agencies had printed and distributed information leaflets, and enlisted 33 000 volunteers to educate people on preparing for cyclones (e.g. bury food to eat after the event) – these volunteers cycled around remote areas giving the warning by megaphone (Figure 16.45).

QUESTIONS

1

(Page 282)

a Copy and complete the diagram to show a glacier system. (8)

b i) Describe the process of freeze–thaw weathering. (2)

ii) Describe two processes by which a glacier can erode the land. (4)

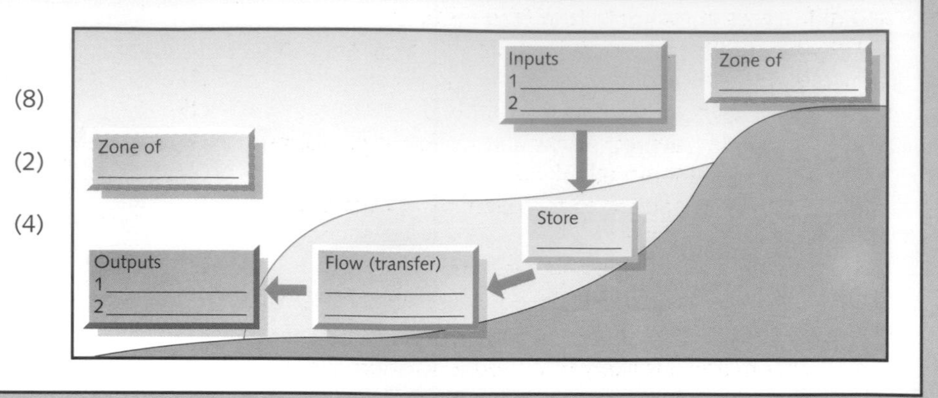

2

(Page 283)

Rearrange the following eleven statements to give a description of the formation of a cirque (corrie or cwm). (11)

ice melts to leave a corrie lake
snow collects in north-facing hollows
freeze–thaw still continues producing scree
moraine left at the end of the glacier
decrease in erosion leaves a rock lip
freeze–thaw weathering loosens rock
glacier moves downhill under gravity
snow is compressed until it becomes ice
corrie floor is overdeepened by abrasion
ice with all air (oxygen) squeezed out turns blue
plucking removes rocks from the backwall

3

(Pages 283–285)

The fieldsketch shows 10 glacial landforms found in the Snowdon area of North Wales. These 10 features, labelled **a** to **j**, include (but not in this order) two cirques, two arêtes, a pyramidal peak, a glacial trough, two hanging valleys, a truncated spur, and a ribbon lake.

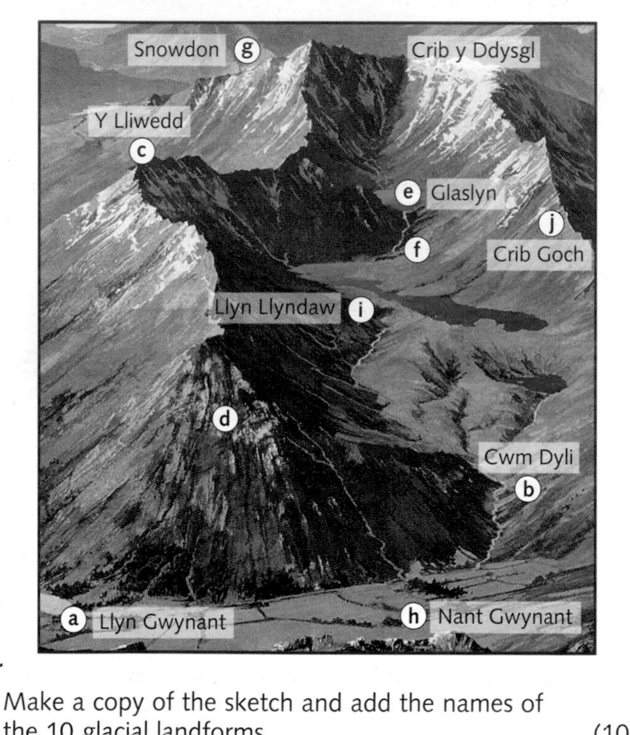

Either

a Make a list matching the landforms with their appropriate letter. (10)

Or

b Make a copy of the sketch and add the names of the 10 glacial landforms. (10)

c With the help of well-labelled diagrams describe the formation of a:

i) pyramidal peak
ii) hanging valley
iii) glacial trough
iv) terminal moraine. (12)

4

(Page 286)

a i) How do large waves form? (1)

ii) Why do waves break? (1)

b What is the difference between:

i) a constructive wave and a destructive wave

ii) the swash and the backwash? (4)

c Describe four processes by which waves can erode the land. (4)

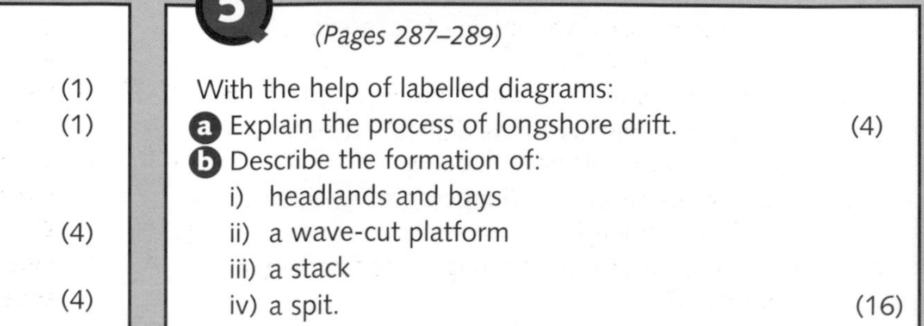

5

(Pages 287–289)

With the help of labelled diagrams:

a Explain the process of longshore drift. (4)

b Describe the formation of:

i) headlands and bays
ii) a wave-cut platform
iii) a stack
iv) a spit. (16)

6

(Pages 287–289)

The map shows part of the Yorkshire coast. The landforms found here are the result of erosion, deposition and differences in rock structure (strength).

a Why have cliffs formed at Scarborough, Filey Brigg and Flamborough Head? (1)

b Why have bays formed at Filey and Bridlington? (1)

c Why can caves, arches, stacks and wave-cut platforms be found at Flamborough Head? (2)

d i) What has happened to the coastline at Holderness since Roman times?
ii) Why has this change taken place? (2)

e Why have groynes been built along the Holderness coast? (1)

f Spurn Head is a spit.
i) What material is likely to be found at Spurn Head?
ii) From which direction must this material have come?
iii) Why was this material deposited here?
iv) Why has Spurn Head a curved (hooked) end?
v) Why is it unlikely that Spurn Head will grow across the Humber Estuary? (5)

g Make a large copy of the map. On it label the following coastal features:
three headlands; three bays; three areas with resistant cliffs; caves, arches and stacks; one area with easily eroded cliffs; groynes; longshore drift (with an arrow to show its direction); spit; sand-dunes. (15)

h Re-draw the map to show the possible shape of this stretch of coastline in several thousand years' time. (3)

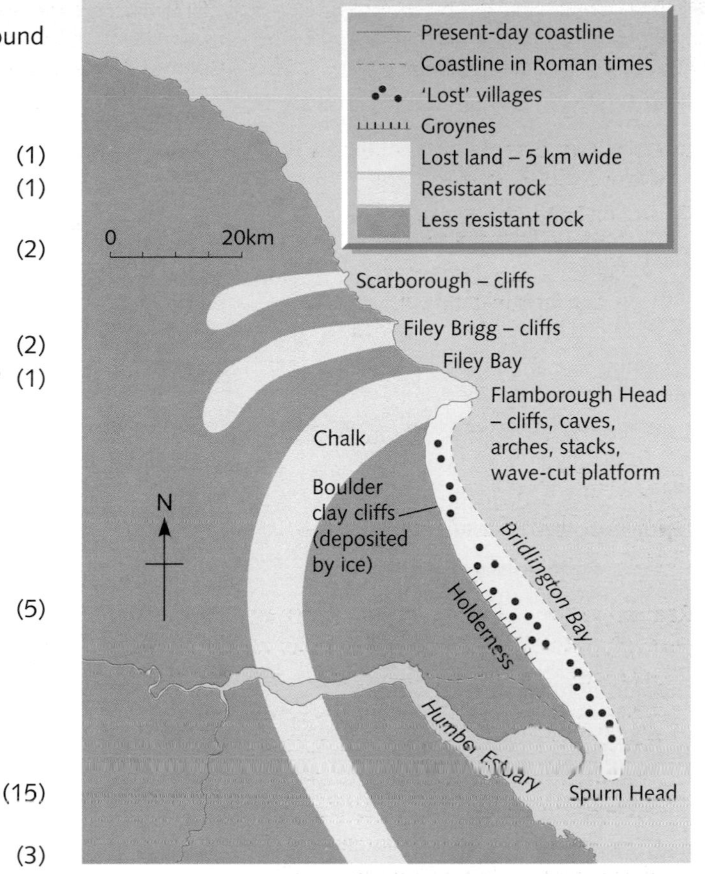

7

(Pages 290–293)

a With reference to the New Forest coastline, name one area with cliffs, one shingle spit, one area of marsh and one river estuary. (4)

b Describe how each of the factors shown on the star diagram adds pressure to the New Forest coastline. (12)

c i) Describe eight of the objectives of the New Forest District Council's management plan. (8)
ii) In what ways do the Council accept that there will be conflicts in implementing the proposals? (3)

d i) What is being, or has been, done to protect:
- the cliffs
- the spit
- wildlife habitats? (6)

ii) What provision has there been to encourage visitors to the area? (4)

Weathering and mass movement
Waves and tides
Industry
NEW FOREST COASTLINE
Recreation
Housing development
Water pollution

8

(Pages 294–297)

a i) Give three reasons why much of Bangladesh is vulnerable to flooding. (3)
ii) Why do so many people live in this flood-prone area? (2)

b i) What are the primary (immediate/short-term) problems that result from flooding? (4)
ii) What are the secondary (longer-term) problems caused by flooding? (4)

c i) Why can each of the following cause flooding in Bangladesh?
- Monsoon rain
- Tropical cyclones (storms)
- Global warming (6)

ii) With the help of a labelled diagram, describe the causes and effects of a storm surge. (5)

d i) What can be done in Bangladesh to try to reduce the risk of:
- flooding by rivers
- flooding by the sea? (6)

ii) Why is it difficult to implement flood prevention schemes in countries like Bangladesh? (4)

Glossary

abrasion erosion caused by the rubbing and scouring effect of material carried by rivers, glaciers, waves and the wind.

acid rain rainwater containing chemicals that result from the burning of fossil fuels.

ageing population the increase in the number of elderly people.

age–sex structure see *population pyramids/structure.*

aid the giving of resources by one country, or an organisation, to another country.

anticyclone an area of high pressure usually associated with settled weather.

appropriate technology technology suited to the area where it is used.

aquifer a rock capable of either storing water or transferring it to another place.

arch a coastal feature formed when waves erode through a small headland.

arête a narrow, knife-edged ridge formed by glacial erosion when two adjacent cirques erode towards each other.

attrition erosion caused when rocks and boulders, transported by rivers and waves, bump into each other and break up into smaller pieces.

biological weathering the breakdown of rock through the actions of plants and animals.

biomass the total amount of organic matter contained by plants and animals in a given area.

biome an ecosystem, such as the tropical rainforest, occurring at a global scale.

birth rate the number of live births per 1000 people per year.

brown earth soil a soil typical of deciduous forests in temperate latitudes.

Burgess model an urban land use model showing five concentric zones, based upon age of houses and wealth of their inhabitants.

central business district (CBD) the commercial and business centre of a town or city where land values are at their highest.

central place any settlement that provides goods and services for smaller neighbouring settlements.

chemical weathering the decomposition of rock caused by a chemical change within that rock.

cirque (corrie) a deep, steep-sided, rounded or semi-circular hollow, often with a lake, found in glaciated highlands.

climate the average weather conditions for a place taken over a period of time.

coastal management the protection of the coastline and its wildlife, the improvement in the quality of life of local residents and the provision of recreational facilities for visitors.

collision margin a boundary between two plates moving together where, as both consist of continental crust, fold mountains form.

Common Agricultural Policy (CAP) the system of agricultural support adopted by the European Union for member states.

commuting the process by which people living in one place, travel to another place for work.

conservative margin a boundary between two plates that are sliding past each other and where crust is being neither formed nor destroyed.

constructive margin a boundary between two plates that are moving apart and where new crust is being formed.

core–periphery model economic growth tends to be most rapid in one part of a country (the core), leaving other places less developed and less well-off (the periphery).

corrasion erosion caused by the rubbing and scouring effect of material carried by rivers, glaciers, waves and the wind.

corrosion erosion caused by acids in rivers and waves dissolving rocks by chemical action.

counterurbanisation the movement of people and employment away from large cities to smaller settlements within more rural areas.

crust the thin outer layer of the Earth.

death rate the number of deaths per 1000 people per year.

deforestation the complete clearance of forested land.

delta an area of silt deposited by a river where it enters the sea or a lake.

demographic transition model a model that tries to show how changes in birth and death rates over a period of time may be related to different stages of economic development.

dependency ratio the ratio between those in the non-economically active age group (taken to be children under 15 and adults over 65) and those in the economically active age group (15 to 64).

deposition the laying down of material previously transported by mass movement, water, glaciers, waves and the wind.

depression an area of low pressure in temperate latitudes usually associated with cloud, rain and strong winds.

desertification the turning of land, often through physical processes and human mismanagement, into desert.

destructive margin a boundary between two plates that are moving together and where one, consisting of oceanic crust, is forced downwards and destroyed.

development the level of economic growth of a country or region and the processes of change taking place within it.

dew point the temperature at which air becomes saturated and water vapour within it condenses to form water droplets.

diet the amount (measured in calories) and quality (the balance between proteins, carbohydrates and vitamins) of food needed to keep a person healthy and active.

discharge the volume of water in a river at a given time, usually measured in cumecs (cubic metres per second).

drainage basin the area of land drained by a main river and its tributaries.

drainage density the total length of all the streams and rivers in a drainage basin divided by the total area of the drainage basin.

drought a prolonged period of dry weather.

drumlin a smooth, elongated mound of material deposited by a glacier and stream-lined in the direction of ice movement.

earthquake a sudden movement within the Earth's crust, usually close to a plate boundary.

ecosystem a natural system in which plants (flora) and animals (fauna) interact with each other and the non-living environment.

ecotourism a sustainable form of 'green' tourism aimed, unlike mass tourism, at protecting the environment and local cultures.

emigrant a person who leaves one country to live in another country.

employment structure the division of jobs into, traditionally, primary, secondary and tertiary sectors, together with, more recently, the quaternary sector.

energy flows the transfer of energy through an ecosystem by means of a food chain (food web).

Enterprise Zone (EZ) an area recognised by the UK government in the 1980s as suffering from acute physical and economic decay and in need of urgent regeneration.

environment the surroundings in which plants, animals and people live.

Environment Agency the amalgamation (1996) of several organisations, including the National Rivers Authority, to secure the protection of water, land and air in the UK.

epicentre the place on the Earth's surface immediately above the focus of an earthquake.

erosion the wearing away of the land by material carried by rivers, glaciers, waves and the wind.

erratic a large boulder transported by ice and deposited in an area of totally different rock.

ethnic group a group of people with common characteristics related to race, nationality, language, religion or culture.

eutrophication the process by which fertiliser (mainly nitrate) causes, on reaching rivers and lakes, rapid algae and plant growth and, subsequently, the depletion of oxygen available for fish.

evapotranspiration the loss of moisture from water surfaces and the soil (evaporation) and vegetation (transpiration).

exfoliation a physical weathering process by which, due to extreme changes in temperature, the surface layers of exposed rock peel away.

exports goods transported by one country for sale in a different country.

favela a Brazilian term for an informal, shanty-type settlement.

fiord a long, narrow, steep-sided inlet formed by glaciers and later drowned by a rise in sea-level.

flood a period of either high river discharge (when a river overflows its banks) or, along the coast, an extremely high tide.

floodplain the wide, flat valley floor of a river where silt is deposited during times of flood.

fold mountains a range of high mountains formed by two plates moving together at a collision margin.

food chain/web the transfer of energy through an ecosystem from primary producers to consumers and decomposers.

footloose industry an industry which, as it is not tied to raw materials, has a free choice of location.

formal and informal sector the difference between employment controlled by large companies and the government, and employment dependent upon the initiative of individuals.

fossil fuels non-renewable forms of energy which, when used, release carbon.

fragile environment an environment which, if not carefully managed, may be irretrievably damaged.

freeze–thaw weathering a process of physical weathering by which rock disintegrates due to water in cracks repeatedly freezing and thawing.

front the boundary between two air masses which have different temperature and humidity characteristics.

GATT an international 'General Agreement on Trade and Tariffs'.

geothermal energy a renewable resource of energy using heated rock within the Earth's crust to produce steam and generate energy.

glacial trough a steep-sided, flat-floored glaciated valley with a characteristic U-shape.

global warming the increase in the world's average temperature, believed to result from the release of carbon dioxide and other gases into the atmosphere by the burning of fossil fuels.

green belt an area of land around a large urban area where the development of housing and industry is severely restricted and the countryside is protected for farming and recreation.

Green Revolution the introduction of high-yielding varieties (HYVs) of cereals (rice and wheat) into economically less developed countries.

Gross National Product (GNP) per capita the total value of goods produced and services provided by a country in a year, divided by the total number of people living in that country.

groundwater water stored underground in permeable rocks.

groyne an artificial structure running out to sea to limit longshore drift.

habitat the natural environment (home) of plants and animals.

hanging valley a tributary valley left high above the main valley as its glacier was unable to erode downwards as quickly as the larger glacier in the main valley, and whose river now descends as a waterfall.

hierarchy a ranking of settlements or shopping centres according to their size or the services which they provide.

high-tech industry an industry using advanced information technology and/or processes involving micro-electronics.

honeypot a place of attractive scenery or historic interest which attracts tourists in large numbers.

Hoyt model an urban land use model showing wedges (sectors), based upon main transport routes and social groupings.

Human Development Index (HDI) a social welfare index, adopted by the United Nations as a measure of development, based upon life expectancy (health), adult literacy (education) and real GNP per capita (economic).

humus organic material found in soil derived from the decomposition of vegetation, dead organisms and animal excreta.

hurricane see *tropical cyclone.*

hydraulic action erosion caused by the sheer force of water breaking off small pieces of rock.

hydrograph a graph showing changes in the discharge of a river over a period of time.

hydrological cycle the continuous recycling of water between the sea, air and land.

igneous rock a rock formed by volcanic activity, either by magma cooling within the Earth's crust or lava at the surface.

immigrant a person who arrives in a country with the intention of living there.

impermeable rock a rock that does not let water pass through it.

imports goods bought by a country from another country.

infant mortality the average number of deaths of children under 1 year of age per 1000 live births.

inner city the part of an urban area next to the city centre, characterised by older housing and industry.

interlocking spur one of a series of spurs that project alternately from the sides of a V-shaped river valley.

island arc a curving line of islands formed by volcanic activity at a destructive plate boundary.

karst an area of Carboniferous limestone scenery, characterised by underground drainage.

lag time the period of time between peak rainfall and peak river discharge.

lahar a rapid downhill movement of mud, water and volcanic ash, often the result of a volcanic eruption.

lava molten rock (magma) ejected onto the Earth's surface by volcanic activity.

leaching the downward movement, and often loss, of nutrients (minerals) in solution in the soil.

levée (dyke) an artificial embankment built to prevent flooding by a river or the sea.

life expectancy the average number of years a person born in a particular country might be expected to live.

literacy rate the proportion of the total population able to read and write.

longshore drift the movement of material along a coast by breaking waves.

magma molten rock occurring beneath the Earth's crust.

malnutrition ill-health caused by a diet deficiency, either in amount (quantity) or balance (quality).

mantle that part of the Earth's structure between the crust and the core.

mass movement the downhill movement of weathered material under gravity.

meander the winding course of a river.

metamorphic rock a rock that has been altered by extremes of heat and pressure.

migration the movement of people (and animals/birds) either within a country or between countries, either voluntary or forced.

million city a city with over one million inhabitants.

model a theoretical representation of the real world in which detail and scale are simplified in order to help explain the reality.

moraine material, usually angular, that is transported and later deposited by a glacier.

mudflow a rapid form of mass movement consisting mainly of mud and water.

multinational corporation see *transnational corporation*.

National Park an area set aside for the protection of its scenery, vegetation and wildlife, so that it may be enjoyed by people living and working there at present, by visitors, and by future generations.

natural increase the growth in population resulting from an excess of births over deaths.

New Commonwealth immigrant a person born in a former British colony in Africa, Asia and the Caribbean (as opposed to the Old Commonwealth countries of Canada, Australia and New Zealand) and who has moved to Britain to live.

newly industrialised country (NIC) a country, mainly in the Pacific Rim of Asia, that has undergone rapid and successful industrialisation since the early 1980s.

new town a well-planned, self-contained settlement complete with housing, employment and services.

non-renewable resource a finite resource, such as a fossil fuel or a mineral, which, once used, cannot be replaced.

nutrient cycle the process by which minerals necessary for plant growth are taken up from the soil, and returned when plants shed their leaves or vegetation dies.

overcultivation the exhaustion of the soil by growing crops, especially the same crop, on the same piece of land year after year.

overgrazing the destruction of the protective vegetation cover by having too many animals grazing upon it.

overpopulation when the number of people living in an area exceeds the amount of resources available to them.

ox-bow lake a crescent-shaped lake formed after a river cuts through the neck of, and later abandons, a former meander.

ozone either a layer of gas found in the atmosphere which protects the Earth from the damaging effects of ultraviolet radiation from the sun, or a harmful gas emitted by vehicle exhausts.

permeable rock a rock that allows water to pass through it.

photosynthesis the process by which green plants (primary producers) take in sunlight, carbon dioxide and water to produce energy and oxygen.

physical weathering the disintegration of rock by mechanical processes without any chemical changes within the rock.

plate margin the boundary between two plates which may be moving towards, away from or sideways past each other creating volcanoes, fold mountains, island arcs and earthquakes.

plate tectonics the theory that the surface of the Earth is divided into a series of plates, consisting of continental and oceanic crust.

plucking a process of glacial erosion by which ice freezes onto weathered rock and, as it moves, pulls pieces of rock with it.

podsol a soil type typical of coniferous forests.

population density the number of people living within a given area (usually a square kilometre).

population pyramids/structure the proportion of males and females within selected age groups, usually shown as a pyramid.

porous rock a rock containing tiny pores through which water can either pass or be stored.

precipitation that part of the hydrological cycle where atmospheric moisture is deposited at the Earth's surface as rain, hail, snow, sleet, dew, frost or fog.

prevailing wind the direction from which the wind usually blows.

primary industry an industry, such as farming, fishing, forestry and mining, that extracts raw materials directly from the land or sea.

pyramidal peak (horn) a triangular-shaped mountain formed by three or more cirques cutting backwards, and with arêtes radiating from the central peak.

quality of life the satisfaction of people with their environment and way of life.

quaternary industry an industry, such as micro-electronics, that provides information and expertise.

range of goods the maximum distance that people are prepared to travel for a specific service.

refugees people forced to move from an area where they lived, and made homeless.

renewable resource a sustainable resource, such as solar energy or water power, which can be used over and over again.

resource a feature of the environment that is needed and used by people.

retailing the sale of goods, usually in shops, to the general public.

ria a river valley drowned by a rise in sea-level.

ribbon lake a long, narrow lake found on the floor of a glaciated valley.

Richter scale the scale used to measure the magnitude of earthquakes.

Rostow model a theory suggesting a five-stage model of economic development, based mainly upon technological innovation.

runoff the surface discharge of water derived mainly from excessive rainfall or melting snow.

rural–urban fringe a zone of transition between the built-up area and the countryside, where there is often competition for land use.

rural–urban migration the movement of people from the countryside to towns and cities where they wish to live permanently.

saltation a process of transportation by rivers in which small particles bounce along the bed in a 'leap-frog' movement.

satellite image a photograph, sometimes using false colours, taken from space and sent back to Earth.

science park/city an estate, often with an edge-of-city location, or a newly planned city, with high-tech industries and a university link.

secondary industry an industry that processes or manufactures primary raw materials (such as steelmaking) assembles parts made by other industries (such as cars) or is part of the construction industry.

sedimentary rock a rock that has been laid down in layers, often as sediment derived from the erosion and transport of an older rock.

self-help housing scheme groups of people, especially in developing countries, are encouraged to build their own homes using materials provided by the local authority.

service industry see *tertiary industry.*

settlement function the main activity, usually economic or social, of a place.

settlement pattern the shape and spacings of individual settlements, usually dispersed, nucleated or linear.

shanty town an area of poor-quality housing, lacking in amenities such as water supply, sewerage and electricity, which often develops spontaneously and illegally (as a squatter settlement) in a city in a developing country.

site the actual place where a settlement (or farm/factory) is located.

situation the location of a settlement in relation to places (physical and human) surrounding it.

soil the thin, loose, surface layer of the Earth which provides a habitat for plants and which consists of weathered rock, water, gases (air), living organisms (biota) and decayed plant matter (humus).

soil creep the slowest type of downhill movement (mass movement) of soil due to gravity.

soil erosion the wearing away and loss of soil due to the action of rain, running water and strong winds, often accelerated by human activity.

soil profile a vertical section of soil showing its different layers (horizons).

solar energy the prime source of energy on Earth, taken into the food chain by photosynthesis in plants, or used by people as a source of electricity.

solution a type of chemical weathering in which water dissolves minerals in rocks.

sphere of influence the area served by a settlement, shop or service.

spit a long, narrow accumulation of sand or shingle formed by longshore drift, with one end attached to the land and the other, projecting out to sea, often with a curved (hooked) end.

stack an isolated piece of rock detached from the mainland by wave erosion.

stalactite and stalagmite formed by water containing calcium carbonate in solution, evaporating in limestone caverns to leave an icicle-shaped feature hanging from the roof (stalactite) or a more rounded feature on the floor (stalagmite).

storm surge a rapid rise in sea-level caused by storms, especially tropical cyclones, forcing water into a narrowing sea area.

subduction zone occurs at a destructive plate margin where oceanic crust, moving towards continental crust, is forced downwards into the mantle and destroyed.

subsidies grants of money made by governments to maintain the price of a specific industrial or agricultural product, e.g. milk.

subsistence farming where all farm produce is needed by the farmer's family or village, and where there is no surplus for sale.

suburbanised village a village that has increasingly adopted some of the characteristics (new housing estates, more services) of urban areas.

suspension a process of transportation by rivers in which material is picked up and carried along within the water itself.

sustainable development a way of improving people's standard of living and quality of life without wasting resources or harming the environment.

swallow hole/sink a hole in the surface of a limestone area, usually formed by solution, down which a river may disappear.

synoptic chart a map showing the state of the weather at a given time.

tariff a customs duty charged on goods imported into a country.

tectonic process a movement within the Earth's crust.

tertiary industry an occupation, such as health, education, transport and retailing, which provides a service for people.

threshold population the minimum number of people needed to ensure that a specific service (shop, school, hospital) will be able to operate economically.

traction a process of transportation by rivers in which material is rolled along the bed.

trade the movement and sale of goods from one country (the producer/exporter) to another country (the consumer/importer).

trading group/bloc a group of countries that have joined together for trading purposes.

transnational corporation a company which, by having factories and offices in several countries, is global in that it operates across national boundaries.

transpiration the loss of moisture from vegetation into the atmosphere.

transportation the movement of material by rivers, glaciers, waves and the wind.

tropical cyclone a severe tropical storm, characterised by low pressure, heavy rainfall and winds of extreme strength which are capable of causing widespread damage.

tropical red earth a soil type typical of the tropical rainforest.

truncated spur a former interlocking spur in a pre-glacial V-shaped valley which, during a later period of glaciation, had its end removed by a glacier.

urban development corporation (UDC) created by the UK government to promote new industrial, housing and community developments in an urban area with large amounts of derelict land and buildings.

urbanisation the increase in the proportion of people living in towns and cities.

urban redevelopment the total clearance of parts of old inner city areas and starting afresh with new houses, especially high-rise flats.

urban regeneration/renewal the improvement of old houses and the addition of amenities in an attempt to bring new life to old inner city areas.

urban sprawl the unplanned, uncontrolled growth of urban areas into the surrounding countryside.

volcano a mountain or hill, often cone-shaped, through which lava, ash and gases may be ejected at irregular intervals.

V-shaped valley a narrow, steep-sided valley formed by the rapid vertical erosion of a river.

water cycle see *hydrological cycle.*

waterfall a vertical, or near vertical, drop of water resulting from a sudden change in the gradient of a river.

watershed a ridge of high land that forms the boundary between two adjacent drainage basins.

water table the upper limit of the zone of saturation found in a porous or permeable rock, or soil.

wave-cut notch an indentation at the foot of a cliff caused by wave erosion.

wave-cut platform a gently sloping, rocky platform found at the foot of a retreating cliff and exposed at low tide.

weather the hour-to-hour, day-to-day state of the atmosphere in relation to temperature, sunshine, precipitation and wind.

weathering the breakdown of rocks *in situ* by either mechanical processes (physical weathering) or chemical changes (chemical weathering).

wetland an ecosystem forming a transition zone between water and dry land.

Index